D1334441

Livestock Handling and Transport

3rd Edition

Livestock Handling and Transport

3rd Edition

Edited by

T. Grandin

Professor, Livestock Handling and Behavior
Department of Animal Sciences
Colorado State University
Fort Collins, Colorado 80523-1171
USA

www.cabi.org

CABI is a trading name of CAB International

CABI Head Office
Nosworthy Way
Wallingford
Oxfordshire OX10 8DE
UK

Tel: +44 (0)1491 832111
Fax: +44 (0)1491 833508
E-mail: cabi@cabi.org
Website: www.cabi.org

CABI North American Office
875 Massachusetts Avenue
7th Floor
Cambridge, MA 02139
USA

Tel: +1 617 395 4056
Fax: +1 617 354 6875
E-mail: cabi-nao@cabi.org

A catalogue record for this book is available from the British Library, London, UK.

A catalogue record for this book is available from the Library of Congress, Washington, DC.

Library of Congress Cataloging-in-Publication Data

Livestock handling and transport / edited by T. Grandin. -- 3rd ed.
    p. cm.
    Includes bibliographical references and index.
    ISBN-13: 978-1-84593-219-0 (alk. paper)
    ISBN-10: 1-84593-219-6 (alk. paper)
    1. Livestock--Handling. 2. Livestock--Transportation. 3.
Livestock--Effect of stress on. I. Grandin, Temple. II. Title.

  SF88.L58 2007
  636.08′3--dc22

                                                                2006036527

ISBN-13: 978 1 84593 219 0

Typeset by AMA DataSet Ltd., Preston, UK.
Printed and bound in the UK by Biddles, King's Lynn.

Contents

Contributors

Jack L. Albright, *Purdue University, West Lafayette, IN 47907, USA.*

Adrian Barber, *Department of Agriculture, PO Box 81, Keith, South Australia, Australia.*

K.E. Belk, *Department of Animal Sciences, Center for Red Meat Safety, Colorado State University, Fort Collins, CO 80523-1171, USA.*

Donald M. Broom, *Centre for Animal Welfare and Anthrozoology, Department of Veterinary Medicine, University of Cambridge, Madingley Road, Cambridge, CB3 0ES, UK.*

Michael S. Cockram, *Division of Veterinary Clinical Sciences, Royal (Dick) School of Veterinary Studies, University of Edinburgh, Easter Bush Veterinary Centre, Easter Bush, Roslin, Midlothian, EH25 9RG, UK.*

Lorna Coppinger, *School of Cognitive Science, Hampshire College, Amherst, MA 01002, USA.*

Raymond Coppinger, *School of Cognitive Science, Hampshire College, Amherst, MA 01002, USA.*

Roger Ewbank, *formerly Director of Universities Federation for Animal Welfare, 19 Woodfield Road, Ealing, W5 1SL, UK.*

Robert B. Freeman, *Agricultural Engineering Section, University of Melbourne, Melbourne, Victoria, Australia.*

Wendy K. Fulwider, *Colorado State University, Fort Collins, CO 80523, USA.*

Carmen Gallo, *Facultad de Ciencias Veterinarias, Instituto de Ciencia Animal y Tecnologia de Carnes, Universidad Austral de Chile, Casilla 567, Valdivia, Chile.*

Temple Grandin, *Department of Animal Sciences, Colorado State University, Fort Collins, CO 80523, USA.*

W.B. Gross, *Department of Large Animal Clinical Sciences, Virginia Polytechnic Institute and State University, Blacksburg, VA 24061, USA.*

P.H. Hemsworth, *Animal Welfare Science Centre, The University of Melbourne and the Department of Primary Industries (Victoria), Parkville, Victoria 3010, Australia.*

K.A. Houpt, *Department of Clinical Sciences, College of Veterinary Medicine, Cornell University, Ithaca, NY 14853-6401, USA.*

G.D. Hutson, *Clifton Press, PO Box 1325, Kensington, Victoria 3031, Australia.*

Toby Knowles, *School of Veterinary Science, University of Bristol, Langford, Bristol, BS40 5DU, UK.*

E. Lambooij, *Animal Sciences Group of Wageningen UR, Animal Production Division, PO Box 65, 8200 AB Lelystad, The Netherlands.*

L.R. Matthews, *Animal Behaviour and Welfare Research Centre, Ruakura Agricultural Centre, Hamilton, New Zealand.*

Miriam Parker, *Livestockwise, The New House, Bausley, Crew Green, Shrewsbury, Shropshire, SY5 9BW, UK.*

J.A. Scanga, *Department of Animal Sciences, Center for Red Meat Safety, Colorado State University, Fort Collins, CO 80523-1171, USA.*

P.B. Siegel, *Department of Animal and Poultry Sciences, Virginia Polytechnic Institute and State University, Blacksburg, VA 24061, USA.*

Joseph M. Stookey, *Western College of Veterinary Medicine, University of Saskatchewan, Saskatoon, Canada.*

Paul Warriss, *School of Veterinary Science, University of Bristol, Langford, Bristol, BS40 5DU, UK.*

Jon M. Watts, *Western College of Veterinary Medicine, University of Saskatchewan, Saskatoon, Canada.*

Claire A. Weeks, *University of Bristol, Department of Clinical Veterinary Science, Langford, Bristol, BS40 5DU, UK.*

Preface

————————

The purpose of this book is to serve as a source of the latest scientific research information and as an archive of practical information. In this third edition, the goal is to bring together in one book the latest research data and practical information on animal handling, the design of facilities and transport. Some of the most valuable contributions to the knowledge of animal handling and transport are often located in producer publications that are difficult to obtain. The extensive reference lists in each chapter will help preserve important knowledge that may not be available on the Internet. At the end of the book there is an index of useful web pages on handling, behaviour and transport.

The first edition was published in 1993 and this third edition was published 14 years later. It is fully updated with the latest research. An extensively revised introductory chapter covers the increasing awareness of animal welfare around the world and outlines the effective auditing programmes of large, corporate meat buyers. Three new authors have chapters on sheep transport, biosecurity and low-stress methods for sorting cattle and weaning calves.

To provide an additional perspective on livestock management in South America, Asia, India and other regions, two new co-authors have been added to the chapters on cattle transport and the handling of cattle raised in close association with people.

The best of the older material – including all the popular handling system layouts and behaviour diagrams – has been kept. Many readers reported that they found these diagrams useful.

All aspects of animal handling are covered, such as handling for veterinary and husbandry procedures, stress physiology, restraint methods, transport, corral and stockyard design, handling at slaughter plants and welfare. Principles of animal behaviour are covered for cattle, sheep, pigs, horses, deer and poultry.

1 Introduction: Effect of Customer Requirements, International Standards and Marketing Structure on the Handling and Transport of Livestock and Poultry

Temple Grandin

Department of Animal Sciences, Colorado State University, Fort Collins, USA

Introduction

Since the 2000 edition, there have been great changes in the industry that have brought about improvements in handling and transport of livestock. Both new international standards and animal handling audits by major meat-buying customers have been drivers of these improvements. Large companies such as McDonald's Corporation, Wendy's International, Tesco Supermarkets and others conduct audits to ensure that meat suppliers maintain high standards.

The author has worked with McDonald's, Wendy's, Burger King and other companies in implementing slaughter plant auditing programmes in the USA, Australia and other countries. These programmes have brought about great improvements since their implementation in 1999 (Grandin, 2001, 2005a). To remain on a customer's approved suppliers list, the plants had to upgrade their practices. Many of the improvements were accomplished by improved equipment maintenance, better training and supervision of employees, and by simple, inexpensive modifications.

Animal handling and welfare auditing programmes are now being conducted in many countries around the world, ranging from South America to Asia. In the USA, Canada, Australia and New Zealand, approximately 90% of the large beef and pork slaughter plants are audited by major customers. The author has observed that the places with the worst practices are slaughter plants that are not audited by customers.

Another significant development since the 2000 edition has been the development of animal welfare guidelines by the OIE (World Organization for Animal Health) in Paris (OIE, 2005a). Due to the increasing globalization of the entire the livestock industry, OIE guidelines are being used in more and more countries to determine standards for trade. These guidelines cover welfare during both slaughter and transport of cattle, pigs, sheep, goats and all types of poultry (OIE, 2005b, c). OIE guidelines are *minimum* worldwide standards for animal welfare. The welfare requirements of major meat-buying customers are usually more strict. The European Union has also made regulations of livestock transport more strict: more rest stops are required and truck drivers will be required to take training courses.

A third major factor of increasing importance is the demands for animal identification and source verification by both commercial customers and governments in countries that import meat. Animals have to be able to be traced back to the farm of origin (Smith *et al.*, 2005). Identification and traceback holds producers accountable for losses due to bruises, dark-cutters and disease. A major motivator for improved animal identification has been the advent of BSE (bovine spongiform encephalopathy) and other animal diseases, the dramatic changes having been brought about by requirements from meat buyers. Both government and private companies require producers to adhere to strict guidelines for animal welfare and food safety. The most effective programmes originate from countries and companies that have a large economic influence on the market.

Guidelines and training materials

Livestock producers' associations, governing bodies and animal welfare groups have responded by publishing more guidelines and training materials for handling and transport. Europe has had guidelines for years but, since 2000, the European Commission has published a major report on animal welfare during transport (Broom et al., 2002). The Farm Animal Welfare Council in the UK has also issued reports (FAWC, 2003, 2005). Canada, the USA and South America have published more training guidelines since 2000.

Various guidelines and programmes are now in place: (i) the US Pork Board Truckers Quality Assurance programme for the training of truck drivers; (ii) Canadian guidelines on transport of unfit animals (Fisher et al., 2005; Mason, 2005; OFAC, 2005; Ontario Beef, 2005); and (iii) South American guidelines for the training of those people handling livestock (Barros and Castro, 2004; Gallo and Stegmaier, 2005). Australia has National Animal Welfare Standards on handling and transport (Edge et al., 2005). The guidelines used by many restaurant companies for auditing meat plants have been updated and now have a standardized audit form (Grandin, 2005b). Retailers are also using standards developed by producer groups to audit farms in many countries.

Both the Canadian and the US governments have new restrictions on the transport and slaughtering of non-ambulatory animals: in the USA, non-ambulatory cattle are not allowed to enter the food supply. To guide USDA (US Department of Agriculture) veterinarians on humane handling enforcement, the Food Safety Inspection Service started HIKE (Humane Interactive Knowledge Exchange), which provides meat inspectors with case histories of handling and stunning problems that have occurred in actual plants. These real-life scenarios provide easy-to-understand instructions on how to enforce the various regulations (see http://www.fsis.usda.gov).

Effect of Customer Audits on Animal Handling

Audits conducted by the large restaurant companies have been one of the most significant factors that have improved animal handling and stunning in the USA and many other countries. The most effective audit programmes encourage suppliers to continuously improve, and plants that repeatedly fail audits are removed from the customer's approved supplier list. This has the effect of making plant management take the audit seriously. Table 1.1 compares measures of stunning and handling before and after the audit programmes started in 57 US beef plants (Grandin, 1997a, 2002, 2005a). Pork plants have also greatly improved. The incidence of cattle or pigs falling down due to slippery floors or rough handling has almost been eliminated in plants that have been in a customer auditing programme for 2 or more years.

Poultry plants that have been in a strict restaurant welfare auditing programme, where they are required to correct deficiencies, also have much better standards of treatment compared with other plants. Twenty-six poultry complexes that had been in a strict handling and stunning audit programme for 3 or more years had no acts of severe abuse observed during an audit. However, 18 complexes that were not part of this programme had acts of severe abuse in 28% of the complexes. Abusive handling included throwing and kicking chickens, putting live chickens in the trash and scalding live chickens. The plants in the strict restaurant audit programme also had superior plant scores on stunning, broken wings and transport cage repair (see Table 1.2).

Constant vigilance is required by both the customer's auditors and plant management to keep standards high. The best beef and pork plants have their own internal audits, but a few plants have had continuous problems and have let their standards slip. The key is plant management

Table 1.1. Improvements in cattle stunning and handling in slaughter plants that were audited by a major customer with strict requirements.

	Cattle rendered insensible with one shot from a captive bolt (%)		Plants passing the stunning audit with 95% or more cattle stunned with one shot (%)	Handling stress indicator – cattle that vocalize (moo or bellow in the stun box and during movement into the stun box) (%)		Plants passing the vocalization audit with 3% or less of animals vocalizing (%)
	Average of all plants	Worst plant		Average of all plants	Worst plant	
Baseline – before customer audits started (*n* = ten plants for stunning)	89	80	10	8	32	43
Fourth year of being audited by three major customers[a] (*n* = 53 plants)	97	86[b]	94	2	6	91

[a] These customers were very strict and plants that failed to correct deficiencies were removed from the approved supplier list.
[b] Cull cows and bulls. One plant has fluctuated from year to year, passing and failing audits due to poor management.

(Grandin, 1988). In several large audited plants, a change in management solved their problems with failed audits. In another case, when a plant lost a good manager it started to fail audits.

In most US, Australian, European and Canadian beef and pork plants, good standards could be achieved and maintained by improvements in stunner maintenance, installation of non-slip flooring and the elimination of distractions such as shiny reflections that cause animals to baulk and refuse to move (Grandin, 1996, 2005b, 2007; Grandin and Johnson, 2005). They did not have to build a new facility. Often, changing the lighting or adding solid sides on a race or stun box was all that was needed to improve animal movement. Moving a lamp will often remove a reflection, and the installation of additional indirect lighting will attract all species into dark stun boxes, restrainers or races.

However, plants in other countries that did not previously have proper stunning equipment have now installed stunners and thereby improved their handling facilities. Since the year 2000, there has been a boom in remodelling and renovation of plants in South America and Asia. This was done to fulfil customer and international requirements for both animal welfare and food safety.

Importance of outcome-based objective standards

A well-written standard that can be consistently applied avoids vague terms such as adequate or proper. What one person may consider proper handling another may consider abusive. An example of a clearly worded standard is: 'all pigs

Table 1.2. Improvements in poultry handling and stunning in slaughter plants that are audited by a major customer with strict requirements.

	Plants having 3% or less broken or dislocated wings (per bird basis, %)	Plants passing the stunning audit with 99% or more of the chickens stunned in a waterbath stunner (%)	Plants passing the audit with 95% or more of the transport cages in good repair (%)
Plants that were not part of a strict restaurant audit programme (*n* = 12 plants for stunning and broken wings; *n* = 18 for level of cage repair)	58	42	88
Plants that were in a strict restaurant audit programme for at least 3 years (*n* = 26 plants[a])	100	96	92

[a]This customer was very strict and plants that failed to correct deficiencies were removed from the approved supplier list.

must have enough space so that they can all lie down at the same time without being on top of each other'.

One of the reasons for the success of the restaurant programmes is that they have used a simple, objective, outcome-based numerical scoring system that minimized subjective judgment. Outcome-based audits that are based on activities and things that auditors can directly observe are much more effective than audits based on the examination of paperwork. Plant management knew exactly what was expected and the objective nature of the scoring system produced similar results with different auditors.

The principle here is to measure relatively few really important outcome-based critical control points (core criteria) that measure numerous problems. For example, animals falling down is a sensitive indicator of either poorly trained handlers or slippery floors. Vocalization (bellowing or squealing) is another indicator of problems; this can be caused by a broken stunner, slipping on the floor, excessive pressure from a restraint device, use of an electric goad or by leaving an animal in the stun box too long.

The five numerically scored core criteria are:

- percentage of animals rendered completely insensible with one application of the stunner;

- percentage that remain insensible (must be 100% for pass);
- percentage that fall down during handling;
- percentage that vocalize during stunning and handling in the stunning box and lead-up race; and
- percentage prodded with an electric goad.

Each animal is scored with simple yes/no scoring. For example, was an animal touched with an electric goad or not touched? For vocalization, each bovine is scored as being either silent or vocalizing. Scoring of cattle vocalization (moo or bellow) during handling is a sensitive indicator of distress. Grandin (1998a) found that 98% of all cattle vocalization in the stun box or in the race leading up to the stun box was associated with an aversive event such as missed stuns, slipping or falling, excessive pressure from a restraint device, single bovine isolated too long or electric goading. Vocalization is associated with stress (Dunn, 1990; Warriss et al., 1994; White et al., 1995; Weary et al., 1998).

There are also five acts of abuse that would lead to automatic audit failure:

- the dragging or throwing of sensible animals;
- the prodding of a goad into sensitive parts of animals;
- deliberate beating of animals;

- the slamming of gates on animals; and
- the intentional driving of one animal over the top of another.

A more complete description can be found in the following: Grandin, 1998b, 2005a, b, 2007. Abusive methods of restraint and handling that are common in developing countries would also lead to automatic audit failure. The following practices would be banned – shackling and hoisting of live animals by the leg, poking out eyes, cutting tendons and puntilla. Another problem area in some countries is the lack of truck loading ramps: animals are pushed or thrown off vehicles. Ramps are easy to build (see Fig. 1.1) and should be one of the required standards.

To pass the audit, a plant must have an acceptable score on all five core criteria and exhibit no acts of abuse. Measurement with numerical scoring enables management to determine if handling has improved or has become worse. Continuous measurement is required to prevent a return of rough handling. The scoring method is objective, but what the minimum acceptable scores would be is determined by either the customer or trading partner. Restaurant audits typically have the following requirements:

no more than 1% of the animals falling and 75% or more moved with no electric goad. The minimum acceptable scores are 95% rendered insensible with one captive bolt shot, 99% correct placement for electrical stunning and 5% or less of cattle vocalizing. All animals must be insensible prior to hoisting. Performing any dressing or slaughter procedure on a sensible animal results in an automatic audit failure.

Auditing of truck loading and unloading

A similar scoring system can be used for monitoring animal handling on farms and during truck loading. The percentage of cattle, pigs and sheep falling down, percentage goad-prodded and the percentage that run into fences can be easily measured. Maria et al. (2004) have developed an effective scoring system for determining stress during loading and unloading of trucks. Higher scores were associated with higher physiological measurements.

Alvaro Barros-Restano (personal communication, 2006) is achieving good results measuring truck loading and unloading and handling in Uruguayan markets. He measures all of the

Fig. 1.1. Well-designed truck-loading ramp that could easily be constructed in a developing country by local people using readily available materials.

previously stated measures plus the percentage of cattle running. Moving animals at a walk or trot is an indicator of good handling, and running is usually an indicator of rough handling. He audited 1200 cattle at a large market, and 99.5% were moved at a walk or trot. Continuous measurement improves handling. Other outcome variables that can be measured are percentage of downed, non-ambulatory animals and dead animals in a truck.

Trucks can also be monitored for waiting time to unload and stocking density. Poor truck scheduling that increases the time pigs have to wait in trucks before unloading will increase the number of dead (Ritter et al., 2005). For poultry, the percentage of birds with broken wings is an effective measure to access chicken handling standards. Scoring of broken and dislocated wings is being used by customer auditors in many different countries. Other effective measures for poultry transport are the percentage of damaged transport cages, percentage of overstocked cages and birds dead on arrival.

The use of measurement to reduce losses

Programmes that reward animal handlers and truck drivers for low levels of damage to animals can be very effective. Hartung et al. (2003) reviewed many transportation studies and recommended paying truck drivers based on reducing losses. These programmes must have accurate measurements of losses. In the poultry industry, paying chicken-catching teams an extra US$30 per person per week for low levels of damage greatly reduced broken wings. Poultry industry data collected by the author showed that incentive pay, combined with measurement of broken wings, reduced damage levels from 5–6% of birds to 1–2%.

Progressive managers use measurement programmes for accessing the percentage of dead pigs or chickens by loading team and by truck drivers. McGlone (2006) found that some truck drivers were linked with twice as many dead or non-ambulatory pigs. Both measurement and the holding of people accountable brought about a 48% reduction in dead pigs (Hill, 2005). Careful measurements also revealed that worker fatigue is a big factor (Ritter et al., 2005). Hill (2005), at Premium Standard Farms in the USA,

found that truck-loading crews became fatigued and the percentage of dead pigs increased after five or six large trucks had been loaded. To reduce death rates, the workload was reduced to six trucks per shift.

In both the poultry and the pork industries, internal data have shown that it pays to work a truck-loading crew for no more than 6 h. Abuse is more likely to occur when handlers are fatigued or the equipment is either poorly designed or broken. Hill (2005) had to record data on many trucks to discover that it would pay not to overwork loading crews. Some of the most useful information comes from studies where large numbers of animals are monitored. Lewis et al. (2005) used a statistical power analysis to determine that over 200 truckloads were required to reliably determine if a new practice made a difference. Another factor is truck driver fatigue: Jennifer Woods, a livestock handling consultant, states that fatigue is a major contribution to livestock truck accidents.

Auditing transport stock density and losses

There are a lot of conflicting data on the proper stocking density for trucks. For both pigs and cattle there is evidence that more space is required for longer trips and those undertaken in hot weather. Pigs will remain standing if the trip is 3 h or less (Guise et al., 1998). After 3 h they will need additional space to lie down. For short journeys, there was little evidence of detrimental effects with a loading density of 281 kg/m^2; this is equivalent to 0.35 m^2 per 100 kg pig (Guise et al., 1998). Similar results have been reported by Ritter et al. (2006a).

Very high stocking densities of 0.39 m^2 for 129 kg pigs resulted in a significant increase in both dead and non-ambulatory animals (Ritter et al., 2006b). The percentage of non-ambulatory pigs and death losses is highly correlated (Hamilton et al., 2003). Similar results have been reported in cattle. On shorter journeys where animals remain standing, they can be stocked more tightly. Filling a vehicle so tightly that closing the gates becomes difficult is a bad practice that should be banned. The author suggests auditing transportation with outcome variables such as death losses, bruising, PSE, dark-cutters, leg injuries and the number of non-ambulatory animals. These data could be used to determine stocking densities for varying weather conditions, journey times and vehicle types.

Importance of stockmanship

A mistake made by many managers is in assuming that technology such as a mechanical chicken harvesting machine or a fancy new fan-ventilated truck will automatically solve all handling problems. Good equipment makes it easier to handle animals but its use must be managed and supervised. During a 35-year career, the author has observed severe animal abuse in poorly managed state-of-the-art facilities. Technology should never be used as a substitute for good management. Audits and financial incentives are powerful tools for the improvement of animal treatment and the reduction of losses.

Paul Hemsworth, an Australian researcher, has clearly shown that good stockmanship and careful, quiet handling pays dividends. Pigs and dairy cattle that are roughly treated and fear people produce fewer progeny, have lower weight gains and produce less milk (Hemsworth and Coleman, 1994; Hemsworth et al., 2000). The attitude of the stock person is also important: animals perform better when they are handled and raised by people who like animals (Hemsworth et al., 1994).

Observations by the author in numerous feedlots and slaughter plants indicate that when the electric prod was no longer the person's primary driving tool, the worker's attitude improved and they were less likely to yell or hit animals. Since audits started, non-electric driving aids such as flags, plastic bags and plastic paddles are now the main tools. The electric goad is picked up only when a stubborn animal refuses to move.

In the best slaughter plants where distractions that cause backing-up and baulking have been removed, 95% or more of cattle or pigs can be moved easily into a stun box or restrainer with no electric goad. For on-farm pigs and sheep, electric goads should not be used. If animals constantly back up, baulk or turn back, distractions in the facility must be eliminated (Grandin, 1996). This is essential for reduction in use of the electric goad.

Further research by Coleman et al. (2003) also shows the importance of removing the electric goad as a person's main handling tool. Measurable improvements in the handler's attitude occurred when they used electric goads with the power turned off. However, there are times when the electric goad is needed. An electric shock is preferable to tail-twisting or beating an animal. Continuous measurement of handling with numerical scoring will help confine use of the electric goad to a very low level.

The quality of stockmanship will have a huge effect on the reduction of dead or non-ambulatory pigs. Multiple shocks with an electric prod and rough handling greatly increased the number of non-ambulatory pigs, and serum lactate levels were greatly elevated (Benjamin et al., 2001). McGlone (2005) conducted observations of hundreds of 115 kg pigs at a large commercial slaughter plant and found that for every 14 pigs electrically prodded, one pig became fatigued and non-ambulatory. Non-ambulatory pigs were approximately four times more numerous when the electric prod was used on over 60% of the pigs compared to the numbers when it was used on fewer than 10%.

Unpublished industry data have also shown that careful truck-driving, with smooth starts and stops, will reduce numbers of non-ambulatory pigs, bruising on cattle and dark-cutters. Economic incentives are powerful motivations for good stockmanship. Large, vertically integrated pork and poultry companies often have a combination of contract farms and company-owned farms. On company-owned farms, hired employees care for the animals; on contract farms, the producer owns the farm and has a bigger financial stake in how well the animals perform. Unpublished internal records from two large companies indicate that contract farms outperform the company operations.

Economic losses from bruising

Smith et al. (1995) and Boleman et al. (1998) reported a bruised carcass level of 48% in US fed steers and heifers. More recent data taken during 2005 – after the restaurant audits had started – indicated that the percentage of bruised fed cattle had dropped to 35% (Smith et al., 2006). Improvements in handling that were required by the restaurant companies helped to reduce bruising. Observations in the USA by the author and Vogel (2006) in large beef plants indicated that facilities where management had worked hard to train truckers had much lower bruising rates than the 35% industry average. In these plants, the level of commercially significant bruising was similar to the 4.1% level reported in a British survey (Weeks et al., 2002). Bruising causes huge economic losses (Marshall, 1977; Blackshaw et al., 1987). Bruised meat must be trimmed out and cannot be used for human consumption.

Sheep and cattle sold through markets had a higher level of bruising than livestock sold directly to the slaughter plant (Cockram and Lee, 1991; McNally and Warriss, 1996; Hoffman et al., 1998; Weeks et al., 2002). A Canadian study showed that 15% of the cattle had severe bruising and 78% of carcasses were bruised (Van Donkersgoed et al., 1997). Smith et al. (1995) found that 22% of cull cows in the USA had severe bruising and 2.2% of these animals had extreme bruising that had destroyed major portions of the carcass. More than 50% of the meat may be destroyed if a cow falls down in a truck and is trampled by other cattle.

Selling cull cows when they are still in good body condition will provide the greatest economic benefit (Apple et al., 1999a, b; Roeber et al., 2001). A survey of cull sows in Minnesota indicated that 67% had foot lesions and 4.6% had shoulder lesions (Ritter et al., 1999). A more recent survey of sows in two large slaughter plants – by Iowa State University – indicated that 12.5% had shoulder lesions and 4.8% of these lesions were open sores (Knauer et al., 2006). The incidence of foot pad lesions was 67.5% (Knauer et al., 2006). Shoulder lesions that occur in sows housed in stalls cause extensive meat damage.

Stress-induced meat quality problems, such as dark-cutters, cause even greater losses. The National Beef Quality Audit estimates that dark-cutters cost the beef industry US$6.08 for every fed animal slaughtered (Boleman et al., 1998). In fed beef, approximately 2% of steers and heifers were found to be dark-cutters (McKenna et al., 2002; Smith et al., 2006). Dark-cutting beef is darker and drier than normal and has a shorter shelf life. Informative reviews on dark-cutting beef can be found in Hood and Tarrant (1981), Fabiansson et al. (1988) and Scanga et al. (1998). The Scanga et al. (1998) study is especially valuable because thousands of cattle were observed.

Weight loss and death losses in cattle

Research at Oklahoma State University (1999) has indicated that the withdrawal of feed from fed feedlot cattle for 24 h prior to slaughter resulted in a loss of $5.00 per animal due to carcass shrinkage and an increased level of dark-cutters. Feedlot managers sometimes do this so that an extra steer can be transported without violating truck weight limitations, but it is a false economy.

Carcass shrinkage (loss of weight) due to rough handling or long hours in transport causes additional losses. Shorthose and Wythes (1988) reviewed numerous studies that quantify shrinkage in cattle and sheep. Large economic losses also occur due to death losses and morbidity in calves that are transported long distances (Hails, 1978). Death losses in US cattle amount to approximately 1% of fed cattle (Jensen et al., 1976; Irwin et al., 1979; Bartlett et al., 1987; Loneragen et al., 2001).

A high percentage of death loss is due to shipping fever, a respiratory disease caused by a combination of shipping stress and viral and bacterial agents. Shipping fever (bovine respiratory disease) costs the US cattle industry US$624 million annually (National Agricultural Statistics Service, 1992–1998). Sickness occurs in about 5% of yearlings (Jensen et al., 1976) and in 14–15% of calves (Bartlett et al., 1987; Snowder et al., 2006). In another study, 22% of steers required medical treatment for sickness (Waggoner et al., 2006).

Dr Dan Thomas, feedlot specialist at Kansas State University, states that in feedlot calf programmes, death losses may rise to 5–10% if the calves have not been pre-weaned and vaccinated prior to arrival (Ishmael, 2005). About 70% of all death losses occur in calves weighing < 225 kg (Noon et al., 1980).

Preconditioning, which consists of weaning and vaccinating 35–45 days prior to shipment to a feedlot, resulted in the reduction of death losses due to respiratory disease (shipping fever) at the feedlot from 0.98 to 0.16% (National

Cattlemen's Association, 1994). A combination of pre-shipment vaccination and good trucking practices can keep death losses on 35-h non-stop trips to < 0.1% (Mills, 1987; Grandin, 1997b). Savings in medical costs and losses from reduced weight gains in sick cattle would be even higher.

Progressive cattle feedlot operators have found that quiet handling during vaccinating enables cattle to go back on feed more quickly. Low-stress handling can cut doctoring and medical expenses by 50% (Maday, 2005). Steve Cote, a specialist in low-stress cattle handling, states that quiet handling can reduce disease from 10 to 1% in calves after they have arrived at a feedlot. Cattle which become agitated while being handled in a squeeze chute will have lower weight gains and yield tougher meat (Voisinet et al., 1997a, b). In another study, Angus cattle that had become excited during handling had lower marbling scores and poorer meat quality compared to those of calm steers (Vann et al., 2006).

Improvements in vaccination and handling practices to reduce sickness can improve profitability. Researchers at Texas A & M University (1998–1992) found that healthy feedlot cattle were more profitable and provided US$49.55–117.42 more profit per animal. Fed cattle that get sick lose some of their marbling and are given a lower-quality grade. A more recent study showed that feedlot cattle that became ill were worth US$69–254 less than healthy steers (Waggoner et al., 2006).

Meat quality losses in pigs

In pigs, deaths during transport, and pale, soft, exudative (PSE) meat cause large financial loss. Rademacher and Davis (2005) found that overloading trucks doubled the number of pigs that died shortly after unloading. PSE is a pork quality defect which is caused by a combination of factors, such as pigs with stress-susceptible genes, rough handling shortly before slaughter and poor carcass chilling. Good reviews on PSE and handling can be found in Smulders (1983), Grandin (1985), Sather et al. (1991), Tarrant (1993), Warriss (2003) and Terlouw (2005).

A Canadian study showed that, even when the stress gene had been bred out of 90% of the pigs, there was still a PSE level of 14.8% (Murray and Johnson, 1998). The authors visited the plant where this study was conducted and observed extremely excessive use of electric prods. In another Canadian plant with good handling, similar pigs demonstrated only 4% PSE. The most recent survey in the USA indicated that 3.34% of the pigs had poor-quality meat having all three of the PSE traits of pale colour, softness and watery texture (VanSickle, 2006).

Pork from heavyweight pigs with the stress gene was judged by a taste panel to be tougher and drier than pork from pigs free of the stress gene (Monin et al., 1999). The author has observed that ultra-lean pigs selected for large muscles have much tougher and drier meat compared to slower-growing pigs with more marbling. Morgan et al. (1993) reported that 9.1% of all hams and loins processed in the USA had PSE. In Denmark, pig breeding and handling are closely monitored: PSE levels in Denmark are at a level of about 2% (Barton-Gade, 1989). (See Chapter 19.)

Marketing System Structure and Losses

The structure of the marketing system can provide either an incentive or a disincentive for reduction of losses. The pork industry in both Europe and the USA has used improvements in genetics, handling and transport to reduce PSE. Pigs marketed through a vertically integrated system usually have less PSE because the farms grow pigs to the customer's specifications. They insist on low PSE genetics and improve handling and transport.

Small producers who grow organic cattle are now part of an integrated chain. They contract directly with the meat-buying customer and no longer go through dealers and middlemen. They have to grow their livestock to meet the customer's specifications. However, some pigs and cattle in the USA are still sold on a live weight basis, where the animals are paid for prior to slaughter. Losses due to bruising, dark-cutters, deaths and PSE are absorbed by the slaughter plant.

A survey by Grandin (1981) indicated that bruises on cattle were greatly reduced when producers switched to a carcass-based selling system where bruise damage was deducted from their payments. Supermarket audits in Brazil – where transporters were held accountable for bruising – reduced bruises on cattle from 20% of all animals to 1.3%. When the frequency of the audits was reduced, bruising rose back to 9% (Paranhas de Costa, personal communication, 2006).

Plants that charge a US$20 fine for non-ambulatory downer pigs have fewer downers than plants without a fining system in place. Marketing systems that allow losses to be passed on to the next buyer provide little incentive to reduce losses. Segmented marketing systems where cattle pass through one or more middlemen, brokers or order buyers prior to reaching their final destination still contribute to substantial losses in both the USA and Mexico. This problem will be greatly reduced when livestock identification and source verification becomes mandatory. DNA fingerprinting could also be used to facilitate traceback of animals (Davis et al., 2006).

Problems with fraud and tag counterfeiting will need to be addressed, especially in local markets. The meat companies which are exporting to premium markets will be motivated to self-police because they will not want to lose their customers, but local dealers who sell old breeding animals will not have this incentive. To prevent cheating, anti-fraud regulations will need to be enforced. During a 30-year career, the author has observed many unethical practices by livestock dealers, but tampering with the weighing scale seldom occurred because US federal laws on scale tampering were enforced. Penalties for transgression are severe.

Handling of pigs greatly improved when the USA entered the Japanese export trade. When slaughter plant managers watched a Japanese grader reject up to 40% of their pork loins due to PSE, a strong economic incentive was created for the improvement of handling. Observations in three different plants showed that simple changes in handling procedures, such as showering, reducing electric prod usage and the resting of pigs, enabled 10% more pork to be exported to Japan. Handling shortly prior to stunning is critical. Stressful handling, use of the electric goad and excitement immediately prior to slaughter are all likely to increase levels of PSE (D'Souza et al., 1999; Hambrecht et al.,

2005; Kuchenmeister et al., 2005; Verstegen and Den Hartog, 2005).

Economic incentives for producers and transporters

A cattle producer is not motivated to vaccinate their calves unless they receive a premium price. There are still many calves entering US feedlots that have not been pre-weaned and vaccinated at the ranch of origin. Only 47% of ranchers had followed recommended guidelines on the weaning and vaccination of calves 45 days before they had left the ranch (Suther, 2006). This bad practice that causes increased illness is gradually decreasing as more producers contract their calves to source-verified beef programmes that require pre-weaned and vaccinated calves.

Fortunately, progressive producer groups are working together to produce truckload lots of calves that have been pre-weaned and vaccinated 5 weeks prior to selling. These calves are being sold at premium prices because buyers know that they will be less likely to become ill, and they also meet strict export requirements for source verification. Powell (2003) and Troxel et al. (2006) reported that cattle producers who had sold preconditioned calves at a special preconditioned calf sale made US$20 more per calf. A total of 52,401 beef breed calves was observed at 15 auctions in Arkansas (Troxel et al., 2006).

Insurance payments for livestock transport must be structured to motivate good practice. If an insurance policy pays for all bruising and deaths, a truck driver has little incentive to reduce losses. Insurance policies should protect a trucking company from a catastrophic loss, such as tipping a truck over, but the policies should not cover one or two dead pigs. People handling livestock or poultry should never be paid based on the number of animals that can run through a race or the number of trucks they can load. This will result in careless work and increased injuries because it provides the wrong incentive. Payment should always be based on the quality of work.

Contracts for buying and selling livestock should have built-in incentives to reduce losses. The Australian sheep-shipping industry provides three examples of a lack of financial incentives to reduce losses. In Australia, death losses on sheep ships sailing to the Middle East average 1–2.5%, but can rise to 6% (Higgs, 1991; Higgs et al., 1991). Grandin (1983) reported that ships' officers stated that very low death losses of 0.47% were possible if sheep were carefully acclimatized in assembly feedlots and prepared prior to loading.

A contributing factor to high death losses is contracts based on the number of live sheep loaded instead of the number of live sheep delivered at the destination (Grandin, 1983). There was little economic incentive to prepare sheep properly and train them to eat pelleted feed prior to transport or to identify the groups of sheep that were likely to have high death losses. Some lines of sheep have very high death losses and overall death losses could be greatly reduced if susceptible sheep could be identified (Norris et al., 1989a, b).

One of the main contributors to ovine death is refusal to eat prior to loading (Norris and Richards, 1989; Norris et al., 1989a, b; Higgs et al., 1991). Death losses during loading are very low, but death losses during unloading may reach 20% (Norris et al., 1990). High discharge death rates occur at ports which have poor facilities and slow unloading, because it is difficult to keep the sheep cool when the ship is stationary. If the people receiving the sheep were required to pay for shipboard losses, they would be motivated to install better unloading facilities. Sheep deaths have also increased when oil prices are high, because ships sail at a slower speed in order to save fuel (Gregory, 1992). This is an unfortunate example of an economic incentive that has increased death losses.

Genetic and Production Problems

Overselection of animals for traits such as rapid weight gain or increased milk production can cause serious welfare problems in both livestock and poultry. Increased selection for rapid growth and a high percentage of lean meat has resulted in weaker pigs, where more are susceptible to death during transport (Grandin and Deesing, 1998). In poultry, selection for rapid growth has led to increases in heart and metabolic problems

(Parkdel et al., 2005). Very lean pigs which have the halothane stress gene will have higher death losses. Murray and Johnson (1998) report that death losses are 9.2% in homozygous-positive pigs, 0.27% in carriers and 0.05% in homozygous-negative pigs. Similar results based on hundreds of truckloads were reported by Holtcamp (2000). Death losses were 0.27% in pigs that were carriers of the stress gene and 0.1% when the gene was removed.

Market pigs grown to very heavy slaughter weights of 129 kg had death losses of 0.23% (Ritter et al., 2005). Heavy pigs with weights over 130 kg tended to have higher death losses compared with lighter pigs (Rademacher and Davis, 2005).

The author has observed that lean hybrids selected for rapid growth and heavy muscling often have double and triple death losses when grown to heavy weights. British pigs which are slaughtered at lighter weights of 100 kg and taken directly from the farm to the slaughter plant have an average death loss of only 0.072% during transport and lairage (Warriss and Brown, 1994). The best average death loss percentage in a British slaughter plant was 0.045% (Warriss and Brown, 1994). In Denmark, the average death loss during transport in pigs free of the stress gene was 0.012% (Barton-Gade et al., 2003). Instances of fatigued pigs that become non-ambulatory without showing symptoms of the porcine stress syndrome often occur: one likely cause is the growing of pigs to heavier and heavier weights.

Finishing pigs that weigh 120–130 kg are common in North America. Rapidly grown heavy pigs need space to lie down on a truck because they are unable to stand as long as lighter pigs. A survey of 42 truckloads of heavy pigs indicated that overloading the truck could cause death losses of > 1% (Ritter et al., 2006a). Another factor is selection for leanness: Durocs selected for lean growth efficiency have significantly higher lactate levels than non-selected pigs from the same genetic lines (Lonergan et al., 2001). High lactate levels are correlated with fatigued downer pigs. Some genetic lines free of the halothane gene may have greater sensitivity to becoming fatigued (Marr et al., 2004).

Another possible problem area is the use of feed additives such as ractopamine. Feeding too much of this beta-agonist can increase problems with fatigued downer pigs. A study carried out by Marchant-Forde *et al.* (2003) showed that pigs on ractopamine were more difficult to drive and likely to become fatigued and go down. The detrimental effect is dose dependent. The author has observed a great increase in fatigued non-ambulatory pigs at the slaughter plants when producers were allowed to feed high doses of this additive.

In fed feedlot beef, feeding at 200 mg/day had a slight effect on handling in the squeeze chute (Baszczak *et al.*, 2006). A higher dose may have a detrimental effect on behaviour. Feedlot workers have reported that the additive may increase respiration during hot weather and increase heat stress: the effect appears to be highly variable. Heavy, black-hided Angus-type feedlot cattle are more prone to heat stress. Mader *et al.* (2002) reported that black-hided *Bos taurus* feedlot cattle panted more during hot weather compared to light-coloured cattle. It is possible that indiscriminate use of ractopamine might increase heat-related death losses in these cattle.

Reduced Disease Resistance

There is evidence that selection for greater and greater growth and yield in pigs has resulted in decreased disease resistance (Meeker *et al.*, 1987; Rothschild, 1998). Continuous selection for greater and greater yields of meat and milk provides economic benefits in the short term, but it may ultimately cause a disaster when an epidemic occurs in high-producing animals with a weakened immune system. In the USA, porcine respiratory and reproductive syndrome (PRRS) became a big problem for producers shortly after the introduction of lean pigs that rapidly gained weight.

Australian chicken producers have reported that disease problems increased when new lines of rapidly growing chickens were introduced. Halibur *et al.* (1998) reported that there were genetic effects on the incidence of infection with PRRS. A team of scientists at the University of Nebraska and the USDA Agricultural Research Service found that pigs selected for lean growth were more susceptible to PRRS than pigs selected for large litters (Johnson *et al.*, 2005).

High-producing Holsteins in the dairy industry have high percentages of lameness and poor reproduction. Zwald *et al.* (2004a, b) reported that data recorded on the farm could be used to select against common health problems. The situation had become so severe that some dairies crossed Holsteins with Jerseys and other dairy breeds to produce more durable animals. The author is concerned that in the future some of the worst animal welfare and disease problems may be caused by over-selection for a narrow range of production traits.

Quality and quantity of meat are two opposing goals. Using either feed additives, hormones or genetics to produce the biggest, fastest-growing cattle or pigs will often reduce quality by reducing tenderness and juiciness. Other quality problems such as dark-cutting beef and PSE pork are also likely to increase. This problem can be avoided if customers pay for quality instead of quantity. Providing the right financial incentives is a major key for the improvement of both animal welfare and meat quality.

Conclusions

Audits by major meat-buying customers have resulted in dramatic improvements in animal handling and transport. The use of objective numerical scoring of handling variables that can be directly observed is more effective than the examination of paperwork. Examples of the handling variables that can be measured in cattle and pigs are the percentages of animals that fall, that vocalize and that are prodded with an electric goad. Losses such as fatalities, bruising and the numbers of downed, non-ambulatory animals should also be measured. Other serious welfare problems that can be easily measured are lameness, emaciated body condition, heat stress symptoms, dirty animals and neglected health problems. In poultry, measuring the percentage of birds with either broken or dislocated wings is an effective method for monitoring handling during catching.

Incentive pay for animal handlers is another powerful tool for the improvement of animal handling. Bonuses based on animal performance and low levels of either sickness or injuries motivate people to handle animals carefully. Another factor that will improve handling is the requirements of both meat-buying customers and international governments for identification and source verification. Holding people accountable for losses will reduce those. Losses are highest in a segmented marketing system where the financial loss can be passed on to the next buyer.

References

Apple, J.K., Davis, J.C., Stephenson, J., Hankins, J.F., Davis, J.R. and Beaty, S.L. (1999a) Influence of body condition scores on the carcass characteristics of subprimal yield in cull beef cows. *Journal of Animal Science* 77, 2660–2669.

Apple, J.K., Davis, J.C. and Stephenson, J. (1999b) Influence of body condition score on by-product yield and value from cull beef cows. *Journal of Animal Science* 77, 670–679.

Barros, A. and Castro, L. (2004) *Bienestar Animal*. Buenas Practicas Operacionales, Institute Nactional de Cavner, Serie Tecnica No. 34, Montevideo, Uruguay.

Bartlett, B.B., Rust, S.R. and Ritchie, H.D. (1987) Survey results of the effects of a pre-shipment vaccination program on sale price, morbidity and morbidity of feeder cattle in northern Michigan. *Journal of Animal Science* 65 (Suppl. 1), 106 (abstract).

Barton-Gade, P. (1989) Pre-slaughter treatment and transportation research in Denmark. In: *35th International Congress of Meat Science and Technology*, Danish Meat Research Institute, Roskilde, Denmark, pp. 140–145.

Barton-Gade, P., Christensen, L., Batzer, M., Fertin, C. and Valentin-Petersen, J. (2003) Investigations into the causes of transport and lairage mortality in Danish slaughter pigs. *International Congress of Meat Science and Technology*, Brazil.

Baszczak, J.A., Grandin, T., Gruber, S.L., Engle, T.E., Platter, W.J., Laudert, S.B., Schroeder, A.L. and Tatum, J.D. (2006) Effects of ractopamine supplementation on behavior of British, Continental, and Brahman crossbred steers during routine handling. *Journal of Animal Science* 84, 3410–3414.

Benjamin, M.E., Gonyou, H.W., Ivers, D.L., Richardson, L.F., Jones, D.J., Wagner, J.R., Seneriz, R. and Anderson, D.B. (2001) Effect of handling method on the incidence of stress response in market swine in a model system. *Journal of Animal Science* (Suppl. 1) 79, 279 (abstract).

Blackshaw, J.K., Blackshaw, A.W. and Kusano, T. (1987) Cattle behavior in a sale yard and its potential to cause bruising. *Australian Journal of Experimental Agriculture* 27, 753–757.

Boleman, S.L., Boleman, S.J., Morgan, W.W., Hale, D.S., Griffin, D.B. *et al.* (1998) National Beef Quality Audit – 1995: survey of producer-related defects and carcass quality and quantity attributes. *Journal of Animal Science* 76, 96–103.

Broom, D.M. *et al.* (2002) *The Welfare of Animals during Transport (Details for Horses, Pigs, Sheep and Cattle)*. Report of the Committee on Animal Health and Animal Welfare, European Commission, Directorate and Scientific Options, Brussels, pp. 1–120.

Cockram, M.S. and Lee, R.A. (1991) Some preslaughter factors affecting the occurrence of bruising in sheep. *British Veterinary Journal* 147, 120–125.

Coleman, G.J., McGregory, M., Hemsworth, P.H., Boyce, J. and Dowling, S. (2003) The relationship between beliefs, attitudes and observed behaviours in abattoir personnel in the pig industry. *Applied Animal Behavior Science* 82, 189–200.

Davis, J., Heller, J., Johnston, E., Fantin, D., Cunningham, W. and Tatum, J. (2006) Case study: use of DNA fingerprinting for verifying identity of individual cattle within a forty-eight hour response period. *Professional Animal Scientist* 22, 139–143.

D'Souza, D.N., Dunshea, F.R., Levry, B.J. and Warner, R. (1999) Effect of mixing boars during lairage and pre-slaughter handling on meat quality. *Australian Journal of Agricultural Research* 50, 109–113.

Dunn, C.S. (1990) Stress reactions of cattle undergoing ritual slaughter using two methods of restraint. *Veterinary Record* 126, 522–525.

Edge, M.K., Maguire, T., Dorian, J. and Barnett, J.L. (2005) *National Animal Welfare Standards for Livestock Processing Establishments*. Australian Meat Industry Council, Crows Nest, New South Wales, Australia, http://www.amic.org.au

Fabiansson, S.U., Shorthose, W.R. and Warner, R.D. (eds) (1988) Dark cutting in cattle and sheep. In: *Proceedings of an Australian Workshop*. Australia Meat and Livestock Development Corporation, Sydney, Australia.

FAWC (2003) *Report on the Welfare of Farmed Animals at Slaughter or Killing*. Farm Animal Welfare Council, London, UK, http://www.fawk.org.uk

FAWC (2005) *Report on the Welfare Implications of Farm Assurance Schemes*. Farm Animal Welfare Council, London, UK, http://www.fawk.org.uk

Fisher, M., Simmons, N. and Follensbee, M. (2005) *Humane Handling of Swine*. Alberta Pork, Edmonton, Canada.

Gallo, C. and Stegmaier, M.V. (2005) *En Bienestar Animal para el Manejo de Bovinos en Predious, Ferias, Medios, de Transporte y Plantas Faenadoras*. Commission National de Buenas, Practicar Agricolas, Ministerio de Agricultura, Chile.

Grandin, T. (1981) Bruises on southwestern feedlot cattle. *Journal of Animal Science* 53 (Suppl. 1), 213 (abstract).

Grandin, T. (1983) *A Survey of Handling Practices and Facilities Used in the Export of Australian Livestock*. Department of Primary Industry, Australian Bureau of Animal Health, Australian Government Publishing Service, Canberra, Australia (also reprinted as Paper No. 84-6533, American Society of Agricultural Engineers, St Joseph, Michigan).

Grandin, T. (1985) *Improving Pork Quality through Handling Systems. Animal Health and Nutrition*. Watt Publishing, Mount Morris, Illinois, July/August, pp. 14–26.

Grandin, T. (1988) Behavior of slaughter plant and auction employees towards animals. *Anthrozoos* 1, 205–213.

Grandin, T. (1996) Factors that impede animal movement at slaughter plants. *Journal of the American Veterinary Medical Association* 209, 757–759.

Grandin, T. (1997a) *Survey of Stunning and Handling in Federally Inspected Beef, Veal, Pork and Sheep Slaughter Plants*. USDA/Agricultural Research Service Project 3602-32000-002-08G, Beltsville, Maryland.

Grandin, T. (1997b) Assessment of stress during transport and handling. *Journal of American Science* 75, 249–257.

Grandin, T. (1998a) The feasibility of vocalization scoring as an indicator of poor welfare during slaughter. *Applied Animal Behavior Science* 56, 121–128.

Grandin, T. (1998b) Objective scoring of animal handling and stunning practices at slaughter plants. *Journal of the American Veterinary Medical Association* 212, 36–39.

Grandin, T. (2001) Welfare of cattle during slaughter and the prevention of non-ambulatory (downer) cattle. *Journal of the American Veterinary Medical Association* 219, 1377–1382.

Grandin, T. (2002) 2001 *Restaurant Audits of Stunning and Handling in Federally Inspected Beef and Pork Slaughter Plants*. http://www.grandin.com

Grandin, T. (2005a) Maintenance of good animal welfare standards in beef slaughter plants by use of auditing programs. *Journal of the American Veterinary Medical Association* 226, 370–373.

Grandin, T. (2005b) *Recommended Animal Handling Guidelines of Audit Guide, 2005 edn*. American Meat Institute Foundation, Washington, D.C. (http://www.animalhandling.org.www.grandin.com).

Grandin, T. (2007) Implementing effective animal welfare auditing programs. In: Gregory, N. (ed.) *Animal Welfare and Meat Science*. CAB International, Wallingford, UK.

Grandin, T. and Deesing, M. (eds) (1998) *Genetics and the Behavior of Domestic Animals*. Academic Press, San Diego, California.

Grandin, T. and Johnson, C. (2005) *Animals in Translation*. Scribner, New York (now published by Harcourt, New York).

Gregory, N.G. (1992) Pre-slaughter handling and stunning. In: *38th International Congress of Meat Science and Technology*, Danish Meat Research Institute, Roskilde, Denmark, pp. 133–139.

Guise, H.J., Riches, H.I., Hunter, E.J., Jones, T.A., Warriss, P.D. and Kettlewell, P.J. (1998) The effect of stocking density in transport on carcass quality and welfare of slaughter pigs. 1. Carcass measurements. *Meat Science* 50, 439–446.

Hails, M.R. (1978) Transportation stress in animals: a review. *Animal Regulation Studies* 1, 282–343.

Halibur, P.G., Rothschild, M.F., Paul, P.S., Thacker, B.J. and Meng, X.J. (1998) Differences in susceptibility of Duroc, Hampshire and Meishan pigs to infection with a high virulence strain (VR2385) of porcine respiratory and reproductive syndrome virus (PRRSV). *Journal of Animal Breeding and Genetics* 115, 181–189.

Hambrecht, E., Eissen, J.J., Newman, D.J., Verstegen, M.W. and Hartog, L.A. (2005) Preslaughter handling affects pork quality and glycolytic potential in two muscles differing in fiber type organization. *Journal of Animal Science* 83, 900–907.

Hamilton, D.N., Ellis, M., Bresser, G.E., Wolter, B.F., Jones, D.J. and Watkins, L.E. (2003) Relationship between environmental conditions on trucks and losses during transport to slaughter in finishing pigs. *Journal of Animal Science* 81 (Suppl. 2), 53 (abstract).

Hartung, J., Marahrens, M. and con Holleben, K. (2003) Recommendations for future development of cattle transport in Europe. *Deutsch Tierarztl Wochenscher* 110, 128–130.

Hemsworth, P.H. and Coleman, G.J. (1994) *Human Livestock Interactions*. CAB International, Wallingford, UK.

Hemsworth, P.H., Coleman, G.J. and Barnett, J.L. (1994) Improving the attitude of behavior of stockpeople towards pigs and the consequences on the behavior and reproductive performance of commercial pigs. *Applied Animal Behavior Science* 39, 349–362.

Hemsworth, P.H., Coleman, G.J., Barnett, J.L. and Borg, S. (2000) Relationships between human–animal interactions and productivity on commercial dairy farms. *Journal of Animal Science* 78, 2821–2831.

Higgs, A.R. (1991) *National Recording Systems for the Livestock Sheep Export Industry*. Miscellaneous Publication 12/91, NDRS, Canberra, Australia.

Higgs, A.R., Norris, R.T. and Richards, R.B. (1991) Season, age and adiposity influence death rates in sheep exported by sea. *Australian Journal of Agricultural Research* 42, 205–214.

Hill, J. (2005) Animal Transport and Handling. Premium Standard Farms, Princeton, Missouri. In: *Animal Handling and Care Conference*, American Meat Institute, Kansas City, Missouri, 9–10 February.

Hoffman, D.E., Spire, M.F., Schwenke, J.R. and Unrah, J.A. (1998) Effect of source of cattle and distance transported to a commercial slaughter facility on carcass bruises in mature beef cows. *Journal of Animal Science* 212, 668–672.

Holtcamp, A. (2000) Gut edema: clinical signs, diagnosis and control. In: *Proceedings of the American Association of Swine Practitioners*, Indianapolis, Indiana, 11–14 March, pp. 337–340.

Hood, D.E. and Tarrant, P.V. (eds) (1981) *The Problem of Dark Cutting Beef*. Martinus Nijhoff, Boston, Massachusetts.

Irwin, M.R., McConnell, S., Coleman, J.D. and Wilcox, G.E. (1979) Bovine respiratory disease complex: a comparison of potential predisposing and etiological factors in Australia and in the United States. *Journal of the American Veterinary Medical Association* 175, 1095–1099.

Ishmael, W. (2005) More than money. *Beef Magazine*, September, 32–33.

Jensen, R., Peirson, R.E., Braddy, P.M., Saari, D.A., Lauerman, L.H., England, J.J., Horton, D.P. and McChesney, A.E. (1976) Diseases of yearling cattle in Colorado. *Journal of the American Veterinary Medical Association* 169, 497–499.

Johnson, R., Petry, D. and Lurney, J. (2005) Genetic resistance to PRRS studied. *National Hog Farmer, 2005 Swine Research Review*, 15 December, 19–20.

Knauer, M., Karriker, L. and Stalder, K. (2006) Understanding why sows are culled. *National Hog Farm Blueprints*, April 15, 27–29.

Kuchenmeister, V., Kuhn, G. and Ender, K. (2005) Preslaughter handling of pigs and the effect on heart rate, meat quality, including tenderness and sarcoplasmic reticulum Ca^{2+} transport. *Meat Science* 71, 690–695.

Lewis, C., Krebs, N., Hubert, L. and McGlone, J. (2005) A model for the study of dead and down pigs associated with transport: effect of maternal pheromone in pigs in transit. *Journal of Animal Science* 83 (Suppl. 1), 324 (abstract).

Loneragen, G.H., Dargartz, D.A., Morley, P.S. and Smith, M.A. (2001) Trends in mortality ratios among cattle in U.S. feedlots. *Journal of the American Veterinary Medical Association* 219, 1122–1127.

Lonergan, S.M., Huff-Lonergan, E., Rowe, L.J., Kuhlers, D.L. and Jungst, S.B. (2001) Selection for lean growth efficiency in Duroc pigs influences pork quality. *Journal of Animal Science* 79, 2075–2085.

Maday, J. (2005) Put out the welcome mat. *Drovers (Feeder Management Suppl.)*, 2–6 September.

Mader, T.L., Hott, G.L., Hahn, M.S., Davis, M.S and Spiers, D.E. (2002) Feeding strategies for managing heat load in feedlot cattle. *Journal of Animal Science* 80, 2373.

Marchant-Forde, J., Lay, D., Pajor, E. and Schinckel, A. (2003) The effects of ractopamine on the behavior and physiology of finishing pigs. *Journal of Animal Science* 81, 416–422.

Maria, G.A., Villaroel, M., Chacon, G. and Gebresenbet, G. (2004) Scoring system for evaluating stress to cattle during commercial loading and unloading. *Veterinary Record* 154, 818–821.

Marr, A.L., Allison, C.P., Berry, N.L., Anderson, D.B., Ivers, D.J., Richardson, L.F., Keffaber, K., Johnson, R.C. and Doumit, M.E. (2004) Impact of halothane sensitivity on mobility status and blood metabolites of HAL-1843-free pigs following an aggressive handling model. *Journal of Animal Science* 82 (Suppl. 2), 33.

Marshall, B.L. (1977) Bruising in cattle presented for slaughter. *New Zealand Veterinary Journal* 25, 83–86.

Mason, S. (2005) *Humane Handling of Dairy Cattle*. Western Dairy Science, Inc., Calgary, Alberta, Canada.

McGlone, J. (2005) Reducing market hog fatigue. *Pork*, December, http://www.pork.mag.com

McGlone, J. (2006) Fatigued pig: the transportation link. *Pork*, 14–16 February.

McKenna, D.R., Roeber, D.L., Bates, P.K., Schmidt, T.R., Hale, D.S., Griffin, D.B., Savell, J.W., Brooks, J.K., Morgan, J.E., Montgomery, T.H., Belk, K.E. and Smith, G.C. (2002) National Beef Quality Audit 2000 Survey of targeted cattle and carcass characteristics related to quality, quantity, and value of fed steers and heifers. *Journal of Animal Science* 80, 1212–1222.

McNally, P.W. and Warriss, P.D. (1996) Recent bruising in cattle at abattoirs. *Veterinary Record* 138 (6), 126–128.

Meeker, D.L., Rothschild, M.F., Christian, L.L., Warner, C.M. and Hill, H.T. (1987) Genetic control of immune response to pseudorabies and atrophic rhinitis vaccines. *Journal of Animal Science* 64, 407–413.

Mills, B. (1987) Arrive alive. *Beef Today*, October, 19–20.

Monin, G., Larzul, C., LeRoy, P., Culioli, J., Mourot, J., Rousset-Akrim, S., Talmant, A., Touraille, C. and Sellier, P. (1999) Effects of halothane genotype and the slaughter weight on texture of pork. *Journal of Animal Science* 77, 408–415.

Morgan, J.B., Cannon, F.K., McKeith, D., Meeker, D. and Smith, G.C. (1993) *National Pork Chain Audit (Packer, Processor, Distributor)*. Final Report to the National Pork Producers Council, Colorado State University, Fort Collins, Colorado and University of Illinois, Champaign-Urbana, Illinois.

Murray, A.C. and Johnson, C.P. (1998) Importance of halothane gene on muscle quality and preslaughter death in western Canadian pigs. *Canadian Journal of Animal Science* 78, 543–548.

National Agricultural Statistics Service (1992–1998) *Cattle and Calves Death Loss*. United States Department of Agriculture, Washington, D.C.

National Cattlemen's Association (1994) *Strategic Alliances: Field Study in Coordination with Colorado State University and Texas A & M University*. National Cattlemen's Beef Association, Englewood, Colorado.

Noon, T.N., Mare, C.J., Mazure, M., Prouth, F., Trautman, R.J. and Reed, R.J. (1980) *Selected Aspects of Respiratory Disease in New Feeder Calves*. Arizona Cattle Feeders Day, University of Arizona, Tucson, Arizona.

Norris, R.T. and Richards, R.B. (1989) Deaths in sheep exported from Western Australia – analysis of ship master's reports. *Australian Veterinary Journal* 66, 97–102.

Norris, R.T., Richards, R.B. and Dunlop, R.H. (1989a) An epidemiological study of sheep deaths before and during export by sea from Western Australia. *Australian Veterinary Journal* 66, 276–279.

Norris, R.T., Richards, R.B. and Dunlop, R.H. (1989b) Pre-embarkation risk factors for sheep deaths during export by sea from Western Australia. *Australian Veterinary Journal* 66, 309–314.

Norris, R.T., Richards, B. and Higgs, T. (1990) The live sheep export industry. *Journal of Agriculture Western Australia* 31 (4), 131–148.

OFAC (2005) *Is the Pig Fit to Load or Transport?* Ontario Farm Animal Council, Ontario, Canada.

OIE (2005a) *Animal Welfare: Global Issues, Trends and Challenge, Scientific and Technical Review*. World Organization for Animal Health, Paris, Vol. 24 (2), http://www.oie.int

OIE (2005b) *Terrestrial Animal Health Code, Guidelines for the Slaughter of Animals for Human Consumption*. World Organization for Animal Health, Paris, http://www.oie.int/eng/en_index.htm

OIEa (2005c) *Terrestrial Animal Health Code, Guidelines for the Land Transport of Animals*. World Organization for Animal Health, Paris.

Oklahoma State University (1999) *Department of Animal Science Research Report*. Oklahoma State University, Stillwater, Oklahoma, p. 965.

Ontario Beef (2005) *New Rules for Transporting Livestock*. Ontario Cattlemen's Association, Guelph, Ontario, Canada, pp. 14–15.

Parkdel, A., van Arendonk, J.A., Vereijken, A.L. and Bovenhuis, H. (2005) Genetic parameters of ascites-related traits in broilers: correlations with feed efficiency and carcass traits. *British Poultry Science* 46, 43–53.

Powell, J. (2003) *Preconditioning Programs for Beef Calves*. Cooperative Extension, University of Arkansas, Fayetteville, Arkansas, http://www.uaex.edu

Rademacher, C. and Davis, P. (2005) Factors associated with the incidence of mortality during transport of market hogs. In: Allen, D., *Leman Swine Conference Proceedings*, Minnesota, pp. 186–191.

Ritter, L.A., Xue, J., Dial, G.D., Morrison, R.B. and Marsh, W.E. (1999) Prevalence of lesions and body condition scores among female swine at slaughter. *Journal of the American Veterinary Medical Association* 214, 525–528.

Ritter, M.J., Ellis, M., Brinkman, J., Keffaber, K.K. and Walter, B.F. (2005) Relationship between transport conditions and the incidence of dead and non-ambulatory finishing pigs at the slaughter plant. *Journal of Animal Science* 82 (Suppl. 1), 259 (abstract).

Ritter, M.J., Ellis, M., Bertelsen, R., Bowman, J., Brinkman, J.M., DeDecker, O., Mendoza, C., Murphy, B.A., Peterson, A., Rojo, J.M., Schlipf, J.M. and Wolter, B.F. (2006a) Impact of animal management and transportation factors on transport losses in market pigs at the packing plant. *Journal of Animal Science* 84 (Suppl. 1), 302 (abstract).

Ritter, M.J., Ellis, M., Brinkmann, J.M., de Decker, J.M., Keffaber, K.K., Kocher, M.E., Peterson, B.A., Schlipf, J.M. and Wolter, B.F. (2006b) Effect of floor space during transport of market-weight pigs on the incidence of transport losses at the packing plant and the relationships between transport conditions and losses. *Journal of Animal Science* 84, 2856–2864.

Roeber, D.L., Mies, P.D., Smith, C.D., Field, T.G., Tatum, J.D., Scanga, J.A. and Smith, G.C. (2001) National market cow and bull beef quality audit, 1999. A survey of producer-related defects in market cows and bulls. *Journal of Animal Science* 79, 658–665.

Rothschild, M.F. (1998) Selection for disease resistance in the pig. In: *Proceedings of the National Swine Improvement Federation*, North Carolina State University, Raleigh, North Carolina.

Sather, A.P., Murray, A.C., Zawadski, S.M. and Johnson, P. (1991) The effect of the halothane gene on pork product quality of pigs reared under commercial conditions. *Canadian Journal of Animal Science* 71, 959–967.

Scanga, J.A., Belk, K.E., Tatum, J.D., Grandin, T. and Smith, G.C. (1998) Factors contributing to the incidence of dark cutting beef. *Journal of Animal Science* 76, 2040–2047.

Shorthose, W.R. and Wythes, J.R. (1988) Transport of sheep and cattle. In: *34th International Congress of Meat Science and Technology*, CSIRO Meat Research Laboratory, Cannon Hill, Queensland, Australia.

Smith, G.C., Savell, H.W., Dolezal, H.G., Field, T.G., Gill, D.G., Griffin, D.B., Hale, D.S., Morgan, J.B., Northcutt, S.L., Tatum, J.D., Ames, R., Boleman, S., Gardner, B., Morgan, W., Smith, M., Lamber, C. and Cowman, G. (1995) *Improving the Quality, Consistency, Competitiveness and Market Share of Beef – a Blueprint for Total Quality Management of the Beef Industry*. The final report of the National Beef Quality Audit (1995). Final report to the National Cattlemen's Beef Association, Colorado State University, Fort Collins, Oklahoma State University, Stillwater, and Texas A & M University, College Station, Texas.

Smith, G.C., Tatum, J.D., Belk, K., Scanga, J.D., Grandin, T. and Sofos, J.N. (2005) Traceability from a U.S. perspective. *Meat Science* 71, 174–193.

Smith, G.C., Savell, J.W., Morgan, B. and Lawrence, T.E. (2006) *National Beef Quality Audit*. Colorado State University, Oklahoma State University and Texas A & M University for the National Cattlemen's Beef Association, Englewood, Colorado.

Smulders, F.J.M. (1983) Pre-stunning treatment during lairage and meat quality. In: Eikelenboom, G. (ed.) *Stunning of Animals for Slaughter*. Martinus Nijhoff, The Hague, Netherlands, pp. 90–95.

Snowder, G.D., Van Vleck, L.D., Cundiff, L.V and Bennett, G.L. (2006) Bovine respiratory diseases in feedlot cattle: environmental, genetic and economic factors. *Journal of Animal Science* 84, 1999–2008.

Suther, S. (2006) *CAB (Certified Angus) Drovers Survey Benchmarks for Producer Practices and Opinions*. Drover's Journal, Vance Publishing, Lenexa, Kansas, pp. 18–19.

Tarrant, P.V. (1993) An overview of production, slaughter and processing factors that affect pork quality: general review. In: Puolanne, E., Demeyer, D.I., Ruusunen, M. and Ellis, S. (eds) *Pork Quality: Genetic and Metabolic Factors*. CAB International, Wallingford, UK.

Terlouw, C. (2005) Stress reactions at slaughter and meat quality in pigs: genetic background and prior experience. A brief review of recent findings. *Livestock Production Science* 94, 125–135.

Texas A & M University (1998–1992) *Ranch-to-Rail Statistics*. Department of Animal Science, College State, Texas.

Troxel, T.R., Barham, B.L., Cline, S., Foley, D., Hardgrave, R., Wiedower, R. and Wiedower, W. (2006) Management factors affecting selling prices of Arkansas beef calves. *Journal of Animal Science* 84 (Suppl. 1), 12 (abstract).

Van Donkergoed, J., Jewison, G., Mann, M., Cherry, B., Altwasser, B., Lower, R., Wiggins, K., Dejonge, R., Thorlakson, B., Moss, E., Mills, C. and Grogen, H. (1997) Canadian Beef Quality Audit. *Canadian Journal of Animal Science* 38, 217–225.

Vann, R.C., Randel, R.D., Welsh, T.H., Willard, S.T., Carroll, V.A., Brown, M.S. and Lawrence, T.E. (2006) Influence of breed type and temperament on feedlot growth and carcass characteristics of beef steers. *Journal of Animal Science* 84 (Suppl. 1), 396 (abstract).

VanSickle, J. (2006) Survey shows lower incidence of PSE. *National Hog Farmer Blueprint*, 15 April, 42.

Verstegen, W.M.A. and Den Hartog, L.A. (2005) Negative effects of stress immediately before slaughter on pork quality are aggravated by suboptimal transport and lairage conditions. *Journal of Animal Science* 83, 440–448.

Vogel, K. (2006) In: Wren, G. (ed.) Prevent bruising in cows. *Bovine Veterinarian*, February, 30–32.

Voisinet, B.D., Grandin, T., Tatum, J.D., O'Connor, S.F. and Struthers, J.J. (1997a) *Bos indicus*-cross feedlot cattle with excitable temperaments have tougher meat and a higher incidence of borderline dark-cutters. *Meat Science* 46, 367–377.

Voisinet, B.D., Grandin, T., Tatum, J.D., O'Connor, S.F. and Struthers, J.J. (1997b) Feedlot cattle with calm temperaments have higher average daily gains than cattle with excitable temperaments. *Journal of Animal Science* 75, 892–896.

Waggoner, J.W., Mathis, C.P., Loest, C.A., Sawyer, J.E. and McColum, F.T. (2006) Impact of feedlot morbidity on performance, carcass characteristics and profitability of New Mexico ranch-to-rail steers. *Journal of Animal Science* 84 (Suppl. 1), 12 (abstract).

Warriss, P.D. (2003) Optimal lairage times and conditions for slaughter pigs: a review. *Veterinary Record* 153, 170–176.

Warriss, P.D. and Brown, S.N. (1994) A survey of mortality in slaughter pigs during transport and lairage. *Veterinary Record* 134, 513–515.

Warriss, P.D., Brown, S.N. and Adams, S.J.M. (1994) Relationship between subjective and objective assessment of stress at slaughter and meat quality in pigs. *Meat Science* 38, 329–340.

Weary, D.M., Braithwaite, L.A. and Fraser, D. (1998) Vocal response to pain in piglets. *Applied Animal Behavior Science* 56, 161–172.

Weeks, C.A., McNeilly, P.W. and Warriss, P.D. (2002) Influence of design of facilities at auction markets and animal handling procedures on bruising cattle. *Veterinary Record* 150, 742–748.

White, R.G., deShazer, J.A. and Trassler, C.J. (1995) Vocalization and physiological responses of pigs during castration with and without a local anesthetic. *Journal of Animal Science* 73, 381–386.

Zwald, N.R., Weigel, K.A., Chang, Y.M., Welper, R.D. and Clay, J.S. (2004a) Genetic selection for health traits using producer recorded data: I. Incidence rates, heritability estimates and sire breeding values. *Journal of Dairy Science* 87, 4287–4294.

Zwald, N.R., Weigel, K.A., Chang, Y.M., Welper, R.D. and Clay, J.S. (2004b) Genetic selection for health traits using producer-recorded data II. Genetic correlations, disease probability and relationship with existing traits. *Journal of Dairy Science* 87, 4295–4302.

2 General Principles of Stress and Well-being

P.B. Siegel[1] and W.B. Gross[2]

[1]Department of Animal and Poultry Sciences; [2]Department of Large Animal Clinical Sciences, Virginia Polytechnic Institute and State University, Blacksburg, Virginia, USA

Introduction

At every level of biological organization, be it molecular, cellular, organismic, population or interspecies, equilibrium patterns emerge. The concept of evolutionary stable strategies in natural populations is well accepted in biological circles. Artificial selection, the foundation of livestock and poultry breeding programmes, can disrupt these patterns and can have major implications in production systems. Accordingly, responses of animals to their environments are easier to evaluate when viewed as aims and strategies for survival. Nature is dynamic, and maintenance of variation within and between populations enhances their adaptation to environmental changes and associated stressors. Thus, conservation of genetic variation allows for ecological niches to be filled, in both the short and long term.

In contrast to populations, the genome of an individual remains constant (barring mutation) throughout life. Factors influencing genetic variation of populations include selection, mutation, migration and chance. Although some individuals do not survive or reproduce and there is unequal reproduction among those that do, an individual's aim is to pass its genes on to subsequent generations. Those individuals best adapted to the current environment seem to have the greatest opportunity of accomplishing this objective. Owners of livestock and poultry may prefer uniformity and high productivity

with no morbidity or mortality. These preferences, however, may be in conflict with the maintenance of genetic variation in populations and with allocation of resources by the animal as it passes through various environments during its life.

Genetic variation in animal populations allows some individuals to survive and exploit environmental changes, which results in differential reproduction. Therefore, within a population there is a range of structural, biochemical, behavioural and disease resistance factors that are under varying degrees of genetic influence and whose phenotypic (outward) expression may be modified by the environment.

Performance in a cow–calf setting may differ from that in the feedlot. Over a period of generations, genetic changes within the population can result in an increasing frequency of individuals that adapt to environmental changes. These genetic changes may occur due to differences in husbandry practices influencing selection (e.g. Muir and Craig, 1998) and in variation among individuals in resistance to diseases (e.g. Sarkar, 1992).

Genetic factors also influence the allocation of an animal's resources to its various components. In response to environmental changes, allocation of resources can also be altered by the stress system. Although this chapter explores relationships between stress and well-being of animals with particular emphasis on studies of chickens, the results are also valid for other farm animals (e.g. Broom, 1988). That is, the paradigm is consistent across livestock and poultry, with human intervention greater in poultry because in most production systems parent–offspring behaviour is not relevant.

How Animals Respond to Stress

Stress is a norm in social animals and our understanding of the stress system is based on that of Selye (1950, 1976), where the major components include the cerebrum, hypothalamus, pituitary, adrenals, glucocorticoids and a cascading of responses as animals attempt to respond to stressors. Glucocorticoids (cortisol and corticosterone) are carried by the blood to all cells of the body, where they enter the nucleus. They then regulate the translation of active genes of the cell into mRNA, which migrates to the mitochondria where encoded proteins are produced.

Understanding the stress system, however, is elusive and suffers from confusion and controversy. The stress response may be viewed as a physiological mechanism that links the stressor to a target organ, and target organ effects may be positive or negative. Thus, stress per se may or may not be damaging, and there can be positive aspects of stress (Zulkifli and Siegel, 1995; Hangalapura, 2006). Zulkifli and Siegel (1995) cite the Yerkes-Dodson law that relates degree of stress and performance efficiency. This law states that performance will be enhanced as arousal increases, but only up to a certain point or optimum level: exceeding the optimum leads to inefficiency. This concept of 'optimum stress' is discussed throughout this chapter.

Stress occurs when an animal experiences changes in the environment that stimulate body responses aimed at re-establishing homeostatic conditions (Mumma et al., 2006). Responses to stressors can include anatomical, physiological and/or behavioural changes. Models to study these responses using continuous infusion of adrenocorticotropin (Thaxton and Puvadolpirod, 2000) or the feeding of corticosterone (Post et al., 2003) have provided an index of the multitude of these changes.

Another type of index involves developmental stability, in which comparisons are made of the degree of asymmetry in bilaterally symmetrical traits (Moller and Swaddle, 1997). Although there is a voluminous literature on this topic, reports using these criteria for measuring genetic (e.g. Yang et al., 1997) and environmental stressors (Yalcin et al., 2005) in livestock and poultry remain sparse (e.g. Moller et al., 1995; Yang et al., 1997; Yalcin et al., 2005). This dearth of research is puzzling, because it allows for diagnostic and retrospective analyses of stressors that may have occurred during life.

Although normal values of various criteria differ not only among livestock and poultry species but also among stocks within classes of livestock and poultry, the general patterns of response are similar, with long-term responses resulting in an increase in size of the adrenal glands and a reduction of lymphoid mass. After removal of the stressor, the return to their prior size is rapid (Gross et al., 1980). Animals differ in their threshold of response to stressors and in the degree of response once thresholds are reached (Siegel et al., 1989). These differences were observed in immunoresponsiveness, efficiency of food utilization, growth, feathering and relative weights of liver, spleen, testes, breast muscle and abdominal fat.

Blood plasma levels of cortisol and/or corticosterone are frequently used as criteria for measuring responses to stressors. Because the utilization of a given blood level of corticosterone differs among individuals, it is sometimes difficult to correlate blood levels of glucocorticoids with other manifestations of a stressor (Panretto and Vickery, 1972). Blood levels of the thyroid hormone T_3 have also been used as a measurement of the stress response in calves (Friend et al., 1985) and turkeys (Yahav, 1998). When plasma corticosterone and thyroid hormones were used to measure effects of long-term stress of chickens under various housing conditions, Gibson et al. (1986) concluded that results were equivocal and that these hormones were not especially useful measures for long-term stress. However, cortisol and/or corticosterone are useful measures of the response to acute, short-term stress induced by handling or restraint. Also involved in responses to hormones are differences in receptors.

In response to stressors, plasma and tissue levels of ascorbic acid may be reduced (Kechik and Sykes, 1979), and it is not surprising that effects of ascorbic acid on stress responses have been studied (e.g. Gross, 1988, 1992b). Alternatives to plasma T_3 and glucocorticoid levels as measures of stressors include various physiological or immunological responses. The number of blood lymphocytes decreases and the number of polymorphs increases after a stressful event (Maxwell and Robertson, 1998). The difficulty of large, normal variation in the numbers of leucocytes can be largely circumvented by the use of polymorph:lymphocyte (P:L) ratios. Recent advances in technology have allowed for increased use of changes in endogenous opioids as measures of stress (e.g. Sapolsky, 1992), including those in sows (Zanella et al., 1998).

The stress system allows animals to allocate resources based on their perception of the environment as well as on direct physical insults from the environment. An animal can be stressed by any change in its internal and external environments. Examples include: (i) rate of growth; (ii) reproductive state; (iii) climate; (iv) unusual sound or light; (v) social interactions; (vi) availability of food and water; (vii) handling and moving; (viii) injected materials such as killed bacteria and some vaccines; and (ix) a disease already in progress (e.g. Siegel, 1980; Freeman, 1987; Gross, 1995; Pierson et al., 1997; Zulkifli et al., 2006).

Profiles and criteria of responses may vary with time (Mitchell and Kettlewell, 1998; Mumma et al., 2006). For example, in chickens, exposure to a short-term stressor such as a brief sound resulted in changes in blood profiles such as heterophil:lymphocyte (H:L) ratios within 18 h. The response peaked at about 20 h and returned to normal in about 30 h (Gross, 1990b), whereas peak H:L ratios were observed 4 h after corticosterone administration (Gross, 1992a).

There may be considerable variation among individuals in their perception of the stressfulness of an event, absorption of glucocorticoids from the blood and response of tissues to glucocorticoids. Improved environmental quality influences these associations (Hester et al., 1996a, b) and increases the correlation between the antibody titre responses of individuals to two different red blood cell (RBC) antigens (Gross and Siegel, 1990).

Levels of stress can also be estimated by presence of diseases. When stress levels are too high, viral diseases and other diseases that stimulate a lymphoid response are more common (Gross and Siegel, 1997). Cell-mediated immunity is reduced, resulting in an increased incidence of tumours and coccidiosis (Gross, 1972, 1976).

When levels of stress are too low, bacterial and parasitic diseases are more common and responses to some toxins are more severe (Brown et al., 1986). At an 'optimum stress' level, incidence to essentially all diseases is reduced. 'Optimum stress', however, may vary among genetic stocks and between individuals within a population because of both differences in their background and prior experiences.

Resource Allocations

Resource allocations should convey the concept that at any particular time resources available to an individual are finite (Rauw et al., 1998). Therefore, there will always be competition for resources among body functions such as growth, maintenance, reproduction and health. Added to this mix are responses to stressors which result in a redistribution of resources. A hypothetical example of redistribution of resources is seen in Fig. 2.1, where a comparison between a stressed and a healthy (biologically balanced) animal is presented. In this example, a healthy animal has a reserve of 10% for the maintenance of health and an equal division of resources for growth, reproduction and maintenance.

When it becomes stressed, however, resources allocated for growth and reproduction become nil and there is a reduction in those available for maintenance because resources have been directed toward improvement of health status. For redistribution of resources to occur it may be essential in some cases that changes occur quickly, while in other cases changes may be more gradual. In addition, the magnitude of response to a stressor may be influenced by the animal's perception not only of its current environment, but also of how the current environment differs from prior environments. Therefore, animals may also acquire, allocate and redistribute resources based on past history and their perceptions of the environment.

Biologically balanced

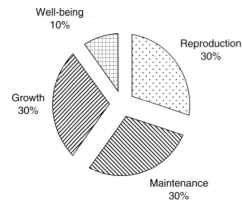

Well-being
10%

Reproduction
30%

Growth
30%

Maintenance
30%

Stressed

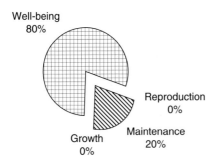

Well-being
80%

Reproduction
0%

Growth
0%

Maintenance
20%

Fig. 2.1. An example of the allocation of resources within a healthy (top) and a stressed (bottom) animal.

The following is an example of resource allocations (Gross and Siegel, 1997). Chickens fed on alternate days exhibit greater resistance to an *Escherichia coli* challenge than those fed *ad libitum*. When subsequently allowed *ad libitum* feeding, their rate of weight gain was greater and their resistance to *E. coli* challenge was less than that of those fed *ad libitum*. It is possible that the restricted feed supply yielded a stress response appropriate for conditions likely to include a bacterial or parasitic challenge. The chickens thus allocated finite resources to defence instead of growth. Adequate feeding then yielded a response appropriate for conditions where a bacterial or parasitic challenge was unlikely. At this point, the need to regain a genetically desired body weight had priority over maintaining a high level of well-being defence. This example is not unlike

that seen with parasitic infections in growing animals (Vanimisetti *et al.*, 2004a, b).

Long-term Higher, Lower and 'Optimum' Stress

'Optimum stress' is a relative term and when we write of higher, lower and optimum we do not wish to imply that if a little stress can be beneficial more is even better. The optimum will vary with genetic stock, prior experience of the animals and the environment. In one experiment with male chickens, Gross and Siegel (1981) characterized environments as high, medium or low social stress.

The low social stress environment consisted of maintaining chickens in individual cages with solid sides and wire fronts, floors and backs with water and feed available continuously. These chickens could hear but not see each other. The high social stress environment consisted of larger cages housing five males. Each day, however, one individual per cage was moved into another cage according to a plan that reduced the possibility of contact with previous individuals.

The medium social stress environment consisted of caging nine males in a series of cages throughout. When continuously exposed for over 3 months to the high social stressor feed consumption was not affected, whereas body weight gain, feed efficiency and the correlation between antibody titres and resistance to *Mycoplasma gallisepticum* challenge infection were reduced. Even though there was a reduction in lymphoid mass, antibody responsiveness to RBCs was not changed and H:L ratios ranged between 0.6 and 1.2. An H:L ratio above 1.3 usually indicates a disease in progress.

When animals are exposed to stressful environments then growth potential is reduced, adaptability increases even though senses are less acute, discrimination is improved and activity is increased. As an animal becomes better adapted to a harsher environment, resources are diverted from growth and reproduction to respond to the stressor. Chickens exposed to the low social stress environment became lethargic, exhibited less preening and their vocalizations suggested contentment. Initially, weight gain and feed efficiency were increased, but after 3 months in this environment feed consumption, growth rate and feed efficiency were greatly reduced. Cockerels maintained in the medium social stress environment maintained their body weight throughout and had the best feed efficiency of the three groups. Also, they had the highest antibody titres to ovine red blood cell antigen. Ranking of the three environments according to the stress hormone corticosterone was high > medium > low.

At low levels of environmental stress, phenotypic variability was reduced and the chickens became unusually susceptible to bacterial infection (Larsen et al., 1985), demonstrating the need for some stress (stimulation) in order to maintain more efficient biochemical activity. Because greater or lesser levels of stress seem to be detrimental, the aim of good husbandry should be to provide an 'optimum stress' level. It is probable that the optimum may vary according to genetic stock and prior experience. Reviews of this topic have been provided by Jones (1987), Zulkifli and Siegel (1995) and Gross and Siegel (1997).

Environmental Stress and Disease Defence

Activation of an animal's defences against disease requires that resources be diverted from growth and reproduction. The stress system allows animals to maintain disease defence at a level commensurate with the risk. When at a low population density, chances of an animal encountering infective levels of bacteria or parasites are reduced, and the need for a phagocytic defence against them is reduced.

As population densities increase, the probability of encountering pathogenic concentrations of bacteria and parasites in the environment also increases. Higher levels of stress enhance defence against bacteria and parasites; however, there is a cost in growth and reproduction. As stress (glucocorticoid) levels increase, levels of superoxide radical in polymorphs increase which, in turn, enhance the ability to destroy bacteria (Som et al., 1983). In contrast, low-stress environments increase susceptibility to opportunistic bacteria such as coliforms, faecal streptococci

Resistant				→ Susceptible	
MG	HH	HL	LH	LL	antibody
EC	HL	HH	LL	LH	heterophil
SV	LH	LL	HH	HL	T-lymphocyte
BT	LL	HL	LH	HH	macrophage

Fig. 2.2. Some results comparing pathogens, parental lines (HH and LL) and their reciprocal crosses (HL and LH). MG, *Mycoplasma gallisepticum*; EC, *Escherichia coli*; SV, splenomeglia virus; BT, *Bacterium tuberculosis*.

and staphylococci (Larsen et al., 1985; Siegel et al., 1987) and to internal and external parasites such as mites and coccidia (Hall et al., 1979).

In higher-stress environments, numbers of chickens susceptible to viral infections and tumours increased (Mohamed and Hanson, 1980; Thompson et al., 1980). In an 'optimal stress' environment chickens are not highly susceptible to bacteria, parasites, viruses or tumours. Once again we emphasize that optimum is a relative term that will vary from flock to flock and herd to herd.

For those interested in further detail, examples from experiments conducted in our laboratory are provided in a summary review (Gross and Siegel, 1997). We would be remiss in not pointing out the conundrum: where there is resistance, the mode of resistance can be an asset in one situation and a liability in another. We compared parental lines and their crosses to four pathogens where the primary body defence differed (Gross et al., 2002). The results showed four different scenarios (see Fig. 2.2) where the combination most resistant to MG was least resistant to TB, and vice versa. The most resistant to E. coli was least resistant to splenomeglia virus and vice versa.

Modifying the Stress Response

Within a population, individuals may differ genetically in their perception of stressors, resulting in considerable variation in the response to the same stressor. Because responses of individuals to different stressors also vary (see Fig. 2.3), it is possible through artificial selection to develop

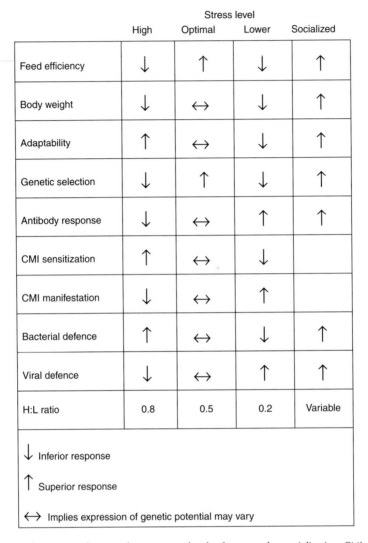

	Stress level			
	High	Optimal	Lower	Socialized
Feed efficiency	↓	↑	↓	↑
Body weight	↓	↔	↓	↑
Adaptability	↑	↔	↓	↑
Genetic selection	↓	↑	↓	↑
Antibody response	↓	↔	↑	↑
CMI sensitization	↑	↔	↓	
CMI manifestation	↓	↔	↑	
Bacterial defence	↑	↔	↓	↑
Viral defence	↓	↔	↑	↑
H:L ratio	0.8	0.5	0.2	Variable

↓ Inferior response

↑ Superior response

↔ Implies expression of genetic potential may vary

Fig. 2.3. A general summary of expected responses to levels of stress and to socialization. CMI, cell-mediated immunity.

populations which have reduced or increased responses to specific stressors (Gross and Siegel, 1985; Faure and Mills, 1998; Jones *et al.*, 2005). This is not unlike what has occurred via natural selection in indigenous livestock and poultry populations.

When animals are repeatedly exposed to the same stressor and the magnitude of each succeeding response is reduced, it can be said that habituation has occurred. Habituation may involve memory and/or physiological manifestations (Siegel, 1989). Prior exposure of young chickens to acute thermal stressors appears to improve heat tolerance later in life. This preconditioning does not have to be instigated by the same stressor, but one requirement is the synthesis and liberation of glucocorticoids (Zulkifli and Siegel, 1995; Zulkifli *et al.*, 1995). These findings are consistent with *in vitro* and *in vivo* studies suggesting that macrophages respond to thermal and non-thermal stressors by producing similar kinds of 'stress proteins' (Miller and Qureshi, 1992).

Effects of stress may be alleviated by chemicals that inhibit the production of adrenal glucocorticoids. One such chemical is ascorbic acid

(Gross *et al.*, 1988b; Gross, 1992b). Others are adrenal-blocking chemicals such as metyrapone (Zulkifli *et al.*, 1995) and 1,1-dichloro-2,2bis/ p-chlorophenyl/ethane (Gross, 1990a). After the administration of an optimal dose of such compounds, the physiological manifestations of stress are reduced. Examples include weight loss associated with transportation, inhibition of viral infections and tumours, increased feed efficiency, reduced effects of heat stress and reduction in the stress inhibition of antibody responsiveness (Gross, 1989, 1990a).

Human–Animal Relationships

Relationships between animals and their handlers can greatly affect responses of those animals to a range of factors and it is not surprising that the roles of stockpersons and veterinarians have been much studied in numerous livestock and poultry species (e.g. Hemsworth and Barnett, 1987; Gross and Siegel, 1997; Hemsworth and Gonyou, 1997; Odendaal, 1998; Sambraus, 1998). Socialized animals welcome the presence of their handlers. The process of socialization may be accomplished by frequent exposure to kind care and handling, beginning at the earliest possible age.

Long-term effects of the human–animal relationship (Jones, 1987; Barnett *et al.*, 1992), coupled with background genotype and prior experiences, influence subsequent responses to various situations (e.g. Gross and Siegel, 1981, 1982; Nicol, 1992). Positive socialization with humans can result in animals approaching caretakers. Negative socialization can result in escape behaviours, and ignored individuals exhibit fear when exposed to humans.

Although gentle handling may exert its strongest influence by facilitating habituation to humans (Jones and Waddington, 1992), feed efficiencies, body weights and antibody responses to RBC antigens are higher for positively socialized chickens than those held under similar environments and ignored. Responses to stressors and to the administration of corticosterone are greatly reduced (Gross and Siegel, 1982) and resistance to most diseases is enhanced in socialized chickens. Both the stressfulness of the environment and socialization influence the responses of chickens (see Fig. 2.3). Socialized chickens,

which are in an 'optimal stress' environment, seem to exhibit the most favourable responses.

Genetic Factors

Interactions with the environment

The response of an animal to environmental factors is determined by its genetic background, as modified by prior environmental experiences. The first week after birth on the human–animal relationship can be very important. Although the expression of traits may differ between stocks, in an environment where stress is optimal genetic differences may be more evident because exposure to environmental stressors may modify genetic influences on expression of traits (Gross *et al.*, 1988a).

Genotype–environment interactions occur when, relative to each other, a series of genotypes do not respond similarly in a series of environments. An example of this was discussed previously and is shown in Fig. 2.2. Numerous reports appeared in the 1970s demonstrating behavioural involvement with genotype–environment interactions. Implications of these interactions can be considerable when viewed in the context of well-being (e.g. Mathur and Horst, 1994).

In experiments measuring production and disease resistance in lines of chickens selected for high- or low-corticosterone response to social strife, extreme divergence of responses were observed between the high line in a higher social strife environment and the low line in a lower social strife environment (Siegel, 1989). The low–low combination was more susceptible to infections from endemic bacteria and external parasites, and the high–high combination to viral infections (Gross and Siegel, 1997). Whether or not an extreme response was advantageous depended upon the measurement criteria. These data are consistent with the view that overall well-being may be at an optimal level when the animal is neither under- nor over-stressed (Zulkifli and Siegel, 1995).

Relationship with well-being

Many components are involved in the development of an animal's well-being. One is adequate food and water; another is protection from environmental insults such as adverse weather. Protection from pathogenic organisms and predators is also important. Stable social (between animals) and physical environments are valuable and contribute to an optimum stress level. These needs are met by good husbandry, which is rewarded by increased productivity and more uniform responses to experimental procedures.

Summary

Although many of the examples in this chapter have come from research conducted with the chicken, our experimental animal, the implications are relevant to all forms of livestock and poultry that are husbanded in flocks and herds. In our opinion, the most important factor affecting the well-being of livestock and poultry is their relationship with their human associates. Kind care (socialization) has many well-being and production benefits. Socialized animals are easier to work with, have enhanced productivity, are more adaptable to adverse environments, are more resistant to diseases and produce better immunity. These factors make genetic selection easier. The responses of individuals within groups are more uniform, thus reducing the number of animals needed for research. Socialization can be easily applied to large groups of animals.

All people who work with animals should be aware of physical, nutritional and behavioural needs and should be able to relate positively to the feelings of the animals. The attitude of handlers is an essential factor in determination of that level of stress which enhances the animal's well-being.

Acknowledgement

Appreciation is extended to C.F. Honaker for assistance in manuscript preparation.

References

Barnett, J.L., Hemsworth, P.H. and Newman, E.A. (1992) Fear of humans and its relationship with productivity in laying hens at commercial farms. *British Poultry Science* 33, 699–710.

Broom, D.M. (1988) Relationship between welfare and disease susceptibility in farm animals. In: Gibson, T.E. (ed.) *Animal Disease – a Welfare Problem*. British Veterinary Association Animal Welfare Foundation, London, pp. 20–29.

Brown, C.W., Gross, W.B. and Ehrich, M. (1986) Effect of social stress on the toxicity of malathion in young chickens. *Avian Diseases* 30, 679–682.

Faure, J.M. and Mills, A.D. (1998) Improving adaptability of animals by selection. In: Grandin, T. (ed.) *Genetics and the Behavior of Domestic Animals*. Academic Press, San Diego, California, pp. 235–264.

Freeman, B.M. (1987) The stress syndrome. *World's Poultry Science Journal* 43, 15–19.

Friend, T.H., Dellomeier, G.R. and Gbur, E.E. (1985) Comparison of four methods of calf confinement. 1. Physiology. *Journal of Animal Science* 60, 1095–1101.

Gibson, S.W., Hughes, B.O., Harvey, S. and Dun, P. (1986) Plasma concentrations of corticosterone and thyroid hormones in laying fowl from different housing systems. *British Poultry Science* 27, 621–628.

Gross, W.B. (1972) Effect of social stress on the occurrence of Marek's disease in chickens. *American Journal of Veterinary Research* 33, 2275–2279.

Gross, W.B. (1976) Plasma steroid tendency, social environment, and *Eimeria necatrix* infection. *Poultry Science* 55, 1508–1512.

Gross, W.B. (1988) Effect of environmental stress on the responses of ascorbic acid-treated chickens to *Escherichia coli* challenge infection. *Avian Diseases* 32, 432–436.

Gross, W.B. (1989) Effect of adrenal blocking chemicals on viral and respiratory infections of chickens. *Canadian Journal of Veterinary Research* 53, 48–51.

Gross, W.B. (1990a) Effect of adrenal blocking chemicals on the responses of chickens to environmental stressors and ACTH. *Avian Pathology* 19, 295–304.

Gross, W.B. (1990b) Effect of exposure to a short duration sound on the stress response of chickens. *Avian Diseases* 34, 759–761.

Gross, W.B. (1992a) Effect of short term exposure to corticosterone on resistance to challenge exposure with *Escherichia coli* and antibody response to sheep erythrocytes. *American Journal of Veterinary Research* 53, 291–293.

Gross, W.B. (1992b) Effects of ascorbic acid on stress and disease resistance in chickens. *Avian Diseases* 36, 388–392.

Gross, W.B. (1995) Relationship between body-weight gain after movement of chickens to an unfamiliar cage and response to *Escherichia coli* challenge infection. *Avian Diseases* 39, 636–637.

Gross, W.B. and Siegel, P.B. (1981) Long term exposure of chickens to three levels of social stress. *Avian Diseases* 25, 312–325.

Gross, W.B. and Siegel, P.B. (1982) Socialization as a factor in resistance to infection, feed efficiency, and response to antigens in chickens. *American Journal of Veterinary Research* 43, 2010–2012.

Gross, W.B. and Siegel, P.B. (1985) Selective breeding of chickens for corticosterone response to social stress. *Poultry Science* 64, 2230–2233.

Gross, W.B. and Siegel, P.B. (1990) Genetic–environmental interactions and antibody response to two antigens. *Avian Diseases* 34, 843–847.

Gross, W.B. and Siegel, P.B. (1997) Why some get sick. *Journal of Applied Poultry Science* 6, 453–460.

Gross, W.B., Siegel, P.B. and DuBose, R.T. (1980) Some effects of feeding corticosterone. *Poultry Science* 59, 516–522.

Gross, W.B., Domermuth, C.H. and Siegel, P.B. (1988a) Genetic and environmental effects on the response of chickens to avian adenovirus group II infection. *Avian Pathology* 17, 767–774.

Gross, W.B., Jones, D. and Cherry, J. (1988b) Effect of ascorbic acid on the disease caused by *Escherichia coli* challenge infection. *Avian Diseases* 32, 407–408.

Gross, W.B., Siegel, P.B. and Pierson, F.W. (2002) Effects of genetic selection for high or low antibody response on resistance to a variety of disease challenges and the relationship of resource allocation. *Avian Diseases* 46, 1007–1010.

Hall, R.D., Gross, W.B. and Turner, E.C. (1979) Population development of *Ornithonyssus sylvarum* on leghorn roosters inoculated with steroids or subjected to extreme social interaction. *Veterinary Parasitology* 5, 287–297.

Hangalapura, B.N. (2006) Cold stress and immunity. PhD dissertation, Wageningen University, Wageningen, Netherlands.

Hemsworth, P.H. and Barnett, J.L. (1987) Human–animal interactions. In: Price, E.V. (ed.) *Farm Animal Behavior*. W.B. Saunders, Philadelphia, pp. 339–356.

Hemsworth, P.H. and Gonyou, H.W. (1997) Human contact. In: Appleby, M.C. and Hughes, B.O. (eds) *Animal Welfare*. CAB International, Wallingford, UK, pp. 205–217.

Hester, P.Y., Muir, W.M., Craig, J.V. and Albright, J.L. (1996a) Group selection for adaptation to multiple-hen cages: hematology and adrenal function. *Poultry Science* 75, 1295–1307.

Hester, P.Y., Muir, W.M. and Craig, J.V. (1996b) Group selection for adaptation to multiple-hen cages: humoral immune response. *Poultry Science* 75, 1315–1320.

Jones, R.B. (1987) Social and environmental aspects of fear. In: Zayan, R. and Duncan, I.J.H. (eds) *Cognitive Aspects of Social Behaviour in Domestic Fowl*. Elsevier, Amsterdam, pp. 82–149.

Jones, R.B. and Waddington, D. (1992) Modification of fear in domestic chicks, *Gallus gallus domesticus*, via regular handling and early environmental enrichment. *Animal Behaviour* 43, 1021–1033.

Jones, R.B., Marin, R.H. and Satterlee, D.G. (2005) Adrenocortical responses of Japanese quail to a routine weighing procedure and to tonic immobility induction. *Poultry Science* 84, 1675–1677.

Kechik, I.T. and Sykes, A.H. (1979) Effect of intestinal coccidiosis (*Eimeria acervulina*) on blood and tissue ascorbic acid concentration. *British Journal of Nutrition* 42, 97–103.

Larsen, C.T., Gross, W.B. and Davis, J.W. (1985) Social stress and resistance of chickens and swine to *Staphylococcus aureus*. *Canadian Journal of Comparative Medicine* 48, 208–210.

Mathur, P.K. and Horst, P. (1994) Methods for evaluating genotype–environment interactions illustrated by laying hens. *Journal of Animal Breeding and Genetics* 111, 265–288.

Maxwell, M.H. and Robertson, G.W. (1998) The avian heterophil leucocyte system: a review. *World's Poultry Science Journal* 54, 155–178.

Miller, L. and Qureshi, M.A. (1992) Heat shock protein synthesis in chicken macrophages: influence of *in vivo* and *in vitro* heat shock, lead acetate and lipopolysaccharide. *Poultry Science* 71, 980–998.

Mitchell, M.A. and Kettlewell, P.J. (1998) Physiological stress and welfare of broiler chickens in transit: solutions not problems. *Poultry Science* 77, 1803–1814.

Mohamed, M.A. and Hanson, R.P. (1980) Effect of social stress on Newcastle disease virus (LaSota) infection. *Avian Diseases* 24, 908–915.

Moller, A.P. and Swaddle, J.P. (1997) *Assymetry, Developmental Stability, and Evolution*. Oxford University Press, New York.

Moller, A.P., Sanorta, G.S. and Vestergaard, K.S. (1995) Developmental stability in relation to population density and breed of chickens *Gallus gallus*. *Poultry Science* 74, 1761–1771.

Muir, W.M. and Craig, J.V. (1998) Improving animal well-being through genetic selection. *Poultry Science* 77, 1781–1788.

Mumma, J.O., Thaxton, J.P., Vizzer-Thaxton, Y. and Dodson, W.L. (2006) Physiological stress in laying hens. *Poultry Science* 85, 761–769.

Nicol, C.J. (1992) Effects of environmental enrichment and gentle handling on behaviour and fear responses of transported broilers. *Applied Animal Behaviour Science* 33, 367–380.

Odendaal, J.S.J. (1998) The practising veterinarian and animal welfare as a human endeavour. *Applied Animal Behaviour Science* 59, 85–91.

Panretto, B.A. and Vickery, M.R. (1972) Distribution of cortisol and its rate of turnover in normal and cold-stressed sheep. *Journal of Endocrinology* 55, 519–531.

Pierson, F.W., Larsen, C.T. and Gross, W.B. (1997) The effect of stress on the response of chickens to coccidiosis vaccination. *Veterinary Parasitology* 73, 177–180.

Post, J., Rebel, M.J. and Ter Huure, H.M. (2003) Physiological effects of elevated plasma corticosterone concentrations in broiler chickens. An alternative means by which to access physiological effects of stress. *Poultry Science* 82, 1313–1318.

Rauw, W.M., Kranis, E., Noordhuizen-Stassen, E.N. and Grommers, F.J. (1998) Undesirable side effects of selection for high production efficiency in farm animals: a review. *Livestock Reproduction Science* 56, 15–33.

Sambraus, H.H. (1998) Applied ethology – its task and limits in veterinary practice. *Applied Animal Behaviour Science* 59, 39–48.

Sapolsky, R.M. (1992) Neuroendocrinology of the stress-response. In: Becker, J.B., Breedlove, S.M. and Crews, D. (eds) *Behavioral Endocrinology*. MIT Press, Cambridge, Massachusetts, pp. 287–324.

Sarkar, S. (1992) Sex, disease and evolution-variations on the theme from J.B.S. Haldane. *BioScience* 42, 448–454.

Selye, H. (1950) *Stress: the Physiology and Pathology of Exposure to Stress*. Acta, Montreal, Canada.

Selye, H. (1976) *The Stress of Life*. McGraw-Hill Book Co., New York.

Siegel, H.S. (1980) Physiological stress in birds. *BioScience* 30, 529–534.

Siegel, P.B. (1989) The genetic–behavioural interface and well-being of poultry. *British Poultry Science* 30, 3–13.

Siegel, P.B., Katanbaf, M.N., Anthony, N.B., Jones, D.E., Martin, A., Gross, W.B. and Dunnington, E.A. (1987) Response of chickens to *Streptococcus faecalis*: genotype–housing interactions. *Avian Diseases* 31, 804–808.

Siegel, P.B., Gross, W.B. and Dunnington, E.A. (1989) Effect of dietary corticosterone in young leghorn and meat-type cockerels. *British Poultry Science* 30, 185–192.

Som, S., Raha, C. and Chatterjee, I.B. (1983) Ascorbic acid: a scavenger of superoxide radical. *Acta Vitamineral Enzymology* 5, 243–250.

Thaxton, J.P. and Puvadolpirod, S. (2000) Model of physiological stress in chickens. Quantitative evaluation. *Poultry Science* 79, 391–395.

Thompson, D.L., Elgert, K.D., Gross, W.B. and Siegel, P.B. (1980) Cell-mediated immunity in Marek's disease virus-infected chickens genetically selected for high and low concentrations of plasma corticosterone. *American Journal of Veterinary Research* 41, 91–96.

Vanimisetti, H.B., Greiner, S.P., Zajac, A.M. and Notter, D.R. (2004a) Performance of hair sheep composite breeds: resistance of lambs to *Haemonchus contortus*. *Journal of Animal Science* 82, 595–604.

Vanimisetti, H.B., Andrew, S.L., Notter, D.R. and Zajac, A.M. (2004b) Inheritance of fecal egg count and packed cell volume and their relationship with production traits in sheep infected with *Haemonchus contortus*. *Journal of Animal Science* 82, 1602–1611.

Yahav, S. (1998) Physiological responses of chickens and turkeys to hot climate. *Proceedings of the 10th European Poultry Conference 1*, Jerusalem, World Poultry Science Association, pp. 84–91.

Yalcin, S., Ozkan, S., Cabuk, M., Buyse, J., Decuypere, E. and Siegel, P.B. (2005) Pre- and postnatal induced thermotolerance on body weight, physiological responses and relative asymmetry of broilers originating from young and old breeder flocks. *Poultry Science* 84, 967–976.

Yang, A., Dunnington, E.A. and Siegel, P.B. (1997) Developmental stability in stocks of White Leghorn chickens. *Poultry Science* 76, 1632–1636.

Zanella, A.J., Brunner, P., Unshelm, J., Mendl, M.T. and Broom, D.M. (1998) The relationship between housing and social rank, β-endorphin and dynorphin (1–13) secretion in sows. *Applied Animal Behaviour Science* 59, 1–10.

Zulkifli, I. and Siegel, P.B. (1995) Is there a positive side to stress? *World's Poultry Science Journal* 51, 63–76.

Zulkifli, I., Siegel, H.S., Mashaly, M.M., Dunnington, E.A. and Siegel, P.B. (1995) Inhibition of adrenal steroidogenesis, neonatal feed restriction, and pituitary–adrenal axis response to subsequent fasting in chickens. *General and Comparative Endocrinology* 97, 49–56.

Zulkifli, I., Norazilna, I., Htin, N.N. and Juriah, K. (2006) Physiological and behavioural responses of laying hens to repeated feed deprivation. *Archiv fur Geflugelkunde* 70, 22–27.

3 Causes of Poor Welfare and Welfare Assessment during Handling and Transport

Donald M. Broom

Centre for Animal Welfare and Anthrozoology, Department of Veterinary Medicine, University of Cambridge, Cambridge, UK

Introduction

The handling, loading, transporting and unloading of animals can have very substantial effects on their welfare. The welfare of an individual can be defined as its state with regard to its attempts to cope with its environment (Broom, 1986) and includes both the extent of failure to cope and the ease or difficulty in coping. Health is an important part of welfare, whilst feelings such as pain, fear and various forms of pleasure are components of the mechanisms for attempting to cope, and so should be evaluated where possible in welfare assessment (Broom, 1998, 2001b, 2006).

Where an individual is failing to cope with a problem, it is said to be stressed. Stress is an environmental effect on an individual which overtaxes its control systems and reduces its fitness or appears likely to do so (Broom and Johnson, 2000). If the effect of the environment is just stimulation, useful experience or an adrenal cortex response which has no adverse consequences, the individual is not stressed. Animal protection is a human activity which is directed towards the prevention of poor welfare in the animals. All stress involves poor welfare but there can be poor welfare without stress because there are no long-term consequences – for example, temporary pain or distress. All of these issues are discussed further in several papers in Broom (2001a).

In this chapter the factors which may result in stress during transport are first introduced. The methodology for assessment of the welfare of the animals during handling and transport is then explained. Finally, some of the various factors that affect the likelihood of stress are discussed, with examples from work on cattle and sheep.

Factors that may Result in Stress during Animal Handling and Transport

People are sometimes cruel to one another but generally believe that other people are aware and sentient so are likely to feel some guilt if they have been cruel. Non-human animals are regarded as aware and sentient by some people but as objects valued only according to their use by others. Hence there is a wide range of attitudes to animals, and these have major consequences for animal welfare.

During handling and transport, these attitudes may result in one person causing high levels of stress in the animals whilst another person doing the same job may cause little or no stress. People may hit animals and cause substantial pain and injury because of selfish financial considerations, because they do not consider that the animals are subject to pain and stress or because of lack of knowledge about animals and their welfare. Training of staff can substantially alter attitudes to, and treatment of, all animals.

Laws can have a significant effect on the ways in which people manage animals. Within the European Union (EU), the Council Regulation (EC) No 1/2005 'On the protection of animals during transport and related operations' takes up some of the recommendations of two separate reports: (i) the EU Scientific Committee on Animal Health and Animal Welfare Report *The Welfare of Animals during Transport (Details for Horses, Pigs, Sheep and Cattle) (March 2002)*; and (ii) the European Food Safety Authority *Report on the Welfare of Animals during Transport (2004)*, which deals with the other species. Laws have effects on animal welfare provided that they are enforced, and the mechanisms for enforcement within EU Member States are the subject of current discussion (2006).

Codes of practice can also have significant effects on animal welfare during transport. The most effective of these, sometimes just as effective as legislation, are retailer codes of practice, since retail companies need to protect their reputation by enforcing adherence to their codes (Broom, 2002).

Some animals are much better able to withstand the range of environmental impacts associated with handling and transport than are others. This may be because of genetic differences associated with the breed of the animal or with selection for production characteristics. Differences between individuals with regard to coping ability also depend on housing conditions and with the extent and nature of contact with humans and conspecifics during rearing.

Since physical conditions within vehicles during transport can affect the extent of stress in animals, the selection of an appropriate vehicle for transport is important in relation to animal welfare. Similarly, the design of loading and unloading facilities is of great importance. The person who designs the vehicle and facilities has a substantial influence, as does the person who decides which vehicle or equipment to use.

Before a journey starts, there must be decisions about the stocking density, grouping and distribution of animals on the vehicle. If there is withdrawal of food from those animals to be transported, this can affect welfare. For all species, tying of animals on a moving vehicle can lead to major problems, and for cattle and pigs any mixing of animals can result in very poor welfare.

The behaviour of drivers towards animals whilst loading and unloading and the way in which people drive vehicles are affected by the method of payment. If personnel are paid more for loading or driving quickly welfare standards will be worse, so such methods of payment should not be permitted. Payment of handling and transport staff at a higher rate for ensuring that the incidences of injury and poor meat quality are low improves welfare. Insurance against bad practice resulting in injury or poor meat quality should not be permitted.

All of the factors mentioned so far should be taken into account in the procedure of planning for transport. Planning should also take account of temperature, humidity and the risks of disease transmission. Disease is a major cause of poor welfare in transported animals. Planning of routes should take account of the needs of the animals for rest, food and water. Drivers or other persons responsible should have plans for emergencies, including a series of emergency contact numbers for obtaining veterinary assistance in the event of injury, disease or other welfare problems during a journey.

The methods used during handling, loading and unloading can have a major effect on animal welfare. The quality of driving can result in poor levels of welfare because of the animals' difficulty in maintaining balance, motion sickness, injury, etc. The ambient conditions – such as temperature and humidity – may change during a journey and require action on the part of the person responsible for the animals. A journey of long duration will have a much greater risk of poor welfare, and some situations inevitably lead to problems. Hence, good monitoring of the animals with inspections of adequate frequency – and in conditions that allow thorough inspection – are important.

Assessment of Welfare

A variety of welfare indicators that can be used to assess the welfare of animals being handled or transported are listed below. Some of these assessments are of short-term effects whilst others are more relevant to prolonged problems. With regard to animals being transported to slaughter, it is mainly the assessment of short-term effects such as behavioural aversion or increased heart rate that is used, but some animals are kept for a long period after transport and assessments such as increased disease incidence or suppression of normal development give information about the effects of the journey on welfare.

Assessments of welfare standards may incorporate the following (from Broom, 2000):

- Physiological indicators of pleasure.
- Behavioural indicators of pleasure.
- Extent to which strongly preferred behaviours can be shown.
- Variety of normal behaviours shown or suppressed.
- Extent to which normal physiological processes and anatomical development are possible.
- Extent of behavioural aversion shown.
- Physiological attempts at coping.
- Immunosuppression.
- Disease prevalence.
- Behavioural attempts at coping.
- Behaviour pathology.
- Brain changes, e.g. those indicating self-narcotization.
- Body damage prevalence.
- Reduced ability to grow or breed.
- Reduced life expectancy.

Details of these and other measures may be found in Broom (1998), Fraser and Broom (1997) and Broom and Johnson (2000).

Behavioural assessment

Changes in behaviour are obvious indicators that an animal is having difficulty coping with handling or transport, and some of these help to show which aspect of the situation is aversive. The animal may stop moving forward, freeze, back off, run away or vocalize. The occurrence of each of these can be quantified in comparisons of responses to different races, loading ramps, etc. Examples of behavioural responses – such as cattle stopping when they encounter dark areas or sharp shadows in a race and pigs freezing when hit or subjected to other disturbing situations – may be found in Grandin (1980, 1982, 1989, 2000).

Behavioural responses are often demonstrated during painful or otherwise unpleasant situations. Their nature and extent vary from one species to another according to the selection pressures that have acted during the evolution of the mechanisms controlling behaviour. Human approach and contact may elicit anti-predator behaviour in farm animals. However, with experience of handling, these responses can be greatly reduced in cattle (Le Neindre et al., 1996).

Social species that can collaborate in defence against predators, such as pigs or man, vocalize a lot when caught or hurt. Species which are unlikely to be able to defend themselves, such as sheep, vocalize far less when caught by a predator, probably because such an extreme response merely gives information to the predator that the animal attacked is severely injured and hence unlikely to be able to escape.

Cattle can also be relatively undemonstrative when hurt or severely disturbed. Human observers sometimes wrongly assume that an animal which is not squealing is not hurt or disturbed by what is being done to it. In some cases, the animal is showing a freezing response and, in most cases, physiological measures must be used to find out the overall response of the animal.

Within species, individual animals may vary in their responses to potential stressors. The coping strategy adopted by the animal can have an effect on responses to the transport and lairage situation. For example, Geverink et al. (1998) showed that those pigs that were most aggressive in their home pen were also more likely to fight during pre-transport or pre-slaughter handling, but pigs driven for some distance prior to transport were less likely to fight and hence cause skin damage during and after transport. This fact can be used to design a test that reveals whether or not animals are likely to be severely affected by the transport situation (Lambooij et al., 1995).

The procedures of loading and unloading animals into and out of transport vehicles can have very severe effects on animals, and these effects are revealed in part by behavioural responses. Species vary considerably in their responses to loading procedures. Any animal which is injured or frightened by humans during the procedure can show extreme responses. However, in most efficient loading procedures, sheep are not greatly affected and cattle are only sometimes affected. Broom *et al.* (1996) and Parrott *et al.* (1998b) showed that sheep show largely physiological responses and these are associated with the novel situation encountered in the vehicle rather than with the loading procedure.

Once a journey starts, some species of farm animals explore the compartment in which they are placed and try to find a suitable place to sit or lie down. Sheep and cattle try to lie down if the situation is not disturbing, but stand if it is. After a period of acclimatization of sheep and cattle to the vehicle environment, during which time sheep may stand for 2–4 h looking around at intervals and cattle may stand for rather longer, most animals will lie down if the opportunity arises. Unfortunately for the animals, many journeys involve so many lateral movements or sudden brakings or accelerations that they cannot lie down.

One important behavioural measure of welfare when animals are transported is the amount of fighting which they show. When male adult cattle are mixed during transport or in lairage they may fight, and this behaviour can be recorded directly (Kenny and Tarrant, 1987). Calves of 6 months of age may also fight (Trunkfield and Broom, 1991). The recording of such behaviour should include the occurrence of threats, as well as the contact behaviours that might cause injury.

One further, valuable, method of using behavioural studies in the assessment of farm animal welfare during handling and transport involves using the fact that the animals remember aversive situations in experimentally repeated exposures to such situations. Any stockkeeper will be familiar with the animal that refuses to go into a crush after having received painful treatment there in the past, or hesitates about passing a place where a frightening event such as a dog threat has previously occurred once.

These observations give us information about both the past and present welfare of the animal. If the animal tries not to return to a place where it previously had an experience, then that experience was clearly aversive. The greater the reluctance of the animal to return, the greater the previous aversion must have been.

This principle has been used by Rushen (1986a, b) in studies with sheep. Sheep that had been driven down a race to a point where gentle handling occurred traversed the race as

rapidly or more rapidly on a subsequent day. Sheep that had been subjected to shearing at the end of the race on the first day were harder to drive down the race subsequently, and those subjected to electro-immobilization at the end of the race were very difficult to drive down the race on later occasions. Hence, the degree of difficulty in driving and the delay before the sheep could be driven down the race are both measures of the current fearfulness of the sheep, and this in turn reflects the aversiveness of the treatment when it was first experienced.

Some behavioural measures are clear indicators that there will be a long-term effect on the animal which will harm it, so these indicate stress. Other behavioural measures provide evidence of good or poor welfare but not necessarily of stress.

Physiological assessment

The physiological responses of animals to adverse conditions – such as those which they may encounter during handling and transport – will be affected by the anatomical and physiological constitution of the animal, as mentioned below. Some physiological assessment criteria are detailed in Table 3.1.

Whenever physiological measurement is to be interpreted, it is important to ascertain the basal level for that measure and how it fluctuates over time (Broom, 2000). For example, plasma cortisol levels in most species vary during the day, tending to be higher before than after midday. A decision must be taken for each measure concerning whether the information required is the difference from baseline or the absolute value. For small effects, e.g. a 10% increase in heart rate, the difference from baseline is the key value to use. With regard to major effects where the response reaches the maximal possible level – for example, cortisol in plasma during very frightening circumstances – the absolute value should be used.

In order to explain this, consider an animal severely frightened during the morning and showing an increase from a rather high baseline of 160 nmol/l but in the afternoon showing the same maximal response, which is 200 nmol/l above the lower afternoon baseline. It is the actual value that is important here rather than a

Table 3.1. Commonly used physiological indicators of poor welfare during transport (modified after Knowles and Warriss, 2000).

Stressor	Physiological variable
Measured in blood or other body fluids	
Food deprivation	↑ FFA, ↑ β-OHB, ↓ glucose, ↑ urea
Dehydration	↑ Osmolality, ↑ total protein, ↑ albumin, ↑ PCV
Physical exertion	↑ CK, ↑ lactate
Fear, lack of control	↑ Cortisol, ↑ PCV
Motion sickness	↑ Vasopressin
Measured otherwise	
Fear, physical effects	↑ Heart rate, ↑ heart rate variability, ↑ respiration rate
Hypo-/hyperthermia	Body temperature, skin temperature

FFA, free fatty acids; β-OHB, β-hydroxybutyrate; PCV, packed cell volume; CK, creatine kinase.

difference whose variation depends on baseline fluctuations. In many studies, the value obtained after the treatment studied can usefully be compared with the maximum possible response for that measure. A very frightened animal may show the highest response of which it is capable.

Some of the parameters useful in the assessment of stress will now be described.

Heart rate

Heart rate can decrease when animals are frightened but, in most farm animal studies, tachycardia – increased heart rate – has been found to be associated with disturbing situations. Heart rate increase is not just a consequence of increased activity: heart rate can be increased in preparation for an expected future flight response. Baldock and Sibly (1990) obtained basal levels for heart rate during a variety of activities by sheep and then took account of these when calculating responses to various treatments. Social isolation caused a substantial response, but the greatest heart rate increase occurred when the sheep were approached by a man with a dog. The responses to handling and transport are clearly much lower if the sheep

have previously been accustomed to human handling.

Heart rate is a useful measure of welfare, but only for short-term problems such as those encountered by animals during handling, loading on to vehicles and certain acute effects during the transport procedure itself. However, some adverse conditions may lead to an elevated heart rate for quite long periods. Parrott *et al.* (1998a) showed that heart rate increased from about 100 to about 160 beats/min when sheep were loaded on to a vehicle, and the period of elevation of heart rate was at least 15 min. During transport of sheep, heart rate remained elevated for at least 9 h (Parrott *et al.*, 1998b). Heart rate variability has also been found to be a useful welfare indicator in cattle and other species (van Ravenswaaij *et al.*, 1993).

Respiratory rate

Observation of animals can provide information about physiological processes without any attachment of recording instruments or sampling of body fluids. Breathing rate can be observed directly or from good-quality video recordings. The metabolic rate and level of muscular activity are major determinants of breathing rate, but an individual animal that is disturbed by events in its environment may suddenly start to breathe more rapidly.

Other directly observable responses

Muscle tremor can be directly observed and is sometimes associated with fear. Foaming at the mouth can have a variety of causes, so care is needed in interpreting the observations, but its occurrence may provide some information about welfare.

Hormones and metabolites

ADRENAL MEDULLARY HORMONES. Changes in the adrenal medullary hormones adrenaline (epinephrine) and noradrenaline (norepinephrine) occur very rapidly, and measurements of these hormones have not been much used in the assessment of welfare during transport. However, Parrott *et al.* (1998a) found that both hormones increased more during loading of sheep by means of a ramp than when loading with a lift.

ADRENAL CORTICAL HORMONES.   Adrenal cortical changes occur in most of the situations which lead to aversion behaviour or increased heart rate, but the effects take a few minutes to become evident and last for 15–120 min, or a little longer. An example comes from work on calves (Kent and Ewbank, 1986; Trunkfield and Broom, 1990; Trunkfield et al., 1991). Plasma or saliva glucocorticoid levels gave information about treatments lasting up to 2 h, but were less useful for journeys lasting longer than this.

Saliva cortisol measurement is useful in cattle, sheep and pigs (Bradshaw et al., 1996a). In the plasma, most cortisol is bound to protein but it is the free cortisol that acts in the body. Hormones such as testosterone and cortisol can enter the saliva by diffusion in salivary gland cells. The rate of diffusion is high enough to maintain an equilibrium between the free cortisol in plasma and in saliva. The level is ten or more times lower in saliva, but stimuli that cause plasma cortisol increases also cause comparable salivary cortisol increases in humans (Riad-Fahmy et al., 1982), sheep (Fell et al., 1985), pigs (Parrott et al., 1989) and in some other species.

The injection of pilocarpine and sucking of citric acid crystals, which stimulate salivation, have no effect on the salivary cortisol concentration. However, any rise in salivary cortisol levels following some stimuli is delayed by a few minutes as compared with the comparable rise in plasma cortisol concentration.

Animals demonstrating substantial adrenal cortical responses during handling and transport also show increased body temperature (Trunkfield et al., 1991). This increase is usually of the order of about 1°C, but the actual level at the end of the journey will depend upon the extent to which any adaptation of the initial response has occurred. Body temperature can be recorded in transit by implanted or superficially attached monitors linked directly or telemetrically to a data storage system.

Parrott et al. (1999) described deep body temperature in eight sheep. When the animals were loaded on to a vehicle and transported for 2.5 h body temperature increased by about 1°C, and in males was elevated by 0.5°C for several hours. Exercise for 30 min resulted in a 2°C increase in core body temperature, which returned rapidly to baseline after exercise. It would seem that prolonged increases in body temperature are an indicator of poor standards of welfare.

PITUITARY HORMONES.   The measurement of oxytocin has not been of particular value in animal transport studies (e.g. Hall et al., 1998b). However, plasma β-endorphin levels have been shown to increase during loading (Bradshaw et al., 1996b). The release of corticotrophin-releasing hormone (CRH) in the hypothalamus is followed by release of pro-opiomelanocortin (POMC) in the anterior pituitary, which quickly breaks down into different components – including adrenocorticotrophic hormone (ACTH), which travels in the blood to the adrenal cortex, and beta-endorphin.

A rise in plasma β-endorphin often accompanies ACTH increases in plasma but it is not yet clear what its function is. Although β-endorphin can have analgesic effects via mu-receptors in the brain, this peptide hormone is also involved in the regulation of various reproductive hormones. Measurement of β-endorphin levels in blood is useful as a back-up for ACTH or cortisol measurement.

Metabolites

Creatine kinase is released into the blood when there has been muscle damage, e.g. bruising, or vigorous exercise. It is clear that some kinds of damage affecting welfare result in creatine kinase release, so this can be used in conjunction with other indicators as a welfare measure. Lactate dehydrogenase (LDH) also increases in the blood after muscle tissue damage, but increases can occur in animals whose muscles are not damaged. Deer that are very frightened by capture show large LDH increases (Jones and Price, 1992). The isoenzyme of LDH, which occurs in striated muscle (LDH5), leaks into the blood when animals are very disturbed, so the ratio of LDH5 to total LDH is of particular interest.

On long journeys, animals will have been unable to drink for much longer than the normal interval. This lack of control over interactions with the environment may be disturbing to the animals and there are also likely to be physiological consequences. The most obvious and straightforward way to assess this is to measure the osmolality of the blood (Broom et al., 1996).

When food reserves are used up there are various changes evident in the metabolites present in the blood. Several of these – for example beta-hydroxy butyrate – can be measured and indicate the extent to which the food reserve depletion is serious for the animal.

Plasma studies in chickens reared for meat production deprived of food for 10 h prior to 3 h of transport, when compared with those in non-deprived birds, showed higher thyroxine and lower tri-iodothyronine, triglyceride, glucose and lactate concentrations, indicating negative energy balance and poor welfare (Nijdam et al., 2004).

Another measure that gives information about the significance for the animal of food deprivation is the time interval since the previous meal. Most farm animals are accustomed to feeding at regular times and if feeding is prevented, especially when high rates of metabolism occur during journeys, the animals will be disturbed by this. Behavioural responses when allowed to eat or drink (e.g. Hall et al., 1997) also give important information about problems of deprivation.

Haematocyte measures

The haematocrit (percentage of red blood cells in blood) is altered when animals are transported. If animals encounter a problem, such as might occur when they are handled or transported, there can be a release of blood cells from the spleen and therefore a higher haematocrit (Parrott et al., 1998b). More prolonged problems, however, are likely to result in a reduced haematocrit (Broom et al., 1996).

Increased adrenal cortical activity can lead to immunosuppression. One or two studies in which transport affected T-cell function were reviewed by Kelley (1985), but such measurements are likely to be of most use in the assessment of more long-term welfare problems. The ability of the animal to react effectively to antigen challenge will depend upon the numbers of lymphocytes and the activity and efficiency of these lymphocytes.

Measurements of the ratios of various white blood cells – for example the heterophil:lymphocyte ratio – are affected by a variety of factors, but some kinds of restraint seem to affect the ratio consistently and so can provide some information

about welfare. Studies of T-cell activity – e.g. in vitro mitogen-stimulated cell proliferation – give information about the extent of immunosuppression resulting from the particular treatment. If the immune system is working less efficiently because of handling/transport treatment, the animal is coping less well with its environment and the welfare is poorer than in an animal that is not immunosuppressed.

Examples of the immunosuppressive effect of transport are: (i) the reduction in four different lymphocyte subpopulations after 24 h of transport in horses (Stull et al., 2004); and (ii) the reduction in phytohaemagglutinin-stimulated lymphocyte proliferation in Bos indicus steers during the 6 days after they had been transported for 72 h (Stanger et al., 2005).

As with behavioural measures, some physiological measures are good predictors of an earlier death or of reduced ability to breed, whilst others are not measures of stress because the effect will be brief or slight.

Carcass and mortality assessment

Measures of body damage, of a major disease condition or of increased mortality are indicators of long-term adverse effects – and hence stress. However, a slight bruise or cut will result in some degree of poor welfare but not necessarily stress, as the effect may be very brief. Death during handling and transport is usually preceded by a period of poor welfare. Mortality records during journeys are often the only records giving information about welfare during the journey, and the severity of the problems for the animals are often only too clear from such records.

Amongst extreme injuries during transport are broken bones. These are rare in the larger animals, but poor loading or unloading facilities and cruel or poorly trained staff who are attempting to move the animals may cause severe injuries. It is the laying hen, however, which is most likely to have bones broken during transit from housing conditions to point of slaughter (Gregory and Wilkins, 1989), especially if the birds have previously had insufficient exercise in a battery cage (Knowles and Broom, 1990).

Bruising, scratches and other superficial blemishes can be scored in a precise way and, when carcasses are downgraded for these reasons, those in charge of the animals can reasonably be criticized for not making sufficient efforts to prevent poor welfare. There is a cost of such blemishes to the industry, as well as to the animals. This cost, in monetary and animal welfare terms, of dark firm dry (DFD) and pale soft exudative (PSE) meat is huge.

DFD meat is associated with fighting in cattle and pigs, but it may also be evidenced in cattle that have been threatened but not directly involved in fighting (Tarrant, personal communication). PSE meat is in part a consequence of possession of certain genes and occurs more in some strains of pigs than in others, but its occurrence is related in most cases to other indicators of poor welfare (Tarrant, personal communication).

Poultry meat quality can often be adversely affected for similar reasons. In a large-scale study of chickens reared for meat production and transported to slaughter in the Netherlands and Germany, Nijdam et al. (2004) found that the mean mortality was 4.6 and the number with bruises was 22 per thousand birds. The major factors that had increased the mortality rate were: (i) increased stocking density in transport containers; (ii) increased transport time; and (iii) increased time in lairage before slaughter.

When animals are subjected to violent handling and respond by energetic struggling, a possible consequence is capture myopathy. This muscle damage that occurs will impair muscular action in the future, at least in the short term, and is an indicator of poor welfare because it reduces coping ability and may be associated with pain (Ebedes et al., 2002).

Experimental methods of assessment

As Hall and Bradshaw (1998) explain, information on the stress effects of transport is available from five kinds of study:

- Studies where transport, not necessarily in conditions representative of commercial practice, was used explicitly as a stressor to evoke a physiological response of particular interest (Smart et al., 1994; Horton et al., 1996).
- Uncontrolled studies with physiological and behavioural measurements being made before and after long or short commercial or experimental journeys (Becker et al., 1985, 1989; Dalin et al., 1988, 1993; Knowles et al., 1994).
- Uncontrolled studies during long or short commercial or experimental journeys (Lambooij, 1988; Hall 1995).
- Studies comparing animals that were transported, with animals that had been left behind to act as controls (Nyberg et al., 1988; Knowles et al., 1995).
- Studies where the different stressors that impinge on an animal during transport were separated out either by experimental design (Bradshaw et al., 1996c; Broom et al., 1996; Cockram et al., 1996) or by statistical analysis (Hall et al., 1998c).

Each of these methods is of value, because some are carefully controlled but less representative of commercial conditions whilst others show what happens during commercial journeys but are less well controlled.

Discussion of Some Key Factors

Animal genetics and transport

Cattle and sheep have been selected for particular breed characteristics for hundreds of years. As a consequence, there may be differences between breeds in how they react to particular management conditions. For example, Hall et al. (1998a, b) found that introduction of an individual sheep to three others in a pen resulted in a higher heart rate and salivary cortisol concentration if it was of the Orkney breed than if it was of the Clun Forest breed. The breed of animal should be taken into account when planning transport.

Farm animal selection for breeding has been directed especially towards maximization of productivity. In some farm species there are consequences for welfare of such selection (Broom, 1994, 1999). Fast-growing broiler chickens may have a high prevalence of leg disorders and Belgian Blue cattle may be unable to calve unaided or without the necessity of Caesarean section. Some of these effects may affect welfare during handling and transport.

Certain rapidly growing beef cattle have joint disorders that result in pain during transport, and some strains of high-yielding dairy cows are much more likely to have foot disorders. Modern strains of dairy cows, in particular, need much better conditions during transport and much shorter journeys if their welfare is not to be poorer than that of the dairy cattle of 30 years ago.

Rearing conditions, experience and transport

If animals are kept in such a way that they are very vulnerable to injury when handled and transported, this must be taken into account during transporting, or the rearing conditions must be changed. An extreme example of such an effect is osteopenia and vulnerability to broken bones, which is twice as high in battery hens as in hens that are able to flap their wings and walk around (Knowles and Broom, 1990). Calves are much more disturbed by handling and transport if they are reared in individual crates than if they are reared in groups, presumably because of lack of exercise and absence of social stimulation in the rearing conditions (Trunkfield et al., 1991).

Human contact prior to handling and transport is also important. If young cattle have been handled for a short period just after weaning they are much less disturbed by the procedures associated with handling and transport (Le Neindre et al., 1996). All animals can be prepared for transport by appropriate previous treatment.

Mixing of social groups and transport

If pigs or adult cattle are taken from different social groups, whether from the same farm or not, and are mixed with strangers just before transport, during transport or in lairage, there is a significant risk of threatening or fighting behaviour (McVeigh and Tarrant, 1983; Guise and Penny, 1989; Tarrant and Grandin, 2000).

The glycogen depletion associated with threat, fighting or mounting often results in DFD meat, injuries such as bruising and associated poor welfare. The problem is sometimes very severe, in welfare and economic terms, but is solved by keeping animals in groups with familiar individuals rather than by mixing strangers. Cattle might be tethered during loading but should never be tethered when vehicles are moving, because long tethers cause a high risk of entanglement and short tethers cause a high risk of cattle being hung by the neck.

Handling, loading, unloading and welfare

Well-trained and experienced stockpeople know that cattle can be readily moved from place to place by human movements that take advantage of the animal's flight zone (Kilgour and Dalton, 1984; Grandin, 2000). Cattle will move forward when a person enters the flight zone at the point of balance, and can be calmly driven up a race by a person entering the flight zone and moving in the opposite direction to that in which the animal intends to go.

Handling animals without the use of sticks or electric goads results in better welfare and less risk of poor carcass quality. Sound knowledge of animal behaviour and good facilities are important for animal welfare during handling and loading.

Ambient temperature and other physical conditions during transport

Extremes of temperature can cause very poor welfare standards in transported animals. Exposure to temperatures below freezing has severe effects on small animals, including domestic fowl.

However, temperatures that are too high are a commoner cause of poor welfare, with poultry, rabbits and pigs being especially vulnerable. For example, de la Fuente et al. (2004) found that plasma cortisol, lactate, glucose, creatine kinase, lactate dehydrogenase and osmolarity were all higher in warmer summer conditions than in cooler winter conditions in transported rabbits. In each of these species, and particularly in chickens reared for meat production, stocking density must be reduced in temperatures of 20°C or higher, or there is a substantial risk of high mortality and poor welfare.

Vehicle-driving methods, stocking density and welfare

When humans are driven in a vehicle, they can usually sit on a seat or hold on to some fixture.

Cattle standing on four legs are far less able to deal with sudden movements such as those caused by swinging around corners or sudden braking. Cattle always endeavour to stand in a vehicle in such a way that they brace themselves to minimize the chance of being thrown around and to avoid making contact with other individuals. They do not lean on other individuals and are substantially disturbed by too much movement or too high a stocking density.

In a study of sheep during driving on winding or straight roads, Hall et al. (1998c) found that plasma cortisol concentrations were substantially higher on winding than on straight roads. Tarrant et al. (1992) studied cattle at a rather high, at an average and at a low commercial stocking density and found that falls, bruising, cortisol and creatine kinase levels all increased with stocking density. Careful driving and a moderate stocking density are crucial for good standards of welfare.

Disease, welfare and transport

The transport of animals can lead to increased disease – and hence to poorer welfare – in a variety of ways: (i) tissue damage and malfunction; (ii) pathological effects which would not otherwise have occurred resulting from pathogens already present; (iii) disease from pathogens transmitted from one transported animal to another; and (iv) disease in non-transported animals because of pathogen transmission from transported animals. Exposure to pathogens does not necessarily result in infection or disease in an animal. Factors influencing this process include the virulence and the dose of pathogens transmitted, route of infection and the immune status of the animals exposed (Quinn et al., 2002).

Enhanced susceptibility to infection and disease as a result of transport has been the subject of much research (Broom and Kirkden, 2004; Broom, 2006). Many reports describing the relationship between transport and incidence of specific diseases have been published. As an example, 'shipping fever' is a term commonly used for a specific transport-related disease condition in cattle. It develops between a few hours and 1–2 days following transport.

Several pathogens can be involved, such as Pasteurella species, bovine respiratory syncytial virus, infectious bovine rhinotracheitis virus and several other herpes viruses, para-influenza 3 virus and a variety of pathogens associated with gastrointestinal diseases, such as rotaviruses, Escherichia coli and Salmonella spp. (Quinn et al., 2002). Transport in general has been shown to result in increased mortality in calves and sheep (Brogden et al., 1998; Radostits et al., 2000), salmonellosis in sheep (Higgs et al., 1993) and horses (Owen et al., 1983). In calves, it can cause pneumonia and subsequent mortality associated with bovine herpes virus–1 (Filion et al., 1984), as a result of a stress-related reactivation of herpes virus in latently infected animals (Thiry et al., 1987).

In some cases, particular aspects of the transport situation can be linked to disease. For example, fighting caused by the mixing of different groups of pigs can depress anti-viral immunity in these animals (de Groot et al., 2001). The presence of viral infection increases the susceptibility to secondary bacterial infection (Brogden et al., 1998).

Transmission of a pathogenic agent begins with shedding from the infected host through oro-nasal fluids, respiratory aerosols, faeces or other secretions or excretions. The routes of shedding vary between infectious agents. Stress related to transport can increase the amount and duration of pathogen shedding and thereby result in increased infectiousness. This is described for Salmonella in various animal species (Wierup, 1994).

The shedding of pathogens by the transported animals results in contamination of vehicles and other transport-related equipment and areas, e.g. in collecting stations and markets. This may result in indirect and secondary transmission. The more resistant an agent is to adverse environmental conditions, the greater the risk that it will be transmitted by indirect mechanisms.

Many infectious diseases may be spread as a result of animal transport. Outbreaks of classical swine fever in the Netherlands and of foot and mouth disease in the UK were much worse than they might have been because animals had been transported and, in some cases, had transmitted the disease at staging points or markets. Schlüter and Kramer (2001) summarized the outbreaks in the EU of foot and mouth disease and classical swine fever and found that, once this latter disease was in the farm stock, 9% of further spread occurred as a result of transport. In a recent epidemic of Highly Pathogenic Avian Influenza virus in Italy it was found that the movement of birds by contaminated vehicles and equipment created a significant problem in the control of this epizootic.

Major disease outbreaks constitute very important animal welfare as well as economic problems, and regulations concerning the risks of disease are necessary on animal welfare grounds. If stress and the mixing of animals and their products are minimized, disease – and hence poor welfare standards – can be prevented or rendered less likely.

References

Baldock, N.M. and Sibly, R.M. (1990) Effects of handling and transportation on heart rate and behaviour in sheep. *Applied Animal Behaviour Science* 28, 15–39.

Becker, B.A., Neinaber, J.A., Deshazer, J.A. and Hahn, G.L. (1985) Effect of transportation on cortisol concentrations and on the circadian rhythm of cortisol in gilts. *American Journal of Veterinary Research* 46, 1457–1459.

Becker, B.A., Mayes, H.F., Hahn, G.L., Neinaber, J.A., Jesse, G.W., Anderson, M.E., Heymann, H. and Hedrick, H.B. (1989) Effect of fasting and transportation on various physiological parameters and meat quality of slaughter hogs. *Journal of Animal Science* 67, 334.

Bradshaw, R.H., Hall, S.J.G. and Broom, D.M. (1996a) Behavioural and cortisol responses of pigs and sheep during transport. *Veterinary Record* 138, 233–234.

Bradshaw, R.H., Parrott, R.F., Forsling, M.L., Goode, J.A., Lloyd, D.M., Rodway, R.G. and Broom, D.M. (1996b) Stress and travel sickness in pigs: effects of road transport on plasma concentrations of cortisol, beta-endorphin and lysine vasopressin. *Animal Science* 63, 507–516.

Bradshaw, R.H., Parrott, R.F., Goode, J.A., Lloyd, D.M., Rodway, R.G. and Broom, D.M. (1996c) Behavioural and hormonal responses of pigs during transport: effect of mixing and duration of journey. *Animal Science* 62, 547–554.

Brogden, K.A., Lehmkuhl, H.D. and Cutlip, R.C. (1998) *Pasteurella haemolytica*-complicated respiratory infections in sheep and goats. *Veterinary Research* 29, 233–254.

Broom, D.M. (1986) Indicators of poor welfare. *British Veterinary Journal* 142, 524–526.

Broom, D.M. (1994) The effects of production efficiency on animal welfare. In: Huisman, E.A., Osse, J.W.M., van der Heide, D., Tamminga, S., Tolkamp, B.L., Schouten, W.G.P., Hollingsworth, C.E. and van Winkel, G.L. (eds) *Biological Basis of Sustainable Animal Production: Proceedings of the 4th Zodiac Symposium*. EAAP Publication No. 67, Wageningen, Netherlands, pp. 201–210.

Broom, D.M. (1998) Welfare, stress and the evolution of feelings. *Advances in the Study of Behavior* 27, 371–403.

Broom, D.M. (1999) The welfare of dairy cattle. In: Aagaard, K. (ed.) *Proceedings of the 25th International Dairy Congress; III, Future Milk Farming*, Aarhus, Netherlands, 1998. Danish National Committee of I.D.F., Aarhus, Netherlands, pp. 32–39.

Broom, D.M. (2000) Welfare assessment and problem areas during handling and transport. In: Grandin, T. (ed.) *Livestock Handling and Transport*, 2nd edn. CABI, Wallingford, UK, pp. 43–61.

Broom, D.M. (ed.) (2001a) *Coping with Challenge: Welfare in Animals including Humans*. Dahlem University Press, Berlin, 364 pp.

Broom, D.M. (2001b) Coping, stress and welfare. In: Broom, D.M. (ed.) *Coping with Challenge: Welfare in Animals including Humans*. Dahlem University Press, Berlin, pp. 1–9.

Broom, D.M. (2002) Does present legislation help animal welfare? *Landbauforschung Völkenrode* 227, 63–69.

Broom, D.M. (2006) Behaviour and welfare in relation to pathology. *Applied Animal Behaviour Science* 97, 71–83.

Broom, D.M. and Johnson, K.G. (2000) *Stress and Animal Welfare*. Kluwer, Dordrecht, Netherlands.

Broom, D.M. and Kirkden, R.D. (2004) Welfare, stress, behavior, and pathophysiology. In: Dunlop, R.H. and Malbert, C.-H. (eds) *Veterinary Pathophysiology*. Blackwell, Ames, Iowa, pp. 337–369.

Broom, D.M., Goode, J.A., Hall, S.J.G., Lloyd, D.M. and Parrott, R.F. (1996) Hormonal and physiological effects of a 15-hour journey in sheep: comparison with the responses to loading, handling and penning in the absence of transport. *British Veterinary Journal* 152, 593–604.

Cockram, M.S., Kent, J.E., Goddard, P.J., Waran, N.K., McGilp, I.M., Jackson, R.E., Muwanga, G.M. and Prytherch, S. (1996) Effect of space allowance during transport on the behavioural and physiological responses of lambs during and after transport. *Animal Science* 62, 461–477.

Dalin, A.M., Nyberg, L. and Eliasson, L. (1988) The effect of transportation/relocation on cortisol, CBG and induction of puberty in gilts with delayed puberty. *Acta Veterinaria Scandinavica* 29, 207–218.

Dalin, A.M., Magnusson, U., Haggendal, J. and Nyberg, L. (1993) The effect of transport stress on plasma levels of catecholamines, cortisol, corticosteroid-binding globulin, blood cell count and lymphocyte proliferation in pigs. *Acta Veterinaria Scandinavica* 34, 59–68.

de Groot, J., Ruis, M.A., Scholten, J.W., Koolhaas, J.M. and Boersma, W.J. (2001) Long-term effects of social stress on anti-viral immunity in pigs. *Physiology and Behavior* 73, 143–158.

de la Fuente, J., Salazar, M.I., Ibáñez, M. and Gonzalez de Chavarri, E. (2004) Effects of season and stocking density during transport on live-weight and biochemical measurements of stress, dehydration and injury of rabbits at time of slaughter. *Animal Science* 78, 285–292.

Ebedes, H., Van Rooyen, J. and Du Toit, J.G. (2002) Capturing wild animals. In: Bothma, J. du P. (ed.) *Game Ranch Management*. Van Scheik, Pretoria, South Africa, pp. 382–440.

Fell, L.R., Shutt, D.A. and Bentley, C.J. (1985) Development of salivary cortisol method for detecting changes in plasma 'free' cortisol arising from acute stress in sheep. *Australian Veterinary Journal* 62, 403–406.

Filion, L.G., Willson, P.J., Bielefeldt-Ohmann, M.A. and Thomson, R.G. (1984) The possible role of stress in the induction of pneumonic pasteurellosis. *Canadian Journal of Comparative Medicine* 48, 268–274.

Fraser, A.F. and Broom, D.M. (1997) *Farm Animal Behaviour and Welfare*. CABI, Wallingford, UK.

Geverink, N.A., Bradshaw, R.H., Lambooij, E., Wiegant, V.M. and Broom, D.M. (1998) Effects of simulated lairage conditions on the physiology and behaviour of pigs. *Veterinary Record* 143, 241–244.

Grandin, T. (1980) Observations of cattle behaviour applied to the design of cattle handling facilities. *Applied Animal Ethology* 6, 19–31.

Grandin, T. (1982) Pig behaviour studies applied to slaughter plant design. *Applied Animal Ethology* 9, 141–151.

Grandin, T. (1989) Behavioural principles of livestock handling. *Professional Animal Scientist* 5 (2), 1–11.

Grandin, T. (2000) Behavioural principles of handling cattle and other grazing animals under extensive conditions. In: Grandin, T. (ed.) *Livestock Handling and Transport*, 2nd edn., CABI, Wallingford, UK.

Gregory, N.G. and Wilkins, L.J. (1989) Broken bones in domestic fowl: handling and processing damage in end-of-lay battery hens. *British Poultry Science* 30, 555–562.

Guise, J. and Penny, R.H.C. (1989) Factors affecting the welfare, carcass and meat quality of pigs. *Animal Production* 49, 517–521.

Hall, S.J.G. (1995) Transport of sheep. *Proceedings of the Sheep Veterinary Society* 18, 117–119.

Hall, S.J.G. and Bradshaw, R.H. (1998) Welfare aspects of transport by road of sheep and pigs. *Journal of Applied Animal Welfare Science* 1, 235–254.

Hall, S.J.G., Schmidt, B. and Broom, D.M. (1997) Feeding behaviour and the intake of food and water by sheep after a period of deprivation lasting 14 h. *Animal Science* 64, 105–110.

Hall, S.J.G., Broom, D.M. and Kiddy, G.N.S. (1998a) Effect of transportation on plasma cortisol and packed cell volume in different genotypes of sheep. *Small Ruminant Research* 29, 233–237.

Hall, S.J.G., Forsling, M.L. and Broom, D.M. (1998b) Stress responses of sheep to routine procedures: changes in plasma concentrations of vasopressin, oxytocin and cortisol. *Veterinary Record* 142, 91–93.

Hall, S.J.G., Kirkpatrick, S.M., Lloyd, D.M. and Broom, D.M. (1998c) Noise and vehicular motion as potential stressors during the transport of sheep. *Animal Science* 67, 467–473.

Hall, S.J.G., Kirkpatrick, S.M. and Broom, D.M. (1998d) Behavioural and physiological responses of sheep of different breeds to supplementary feeding, social mixing and taming, in the context of transport. *Animal Science* 67, 475–483.

Higgs, A.R.B., Norris, R.T. and Richards, R.B. (1993) Epidemiology of salmonellosis in the live sheep export industry. *Australian Veterinary Journal* 70, 330–335.

Horton, G.M.J., Baldwin, J.A., Emanuele, S.M., Wohlt, J.E. and McDowell, L.R. (1996) Performance and blood chemistry in lambs following fasting and transport. *Animal Science* 62, 49–56.

Jones, A.R. and Price, S.E. (1992) Measuring the response of fallow deer to disturbance. In: Brown, R.D. (ed.) *The Biology of Deer*. Springer Verlag, Berlin.

Kelley, K.W. (1985) Immunological consequences of changing environmental stimuli. In: Moberg, G.P. (ed.) *Animal Stress*. American Physiological Society, Betheda, Maryland, pp. 193–223.

Kenny, F.J. and Tarrant, P.V. (1987) The reaction of young bulls to short-haul road transport. *Applied Animal Behaviour Science* 17, 209–227.

Kent, J.F. and Ewbank, R. (1986) The effect of road transportation on the blood constituents and behaviour of calves. III. Three months old. *British Veterinary Journal* 142, 326–335.

Kilgour, R. and Dalton, C. (1984) *Livestock Behaviour: a Practical Guide.* Granada Publishing, St Albans, UK.

Knowles, T.G. and Broom, D.M. (1990) Limb bone strength and movement in laying hens from different housing systems. *Veterinary Record* 126, 354–356.

Knowles, T.G. and Warriss, D. (2000) Stress physiology of animals during transport. In: Grandin, T. (ed.) *Livestock Handling and Transport*, 2nd edn., CABI, Wallingford, UK, pp. 385–407.

Knowles, T.G., Warriss, P.D., Brown, S.N. and Kestin, S.C. (1994) Long-distance transport of export lambs. *Veterinary Record* 134, 107–110.

Knowles, T.G., Brown, S.N., Warriss, P.D., Phillips, A.J., Doland, S.K., Hunt, P., Ford, J.E., Edwards, J.E. and Watkins, P.E. (1995) Effects on sheep of transport by road for up to 24 hours. *Veterinary Record* 136, 431–438.

Lambooij, E. (1988) Road transport of pigs over a long distance: some aspects of behaviour, temperature and humidity during transport and some effects of the last two factors. *Animal Production* 46, 257–263.

Lambooij, E., Geverink, N., Broom, D.M. and Bradshaw, R.H. (1995) Quantification of pigs' welfare by behavioural parameters. *Meat Focus International* 4, 453–456.

Le Neindre, P., Boivin, X. and Boissy, A. (1996) Handling of extensively kept animals. *Applied Animal Behaviour Science* 49, 73–81.

McVeigh, J.M. and Tarrant, V. (1983) Effect of propanolol on muscle glycogen metabolism during social regrouping of young bulls. *Journal of Animal Science* 56, 71–80.

Nijdam, E., Arens, P., Lambooij, E., Decuypere, E. and Stegman, J.A. (2004) Factors influencing bruises and mortality of broilers during catching, transport and lairage. *Poultry Science* 83, 1610–1615.

Nyberg, L., Lundstrom, K., Edfors-Lilja, I. and Rundgren, M. (1988) Effects of transport stress on concentrations of cortisol, corticosteroid-binding globulin and glucocorticoid receptors in pigs with different halothane genotypes. *Journal of Animal Science* 66, 1201–1211.

Owen, R.A., Fullerton, J. and Barnum, D.A. (1983) Effects of transportation, surgery, and antibiotic therapy in ponies infected with Salmonella. *American Journal of Veterinary Research* 44, 46–50.

Parrott, R.F., Misson, B.H. and Baldwin, B.A. (1989) Salivary cortisol in pigs following adrenocorticotrophic hormone stimulation: comparison with plasma levels. *British Veterinary Journal* 145, 362–366.

Parrott, R.F., Hall, S.J.G. and Lloyd, D.M. (1998a) Heart rate and stress hormone responses of sheep to road transport following two different loading responses. *Animal Welfare* 7, 257–267.

Parrott, R.F., Hall, S.J.G., Lloyd, D.M., Goode, J.A. and Broom, D.M. (1998b) Effects of a maximum permissible journey time (31 h) on physiological responses of fleeced and shorn sheep to transport, with observations on behaviour during a short (1 h) rest-stop. *Animal Science* 66, 197–207.

Parrott, R.F., Lloyd, D.M. and Brown, D. (1999) Transport stress and exercise hyperthermia recorded in sheep by radiotelemetry. *Animal Welfare* 8, 27–34.

Quinn, P.J., Markey, B.K., Carter, M.E., Donnelley, W.J. and Leonard, F.C. (2002) *Veterinary Microbiology and Microbial Diseases.* Blackwell, Oxford, UK.

Radostits, O.M., Gay, C.C., Blood, D.C. and Hinchcliff, K.W. (2000) *Veterinary Medicine: a Textbook of the Diseases of Cattle, Sheep, Pigs, Goats and Horses*, 9th edn. W.B. Saunders, London.

Ravenswaaij, C.M.A., van, Kolléè, L.A.A., Hopman, J.C.W., Stoelinga, G.B.A. and van Geijn, H. (1993) Heart rate variability. *Annals of Internal Medicine* 118, 435–437.

Riad-Fahmy, D., Read, G.F., Walker, R.F. and Griffiths, K. (1982) Steroids in saliva for assessing endocrine function. *Endocrinology Review* 3, 367–395.

Rushen, J. (1986a) The validity of behavioural measure of aversion: a review. *Applied Animal Behaviour Science* 16, 309–323.

Rushen, J. (1986b) Aversion of sheep for handling treatments: paired choice experiments. *Applied Animal Behaviour Science* 16, 363–370.

Schlüter, H. and Kramer, M. (2001) Epidemiologische Beispiele zur Seuchenausbreitung. *Deutsche Tierärztliche Wochenschrift* 108, 338–343.

Smart, D., Forhead, A.J., Smith, R.F. and Dobson, H. (1994) Transport stress delays the oestradiol-induced LH surge by a non-opioidergic mechanism in the early postpartum ewe. *Journal of Endocrinology* 142, 447–451.

Stanger, K.J., Ketheesan, N., Parker, A.J., Coleman, C.J., Lazzaroni, S.M. and Fitzpatrick, L.A. (2005) The effect of transportation on the immune status of *Bos indicus* steers. *Journal of Animal Science* 83, 2632–2636.

Stull, C.L., Spier, S.J., Aldridge, B.M., Blanchard, M. and Stott, J.L. (2004) Immunological response to long-term transport stress in mature horses and effects of adaptogenic dietary supplementation as an immunomodulator. *Equine Veterinary Journal* 36, 583–589.

Tarrant, P.V., Kenny, F.J., Harrington, D. and Murphy, M. (1992) Long-distance transportation of steers to slaughter, effect of stocking density on physiology, behaviour and carcass quality. *Livestock Production Science* 30, 223–238.

Tarrant, V. and Grandin, T. (2000) Cattle transport. In: Grandin, T. (ed.) *Livestock Handling and Transport*, 2nd edn. CAB International, Wallingford, UK.

Thiry, E., Saliki, J., Bublot, M. and Pastoret, P.-P. (1987) Reactivation of infectious bovine rhinotracheitis virus by transport. *Comparative Immunology Microbiology and Infectious Diseases* 10, 59–63.

Trunkfield, H.R. and Broom, D.M. (1990) The welfare of calves during handling and transport. *Applied Animal Behaviour Science* 28, 135–152.

Trunkfield, H.R. and Broom, D.M. (1991) The effects of the social environment on calf responses to handling and transport. *Applied Animal Behaviour Science* 30, 177.

Trunkfield, H.R., Broom, D.M., Maatje, K., Wierenga, H.K., Lambooij, E. and Kooijman, J. (1991) Effects of housing on responses of veal calves to handling and transport. In: Metz, J.H.M. and Groenestein, C.M. (eds) *New Trends in Veal Calf Production*. Pudoc, Wageningen, Netherlands, pp. 40–43.

Wierup, M. (1994) Control and prevention of *Salmonella* in livestock farms. *Proceedings of the 16th Conference, OIE Regional Commission, Europe,* Stockholm, 28 June–1 July 1994, pp. 249–269.

4 Behavioural Principles of Handling Cattle and Other Grazing Animals under Extensive Conditions

Temple Grandin

Department of Animal Sciences, Colorado State University, Fort Collins, Colorado, USA

Introduction

More and more ranchers and feedlot managers are adopting low-stress handling methods. Since 2000 there have been several new innovations in herding cows on the range and in receiving procedures for calves arriving in a feedlot. It is likely that these methods may be rediscoveries of the ways of the stockmen of bygone years.

In the late 1800s, cowboys handled and trailed cattle quietly on the great cattle-drives from Texas to Montana. In a cowboy's diary Andy Adams wrote: 'Boys, the secret of trailing cattle is to never let your herd know that they are under restraint. Let everything that is done be done voluntarily by the cattle' (Adams, 1903). Unfortunately, the quiet methods of the early 1900s were forgotten and some more modern cowboys were rough (Wyman, 1946; Hough, 1958; Burri, 1968). There is an excellent review of the history of herding in Smith (1998). Progressive producers of cattle know that reducing stress will improve both productivity and safety.

Motivated by Fear

Cattle and other grazing, herding animals such as horses are prey species. Fear motivates them to be constantly vigilant in order to escape from predators. Fear is a very strong stressor

(Grandin, 1997). Fear stress can raise stress hormones to higher levels than can many physical stressors. When cattle become agitated during handling they are motivated by fear. The circuits in the brain that control fear-based behaviour have been studied and mapped (LeDoux, 1996; Rogan and LeDoux, 1996).

Calm animals are easier to handle and sort than agitated, fearful cattle. Fearful cattle stick together and handling becomes more difficult. The secret to low-stress cattle handling is to keep them calm. If cattle become frightened, it takes 20 min for them to calm down. Feedlot operators who handle thousands of extensively raised cattle have found that quiet handling during vaccination enabled their charges to go back on feed more quickly (Grandin, 1998a).

Voisinet *et al.* (1997) reported that cattle which became highly agitated during restraint in a squeeze chute had lower weight gains than calm cattle that had stood quietly in the chute. Further research has shown that cattle that run rapidly out of the chute are also more susceptible to pre-slaughter stress and yielded tougher meat (Petherick *et al.*, 2002; Vann *et al.*, 2004). King *et al.* (2005) reported that extensively raised yearling steers with an excitable temperament had higher cortisol levels after handling.

Sheep and cattle may have an innate fear of dogs. Sheep were more willing to approach a goat compared to an unfamiliar human or a quiet sitting dog (Beausoleil *et al.*, 2005). The unfamiliar goat may be perceived as a herdmate rather than as a threat. When grazing animals see a novel or potentially threatening thing they will raise their heads in a vigilant posture (Welp *et al.*, 2004). The vigilant posture occurs when they perceive a potential threat. The eyes also provide an indicator of a bovine's stress level. Frightening cattle by suddenly opening an umbrella caused a greater percentage of the white portion of the eye to show (Sandem *et al.*, 2004).

Further studies have shown that the percentage of visible white eye increases when a calf is separated from its mother, while the tranquillizer diazepam reduces it (Sandem *et al.*, 2006). People working with cattle need to improve their handling methods if cattle show the whites of their eyes and their heads are constantly up. When cattle are handled quietly, these signs of fear will be absent.

Perception of grazing animals

Vision

To help in the avoidance of predation, cattle have wide-angle (360°) panoramic vision (Prince, 1977), and vision has dominance over hearing (Uetake and Kudo, 1994). They can discriminate colours (Thines and Soffie, 1977; Darbrowska *et al.*, 1981; Gilbert and Arave, 1986; Arave, 1996). Cattle, sheep and goats are dichromats (only two of the three primary colours can be discerned), with cones that are most sensitive to yellowish green (552–555 nm) and blue–purple light (444–455 nm) (Jacobs *et al.*, 1998). The horse is most sensitive at 539 nm and 428 nm (Carroll *et al.*, 2007).

Pick *et al.* (1994) tested a horse and found that it could discriminate red from grey and blue from grey, but could not discriminate green from grey. In another study, Smith and Goldman (1999) found that most horses could discriminate grey from red, blue, yellow and green, but one horse was not able to distinguish yellow from green. Dichromatic vision may provide better vision at night and aid in detecting motion (Miller and Murphy, 1995). The visual acuity of bulls may be worse than that of younger cattle or sheep (Rehkamper and Gorlach, 1998).

Grazing animals can see depth (Lemmon and Patterson, 1964). Horses are sensitive to visual depth cues in photographs. However, grazing animals may have to stop and put their heads down to see depth. This may explain why they baulk at shadows on the ground. Observations by Smith (1998) indicate that cattle do not perceive objects that are overhead unless they move. Smith (1998) also observed that, due to their horizontal pupil, cattle might see vertical lines better than horizontal lines. It is of interest that most grazing animals have horizontal pupils and most predators have round ones.

Research with horses indicates that they have a horizontal band of sensitive retina, instead of a central fovea as in the human (Saslow, 1999). This enables them to scan their surroundings while grazing. Grazing animals have a visual system that is very sensitive to motion and contrasts of light and dark. They are able to scan the horizon constantly while grazing and they may have difficulty in quickly focusing on nearby objects, due to weak eye muscles (Coulter and Schmidt, 1993). This may explain why grazing animals 'spook' at nearby objects that suddenly move.

Wild ungulates, domestic cattle and horses respect a solid fence and will seldom ram or try to run through a solid barrier. Sheets of opaque plastic can be used to corral wild ungulates (Fowler, 1995), whereas portable corrals constructed from canvas have been used to capture wild horses (Wyman, 1946; Amaral, 1977). Excited cattle will often run into a cable or chain-link fence because they cannot see it. A 30 cm-wide, solid, belly rail installed at eye height or ribbons attached to the fence will enable the animal to see the fence and prevent fence ramming (Ward, 1958). Cattle also have a strong tendency to move from a dimly illuminated area to a more brightly lit one (Grandin, 1980a, b). However, they will not approach a blinding light.

Visual distractions that cause animals to back up and refuse to move must either be removed from a handling facility or blocked by solid walls. Some of the most common distractions are: dangling chains, reflections, shadows, moving people, vehicles and flapping objects (Grandin, 1987, 2006). To locate visual distractions, people need to get into the race and look at it from the point of view of the bovine eye.

Hearing

Grazing animals are very sensitive to high-frequency sounds. The human ear is most sensitive at 1000–3000 Hz, but cattle are most sensitive to 8000 Hz (Ames, 1974; Heffner and Heffner, 1983). Cattle can easily hear up to 21,000 Hz (Algers, 1984).

Heffner and Heffner (1992) found that cattle and goats have a poorer ability to localize sound than most mammals. The authors speculate that, since prey species animals have their best vision directed to nearly the entire horizon, they may not need to locate sounds as accurately as an animal with a narrow visual field. The author has observed that cattle and horses will 'watch' people and other animals with their ears. They will point each ear independently at two different people or animals. Noise is stressful to grazing animals (Price et al., 1993). The sounds of people yelling or whistling were more stressful to cattle than the sounds of gates clanging (Waynert et al., 1999).

Cattle are able to differentiate between the threatening sound of a person yelling at them and a machinery sound that is not directed at them. Shouting close to the ear of a cow may be as aversive as an electric prod (Pajor et al., 2003). Lanier et al. (1999a, 2000) found that cattle that became agitated in an auction ring were more likely to flinch or jump in response to sudden, intermittent movement or sounds. Intermittent movements or sounds appear to be more frightening than steady stimuli.

Talling et al. (1998) found that pigs were more reactive to intermittent sounds compared with steady sounds. High-pitched sounds increased a pig's heart rate more than low-pitched sounds (Talling et al., 1996). Sudden movements have the greatest activating effect on the amygdala (LeDoux, 1996), the part of the brain that controls fearfulness (LeDoux, 1996; Rogan and LeDoux, 1996). See Chapter 5, this volume, for more information on the effects of sound.

Effects of sudden novelty, high visual contrast and rapid movement

Cattle and other ungulates are frightened by novelty when they are suddenly confronted by it.

Animals will baulk at a sudden change in fence construction or floor texture (Lynch and Alexander, 1973). Shadows, drain grates and puddles will also impede cattle movement (Grandin, 1980a). In areas where animals are handled, illumination should be uniform to prevent shadows, and handling facilities should be painted all one colour to avoid contrasts. In indoor handling facilities, white, translucent skylights should be installed in either the walls or roof to let in lots of shadow-free natural light. The ideal illumination should look like a bright, but cloudy day.

Contrasts have such an inhibiting effect on cattle movement that road maintenance departments have stopped cattle from crossing a road by painting a series of white lines across it (*Western Livestock Journal*, 1973). Dairy cattle that are handled every day in the same facility will readily walk over a drain grate or a shadow because it is no longer novel. However, the same dairy cow will baulk and put her head down to investigate a strange piece of paper on the floor of a familiar alley.

The paradoxical aspect of novelty is that it is both frightening and attractive (Grandin and Deesing, 1998). A clipboard on the ground will attract cattle when they can voluntarily approach it, but they will baulk and may refuse to step over it if driven towards it. A prey species must be wary of novelty because novelty can mean danger. For example, Nyala (antelope) in a zoo have little fear of people standing by their fence, but the novelty of people fixing a barn roof provoked an intense flight reaction.

A review of the literature about cattle drives in the 1800s and early 1900s indicated that sudden novelty was the major cause of stampedes. Stampedes were caused by a hat flying in the wind, a horse bucking with a saddle under its belly, thunder, a cowboy stumbling or a flapping raincoat (Harger, 1928; Ward, 1958; Linford, 1977). Stampedes were also more likely to occur at night (Ward, 1958; Linford, 1977). Objects that move quickly are more likely to scare. Rapid motion has a greater activating effect on the amygdala than slow movement (LeDoux, 1996).

Dantzer and Mormede (1983) and Stephens and Toner (1975) both reported that novelty is a strong stressor. Placing a calf in an unfamiliar place is probably stressful (Johnston and

Buckland, 1976). In tame beef cattle, throwing a novel-coloured ball into the pen caused a crouch–flinch reaction in 50% of the animals (Miller *et al.*, 1986). Cattle which had previous handling experience in a livestock market settled down more quickly at the slaughter plant stockyards because it appeared less novel and frightening to cattle that had been in a livestock market (Cockram, 1990).

Handling and weaning of calves will be less stressful if the cows and calves are quietly moved through the chutes and corrals several times before any actual veterinary procedures are done (Humphries, 2006). It is recommended to get cattle accustomed to being handled by people on foot, on horses and in vehicles, in order to prevent the animals from becoming excited by the novelty of handling at a feedlot, auction or slaughter plant.

Zebu cattle reared in the Philippines are exposed to so much novelty that new experiences seldom alarm them. Halter-broken cows and their newborn calves are moved every day to new grazing locations along busy roads full of buses and cars.

Studies on handling stress

There is an old saying: 'You can tell what kind of a stockman a person is by looking at his cattle'. Many cattlemen and women believe that early handling experiences have long-lasting effects (Hassal, 1974). Cattle with previous experience of gentle handling will be calmer and easier to handle in the future than cattle that have been handled roughly (Grandin, 1981). Calves and cattle accustomed to gentle handling at the ranch of origin had fewer injuries at livestock markets because they had become accustomed to handling procedures (Wythes and Shorthose, 1984).

Rough handling can be very stressful. In a review of many different studies, Grandin (1997) found that cortisol levels were two-thirds higher in animals subjected to rough treatment. Rough handling and sorting in poorly designed facilities resulted in much greater increases in heart rate compared with handling in well-designed facilities (Stermer *et al.*, 1981).

The severity and duration of a frightening handling procedure determine the length of time required for the heart rate to return to normal. Over 30 min is required for the heart rate to return to baseline levels after severe handling stress (Stermer *et al.*, 1981). Moving cattle through a handling facility will raise their body temperature, and intake of feed may be reduced for up to 2 days following handling (Mader *et al.*, 2005).

Measurement of cortisol levels has shown that animals can become accustomed to handling procedures. They will adapt to repeated, non-painful procedures, such as moving through a race or having blood samples taken through an indwelling catheter while they are held in a familiar tie stall (Alam and Dobson, 1986; Fell and Shutt, 1986). Wild beef calves can adapt to a non-painful, relatively quick procedure, such as weighing. Peischel *et al.* (1980) reported that daily weighing did not affect weight gain.

Cattle will not readily adapt to severe procedures that cause pain or to a series of rapidly repeated procedures where the animal does not have sufficient time to calm down between procedures. Fell and Shutt (1986) found that cortisol levels did not decrease after repeated trips in a truck where some animals had fallen down and lost their footing. Tame animals are likely to have a milder reaction to an aversive procedure than wild ones. Calves on an experiment station where they were petted by visitors had significantly lower cortisol levels after restraint and handling than calves that had had less contact with people (Boandle *et al.*, 1989).

Training and habituation of animals to handling

Ried and Mills (1962) have suggested that animals can be trained to accept some irregularities in management, which would help reduce violent reactions to novelty. Exposing animals to reasonable levels of music or miscellaneous sounds will reduce fear reactions to sudden, unexpected noises. When a radio is played in a pig barn, pigs have a milder reaction to a sudden noise such as a door slamming. Playing instrumental music or miscellaneous sounds at 75 dB improved weight gains in sheep (Ames, 1974). Louder sound reduced weight gains.

Binstead (1977), Fordyce *et al.* (1985) and Fordyce (1987) reported that training young

Bos indicus heifer calves produced calmer adult animals that were easier to handle. Training of weaner calves involves walking quietly among them in the corrals, working them through races and teaching them to follow a lead horseman (Fordyce, 1987). These procedures are carried out over a period of 10 days.

Becker and Lobato (1997) also found that ten sessions of gentle handling in a race made zebu cross-bred calves calmer and less likely to attempt to escape or charge a person in a small pen. Training bongo antelope to voluntarily cooperate with injections and blood sampling resulted in very low cortisol levels that were almost at baseline (Phillips *et al.*, 1998).

All training procedures must be done gently. Burrows and Dillon (1997) suggest that training may provide the greatest benefit for cattle with an excitable temperament. There are great individual differences in how animals react to handling and restraint. Ray *et al.* (1972) found that cortisol levels varied greatly between individual cattle: in semi-tame beef cattle, one animal had almost no increase in cortisol levels during restraint and blood sampling from the jugular, whereas the other five cattle in the experiments had substantial increases.

Acclimatization of newly arrived cattle at the feedlot

Extensively raised cattle that are not habituated to people often have difficulty settling down and going on feed at the feedlot. Veterinarians Lynn Locatelli and Tom Noffsinger in Nebraska train feedlot receiving crews in methods of reducing the animals' fear and of habituating them to closer contact with people. Handlers approach a group of cattle and, when they first start to react, they back away. Approaching too quickly or too close will cause the animals to run. Gradually they are able to habituate the cattle to people moving closer.

The principle is to gain their trust by relieving pressure at the first sign of a reaction. Over a period of 10–20 min of quietly repeating approach and backing off, most cattle will allow people to move closer without running away. If a pen of calves is milling and bawling, they can be calmed by walking with them in

the *same* direction as they are moving. These methods were developed by Bud Williams and there is further information in Maday (2005).

Genetic differences in temperament

Genetic differences within a breed can affect stress responses during handling. Animals with flighty genetics are more likely to become extremely agitated when confronted by a sudden novel event – such as seeing a waving flag for the first time – than animals with a calmer temperament (Grandin and Deesing, 1998). A basic principle is that animals with flighty, excitable genetics must have new experiences introduced more gradually than animals with calm genetics. One of the major differences between wild and domesticated animals is that the wild species have higher levels of fearfulness and a stronger reaction to environmental change (Price, 1998).

In extensively reared, untrained, wild 260 kg Gelbvieh × Simmental × Charolais cross-bred cattle, behavioural traits were persistent over a series of four monthly handling and restraint sessions (Grandin, 1993). A small group of cattle (9% bulls and 3% steers) became extremely agitated and violently shook the squeeze chute (crush) every time they were handled. Another group (25% bulls and 40% steers) stood very calmly in the squeeze chute every time they were handled. There was also a large group of animals that were sometimes calm and sometimes agitated.

The animals were handled carefully and gently during all the observed restraint sessions. These differences in temperament can probably be explained by a combination of genetic factors and handling experiences as young calves. The behaviour of the few extremely agitated animals did not improve over time. These observations illustrate that the behaviour patterns that are formed at a young age may be very persistent. There was also a tendency for the agitated animals to avoid coming through the race with the first bunches of cattle. Orihuelo and Solano (1994) found that animals first in line in a single-file race moved more quickly through the race compared with animals last in line.

Species such as American bison and antelope are so fearful that they often severely injure themselves when they are restrained. Whereas domestic cattle will tolerate being gently forced into a restraint device, bison and antelope are creatures that need to be trained to cooperate voluntarily (Grandin, 1999). Jennifer Lanier in our research team has had some success in training American bison to move voluntarily through races for feed rewards (Lanier et al., 1999b).

Bison, deer and other flighty species should be moved in small groups. They will remain calmer if only one or two animals are brought from the forcing pen to the restraint device through a short, single-file race. Whereas domestic cattle will stand quietly in a single-file race, many wild ungulates become stressed and agitated if they are made to wait in line. Even with domestic animals, some individuals will habituate to a forced, non-painful procedure and others may respond by getting increasingly stressed. Lanier et al. (1995) found that some pigs habituated quickly to swimming and their adrenaline levels dropped to baseline, whereas others remained frightened and their adrenaline levels remained high.

In Holstein calves, the sire had an effect on cortisol response to transportation stress (Johnston and Buckland, 1976). The sire also has an effect on learning ability and activity levels in Holstein calves (Arave et al., 1992).

The breed has a definite effect on bovine temperament. Brahman-cross cattle became more behaviourally agitated in a squeeze chute compared with Shorthorns (Fordyce et al., 1988). Both Hearnshaw et al. (1979) and Fordyce et al. (1988) reported that temperament is heritable in B. indicus cattle. Stricklin et al. (1980) reported that Herefords were the most docile British breed and Galloways the most excitable. The continental European breeds of Bos taurus were generally more excitable than British breeds. Within a breed the sire was found to have an effect on temperament scores.

LeNeindre et al. (1996) discussed problems associated with taking breeds which were developed for an intensive system and putting them out on an extensive range. For example, a bull can produce daughters that are gentle in an intensive system and aggressive towards handlers when raised on the range. The author has observed that these problems are most likely to occur in excitable, flighty cattle that panic in a novel situation. Some genetic lines of Saler cattle are calm and easy to handle when they are with familiar people, but they panic, kick and charge people when confronted with the noise and novelty of an auction or slaughter plant. These problems are most likely to occur in high-fear breeds such as Saler.

In the USA, the various breed associations have implemented temperament scoring and ranchers are culling the wild cows that become highly agitated during handling. It is important to cull the really 'berserk' cows, but selecting for the absolute calmest is a mistake. Ranchers on extensive rangeland have reported that selection for the most calm may reduce either mothering or foraging ability.

Grandin et al. (1995) and Randle (1998) found that cattle with small, spiral hair whorls above the eyes had a larger flight distance and were more likely to become agitated during restraint than cattle with hair whorls below the eyes. Observations of cattle, horses, dogs and other animals indicated that animals with a slender body and fine bones were more nervous and flighty. Lanier and Grandin (2002) found that cattle that had a smaller diameter cannon bone (foreleg) were more nervous and ran out of the squeeze chute more quickly.

Breed differences in handling patterns

Different breeds of cattle also have different behavioural characteristics that affect handling. Pure-bred B. indicus cattle have a greater tendency to follow people or lead animals. It is sometimes easier to train these cattle to lead instead of driving them. Brahman and Brahman-cross cattle also tend to flock more tightly together when they are alarmed compared with British breeds. Brahman-type cattle are also more difficult to block at gates compared with British breeds (Tulloh, 1961). Salers will bunch more tightly when they get scared than Angus cattle.

Brahman and Brahman-cross cattle are more prone to display tonic immobility during restraint (Fraser, 1960; Grandin, 1980a). Brahman-cross cattle are more likely to lie down in a single-file race and refuse to move compared with British breeds (Grandin, 1980a). Excessive electric prodding of a submissive Brahman can kill it, but it will usually get up if left alone.

Fraser and Broom (1990) stated that an uninjured downed cow will often get up if its environment is changed, such as moving it from inside to outside. Zavy *et al.* (1992) found that Hereford × Brahman crosses and Angus × Brahman crosses had higher cortisol levels during restraint in a squeeze chute compared with Hereford × Angus crosses. Brahman genetics increased cortisol levels and Angus × Brahman crosses had the highest levels.

Flight zone

The concept of flight distance was originally applied to wild ungulates. Hedigar (1968) states: 'By intensive treatment, i.e. by means of intimate and skilled handling of the wild animals, their flight distance can be made to disappear altogether, so that eventually such animals allow themselves to be touched. This artificial removal of flight distance between animals and man is the result of the process of taming'.

This same principle applies to domestic cattle and wild ungulates. Extensively raised wild cows on an Arizona ranch may have a 30 m flight distance, whereas feedlot cattle may have a flight distance of 1.5–7.61 m (Grandin, 1980a). Cattle with frequent contact with people will have a smaller flight distance than cattle that seldom see people. Cattle subjected to gentle handling will usually have a smaller flight zone than cattle subjected to abusive handling.

Excitement will enlarge the flight zone. Totally tame dairy cattle may have no flight zone and people can touch them. The edge of the flight zone can be determined by slowly walking up to a group of cattle. When the flight zone is entered, the cattle will start to move away and, if the person stands still, the cattle will turn and face the handler, but keep their distance. When a person re-enters the edge of the flight zone, the animals will turn around and move away. When the flight zone of a group of bulls was invaded by a moving mechanical trolley, the bulls moved away and maintained a constant distance between themselves and the moving trolley (Kilgour, 1971). The flight distance

was determined by the size of a piece of cardboard attached to the trolley.

Cattle remain further away from a larger object (Smith, 1998). When a person approaches full-face, the flight zone will be larger than when approaching with a small, sideways profile. The author has observed that the principle that large objects are more threatening is even more apparent in flighty antelope. Tame, hand-reared pronghorn antelope panicked and hit the fence of their enclosure when a large, novel object such as a wheelbarrow was brought into their pen. They had to be carefully habituated to each new large object. Small novel objects – such as coffee cups that had been brought into their pens – had no effect.

Cattle can be moved most efficiently if the handler works on the edge of the flight zone (Grandin, 1980b, 1987). The animals will move away when the flight zone is penetrated and stop when the handler retreats. Smith (1998) explains that the edge of the flight zone is not distinct and that approaching an animal quickly will enlarge the flight zone. Excited cattle will have a larger flight zone, and eye contact with the animals will also enlarge the flight zone. If a handler wants an animal to walk past him, he should look away from it. To make an animal move forward, the handler should stand in the shaded area marked A and B in Fig. 4.1 and stay out of the blind spot at the animal's rear.

Point of balance

To make an animal move forward, the handler should stand behind the point of balance at the shoulder and, to make the animal back up, the handler should stand in front of the point of balance (Kilgour and Dalton, 1984). To turn an animal, start on the edge of the flight zone at the point of balance and approach the animal's rear on an angle (Cote, 2003).

Another principle is that grazing animals, either singly or in groups, will move forward when a handler quickly passes the point of balance at the shoulder in the opposite direction of desired movement (Grandin, 1998a; Fig. 4.2). The principle is to move *inside* the flight zone in the opposite direction to the desired movement and *outside* the flight zone in the same direction as the desired movement. Use of the movements

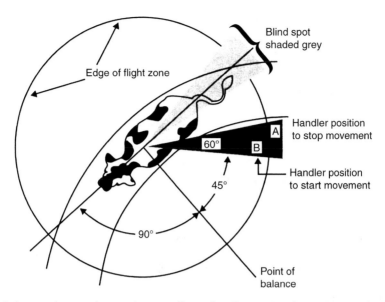

Fig. 4.1. Flight zone diagram showing the most effective handler positions for moving an animal forward.

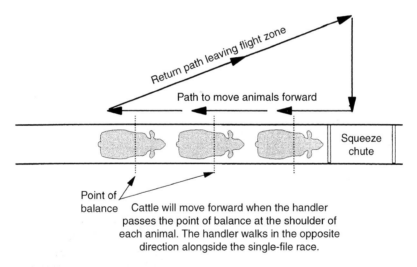

Fig. 4.2. Handler movement pattern to induce cattle to move forward (from Grandin (1998a) In: Gregory, N.G. (ed.) *Animal Welfare and Meat Science*, CABI, Wallingford, UK, p. 47.

to induce cattle to enter a squeeze chute makes it possible to greatly reduce or eliminate electric prod use (see Fig. 4.2).

When an animal is approached head on, it will turn right if the handler moves left, and vice versa (Kilgour and Dalton, 1984). Calm cattle in crowd pens and other confined areas can be easily turned by shaking plastic strips on a stick next to their head (see Fig. 4.3). For example,

when a cow's vision is blocked on the left side by the plastic strips, she will turn right. Handlers should avoid deep penetration of the flight zone, because this may cause cattle to panic.

If an animal rears up in a race, handlers should back up to remove themselves from the animal's flight zone. Handlers should not attempt to push a rearing animal down, because deep penetration of flight zone causes increasing

Fig. 4.3. For turning cattle, plastic streamers on the end of a stick are useful by shaking them alongside the animal's head.

panic and attempts to escape. If cattle attempt to turn back in an alley, the handler should back up and remove him/herself from deep inside the flight zone. The angle of approach and the size of the animal's enclosure will also affect the size of the flight zone.

Experiments with sheep indicated that animals confined in a narrow alley had a smaller flight zone than animals confined in a wider alley (Hutson, 1982). Cattle will have a larger flight zone when they are approached head-on. A basic principle is that the flight zone is smallest along the sides of the animal and greatest in front and behind (Cote, 2003). Extremely tame cattle are often hard to drive because they no longer have a flight zone. These animals should be led. More information on flight zones can be found in Smith (1998).

Moving large groups

Ward (1958) described the methods used on the old cattle drives in the USA to move herds of 1000 cattle, in which many people were required to keep the cattle together. Cattle handling specialist Bud Williams developed movement patterns for moving and gathering cattle. The author

had the opportunity to observe him and develop these guidelines to teach the principles.

All cattle movements are done at a slow walk with no yelling. Figure 4.4 shows the handler movement patterns which will keep a herd moving in an orderly manner. It will work both along a fence and on an open pasture. If a single person is moving cattle, position 2 on Fig. 4.4 shows the handler movement patterns which will keep a herd moving in an orderly manner.

The principle is to alternately penetrate and withdraw from the flight zone. Continuous steady pressure will cause the herd to split. As the herd moves, the handler should keep repeating the movement pattern. For a more complete description, refer to Grandin (1990). Ward (1958) also showed a similar movement pattern. The principle is to move inside the flight zone in the direction opposite to the desired movement and to be outside the flight zone in the same direction as the desired movement.

Figure 4.5 illustrates how to bring the herd back together if it splits. The handler should not act like an attacking predator and run around behind the stragglers to chase them. He/she should move towards the stragglers while gradually impinging on the collective flight zones and

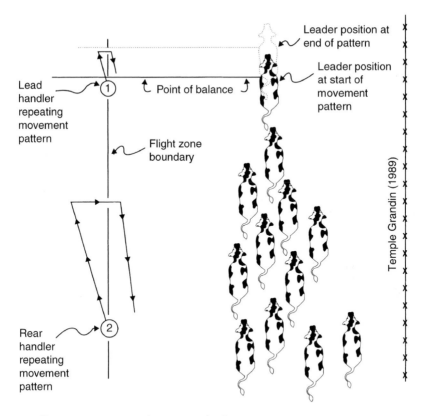

Fig. 4.4. Handler movement patterns for moving a herd.

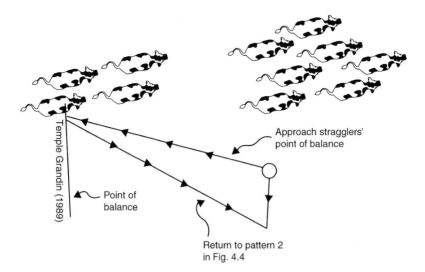

Fig. 4.5. Handler movement pattern for bringing a split herd back together.

stop at the point of balance of the last animal. After the herd closes up, he/she should walk forward at an angle to gradually decrease pressure on the collective flight zone.

Gathering cattle on pasture

Wild and semi-wild cattle can be easily gathered on pasture by inducing their natural behaviour to loosely bunch. Figure 4.6 shows a 'windscreen wiper' pattern, where the handler walks on the edge of the group's collective flight zone. The handler moves at a slow walk and must be careful not to circle around the animals. The handler must also resist the urge to chase stragglers. When the bunching instinct is triggered, the herd will come together and the stragglers will join the other cattle.

Care must be taken to be quiet and keep the animals moving at a walk. The principle is to induce bunching before any attempt is made to move the herd. The animals will move towards the pivotal point of the 'windscreen wipers'. If too much pressure is applied to the collective flight zone prior to bunching, the herd will scatter. More information can be found in Grandin (1998b), Smith (1998) and Ruechel (2006). This method will not work on completely tame animals with little or no flight zone. Leading is often the best way to move really tame cattle, and it is very non-stressful.

Why does it work?

The author speculates that the behavioural principles of moving cattle and other ungulates are

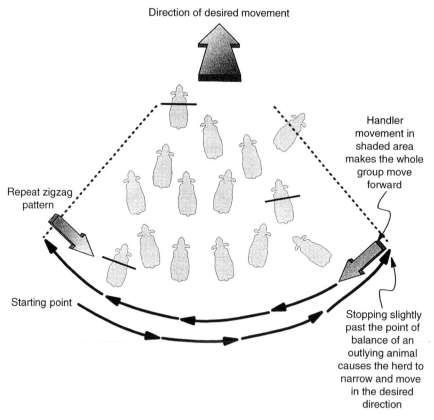

Fig. 4.6. Handler movement pattern to induce bunching for gathering cattle. The handler must zigzag back and forth to keep the herd heading straight. Imagine that the leaders are the pivot point of the 'windscreen wiper' and the handler is out on the end of the blade, sweeping back and forth. As the herd narrows and gets good forward movement, the width of the handler's zigzag narrows.

based on innate, instinctual, anti-predator behaviours (Grandin, 1998c). There appear to be four basic behaviours: (i) turn and orient towards a novel stimulus, but keep a safe distance; (ii) point of balance; (iii) loose bunching; and (iv) milling and circling. The study of many nature programmes on television has indicated that the point-of-balance principle enables an animal to escape a chasing predator.

Cattle bunching behaviour appears to be less intense compared to that in sheep: sheep have extreme bunching behaviour compared to cattle. A flock of sheep will often immediately bunch when they see a dog. Inducing bunching activates a mild anxiety, and the high stress of behaviour (iv) should be avoided. The least stressful handling procedure would be entirely voluntary.

Smith (1998) states that there is no black-and-white dividing line between herding, leading and training. It is likely that cows gathered with the 'windscreen wiper' method may have slight anxiety at first, but then become completely trained and have diminishing anxiety. Bud Williams, cattle-handling specialist, recommends using a straight zigzag motion, instead of the slight curve in the 'windscreen wiper' pattern. The handler must not circle around the cattle and the arc should be very slight.

The principle is to work in a straight line perpendicular from the direction one wants to go (Dylan Biggs, personal communication, 2006). Using these movement patterns probably triggers an instinct to bunch, similar to the behaviour of cattle in bear country, where they graze in tighter bunches.

Training cattle to trust people during herding on the range

Bud Williams has further developed his herding methods so that they no are longer dependent on triggering hard-wired, instinctual behaviour patterns. Cote (2003) has written a book that explains these methods. The principle is to train animals to calmly respond to pressure on the flight zone in a controlled manner. It is a process of training them to move quietly rather than just habituation. Cattle that are calm and trust the handler will move straight and not attempt to circle around and look at that person.

This instinctual behaviour is overridden by trust and learning. Do not work in the blind spot – cattle will turn if they cannot see the handler. All handler movements are done at a walk. The three main principles are:

- Never apply pressure to the flight zone when an animal is doing what you want.
- Release pressure when animals move.
- Reapply pressure only when the cattle slow down.

Working in corrals

Figure 4.7 illustrates the correct positions for emptying a pen and sorting cattle out through a gate (Smith, 1998). The diagram shows the movements for stopping an animal from going out of a gate. Eye contact can be used to hold back animals. The handler should avert his/her eyes away from animals sorted out of the gate. When a pen is emptied, the handler should avoid chasing the cattle out. They should move past the handler at a controlled rate, so that they learn that the handler controls their movements.

Rancher Darol Dickinson states that you need to train cattle (McDonald, 1981). Additional methods for moving and loading cattle are shown in McDonald (1981). One of the most common mistakes is to place too many cattle in a crowd (forcing) pen that leads to a single-file race. An overloaded crowd pen causes problems because the cattle do not have room to turn. For best results fill the crowd pen half-full. To utilize natural following behaviour, handlers should wait for the single-file race to become almost empty before refilling it (Grandin, 1980a).

Many handlers overuse and sometimes abuse electric prods and other persuaders. Electric prods must never be used as a person's primary driving tool. People should not constantly carry electric prods and the only place they may be needed is at the entrance to the squeeze chute. After the electric prod is used to move a stubborn animal, it should be put away. If tail twisting is used to move cattle through a race, pressure on the tail should be instantly released when the cow moves. Breeding cattle quickly learn that they can avoid having their tails twisted if they move promptly when the tail is touched. Gentle tail twisting is less aversive than

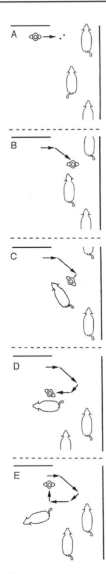

Fig. 4.7. Handler movement patterns for turning an animal back when sorting out a gate (from Smith, 1998).

shouting or use of an electric prod (Pajor *et al.*, 2003).

Handlers should wait until the tailgate of the squeeze chute is open before initiating the movement patterns shown in Fig. 4.2. If the movement pattern fails on the first attempt, walking past the point of balance a second time will often work. At ranches and feedlots, over 90% of the cattle should enter the squeeze chute with no electric prod. Animals learn to associate

the sound of a prod buzzing with the shock (Croney *et al.*, 2000). If one animal baulks, uncooperative behaviour will spread to other cattle. Harger (1928) discusses how one hysterical animal can have a negative influence on the rest of the group.

Cattle are herd animals and they become stressed and upset when they are isolated from their herd mates (Ewbank, 1968). Isolated lone animals that have panicked cause many injuries to both people and cattle. To move a frantic animal some other cattle should be put in with it. Often, the animals most difficult to handle are the last ones in a group to move through a race (Orihuela and Solano, 1994).

Leaders

The natural following behaviour of cattle can be used to facilitate cattle movements. On the old cattle drives in the USA, the value of leaders was recognized. The same leaders would lead a herd of 1000 cattle every day (Harger, 1928). A good leader is usually a sociable cow and is not the most dominant animal. Smith (1998) contains excellent information on the effect of social behaviour on handling.

Excitable, nervous animals that became leaders were usually destroyed and calm leaders were kept (Harger, 1928). If the cattle herd refused to cross a bridge or brook, a calf would be roped and dragged across to encourage the other cattle to follow (Ward, 1958). In Australia, a herd of tame 'coacher' cattle is used to assist in gathering wild feral cattle (Roche, 1988), and similar methods have also been used with wild horses (Amaral, 1977).

Fordyce (1987) also recommends mixing a few quiet, old steers in with *B. indicus* calves to facilitate training to handling procedures. Dumont *et al.* (2005) found that to determine which animal is the true leader one should observe spontaneous long-distance movements to a new feeding site. The leader cannot be determined by watching cattle slowly graze through a field.

Cattle reared under extensive conditions can easily be trained to come when called: the animals learn to associate a vehicle horn with feed (Hasker and Hirst, 1987). In the northern USA, when snow is on the ground cattle will come running when they see the hay truck. However, cattle can become a nuisance by chasing a truck for feed, so the animals should be trained to associate the vehicle horn with feed, then the truck can be driven in a pasture without the animals running after it. Young calves are less likely to become stressed and separated from their mother if their mothers have learned not to constantly chase trucks.

More and more ranchers are adopting intensive grazing systems where cattle are switched to new pasture every few days (Savory, 1978; Smith et al., 1986). The cows quickly learn to make the switch, but calves are sometimes stressed when the cows rapidly run into the new pasture and leave them behind. To avoid calf stress, handlers should either stand near the gate of the new pasture and make the cows walk by them at a controlled rate or lead them slowly into the pasture.

Intensive grazing without fences

There is increasing interest in practising intensive grazing methods without the expense of fences. Herding methods are being used to keep cattle bunched and to move them to different grazing areas. When cows and calves are moved, it is important to move them slowly so that the calves do not become separated from the cows. On long moves the animals should have time to graze. One of the big problems is that some cattle are 'bunch quitters' and do not want to stay with the herd (Nation, 1998). 'Bunch quitters' are most likely to be high-headed, nervous cows.

Selling the 'bunch-quitting' cow is often the best option. Herding works best with uniform groups of cattle that have all been raised on the same ranch. Bunch-quitters are likely to be a problem in cows obtained from several different ranches. The principle of herding without fences is to relieve pressure on the collective flight zone when cows stay where you want them and apply pressure when they go where you do not want them to.

Herders have to spend many hours with their herds and have lots of patience. Low-stress herding is most difficult with older cows from several different ranches that have had completely different previous experiences with herding

and handling (Nation, 1999). More information on pasture herding can be found in Biggs and Biggs (1996, 1997), Herman (1998), Nation (1998), Smith (1998), Williams (1998) and Lanier and Smith (2006a, b).

Genetic and maternal effects on grazing behaviour

Bailey (2004), a grazing specialist in Montana, suggests a combination of herding and other methods to encourage cattle to stay in a new location after they have been moved. When cows with calves are moved long distances, the move should be timed so that they arrive at their destination in the late afternoon when it is time for the calves to bed down. This will encourage the cows to stay in the new location. Tasty supplements placed in the pasture also encourage cows to stay.

Some cattle prefer to graze on the low flatlands and others prefer hills (Bailey et al., 2004, 2005; Rook et al., 2004). Terrain preference may be inherited. Within a breed of cattle, the sire had a significant effect on terrain preference. Breeds developed in the mountains tend to prefer grazing on the hills, but there is also a lot of variation within a breed. Cows sired by Piedmontese bulls, a breed developed in the mountains, preferred steeper rougher terrain than cows sired by Angus bulls, a breed originating from the lowlands (Van Wagoner et al., 2006).

Keeping animals in the right place will be easier if the animal has a preference for the type of terrain you want it on. To determine the terrain preference for a cow, check her location at 7:00 in the morning (Bailey, 2004).

Learning also has a huge effect on grazing behaviour. Livestock prefer the feeds that they ate with their mother when they were little (Provenza, 2003). Bringing adult cows to an area that has unfamiliar feeds can cause big problems. When replacement animals are purchased they should come from areas that have similar pastures (Nation, 2003). Calves can be taught to eat a new feed in the future by feeding it to both the cow and the calf. Cattle that are being brought in from a different part of the country can be taught as calves to eat feeds they will encounter when they grow up.

Herding by pastoral people

The herding methods described in the last section are a relearning of old pastoral herding methods that have been used for thousands of years. In all of these methods, a great deal of time is spent with the animals. Norwegian reindeer herders are in close contact with their animals and the animals associate the smells and noises of camp with serenity (Paine, 1994). The Fulani African tribesmen have no horses, ropes, halters or corrals (Lott and Hart, 1977): their cattle are completely tame and have no flight zone. Instead of chasing the cattle, the herdsman becomes a member of the herd and the cattle follow him (Lott and Hart, 1979).

Bos indicus cattle have a much stronger following instinct than *B. taurus*. Observations by the author indicate that tame, pure-bred Brahmans are difficult to drive and they will often follow a person or a trained lead animal. In Australia they have been trained to follow lead dogs. The nomadic Fulani tribesmen use the animals' natural following, dominance, submission and grooming behaviour to control their overall behaviour. If a bull makes a broadside threat, the herdsman yells and raises a stick. The herdsman charges at the bull and hits it with a stick if it attempts a charge.

Similar methods have also been used successfully in other species. Raising a stick over the handler's head has been used to exert dominance over bull elk (B. Williams, personal communication; Smith, 1998). The stick is never used to hit the bull elk.

The author has used similar methods to control aggressive pigs that exert dominance by biting and pushing against the other pig's neck (Houpt and Wolski, 1982). The aggressive behaviour was stopped by shoving a board against the pig's neck to simulate the bite of another pig. Using the animal's natural method of communication was more effective than slapping it on the rear. Exerting dominance is not beating an animal into submission: the handler uses the animal's own behavioural patterns to become the 'boss' animal.

The aversion cattle have for manure can be used to keep them away from crops, by smearing the borders of a field with faeces (Lott and Hart, 1982). Manure is also smeared on the cow's udder to limit milk intake by the calf. The Fulani stroke their cattle in the same areas where a mother cow licks her calf (Lott and Hart, 1979), so adult cattle will approach and stretch out their necks to be stroked under the chin (Lott and Hart, 1982).

Similar methods are used at the J.D. Hudgins Ranch in Hungerford, Texas, and at the J. Carter Thomas Ranch in Cuero, Texas. Pure-bred Brahmans are led to the corrals and will eat out of the rancher's hand. Cows and bulls in the pasture will come up to Mr Thomas for stroking and brushing (Julian, 1978).

Small herds of zebu cattle raised in the Philippines have no flight zone and they are easily led by small children. The author's observations indicate that taming cross-breeds of Brahman and *B. taurus* is more difficult. This may be partially due to a lower level of inquisitiveness, desire for stroking and following behaviour. The cattle-herding methods of the Fulani are also practised by other African tribes such as the Dinka (Deng, 1972; Schwabe and Gordon, 1988) and the Nuer (Evans-Pritchard, 1940). The less nomadic tribes do use corrals and tethers, but the cattle are still completely tame with no flight zone. Surplus bulls are castrated and kept as steers by all tribes.

The cattle-handling practices of African tribes date back to before the great dynasties of Egypt (Schwabe, 1985; Schwabe and Gordon, 1988). It is also noteworthy that the religion of the Nuer and Dinka tribes centres around cattle (Seligman and Seligman, 1932; Evans-Pritchard, 1940). One factor which makes African tribal handling methods successful is that relatively small herds are handled and each tribe has many herdsmen. Therefore, each herdsman has time to develop an intimate relationship with each animal.

Bull behaviour

Dairy bulls have a bad reputation for attacking humans, possibly due to the differences in the way beef bulls and dairy bulls are raised. Bulls are responsible for about half of the fatal accidents with cattle (Drudi, 2000). Dairy bull calves are often removed from the cow shortly after birth and raised in individual pens, whereas beef bull calves are reared by the cow.

Price and Wallach (1990) found that 75% of Hereford bulls reared in individual pens from

1–3 days of age threatened or attacked the handlers, whereas only 11% threatened handlers when they were hand-reared in groups. These authors also report that they have handled over 1000 dam-reared bulls and have experienced only one attack. Bull calves that are hand-reared in individual pens may fail to develop normal social relations with other animals, and they possibly view humans as a sexual rival (Reinken, 1988).

Both dairy and beef breed bulls will be safer if bull calves are raised on a cow and kept in groups with other cattle. This provides socialization with their own species and they will be less likely to direct attacks towards people. Similar aggression problems have also been reported in hand-reared male llamas (Tillman, 1981). Fortunately, hand-rearing does not cause aggression problems in females or castrated animals: it will make these animals easier to handle. More information on bulls can be found in Smith (1998).

Conclusions

Cattle are animals that fear novelty and become accustomed to a routine. They have a good memory and animals with previous experience of gentle handling will be easier to handle than animals with a history of rough handling. Both genetic factors and experience influence how cattle will react to handling.

An understanding of natural behaviour patterns will facilitate handling. Handlers who use the movement diagrams in this chapter will be able to move both large and small groups of cattle safely and quietly.

To reduce stress, progressive producers should work with their animals to habituate them to a variety of quiet handling methods such as people on foot, riders on horses and vehicles. Training animals to accept new experiences will reduce stress when animals are moved to a new location.

References

Adams, A. (1903) *Log of a Cowboy*. Houghton Mifflin, New York (reissued in 1964 by Bison Book, University of Nevada Press, Reno, Nevada).

Alam, M.G.S. and Dobson, H. (1986) Effect of various veterinary procedures on plasma concentrations of cortisol, luteinizing hormone and prostaglandin F_2 metabolite in the cow. *Veterinary Record* 118, 7–10.

Algers, B. (1984) A note on responses of farm animals to ultrasound. *Applied Animal Behaviour Science* 12, 387–391.

Amaral, A. (1977) *Mustang: Life and Legends of Nevada's Wild Horses*. University of Nevada Press, Reno, Nevada.

Ames, D.R. (1974) Sound stress and meat animals. In: *Proceedings of the International Livestock Environment Symposium*. SP-0174, American Society of Agricultural Engineers, St Joseph, Michigan, p. 324.

Arave, C.W. (1996) Assessing sensory capacity of animals using operant technology. *Journal of Animal Sciences* 74, 1996–2009.

Arave, C.W., Lamb, R.C., Arambel, M.J., Purcell, D. and Walters, J.L. (1992) Behaviour and maze learning ability of dairy calves are influenced by housing, sex and size. *Applied Animal Behaviour Science* 33, 149–163.

Bailey, D.W. (2004) Management strategies for optimal grazing distribution and use of arid rangelands. *Journal of Animal Science* 82 (E. Suppl.), E147–E153.

Bailey, D.W., Keil, M.R. and Rittenhouse, L.R. (2004) Research observation: daily movement patterns of hill climbing and bottom dwelling cows. *Journal of Range Management* 37, 20–28.

Bailey, D.W., Van Wagoner, H.C., Garrick, D.J., Thomas, M.G. and Weinmeister, R. (2005) Sire and dam effects on distribution patterns of cows grazing mountainous rangeland. *Proceedings Western Section, American Society of Animal Science*, 56.

Beausoleil, N.J., Stafford, K.J. and Mellor, D.J. (2005) Sheep show more aversion to a dog than to a human in an arena test. *Applied Animal Behavior Science* 91, 219–232.

Becker, B.G. and Lobato, J.F.P. (1997) Effect of gentler handling on the reactivity of Zebu crossed calves to humans. *Applied Animal Behaviour Science* 53, 219–224.

Biggs, D. and Biggs, C. (1996) Using the magnetic nature of herd movement. *Stockman Grassfarmer*, September. Mississippi Valley Publishing, Ridgeland, Mississippi, pp. 15–17.

Biggs, D. and Biggs, C. (1997) Gatekeepers: low stress tactics for moving cattle. *Stockman Grassfarmer*, September. Mississippi Valley Publishing, Ridgeland, Mississippi, pp. 36–37.

Binstead, M. (1977) Handling cattle. *Queensland Agriculture Journal* 103, 293–295.

Boandle, K.E., Wohlt, J.E. and Carsia, R.V. (1989) Effect of handling, administration of a local anesthetic and electrical dehorning on plasma cortisol in Holstein calves. *Journal of Dairy Science* 72, 2193–2197.

Burri, R. (1968) *The Gaucho*. Crown Publishing, New York.

Burrows, H.M. and Dillon, R.D. (1997) Relationship between temperament and growth in a feedlot and commercial carcass traits of *Bos indicus* crossbreeds. *Australian Journal of Experimental Agriculture* 37, 407–411.

Carroll, J.M., Murphy, C.J., Neitz, M. and Hoeve, J.N. (2007) Photopigment basis for dichromatic vision in the horse. *Journal of Vision* 1, 80–87.

Cockram, M.S. (1990) Some factors influencing behavior of cattle in a slaughter house lairage. *Animal Production* 50, 475–481.

Cote, S. (2003) *Stockmanship: a Powerful Tool for Grazing Management*. USDA Natural Resources Conservation Service, Arco, Idaho.

Coulter, D.B. and Schmidt, G.M. (1993) Special senses 1: vision. In: Swenson, M.J. and Reece, W.O. (eds) *Duke's Physiology of Domestic Animals*, 11th edn. Comstock Publishing, Ithaca, New York.

Croney, C.C., Wilson, L.L., Curtis, S.E. and Cash, E.H. (2000) Effects of handling aids in calf behaviour. *Applied Animal Behavior Science* 69, 1–13.

Dantzer, R. and Mormede, P. (1983) Stress in farm animals: a need for re-evaluation. *Journal of Animal Science* 57, 6–18.

Darbrowska, B., Harmata, W. and Lenkiewiez, Z. (1981) Color perception in cows. *Behaviour Process* 6, 1–10.

Deng, F.M. (1972) *The Dinka of Sudan*. Holt, Rinehart and Winston, New York.

Drudi, D. (2000) Are animals occupational hazards? Compensation and working conditions. *U.S. Dept. of Labor*, Washington, DC, Fall, pp. 15–22.

Dumont, B., Boissey, A. and Archard, C. (2005) Consistency of order in spontaneous group movements allows the measurement of leadership in a group of grazing heifers. *Applied Animal Behaviour Science* 95, 55–66.

Evans-Pritchard, E.E. (1940) *The Nuer*. Clarendon, Oxford, UK.

Ewbank, R. (1968) The behavior of animals in restraint. In: Fox, M.W. (ed.) *Abnormal Behavior in Animals*. W.B. Saunders, Philadelphia, Pennsylvania.

Fell, L.R. and Shutt, D.A. (1986) Adrenal response of calves to transport stress as measured by salivary cortisol. *Canadian Journal of Animal Science* 66, 637–641.

Fordyce, G. (1987) Weaner training. *Queensland Agricultural Journal* 113, 323–324.

Fordyce, G., Goddard, M.E., Tyler, R., Williams, C. and Toleman, M.A. (1985) Temperament and bruising in *Bos indicus* cross cattle. *Australian Journal of Experimental Agriculture* 25, 283–288.

Fordyce, G., Dodt, R.M. and Wythes, J.R. (1988) Cattle temperaments in extensive herds in northern Queensland. *Australian Journal of Experimental Agriculture* 28, 683–687.

Fowler, M.E. (1995) *Restraint and Handling of Wild and Domestic Animals*, 2nd edn. Iowa State University Press, Ames, Iowa.

Fraser, A.F. (1960) Spontaneously occurring forms of tonic immobility in farm animals. *Canadian Journal of Comparative Medicine* 24, 330–333.

Fraser, A.F. and Broom, D.M. (1990) *Farm Animal Behavior and Welfare*. Baillière Tindall, London.

Gilbert, B.J. and Arave, C.W. (1986) Ability of cattle to distinguish among different wavelengths of light. *Journal of Dairy Science* 69, 825–832.

Grandin, T. (1980a) Livestock behavior as related to handling facilities design. *International Journal for the Study of Animal Problems* 1, 33–52.

Grandin, T. (1980b) Observations of cattle behavior applied to the design of cattle handling facilities. *Applied Animal Ethology* 6, 19–31.

Grandin, T. (1981) Bruises on Southwestern feed lot cattle. *Journal of Animal Science* 53 (Suppl. 1), 213 (abstract).

Grandin, T. (1987) Animal handling. In: Price, E.O. (ed.) Farm Animal Behavior. *Veterinary Clinics of North America: Food Animal Practice* 3 (2), 323–338.

Grandin, T. (1990) Forget that rodeo cowboy stuff. *Beef* (Intertec Publishing, Minneapolis, USA) 26 (6), 14–18.

Grandin, T. (1993) Behavioral agitation during handling of cattle is persistent over time. *Applied Animal Behavior Science* 36, 1–9.

Grandin, T. (1997) Assessment of stress during handling and transport. *Journal of Animal Science* 75, 249–257.

Grandin, T. (1998a) Handling methods and facilities to reduce stress on cattle. *Veterinary Clinics of North America: Food Animal Practice* 14, 325–341.

Grandin, T. (1998b) Calm and collected. *Beef* (Minneapolis, Minnesota) March, 74–75.

Grandin, T. (1998c) Can acting like a predator produce low stress cattle handling? *Cattleman* (Winnipeg, Manitoba, Canada) October, 42.

Grandin, T. (1999) Safe handling of large animals. *Occupational Medicine: State of the Art Reviews* 14, 195–212.

Grandin, T. (2005) Maintenance of good animal welfare standards in beef slaughter plants by use of auditing programs. *Journal of the American Veterinary Medical Association* 226, 370–373.

Grandin, T. and Deesing, M.J. (1998) Behavioral genetics and animal science. In: Grandin, T. (ed.) *Genetics and the Behavior of Domestic Animals*. Academic Press, San Diego, California, pp. 1–30.

Grandin, T., Deesing, M.J., Struthers, J.J. and Swinker, A.M. (1995) Cattle with hair whorls above the eyes are more behaviorally agitated during restraint. *Applied Animal Behaviour Science* 46, 117–123.

Harger, C.M. (1928) *Frontier Days*. Macrae Smith, Philadelphia, Pennsylvania.

Hasker, P.J.S. and Hirst, D. (1987) Can cattle be mustered by audio conditioning? *Queensland Agriculture Journal* 113, 325–326.

Hassal, A.C. (1974) Behavior patterns in beef cattle in relation to production in the dry tropics. *Proceedings of the Australian Society of Animal Production* 10, 311–313.

Hearnshaw, H., Barlow, R. and Want, G. (1979) Development of a 'temperament' or handling difficulty score for cattle. *Proceedings of the Australian Association of Animal Breeding and Genetics* 1, 164–166.

Hedigar, H. (1968) *The Psychology and Behavior of Animals in Zoos and Circuses*. Dover Publications, New York.

Heffner, R.S. and Heffner, H.E. (1983) Hearing in large mammals: horse (*Equus caballus*) and cattle (*Bos taurus*). *Behavioral Neuroscience* 97(2), 299–309.

Heffner, R.S. and Heffner, H.E. (1992) Hearing in large mammals: sound-localization acuity in cattle (*Bos taurus*) and goats (*Capra hircus*). *Journal of Comparative Psychology* 106, 107–113.

Herman, J. (1998) Patience and silence are the major virtues in successful stockherders. *Stockman Grass Farmer* (Mississippi Valley Publishing, Ridgeland, Mississippi) October, 55 (10), 1, 8.

Hough, E. (1958) The roundup. In: Neider, C. (ed.) *The Great West*. Coward-McCann, New York.

Houpt, K. and Wolski, T.R. (1982) *Domestic Animal Behavior for Veterinarians and Animal Scientists*. Iowa State University Press, Ames, Iowa.

Humphries, E. (2006) Weaning doesn't have to be wearing cattlemen. *Texas and Southwestern Cattle Raisers Association* (Fort Worth, Texas), March, pp. 38–50.

Hutson, G.D. (1982) Flight distance in Merino sheep. *Animal Production* 35, 231–235.

Jacobs, G.H., Deegan, J.F. and Neitz, J. (1998) Photopigment basis for dichromatic colour vision in cows, goats and sheep. *Visual Neuroscience* 15, 581–584.

Johnston, J.D. and Buckland, R.B. (1976) Response of male Holstein calves from seven sires to four management stresses as measured by plasma corticoid levels. *Canadian Journal of Animal Science* 56, 727–732.

Julian, S. (1978) Gentle on the range. *Western Livestock Journal* (Denver, Colorado), November, 24–30.

Kilgour, R. (1971) Animal handling in works: pertinent behavior studies. In: *Proceedings of the 13th Meat Industry Research Conference*, Hamilton, New Zealand, pp. 9–12.

Kilgour, R. and Dalton, C. (1984) *Livestock Behavior: a Practical Guide*. Granada Publishing, St Albans, UK.

King, D.A., Schuehle, C.E., Randel, R.D., Welsh, T.H., Oliphant, R.A., Curley, K.O., Vann, R.C. and Savell, J.W. (2005) ? of Animal Temperament on the Tenderness of Beef M. Longissimus on the Tenderness of Beef. M. Longissimus Lumborum Steaks. Beef Cattle Research in Texas, Texas A & M University, College Station, Texas, pp. 129–133.

Lanier, E.K., Friend, T.H., Bushong, D.M., Knabe, D.A., Champney, T.H. and Lay, D.G. (1995) Swim habituation as a model for eustress and distress in the pig. *Journal of Animal Science* 73 (Suppl. 1), 126 (abstract).

Lanier, J. and Grandin, T. (2002) The relationship between *Bos taurus* feedlot cattle temperament and cannon bone measurements. *Proceedings, Western Section, American Society of Animal Science* 53, 97–101.

Lanier, J. and Smith, B. (2006a) Slow is fast, low stress livestock handling I. *Stockman Grass Farmer*, Ridgeland, Mississippi, February, 30–31.

Lanier, J. and Smith B. (2006b) Slow is fast, low stress livestock handling II. *Stockman Grass Farmer*, Ridgeland, Mississippi, 18–19.

Lanier, J.L., Grandin, T., Green, R.D., Avery, D. and McGee, K. (1999a) The effect on cattle of sudden inter-mittent movements and sounds associated with an auction ring. In: *1999 Beef Program Report*. Colorado State University, Fort Collins, Colorado, pp. 225–230.

Lanier, J.L., Grandin, T., Chaffin, A. and Chaffin, T. (1999b) Training American bison calves, 1999. *Bison World* 24 (4), 94–99.

Lanier, J.L., Grandin, T., Green, R.D., Avery, D. and McGee, K. (2000) The relationship between reaction to sud-den intermittent movements and sounds to temperament. *Journal of Animal Science* 78, 1467–1474.

LeDoux (1996) *The Emotional Brain*. Simon and Schuster, New York.

Lemmon, W.B. and Patterson, G.H. (1964) Depth perception in sheep: effects of interrrupting the mother–neonate bond. *Science* 145, 835–836.

LeNeindre P., Boivin, X. and Boissy, A. (1996) Handling extensively kept animals. *Applied Animal Behavior Science* 49, 73–81.

Linford, L. (1977) Stampede. *New Mexico Stockman* 42 (11), 48–49.

Lott, D.F. and Hart, B. (1977) Aggressive domination of cattle by Fulani herdsmen and its relation to aggres-sion in Fulani culture and personality. *Ethos* 5, 172–186.

Lott, D.F. and Hart, B. (1979) Applied ethology in a nomadic cattle culture. *Applied Animal Ethology* 5, 309–319.

Lott, D.F. and Hart, B.L. (1982) The Fulani and their cattle: applied behavioral technology in a nomadic cattle culture and its psychological consequences. *National Geographic Research Reports* 14, 425–430.

Lynch, J.J. and Alexander, G. (1973) *The Pastoral Industries of Australia*. University Press, Sydney, Australia, pp. 371–400.

Maday, J. (2005) Put out the welcome mat. *Drovers Feeder Management Supplement*, September 2–12.

Mader, T., Davis, M.S. and Kreikemeier, W.M. (2005) Case study: tympanic membrane temperature and behavior associated with moving feedlot cattle. *Professional Animal Scientist* 21, 339–344.

McDonald, P. (1981) Thinking like a cow makes the job easier. *Beef* (Intertec Publishing, Minneapolis, Min-nesota), October, 81–86.

Miller, C.P., Wood-Gush, D.G.M. and Martin, P. (1986) The effect of rearing systems on the responses of calves to novelty. *Biology of Behaviour* 11, 50–60.

Miller P.E. and Murphy, C.J. (1995) Vision in dogs. *Journal of the American Veterinary Medical Association* 12, 1623–1634.

Nation, A. (1998) Western grazers practice management intensive grazing without fences. *Stockman Grass Farmer*, September (Mississippi Valley Publishing, Ridgeland, Mississippi), 1–12.

Nation, A. (1999) Herding is not as easy as it looks. *Stockman Grassfarmer* (Mississippi Valley Publishing, Ridgeland, Mississippi), September, 1–13.

Nation, A. (2003) Allan's observations. *Stockman Grass Farmer*, November, 8–11.

Orihuelo, J.A. and Solano, J.J. (1994) Relationship between order of entry in slaughterhouse raceway and time to traverse raceway. *Applied Animal Behaviour Science* 40, 313–317.

Paine, R. (1994) *Herds of the Tundra: a Portrait of Saami Reindeer Pastoralism*. Smithsonian Institute Press, Washington, DC.

Pajor, E.A., Rushen, J. and dePaisille, A.M.B. (2003) Dairy cattle choice of handling treatments in a Y maze. *Applied Animal Behaviour Science* 80, 93–107.

Peischel, P.A., Schalles, R.R. and Owenby, C.E. (1980) Effect of stress on calves grazing Kansas Hills range. *Journal of Animal Science* (Suppl. 1), 24–25 (abstract).

Petherick, J.C., Holyroyd, R.G., Doogan, V.J. and Venus, B.K. (2002) Productivity, carcass, meat quality of feedlot *Bos indicus* cross steers grouped according to temperament. *Australian Journal of Experimental Agriculture* 42, 389–398.

Phillips, M., Grandin, T., Graffam, W., Irlbeck, N.A. and Cambre, R.C. (1998) Crate conditioning of Bongo (*Tragelephus eurycerus*) for veterinary and husbandry procedures at Denver Zoological Gardens. *Zoo Biology* 17, 25–32.

Pick, D.F., Lovell, G., Brown, S. and Dail, D. (1994) Equine color vision revisited. *Applied Animal Behavior Science* 42, 61–65.

Price, E.O. (1998) Behavioral genetics and the process of animal domestication. In: Grandin, T. (ed.) *Genetics and the Behavior of Domestic Animals*. Academic Press, San Diego, California, pp. 31–65.

Price, E.O. and Wallach, S.J.R. (1990) Physical isolation of hand reared Hereford bulls increases their aggres-siveness towards humans. *Applied Animal Behaviour Science* 27, 263–267.

Price, S., Sibley, R.M. and Davies, M.H. (1993) Effects of behavior and handling on heart rate in farmed red deer. *Applied Animal Behavior Science* 37, 111–123.

Prince, J.H. (1977) The eye and vision. In: Swenson, M.J. (ed.) *Dukes Physiology of Domestic Animals*. Cornell University Press, New York, pp. 696–712.

Provenza, F.D. (2003) *Foraging Behavior Managing to Survive a World of Change*. USDA, Utah State University, Logan, Utah.

Randle, H.D. (1998) Facial hair whorl position and temperament in cattle. *Applied Animal Behavior Science* 56, 139–147.

Ray, D.E., Hansen, W.J., Theurer, B. and Stott, G.H. (1972) Physical stress and corticoid levels in steers. *Proceedings Western Section, American Society of Animal Science* 23, 255–259.

Rehkamper, G. and Gorlach, A. (1998) Visual identification of small sizes by adult dairy bulls. *Journal of Dairy Science* 81, 1574–1580.

Reinken, G. (1988) General and economic aspects of deer farming. In: Reid, H.W. (ed.) *The Management and Health of Farmed Deer*. Springer, London, pp. 53–59.

Ried, R.L. and Mills, S.C. (1962) Studies of carbohydrate metabolism of sheep. XVI The adrenal response to physiological stress. *Australian Journal of Agricultural Research* 13, 282–294.

Roche, B.W. (1988) Coacher mustering. *Queensland Agricultural Journal* 114, 215–216.

Rogan, M.T. and LeDoux, J.E. (1996) Emotion: systems, cells and synaptic plasticity. *Cell* (Cambridge, Massachusetts) 83, 369–475.

Rook, A.J., Dumont, B., Isselstein, J., Osorok, Wallis deVries, M.F., Parente, G. and Mills, J. (2004) Matching type of livestock to desired biodiversity outcomes in pastures – a review. *Biological Conservation* 119, 137–150.

Ruechel, J. (2006) *Grassfed Cattle*. Story Publishing, North Adams, Massachusetts.

Sandem, A.I. and Braastad, B.O. (2005) Effects of calf separation and visible eye white and behavior in dairy cows – a brief report. *Applied Animal Behavior Science* 95, 233–239.

Sandem, A.I., Janczak, A.M. and Braastad, B.O. (2004) A short note on effects of exposure to a novel stimulus (umbrella) or behaviour and percentage of white eye in the cow. *Applied Animal Behaviour Science* 89, 309–314.

Sandem, A.I., Janczak, A.M., Salle, R. and Braastad, B.O. (2006) The use of diazepam as a pharmacological validation of eye white as an indicator of emotional state in dairy cows. *Applied Animal Behaviour Science* 96, 177–183.

Saslow, C.A. (1999) Factors affecting stimulus visibility in horses. *Applied Animal Behaviour Science* 61, 273–284.

Savory, A. (1978) Ranch and range management using short duration grazing. In: *Beef Cattle Science Handbook*. Agriservices Foundation, Clovis, California, pp. 376–379.

Schwabe, C.W. (1985) A unique surgical operation on the horns of African bulls in ancient and modern times. *Agricultural History* 58, 138–156.

Schwabe, C.W. and Gordon, A.H. (1988) The Egyptian w3s-sceptor and its modern analogue: uses in animal husbandry, agriculture and surveying. *Agricultural History* 62, 61–89.

Seligman, C.G. and Seligman, B.Z. (1932) *Pagan Tribes of the Nilotic Sudan*. Routledge and Kegan Paul, London.

Smith, B. (1998) *Moving 'Em: a Guide to Low Stress Animal Handling*. Graziers Hui, Kamuela, Hawaii.

Smith, B., Leung, P. and Love, G. (1986) *Intensive Grazing Management: Forage, Animals, Men, Profits*. Graziers Hui, Kamuela, Hawaii.

Smith, S. and Goldman, L. (1999) Color discrimination in horses. *Applied Animal Behaviour Science* 62, 13–25.

Stephens, D.B. and Toner, J.N. (1975) Husbandry influences on some physiological parameters of emotional responses in calves. *Applied Animal Ethology* 1, 233–243.

Stermer, R., Camp, T.H. and Stevens, D.G. (1981) *Feeder Cattle Stress During Transportation*. Paper No. 81–6001, American Society of Agricultural Engineers, St Joseph, Michigan.

Stricklin, W.R., Heisler, C.E. and Wilson, L.L. (1980) Heritability of temperament in beef cattle. *Journal of Animal Science* 51 (Suppl. 1), 109 (abstract).

Talling, J.C., Waran, N.K. and Wathes, C.M. (1996) Behavioral and physiological responses of pigs to sound. *Applied Animal Behavior Science* 48, 187–202.

Talling, J.C., Waran, N.K., Wathers, C.M. and Lines, J.A. (1998) Sound avoidance by domestic pigs depends upon characteristics of the signal. *Applied Animal Behavior Science* 58 (3–4), 255–266.

Thines, G. and Soffie, M. (1977) Preliminary experiments on color vision in cattle. *British Veterinary Journal* 133, 97–98.

Tillman, A. (1981) *Speechless Brothers: the History and Care of Llamas.* Early Winters Press, Seattle, Washington.

Tulloh, N.M. (1961) Behavior of cattle in yards. II. A study of temperament. *Animal Behavior* 9, 25–30.

Uetake, K. and Kudo, Y. (1994) Visual dominance over hearing in feed acquisition procedure of cattle. *Applied Animal Behavior Science* 42, 1–9.

Vann, R.C., Paschal, J.C. and Randal, R.D. (2004) Relationships between measures of temperament and carcass traits in feedlot steers. *Journal of Animal Science* 82 (Suppl. 1), 259 (abstract).

Van Wagoner, H.C., Bailey, D.W., Kress, D.D., Anderson, D.C. and Davis, K.C. (2006) Differences among beef sire breeds and relationships between terraine, use and performance when daughters graze foothill rangeland as cows. *Applied Animal Behaviour Science* 97, 105–121.

Voisinet, B.D., Grandin, T., Tatum, J.D., O'Connor, S.F. and Struthers, J.J. (1997) Feedlot cattle with calm temperaments have higher average daily gains than cattle with excitable temperaments. *Journal of Animal Science* 75, 892–896.

Ward, F. (1958) *Cowboy at Work.* Hasting House Publishers, New York.

Waynert, D.E., Stookey, J.M., Schwartzkopf-Gerwein, J.M., Watts, C.S. and Waltz, C.S. (1999) Response of beef cattle to noise during handling. *Applied Animal Behavior Science* 62, 27–42.

Welp, T., Rushen, J., Kramer, D.L., Festa-Bianchet, M. and de Passille, A.M.B. (2004) Vigilance as a measure of fear in dairy cattle. *Applied Animal Behaviour Science* 87, 1–13.

Western Livestock Journal (1973) 'Put on' foils escape minded cattle. *Western Livestock Journal* (Denver, Colorado), July, 65–66.

Williams, E. (1998) How to place livestock and have them stay where you want them. *Stockman Grass Farmer* (Mississippi Valley Publishing, Ridgeland, Mississippi), September, 6–7.

Wyman, W.D. (1946) *The Wild Horse of the West.* Caxton Printers, Caldwell, Idaho.

Wythes, J.R. and Shorthose, W.R. (1984) *Marketing Cattle: its Effect on Live Weight and Carcass Meat Quality.* Australian Meat Research Committee Review No. 46, Australian Meat and Research Corporation, Sydney, Australia.

Zavy, M.T., Juniewicz, P.E., Williams, A.P. and Von Tungeln, D.L. (1992) Effect of initial restraint, weaning and transport stress on baseline and ACTH stimulated cortisol responses in beef calves of different genotypes. *American Journal of Veterinary Research* 53, 552–557.

5   Low-stress Restraint, Handling and Weaning of Cattle

Joseph M. Stookey and Jon M. Watts

Western College of Veterinary Medicine, University of Saskatchewan, Saskatoon, Canada

Introduction

What more can be learned about handling cattle? Our relationship with domestic cattle dates back approximately 10,000 years (Hanotte *et al.*, 2002). There is evidence that our ancestors had a good understanding of the general factors that influence animal movement. Pictographs on Egyptian tombs, for example, depict herdsmen driving and leading cattle across the crocodile-infested Nile River – certainly no easy task – by taking advantage of the maternal instinct of a cow to follow her calf, and the instinct of cattle to follow one another (see Fig. 5.1).

So even as far back as the ancient Egyptians, people knew how to take advantage of natural behaviour to herd and stimulate movement. A modern-day sceptic might even claim that there is nothing new to be learned about cattle behaviour. However, just as the study of ruminant nutrition has been able to tease apart factors that influence forage quality and digestibility, modern behavioural studies have been able to dissect the factors that influence an animal's response to handling.

Genetics, previous experience, the internal motivation of an animal and the surrounding external stimuli all interact to influence an animal's behaviour. The challenge for ethologists has been to determine the relative influence of these various factors on the behaviour we observe in cattle.

Assessment of Behaviour using Movement-measuring Devices

Several researchers have subjectively scored temperament of cattle during their movement through a chute (race) system or during their restraint in a headgate (Tulloh, 1961; Holmes *et al.*, 1972; Heisler, 1979). Subjective scores have been useful in determining the genetic heritability of temperament (Stricklin *et al.*, 1980, pp. 44–48) as it relates to handling. While the scoring is subjective, it still requires an observer to calculate mentally the animal's level of excitement, by taking into account the relative amount of movement and agitation displayed.

However, in experimental work, subjective scores are susceptible to inadvertent observer effects that may compromise their reliability (Lehner, 1996). Arguably, a better way to record an animal's response to handling is to quantify objectively the amount of movement an animal makes while exposed to various stimuli. Indirectly, an electronic scale can serve this purpose, since the voltage output from the load cells will vary as the animal moves on the scale platform.

Stookey *et al.* (1994) used a 'movement-measuring device' (MMD) to collect analogue signals from the load cells of an electronic animal scale. This showed that cattle stood more calmly within sight of a conspecific compared with visually isolated animals. The same device (along with heart rate measurements) was used

Tomb-owner: Ti
Location: Saqqara
Date: late 5th dynasty
Wild, H. Le Tombeau de Ti. Vol. 2, plate cxiv.

Fig. 5.1. Egyptian tomb pictograph depicting the annual movement of cattle herds returning to the Nile valley from the lush Delta region where they had previously been pastured (courtesy of Linda Evans, Department of Ancient History, Macquarie University, Sydney, Australia).

to show that cattle were calmer when exposed to their mirror image compared with cattle in isolation (Piller *et al.*, 1999). Both studies showed that cattle are calmer in the presence of their conspecifics. Moreover, the tools used to quantify these responses have also been used to evaluate the importance of other visual, auditory or olfactory cues that may influence an animal's response to handling.

Responses to Human Beings

The human voice

Waynert *et al.* (1999) recorded heart rate and MMD values to compare the response of British × Continental (*Bos taurus*) cattle exposed to two categories of sounds that are typically present during handling: namely, the vocalizations of humans urging animals to move forward and the noise from metal gates and sides clanging and banging. Not surprisingly, cattle tested individually moved more on the electronic scale and had elevated heart rates when exposed to the recordings of both types of handling noises combined, compared with cattle exposed to silence.

To determine the contribution of each specific type of noise to the animal's response, Waynert *et al.* (1999) conducted a second experiment using another group of animals. In this experiment,

50% of the animals were exposed to the sounds of human vocalizations and the other 50% were exposed only to the chute-banging sounds. Both sounds were adjusted in a studio recording so that they could be played back at the same volume in decibels. The cattle exposed to the human vocalizations were more agitated, based on heart rate and MMD values, compared with cattle exposed to the sounds of clanging metal. It is an interesting finding and suggests that humans play a significant role – not just in their physical visual presence – but also in their vocalizations, in the response of cattle to handling.

In another study, at the Western College of Veterinary Medicine (Clavelle *et al.*, unpublished), researchers attempted to determine whether it was the sound of human vocalizations *per se* that were unsettling to cattle or whether it was the intent on the part of handlers to use their voice in stimulating cattle movement. Pairs of voice recordings were made by several handlers, including men and women. In the first recordings of each pair, handlers attempted to reproduce the tone of voice they would employ to urge cattle to move, and such vocal invective as each individual customarily used.

In the second recordings, each person uttered the same sequence of vocal expressions but in an even, neutral tone of voice, as though he or she was completely indifferent as to whether the animal moved or not. When these pairs of recordings were played to cattle, their overall heart rate and movement responses to the first type of recording were not significantly different from their responses to the second type. Put another way, it does not appear to matter to the animals whether a person uses vocal expression to make them move, or simply talks in their presence. In either situation the sound of the human voice can be inherently upsetting to cattle that are not habituated to it.

Handling

Mainly through the work and writings of Temple Grandin, many cattle producers now appreciate that the sight of humans can have a pronounced effect on cattle during handling. Hence, the common recommendation today is to construct handling facilities with solid sides in order to prevent cattle from inadvertently glimpsing people (Grandin, 1980, 1997). Lay et al. (1992) showed the benefits of using a dark 'breeding box' during the artificial insemination of cattle, perhaps because the box helped eliminate the sight of people (see figures in Chapter 7, this volume).

It is commonly believed that solid walls facilitate movement of cattle through the chute by helping them focus on the only way out and by removing the human factor. Observations in slaughter plants indicate that adding solid sides to prevent cattle from seeing vehicles, conveyors and other moving objects greatly improves the movement of cattle through the chutes (Grandin, 1996). Most solid-sided handling systems incorporate a catwalk on the side of the chute. People can remain out of sight while off the catwalk or 'appear' on the catwalk only when cattle require encouragement to move through the system. Cattle should theoretically remain calmer during handling if they are unable to see people. Unfortunately, the location where humans are closest and most visible to the cattle is in the front of the chute complex, where cattle are caught and restrained in a head gate.

One way of removing the sight of humans at the front is to blindfold the cattle during restraint. Andrade et al. (2001) reported that blindfolded cattle had lower heart rates and yielded lower temperament scores during a 3-min period of restraint compared with controls. Mitchell et al. (2004) reported similar findings, but replaced temperament scoring with an

objective method of quantifying the animal's struggle while it was restrained. Strain gauges were attached to the head gate (as described by Schwartzkopf-Genswein et al., 1997) to measure the exertion force cattle applied against it and found that blindfolded cattle struggled less compared with controls.

Blindfolding cattle during restraint may not be practical during routine procedures that are relatively quick, such as vaccination, ear tagging, etc., but may have application for procedures requiring longer periods of restraint, such as during calving or surgical procedures. How one applies blindfolds in a quick and ergonomic manner deserves some thought, especially if one wishes to use the technique during the processing of large numbers of animals. Mitchell et al. (2004) blindfolded their cattle using a dark towel tucked under a rope halter.

The use of solid sides and curved chutes has swept through the cattle industry as useful and necessary design features that facilitate cattle movement through chute complexes in slaughter facilities, auction markets and feedlots. These systems work very well with low-skilled labour and in areas where there are many distractions outside the facility such as vehicles, people walking by and objects with high visual contrast.

However, there are advocates who claim that open-sided and straight chutes offer distinct advantages under extensive pasture conditions where there are few disturbances outside the chute and where handlers have a keen understanding of cattle behaviour.

North American cattle handlers Bud Williams (2006) and Dylan Biggs (2006) offer workshops for producers and teach low-stress handling techniques. Both instructors demonstrate in their workshops that the positioning of humans is more critical in gaining proper movement of cattle than facility design. Design is less important when highly skilled people who understand behaviour are handling cattle (see Chapter 4, this volume). For example, the rapid movement of humans passing very close to the cattle in the opposite direction (moving from the front towards the back of a group or line of animals) actually speeds up the movement of cattle past the handler, and is more effective than pushing cattle from the rear (see Fig. 5.2). Also, by allowing cattle the opportunity to see through the sides, Williams and Biggs claim that people can apply or release pressure on

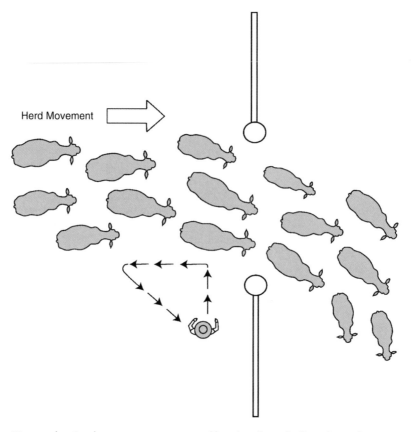

Fig. 5.2.   Diagram showing the movement pattern used by a handler to facilitate forward movement of animals through a gate, by repeatedly moving past the group against the desired direction of movement (adapted from Biggs, 2006).

the cattle by their positioning much more sensitively than if walls are solid.

With solid walls, people on catwalks are physically unable to adjust their proximity to the cattle other than being completely off the catwalk and out of sight, or on the catwalk in close proximity. Another good system is to make one wall of the single file race solid and the other open, but only on the top half. This eliminates the catwalk and the cattle can respond to the position of people on the ground. For nervous cattle (i.e. those with a wilder temperament and a bigger flight zone) humans on catwalks can be too close and can apply too much pressure. This can cause animals to panic and increases the risk of serious injury. Of course, the disadvantage of open sides is the potential distraction of other people or objects outside the chute that negatively influence movement.

Experimental Testing of Cattle Movement

Our research group initiated a study to see if we could combine the best of both principles by allowing cattle to have a greater view of what lies ahead of them, but block their immediate side view of the facility (Stookey *et al.*, unpublished data). A group of cattle, naïve to the facility, were tested once individually to determine their choice between alternative paths leading out of a simple Y-maze. The arms consisted of two identical curved chutes branching off in opposite directions at a 30° angle from the junction of the straight arm (see Fig. 5.3). At the junction of the Y it was impossible for the animal to see the exits at the end of each arm of the maze.

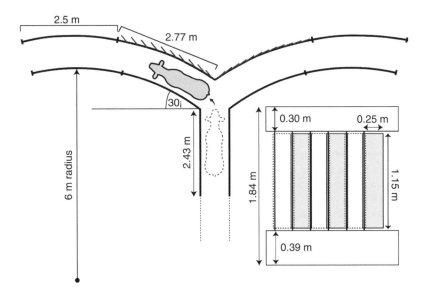

Fig. 5.3. Diagram of Y-maze used to test preference of cattle for varying visual conditions. The chute walls contained overlapping 'louvres' that could be opened or closed: these are located in a 2.77 m section of the outer fence starting at the junction of the Y.

During each test one curved arm had vertical overlapping slats positioned in the open condition within the outside wall. The slats offered the impression of an opening in the outside wall ahead of the animal, while the wall directly beside the animal appeared solid. The opposite arm of the Y-maze had all the slats in the closed position. The arms of the maze with the open and closed slats were randomly assigned, but balanced over the course of the trial.

Animals were free to exit the maze on their own initiative and no humans were visible during their 'choice'. Both exit arms of the maze were raked between test animals to reduce the risk of cues left by the previous animal. No other animal or human was visible to the test animal while it made its choice. Thirty-eight of 57 cattle exited via the arm with open slats ($\chi^2 = 6.366$, $P < 0.02$).

The same Y-maze was used to test other visual conditions that were thought to influence or facilitate movement. In the second experiment, another group of cattle was used to determine the effects of overhead lighting. Twenty-four of 27 cattle exited via the lit arm ($\chi^2 = 17.77$, $P < 0.001$) when the starting position in the straight arm was also lit. When the starting position in the straight arm was dark, 18 of 22 cattle exited via the lit arm ($\chi^2 = 11.88$,

$P < 0.01$). Overall, 42 of 49 cattle (86%) exited via a lit arm ($P < 0.001$), proving that cattle have a very strong preference for moving through a lighted facility.

In a third experiment, we offered cattle choices between different-coloured interior walls of the chute. One arm of the Y-maze was fitted with light beige-coloured interior walls, while the walls of the opposite arm were dark brown. The interior walls were interchangeable. When the starting position in the straight arm was dark brown, only 13 of 22 cattle exited via the lighter-coloured beige arm. Nine of 20 exited via the beige arm when the starting position was also beige. Neither of the results differed from random chance.

From this series of experiments we determined that movement of cattle through a chute complex may be facilitated by the use of vertically slatted openings along the outside wall of a curved chute and with the use of lights, but the relative shade (light versus dark) of the chute walls may not be important.

Moving the Herd

Not all handling and movement of cattle occurs within a chute complex. Cattle are also moved, sorted or regrouped for a variety of reasons. The art of herding cattle and moving cattle in extensive environments has become the subject of popular workshops for cattle producers in North America (Biggs, 2006; Williams, 2006). Many of the techniques taught are somewhat counterintuitive. For example, driving cattle directly from the rear of the herd by pushing straight into them in the desired direction of movement will inevitably cause the leaders of the group to turn to one side, in their attempt to keep the humans in their view.

To drive cattle in a straight line, a more effective method is for the herder to move back and forth, all the way across the rear of the herd and out beyond the herd edges, perpendicular to the intended direction of movement, so that animals on either side of the herd will see the person and turn back into the herd and not veer off left or right.

The workshops teach producers how to initiate cattle movement, control the direction of a herd, and speed them up or slow them down in a calm manner by using what have been called 'low-stress handling techniques'. These techniques focus on proper positioning, understanding cattle behaviour and the application of 'pressure and release' to generate cattle movement without unnecessary fear and stress.

Many people fail to appreciate that the movement of a few animals can stimulate others to follow. Therefore, when trying to empty an entire pen of animals, a person can be successful by gently nudging just a few animals through the opening and then standing back to let the leaders draw out the rest of the group. Often, people mistakenly assume that they must position themselves behind the entire group to empty a pen or pasture; sometimes, that positioning serves only to draw attention away from the opening and back on to themselves.

Some people also tend to apply more and more pressure as the cattle come closer to their destination or gate opening, as if trying to close the sides of an invisible net. Such pressure inadvertently communicates to the cattle that they are moving into a worse destination than they are leaving, causing them to have a stronger urge to turn back and try to escape the pressure. Low-stress handling techniques teach producers to 'reward' cattle for moving in the right direction by releasing pressure. The same principles

work when working cattle through a chute or up a loading ramp. If cattle are moving in the right direction, one should release the pressure.

It is important to appreciate that the responses of cattle to any particular handling situation are inherently plastic. That is to say, their behaviour in response to a particular set of stimuli is likely to alter with experience due to active learning processes. Knowledgeable handlers can exploit the learning abilities of cattle to shape their behaviour. This can make subsequent handling easier and less stressful for both animals and people.

Learning Responses to Handling by Cattle

A simple example is the correct use of tail-twisting to move cattle in a chute. This is not a desirable form of pressure to apply routinely to an animal, since it undoubtedly causes some pain and may cause injury if done too enthusiastically. However, inevitably, some animals that are unwilling to move, either due to fear – or, paradoxically because of over-habituation to people, in other words a lack of fear – will require additional pressure to stimulate movement.

Tail-twisting is properly used as a form of negative reinforcement. Gentle bending force is applied to the vertebrae in the tail, causing discomfort. Any forward movement by the animal, even a tiny amount, should be rewarded by releasing the force *instantly*. If this is done sensitively, the animal will learn – within two or three applications – that escaping the painful stimulus is a consequence of its own behaviour of moving forward. In effect, the handler has taught the animal what is required of it. On subsequent occasions, the handler may only have to gently touch the tail to remind the animal that movement is needed.

The mistake made by many handlers when applying painful stimuli to produce movement (and this also applies to the use of electric prods) is failing to release the stimulus when the desired behaviour occurs. It is tempting to believe that if a little tail-twist causes movement, then prolonged and harder tail twisting should produce more movement. In fact, this is counterproductive and will ultimately make animals harder to move. It is not the pain – but rather it is the rewarding consequence of the animal's action in escaping or avoiding the pain – that causes it to repeat the behaviour of moving.

The above example illustrates how, in handling situations, animals learn associations between their own actions and the consequences they produce. It is also possible for them to mistakenly associate a rewarding outcome with the behaviour they were performing immediately before, when the actual events are causally unrelated. As a consequence, they are more likely to repeat the same actions on subsequent occasions. In the psychological literature this type of behaviour is sometimes called 'superstitious' (Skinner, 1948).

In a recent experiment (unpublished), we tried to determine whether this kind of mistaken attribution could be exploited to help make cattle-handling easier or safer. Cattle released from a headgate may attribute their release to what they were doing immediately before. Thus, an animal that was struggling when released might learn to struggle more during subsequent restraint. Similarly, an animal that was relatively motionless prior to release might try to reproduce this state on a subsequent occasion to bring about release.

In this experiment, we handled beef heifers for 4 consecutive days by restraining them in a headgate fitted with electronic strain gauges, and releasing them only when a specific behavioural criterion had been met. Half of the animals we tried to make calmer during restraint, by releasing them from the headgate only when they had exhibited a period of calm, motionless behaviour. The other half of the group we tried to make wilder, by only releasing them when they showed a period of violent, struggling behaviour.

The animals in the struggling group produced the desired behaviour faster each consecutive day, suggesting that they had learned that this behaviour earned them their release. The calm group did not learn as quickly. On the fifth day, the animals were restrained for an extra period of 30 s while we measured the strain forces, and compared them with those recorded on the first day. Both the calm and struggle treatments exerted less force than at the beginning of the experiment.

We concluded that the behaviour of restrained cattle can indeed be influenced by appropriately timing their release from the headgate.

However, it seems that cattle may learn to struggle to obtain release more readily than they learn to remain calm. By releasing cattle only when they are calm, handlers can at least avoid inadvertently training the animals to struggle more during restraint.

Minimizing the Stress Associated with Sorting and Weaning Beef Cattle

Usually, the purpose or goal of handling cattle is to complete a necessary chore that may in itself be stressful (i.e. routine procedures such as castration, dehorning, branding, vaccination or implanting). Much has been written on the pain and stress associated with these procedures and will not be the subject of discussion here. However, one handling procedure that does result in considerable stress to cattle is the separation and sorting of cows and calves from each other for weaning. The entire procedure is stressful, partly because of the immediate stress associated with handling and sorting, but also because of the breaking of the maternal offspring bond over several days.

One way of minimizing the stress of sorting cows from calves has been developed in Australia under extensive range conditions. Over the course of 24 h, calves and cows can be automatically sorted and separated from each other during their visit to a watering station. The watering location is surrounded by a fence where cows and calves enter and exit the area daily through the use of one-way spear gates (see Fig. 5.4). On the day of separation, the calves' entrance is fitted with a one-way spear gate to prevent calves from exiting. As the calves enter the watering area they are diverted into a separate pen without an exit. This system allows the cattle to be separated automatically without humans being present.

While weaning is a natural phenomenon, the procedure is typically forced on to cattle at an earlier stage than they would wean naturally. This should more accurately be referred to as *artificial* weaning. Artificially weaning cows and calves by abrupt separation is perhaps the greatest psychological stressor imposed on beef cattle during their lifetime (Stookey and Watts, 2004). We know artificial weaning, or abrupt separation, is stressful because of the associated

View A

View B

Fig. 5.4. Photographs of Australian cow–calf separator placed around water location to capture and sort calves from cows. To enter the watering area, cows learn to push open the swinging gate shown at lower left of View B, while calves avoid the cow gate and slip through the smaller opening. The spear gate exit can be seen at left in View A and at right in View B. On capture days, a one-way spear gate is attached to the calf entrance to prevent them from leaving the watering area, and a portable fence divides the cow and calf entrances such that calves are diverted and trapped in a pen separate from adults (Photographs compliments of C. Petherick, Department of Primary Industries & Fisheries, Queensland, Australia).

increases in cortisol, setback in weight gain, increases in morbidity and significant changes in behaviour. Increased vocalization, decreased feeding, decreased time spent lying and increased walking are some of the main behavioural changes that persist for 3–5 days after separation.

One social consequence of artificial weaning is that, following separation from their dams, calves are placed in a situation where no adult cattle are present. Some researchers have attempted to minimize the stress of weaning on calves by adding adult 'trainer' cows into the pens of newly weaned calves. This idea has had, at best, minimal success (Gibb *et al.*, 2000; Loerch and Fluharty, 2000).

One possible explanation for this poor outcome may have been that the adult cattle were unfamiliar with the newly weaned calves in both of these studies. However, using a cross-over design whereby cows were split into two groups and given each other's calves, Nicol (1977) could not show an improvement in weight gain compared to the traditional method of abrupt separation, suggesting that calves treated this way were equally stressed. Apparently, newly weaned or separated cows and calves can not be pacified by the presence of any other cattle, unfamiliar or familiar.

Low-stress weaning methods

Therefore, minimizing the stress of weaning would seem to hinge on allowing some contact between cow and calf until the dependency is over. One obvious solution is fenceline weaning, first described by Nicol (1977). Fenceline weaning prevents nursing, but may allow some physical contact depending upon the structure of the fence. Most importantly, fenceline weaning provides visual and auditory contact between separated pairs. The technique has been successfully used in horses (McCall *et al.*, 1985), elk (Haigh *et al.*, 1997) and beef cattle (Stookey *et al.*, 1997; Price *et al.*, 2003). Fenceline-weaned calves gained more weight than abruptly separated controls during the first week following weaning, and were still heavier than traditionally weaned calves after 10 weeks (Price *et al.*, 2003).

Most recently, Haley *et al.* (2005) found that by weaning in two stages, first by preventing the calf from obtaining its dam's milk and secondly by separating the pair several days later, the entire weaning process can be made significantly less stressful. In our studies, nursing was prohibited by a plastic anti-sucking device, which hung from the calf's nose and prevented

Fig. 5.5. Photograph of a calf wearing an anti-sucking device used to prevent nursing. The device facilitates weaning of cow–calf pairs while kept in the presence of each other, commonly referred to as two-stage weaning.

calves from getting the teat into their mouth (see Fig. 5.5).

In one trial, 12 beef cow–calf pairs were randomly assigned to a control (abrupt weaning treatment) or to the two-stage weaning procedure. Nursing was prevented for 4 days prior to separation for the two-stage calves. Prior to imposing the treatments, baseline information was collected on the amount of nursing and general behaviour patterns of the cow–calf pairs for 4 days. We then observed the animals during the 4 day period when nursing was prevented for the two-stage weaning group and while control groups were allowed to nurse. Finally we observed cows and calves for the 4 days following their separation. All anti-sucking devices were removed on the day of separation.

During the baseline period, vocalizations by both cows and calves were extremely rare. The only behavioural change associated with preventing nursing, during the next 4 days, was

a slight increase in the amount of vocalization (cows = 24 vocalizations/day; calves = six vocalizations/day). However, during the 4 days after cows and calves had been separated, two-stage weaned cows vocalized 80% less than the control cows weaned the traditional way. For calves, the difference was even more remarkable: two-stage weaned calves vocalized 95% less than traditional weaned calves, calling at the same rate as during baseline observations.

Two-stage weaned cows and calves spent over 25% more time eating compared to controls, and two-stage calves spent roughly 50% less time walking than the abruptly weaned calves after separation.

We conducted another trial to determine the relative distance that calves walked before and following separation, using pedometers. We found that controls walked nearly three times further than two-step calves during the first two days after separation (40 versus 15 km,

$P < 0.05$). Two-stage weaning drastically reduces the behavioural indicators of stress associated with the traditional abrupt method of weaning.

In a separate trial, Haley (2006) reported that two-stage calves had significant reduction in vocalization and standing and an increase in time spent lying compared with fenceline-weaned calves, suggesting that two-stage weaning has some distinct advantages over other favourable weaning techniques. We now know that simultaneously removing the mother and the milk – as done in traditional weaning – causes a far greater behavioural response than if the two events occur at separate times (Haley et al., 2005).

One disadvantage of two-stage weaning is that calves must be handled twice: once for inserting the anti-sucking device and once for removal. It means that calves must be sorted twice from the cows. One simple method to facilitate sorting has been developed by Biggs (2006). He moves the entire group of cows and calves into a smaller paddock. He then takes advantage of the natural tendency of cattle to want to exit the pen via the route they had entered, so he stands at the gate and allows cows to exit while diverting calves through a 'lower' calf gate that traps them in a separate pen. Using this technique, a single person can successfully sort hundreds of cows from calves in a quick, calm and stress-free manner. This technique is useful whenever cow–calf pairs need sorting and makes two-stage weaning possible for large herds.

Conclusions

Some understanding of cattle behaviour, cognition and perception is essential for their management, especially during handling and restraint. Behavioural principles have been successfully applied in the design of handling facilities that facilitate movement of animals, even by inexperienced stockpersons. With appropriate training and experience, operators can make use of 'low stress' handling techniques that make successful handling possible in open areas or where facilities are primitive or nonexistent.

Managing what cattle see and hear during handling and restraint is important for the minimization of stress and facilitation of movement. The responses of cattle to handling can change as a result of experience. This learning capability can be used to advantage, to train animals to handle more easily and remain calm during handling.

References

Andrade, O., Orihuela, A., Solano, J. and Galina, C.S. (2001) Some effects of repeated handling and the use of a mask on stress responses in zebu cattle during restraint. Applied Animal Behaviour Science 71, 175–181.

Biggs, D. (2006) Dylan Biggs Stockmanship Clinics: Low Stress Livestock Handlling. http://www.tkranch.com/

Gibb, D.J., Schwartzkopf-Genswein, K.S., Stookey, J.M., McKinnon, J.J., Godson, D.L., Wiedmeier, R.D. and McAllister, T.A. (2000) Effect of a trainer cow on health, behavior, and performance of newly weaned beef calves. Journal of Animal Science 78, 1716–1725.

Grandin, T. (1980) Observations of cattle behavior applied to the design of cattle handling facilities. Applied Animal Ethology 6, 19–31.

Grandin, T. (1996) Factors that impede animal movement at slaughter plants. Journal of the American Veterinary Medical Association 209, 757–759.

Grandin, T. (1997) The design and construction of facilities for handling cattle. Livestock Production Science 49, 103–119.

Haigh, J.C., Stookey, J.M., Bowman, P. and Waltz, C. (1997) A comparison of weaning techniques in farmed wapiti (Cervus elaphus). Animal Welfare 6, 255–264.

Haley, D.B. (2006) Factors affecting the behaviour response of beef cattle to artificial weaning. PhD thesis, University of Saskatchewan, Saskatoon, Saskatchewan.

Haley, D.B., Bailey, D.W. and Stookey, J.M. (2005) The effects of weaning beef calves in two stages on their behavior and growth rate. Journal of Animal Science 83, 2205–2214.

Hanotte, O., Bradley, D.G., Ochieng, J.W., Verjee, Y., Hill, E.W. and Rege, E.O. (2002) African pastoralism: genetic imprints of origins and migrations. *Science* 296, 336–339.

Heisler, C.E. (1979) Breed differences and genetic paramaters of temperament in beef cattle. MS thesis, University of Saskatchewan, Saskatoon, Saskatchewan.

Lay, D.C., Friend, T.H., Grissom, K.K. and Bowers, C.L. (1992) Novel breeding box has variable effects on heart rate and cortisol response in cattle. *Applied Animal Behaviour Science* 35, 1–10.

Lehner, P.N. (1996) *Handbook of Ethological Methods*, 2nd edn. Cambridge University Press, Cambridge, UK, pp. 210–215.

Loerch, S.C. and Fluharty, F.L. (2000) Use of trainer animals to improve performance of newly arrived feedlot calves. *Journal of Animal Science* 78, 539–545.

McCall, C.A., Potter, G.D. and Kreider, J.L. (1985) Locomotor, vocal and other behavioural responses of varying methods of weaning foals. *Applied Animal Behaviour Science* 14, 27–35.

Mitchell, K.D., Stookey, J.M., Laturnas, D.K., Watts, J.M., Haley, D.B. and Hayde, T. (2004) The effects of blind folding on behaviour and heart rate in beef cattle during restraint. *Applied Animal Behaviour Science* 85, 233–245.

Nicol, A.M. (1977) Beef cattle weaning methods. *New Zealand Journal of Agriculture* 134, 17–18.

Noble, J., Stookey, J.M., Watts, J.M., Olenick, D. and Kendall, R. (2006) Handler behaviour can influence the amount of struggling by beef cattle during restraint. *Applied Animal Behaviour Science*, In press.

Pillar, C.A.K., Stookey, J.M. and Watts, J.M. (1999) Effects of mirror image exposure on heart rate and movement in isolated heifers. *Applied Animal Behaviour Science* 63, 93–102.

Price, E.O., Harris, J.E., Borgwardt, R.E., Sween, M.L. and Connor, J.M. (2003) Fenceline contact of beef calves with their dams at weaning reduces the negative effects of separation on behaviour and growth rate. *Journal of Animal Science* 81, 116–121.

Schwartzkoph-Genswein, K.S., Stookey, J.M. and Welford, R. (1997) Behaviour of cattle during hot-iron and freeze branding and the effects on subsequent handling ease. *Journal of Animal Science* 75, 2064–2072.

Skinner, B.F. (1948) 'Superstition' in the pigeon. *Journal of Experimental Psychology* 38, 168–172.

Stookey, J.M. and Watts, J.M. (2004) Production practices and well-being: beef cattle. In: Benson, G.J. and Rollin, B.E. (eds) *The Well-being of Farm Animals: Challenges and Solutions*. Blackwell Publishing Professional, Ames, Iowa, pp. 185–205.

Stookey, J.M., Nickel, T., Hanson, J. and Vandenbosch, S. (1994) A movement-measuring-device for objectively measuring temperament in beef cattle and for use in determing factor that influence handling. *Journal of animal Science* 1, 207, Suppl.

Stookey, J.M., Schwartzkopf-Genswein, K.S., Waltz, C.S. and Watts, J.M. (1997) Effects of remote and contact weaning on behaviour and weight gain of beef calves. *Journal of Animal Science* 75 (Suppl. 1), 157 (abstract).

Stricklin, W.R., Heisler, C.E. and Wilson, L.L. (1980) Heritability of temperament in beef cattle. *Journal of Animal Science* 51 (Suppl. 1), 109–110.

Tulloh, N.M. (1961) Behaviour of cattle in yards II A study of temperament. *Animal Behavior* 9, 25–30.

Waynert, D.F., Stookey, J.M., Schwartzkopf-Genswein, K.S. and Waltz, C.S. (1999) The response of beef cattle to noise during handling. *Applied Animal Behaviour Science* 63, 27–42.

Williams, B. (2006) Bud Williams Stockmanship School: Teaching Low Stress Livestock Handling Methods. http://www.stockmanship.com/

6 Handling Cattle Raised in Close Association with People

Roger Ewbank[1] and Miriam Parker[2]

[1]19 Woodfield Road, Ealing, London, UK (formerly Director of Universities Federation for Animal Welfare); [2]Livestockwise, The New House, Bausley, Crew Green, Shrewsbury, UK

Introduction

Cattle have been domesticated by humans for several thousands of years and they are currently used in very large numbers as sources of meat and/or milk and for draught purposes. As a result of this long and close contact there is great practical knowledge within the agricultural community as to their management, care and handling.

There are two overall behaviour patterns in intensively raised cattle. In the first type, cattle are completely tame and can easily be led with a halter (head collar) and lead rope. These cattle are led by a person and are usually not driven. Cattle that are trained to lead are common in Asia, India, China and France (see Fig. 6.1).

The second type of cattle are not trained to lead, but they usually have a small flight zone. They are raised in groups and they are in close contact with people every day. When they are moved they are either driven or led by handlers and no lead ropes are used. Often, these cattle become very accustomed to many human activities. *Bos indicus* cattle grazed along the roads learn to ignore bicycles and buses (see Fig. 6.2).

The facilities required to handle cattle that have been raised in close association with people can be much simpler than the facilities that are required to handle much wilder, extensively raised cattle that have much less contact with people. In many countries, cattle that are trained to lead can be easily handled without races and crowd pens. In Sudan, the second author has observed skilled herders at a local market who corralled their cattle by using people as fences (see Fig. 6.3). The people positioned themselves around a group of cattle on the boundary of the flight zone.

In the UK and New Zealand, groups of cattle come when called and are led by a handler to a new pasture. However, in abattoirs and in large veterinary facilities, cattle that are not trained to a lead rope should be handled in facilities that are designed for extensively raised cattle (see Chapters 7 and 20, this volume).

Information on the behaviour of British and European cattle – an important component of any understanding of how they can be effectively and sympathetically handled – is to be found in Hafez and Bouissou (1975), Albright and Arave (1997), Hall (2002), Phillips (2002) and Houpt (2005). Grandin (1998a, 2005) and Lawrence (1991) give accounts of behaviour of animals in restraint, and the general behaviour and welfare of farm animals – with much attention to bovines – is discussed in Fraser and Broom (1990).

Fig. 6.1. In much of the world, tame cattle live in close association with people and they are trained to lead. Complex handling facilities are not required for cattle that are trained to lead and have no flight zone.

Fig. 6.2. Cattle in the Philippines are grazed along roads and other places with high human activity. They have lived near cars, bicycles and other activity all their lives. These animals do not react to a bicycle or car because it is no longer novel.

Effects of Genetics and Experience

Many different breeds of cattle have been developed (Epstein and Mason, 1984) and it is generally accepted that these can be grouped into two main types: (i) the humped (zebu) (*Bos indicus*) animals, native to subtropical and tropical areas (see Payne, 1990 for an account of Zebu cattle types and their husbandry); and (ii) the non-humped (*Bos taurus*) cattle of European origin.

There are some behavioural differences between these cattle types (Hafez and Bouissou, 1975) and it is believed – without much real evidence – that they react differently to handling. It is widely held for example, that zebu-type cattle are more difficult to restrain than their European equivalents. This view probably stems from experience with these animals under extensive pastoral conditions. *B. indicus* cattle husbanded/reared in close contact with humans in the family farm as milk or draught animals

Fig. 6.3. In Sudan, cattle are corralled for sale by positioning groups of people around them on the perimeter of their flight zone. The animals are fenced in by the people. They have a small flight zone, but are not trained to lead.

are usually easily handled. When they are completely tamed with no flight zone, the *B. indicus* breeds appear to be more responsive to stroking compared with *B. taurus* breeds. Intensively reared *B. indicus* cattle often actively seek out stroking. This may help form a bond with the handler.

The husbandry systems under which cattle have been reared probably have a more significant effect on flight zone size and tameness than the effects of any genetic differences between the different breeds (Murphey *et al.*, 1980; Kabuga and Appiah, 1992). It is held as a general rule, however, that within European (*B. taurus*) cattle, females of the dairy breeds are quieter to handle than females of the beef breeds, but the converse is true for the males. See Chapters 4 and 8, this volume, for more information on the handling of bulls.

The most important factor which influences the ease or otherwise with which cattle or any other creature can be handled is the extent and severity of their previous experiences of contact with humans. For a general account of the effects of this human–animal relationship in stock farming see Hemsworth and Coleman (1998) and Raussi (2005).

It is widely believed – and there is much field evidence to support this view – that cattle which have had frequent, gentle and early contact with humans are usually tame and easy to handle. This early contact may have to have been over an extended period if it is to have a long-term effect. Handling restricted to the first month of the calf's life (Sato *et al.*, 1984) seemed to have little influence on its subsequent

behaviour; the regular handling of heifers up to the age of 9 months (Bouissou and Boissy, 1988), however, was shown to reduce permanently their fear of humans. A detailed discussion on the effects of early experience on farm livestock is to be found in Creel and Albright (1987).

Young cattle are inquisitive, active and playful, while older animals are more placid, but both types can be readily trained (Albright, 1981; Lemenager and Moeller, 1981; Dickfos, 1991). For general accounts of learning and training in the domesticated species see Kilgour (1987) and Universities Federation for Animal Welfare (1990).

Traditional Family Farm

In the traditional, family-farmed systems of stock keeping still common in many parts of the world, small numbers of cattle are reared and kept in close contact with humans (FAO, 2002). They are grazed under careful supervision in small fields or on tethers and/or are kept indoors in tie-up cow sheds or in covered yards. Their calves are often hand-reared by humans or, if kept on their dams, are habituated to the near presence of humans at an early age. As a general rule, cattle husbanded in these ways are tame, cooperative and easily handled. It seems that they associate handling with particular locations (Rushen *et al.*, 1998) and that they may at times differentiate between different human handlers (Taylor and Davis, 1998).

Oxen and Other Intensively Handled Bovines in Developing Countries

Fifty per cent of the world's population uses draught animals such as cattle or water buffalo to cultivate their crops (Wilson, 2002). Compared to donkeys and horses, bovines can pull more weight (Pearson and Vall, 2004). Their speed is slower, but the greater power of a heavier animal is preferable for many tasks. Additional information on draught animals may be found in Bartholomew *et al.* (1995), Conroy (1999) and Sims *et al.* (2003).

The second author has made many observations during extensive travels in Asia and Africa. She observed that educating people on some basic principles of handling could bring about great improvements in animal welfare. She trained people to use a simple rope halter (head collar) (see Fig. 6.4) instead of tying animals by a nose ring. Tying cattle in a transport vehicle by the nose may result in severe injuries. Below is a list of simple, easy to implement, low-tech recommendations for prevention of injury when animals are tied up by a lead rope.

1. Do not tie with the nose ring; a halter or head collar should be used.
2. Do not tie large and small animals together, or weak and strong. Animals should be grouped by size and weak animals should be in a separate group.
3. Tie horned and dehorned or polled cattle in separate groups.
4. Provide non-slip flooring in the transport vehicle.
5. Provide ramps for trucks unloading. Do not let animals jump off vehicles.

Animals can easily be trained to voluntarily cooperate during minor procedures such as injections. Cattle can be trained to cooperate willingly if stress is greatly reduced (Grandin, 2000). Cattle and other animals kept for special purposes such as show cattle, AI studies and for agricultural trials are usually easier to handle (Grandin, 1998b). Figure 6.5 shows a simple hoof-trimming stand that a tame cow is led into.

Fig. 6.4. A rope halter should be used for tying a tame bovine. Tying by a nose ring or a cord through the nose may cause nasal injuries. The ring or nasal cord should only be used during handling or for steering a draught animal.

Fig. 6.5. A simple stand for holding a tame cow for hoof trimming or veterinary work. This will work well for animals that are trained to lead. The area must have a solid non-slip floor.

Types of Cattle Husbandry Systems

In the extensive systems of stock farming, cattle are kept in considerable numbers in large fields or on free range; their welfare is supervised by humans but the animals are relatively untamed and only occasionally handled, and then mainly in groups in gathering pens (corrals), races and crushes.

The modern moves towards large-scale cattle operations mean that large numbers of animals are looked after by small numbers of attendants. Many of the regular stock tasks, such as feeding, milking and dung-cleaning, are at least partly carried out by machines: calves and young stock are not often as closely handled as they are on the traditional family farm and there is little time for individual attention to any one animal.

As a result of the low need for close physical contact between humans and cattle, there has perhaps been a tolerance of uncooperative behaviour and less incentive to select for quietness in the stock. Whatever the causes, it is widely held that many of the cattle held in the large-scale feedlot and ranch systems are increasingly seen to be nervous, uncooperative and difficult to control.

The equipment and methods used to handle cattle at close quarters vary considerably around the world in their cost, manpower required, level of sophistication and effect on the animal; however, there are four broad approaches.

Methods for Restraining Cattle for Veterinary Procedures

1. In 'linear' groups in a single-file race (see Chapters 7 and 20, this volume); drenching and pregnancy testing is performed while the group of animals is in the race.
2. While being held in their own tie-up stalls or self-locking stanchions; this is a common system on dairies.
3. While being confined closely together as a group in a small space. This works well for calves.
4. Individually:
- In a crush (squeeze chute) that is located at the end of a single file race. A stanchion or other device restrains the animal by the neck (see Fig. 7.6; Chapter 7, this volume).
- An animal is restrained with ropes. This is done when a crush or squeeze chute is not available.

Training the animal to cooperate will greatly reduce stress. The animal is trained to stand while being held with a halter or lead rope.

In linear groups via forcing pens, races and crushes

In many intensive dairy and beef cattle enterprises, the animals are regularly passed through specially designed handling units, which are made up of various combinations of funnels, forcing (gathering) pens, races, crushes (squeeze chutes), shedding gates and automatic cow traps (Chapter 5, this volume, shows an automatic sorting system). Some of the layouts incorporate foot baths, weighing crates and/or loading ramps.

For dairy cattle these facilities are usually placed so that each time the animals leave the milking area they traverse all or part of the handling unit on the way back to the feeding/resting areas or the grazing fields. Beef cattle will often be regularly passed once a week – say through the foot bath and/or weigher. In both types of enterprise, the animals are used to being regularly handled while still being in contact with other members of their social (pen) group.

If the facilities have been well designed and constructed – there are numerous advisory booklets and pamphlets on these matters (e.g. Shepherd, 1972; Graves, 1983; MAFF, 1984; Gilbert, 1991; Bickert, 1998) – and the stockmen and -women are skilled and patient, the handling can often be efficient, effective and relatively stress-free to both human and animal.

The terminal part of the handling unit is usually some sort of so-called cattle crush (restraint device). These devices are essentially a solid or bar-sided box designed and constructed to hold one animal. There is a vertically pivoted release gate on the front and a sliding back gate (or bars) which can be pushed across the race at the rear, to stop the animal backing away.

The animal's neck can either be caught in a yoke (stanchion), which is built into the front gate or held between two vertically pivoted side panels or bars; these are set so that when they swing in they trap the neck of the animal between them as it stands just short of the front of the crush. The yoke can only be used if the animal volunteers – or can be persuaded – to put its head through the gap in the front gate: successful use of the side panel/bar neck restrainer does not depend on this kind of cooperation.

Sometimes, cattle restrained in crushes become very stubborn, put their heads down and refuse to move and even adopt a so-called kneeling, submissive position (Ewbank, 1961) – a possible form of tonic immobility (Fraser, 1960; see also below). Once the head is held, various hinged panels (gates) in the sides of the crush can be swung open to gain acess to the body of the animal. Some of the more complex

handling (squeeze) crates have one of the sides so pivoted along the horizontal axis that it can be moved to hold the animal tight for restraint, and some are even equipped with leg-winders (devices to hold the feet) so that the feet of the animal can be readily restrained for examination and treatment.

At times it is necessary to take a cow out of the crush and handle in the adjacent open area. To do this, a halter should be put on its head, the front gate opened and the animal led forward. This manoeuvre is easy with a side-panel/bar head restraint crush and with a front gate yoke in which the neck of the animal is being held between the gate and the frame of the crush. In both these cases the head can be released while the animal is standing still. In the central yoke gate design, however, the cow has first to be driven back clear of the gate and then the gate swung open.

Cattle, especially those which routinely traverse a group-housing system, will often readily follow each other through the funnels, forcing pens and races. Once an individual animal is held in the cattle crush, however, the forward movement of the group stops and can start again only when the head-held animal is released. It may be difficult to get the next animal in line to move forward and enter the crush. This is especially true if the previously handled animal has become in any way distressed.

To overcome this start-stop-start problem, it is common practice for handlers – who may be standing on a raised catwalk running alongside the run-up race – to attempt to examine or treat animals by reaching over the top sides of the crush and the run-up race. In this technique, only the heads and top lines of the cattle are directly accessible to stockmen and -women. Injections should not be given in the rump of cattle because this will likely cause damage to the most expensive cuts of beef (George et al., 1995, 1996; Roeber et al., 2002).

If necessary, the heads of the animals can be held by hand: the animals usually stand still as they have little space for body movement and they feel relatively secure, being in close contact with other members of their group. If the race has one partially open barred side, pregnancy testing can be performed while the cattle are held snugly in a linear line. Handling under these circumstances is really a linear form of 'being closely confined together as a group in a small space' – with the advantage that the human handler is safely outside the race (pen).

Once the cattle are released from the crush via the front gate, they can be diverted by sorting gates into side pens, which may contain automatic cow traps (or neck yokes) set at the feed trough, or they can be allowed to return to the grazing fields or their housed accommodation.

A modified version of the cattle crush concept can be used in association with automatic captive yokes (cow traps). Once a cow is caught up in a yoke, a wheeled handling box (a moveable crush but without a front restraining yoke or gate) is pushed/positioned around the cow. The animal, in effect, is now restrained as in a crush. This form of restraint is most applicable in the dairy herd which, as a routine mastitis control technique, traps affected cows in automatic yokes as they leave the milking parlour, by putting a small quantity of an attractive food in the front trough of the traps.

Handling while in their own tie-up stalls or self-locking stanchions

This method is mainly used in dairies. The advantages for the stockperson in handling cattle while they are being held in their own tie-up stalls (stanchions) are as follows:

1. The animal is already caught, i.e. it does not have to be chased and captured. The cattle are already accustomed to being restrained when they are fed.
2. The head of the animal is held in either the chain tie or the neck yoke.
3. The animal is in its own living place and is between known companions in the adjacent stalls, i.e. it is usually relaxed and confident.
4. The husbandry system is generally such that the animal is used to frequent physical contact with humans while it is being fed/milked/having its bedding replaced, etc.
5. It is surrounded by bars and pipes to which the various ropes, etc. used in restraint (see later) can be attached.

There are also a number of disadvantages:

1. There may not be much space around the animal. Cattle in adjacent stalls can be moved

away, but this often distresses the one left isolated.
2. The floor is probably hard (concrete) and the cow may damage itself if it falls.
3. It may begin to associate its own stanchion with unpleasantness and become reluctant to enter it again. This is especially true if painful operations, e.g. foot dressing, are repeatedly carried out in the home stall. It may be better to remove the animal to a 'neutral' place for potentially distressing procedures.

Confined closely together as a group in a small pen

In this method of handling the animals, which can be young stock, polled or dehorned dairy cows or intensively housed beef cattle are put into a restricted area (pen) such that they are in close contact with each other and have little space for movement. Once the cattle have settled down, the human handler(s) can enter the pen, push between the animals and catch and/or handle them at will. The animals usually remain quiet. This is possible because: (i) they have little scope for individual movement; and/or (ii) they remain relaxed while in close physical contact with their peers. An animal separated from its group soon becomes agitated and difficult to handle.

Handled individually by means of ropes and specially designed equipment

There are many different descriptions of the way ropes and specially designed pieces of handling equipment can be used to control and manipulate cattle (McNitt, 1983; Holmes, 1991; Battaglia, 1998; Haynes, 2001). There are also a number of veterinary texts (e.g. Stober, 1979; Kennedy, 1988; Fowler, 1995; Leahy, 1995; Hanie, 2006; Sheldon et al., 2006) which give details of the numerous specialized handling and surgical restraint techniques that can be used as appropriate on calves, young stock and adult cattle. Ropes in particular are used across the world to aid in the restraint or handling of cattle; however, there are a multitude of regional and local variations, with few formally documented or researched.

When cattle are being handled individually this way, ideally the animal should be within a 'safe' space such as the centre of a pen or large loose box. However, for some methods or procedures it may be necessary to be positioned beneath overhead attachment points for ropes or slings, or in a tie-up stall, in a cow cubicle, in the space alongside the cattle crush or in the area immediately adjacent to a tubular metal pen side or wooden fenceline.

The use of pressure or force on the body

Once cattle are confined or restrained, for example within a crush, small pen, stall or by means of a halter or ropes, additional steps may be necessary to reduce movement further and/or safely access a part of the animal's body to carry out a procedure. A number of methods have become established over the years that act on the principle of applying a degree of force to the animal's body, and in some instances this has shown to produce both calming and immobilizing effects.

Pressure applied to the whole body

It has long been recognized that the tightening of single or multiple rope loops that have been placed around the bodies of cattle, e.g. udder kinches, chest twitches and Reuff's method of casting, cause the animal to stand still and, if tightened further, to appear to be partly paralysed and go down. If the rope loops are kept tight, the animals stay down. The reason as to why this works is largely unknown.

It may be that this partial paralysis is related to the so-called tonic immobility (fear paralysis) seen in some animals when pressure is applied to their body and/or when they are exposed to fear-inducing situations. This phenomenon has been studied under experimental conditions in chickens, rodents and rabbits, and there is now some understanding (Carli, 1992) of the behavioural and physiological mechanisms involved.

It is known in rodents that sudden and/or aversive stimuli can, at times, lead to tonic immobility and that this is sometimes accompanied by a drop in the animal's heart rate (Hofer, 1970; Steenbergen et al., 1989). It has been noted in cattle (Leigh, 1937, p. 177) that the tightening of a rope round the body slows down the movement of the heart.

The quietness shown by animals both in the squeeze crush (squeeze chute) and when being restrained in close physical contact with others of their kind – as in handling pens, raceways or loose boxes – may be at least partly as a result of this 'pressure on the animal's body' phenomenon (Ewbank, 1961, 1968).

Pressure applied to the legs

A rope noose tightened round the hind leg of a cow just above the hock joint or the application of a specially designed pincer clip device to each side of the tendon that runs down to the point of the hock can result in the leg seemingly becoming temporarily paralysed. The animal itself is not really calmed and is often somewhat distressed, but the leg is easily handled. Occasionally the animal will go down, but this seems more as a result of loss of balance on the three non-affected legs rather than from any general calming/immobilization effect. The removal of the rope noose or the pincer clip soon restores the animal's ability to move and use the leg, although there is sometimes a degree of swelling at the application site and a slight, associated, temporary lameness.

Pressure applied to the tail

The holding of a cow's or calf's tail in a near-vertical position seems to partially restrict the movement of the hind legs and allows a relatively safe examination of the groin or udder region of animals that are otherwise unhandleable. This method is used to make animals stand still. It should never be used as a method for making the animal move forward.

Pressure applied to the nose

The grasping of the nose of the animal by placing a finger and thumb each side of the nasal septum or by the application of blunt-ended, pincer-like tongs (bulldogs) to the same site usually results in the animal standing still and tolerating a certain amount of interference to the rest of its body. The impression is given that the animal's whole attention is concentrated on its nose. It has been suggested that it will only continue to stand still as long as the distress/pain of the interference is equal to or less than the distress/pain of its nose: once the interference distress is greater than the nose distress, control will be lost.

A major disadvantage of using a nose pincer is that cattle remember the pain and they will resist application of the pincer in the future. When a halter is used to hold the head for multiple blood samples collected over a series of days, the animal will often become increasingly more cooperative. When the nose pincer is used it may become progressively more difficult to handle. In most instances, a halter or head collar should be used to restrain the head.

The twitch used on the horse is a more subtle device, in that the loop of cord round the horse's upper lip can be relaxed and tightened at will and, in effect, the stimulus to the lip can be titrated against the stimulus of the interference being carried out on other parts of the animal. In the horse, it has been suggested that the twitch has somewhat similar action to acupuncture (Lagerweiz et al., 1984), with an increase of circulatory endorphins causing possible changes in pain susceptibility and/or mood of the animal. It is not known whether there are increases in endorphin levels in the blood of cattle subjected to pressure/force being applied to all or part of their bodies.

Metal rings or ropes are often permanently placed through the soft anterior parts of the nasal septum in cattle (mainly adult males) to facilitate handling (fig bull ring and Indian cow) (Fig. 6.4). Various staffs, chains, ropes, etc. can be attached to the ring or rope when needed, and probably help in controlling the animals by causing them a degree of discomfort/pain if they do not respond to the pushes and pulls exerted on the rings.

A halter (head collar) should always be used in conjunction with a rope tied to the ring. This will prevent a frightened animal from pulling the ring out. For dairy bulls, a snap attached to a 1 m-long stick can be attached to the ring to enable the handler to keep the bull at a safe distance. This is often used in Holstein bull studs when bulls are handled for semen collection.

Restraint of the head and neck

Animals which have their necks held in yokes or tie-chains in their individual standings (stanchions) are still able to move their head, and are usually capable of some limited movement of their whole body. This is also true of animals detained within the various designs of cattle crushes. Although there is now available a mechanical, rubber-lined scoop device which can be attached to the front of a cattle crush and which allows the head to be held firmly in one position, additional physical control of these animals is generally achieved by the grasping of the head.

This is usually done by a human handler, who should be standing to one side of the animal's head, taking hold of the horn or ear closest to him/her and, at the same time, seizing the nasal septum between a finger and thumb of the other hand. When this is done, cattle will usually keep their heads and bodies still. If the head is pulled sideways and back towards the animal's flank, i.e. rotated around the chain tie/bar of the yoke, the cow will tend to swing away from the handler and, in many instances, steady itself by pushing the side of its body against the adjacent pen division, stall crush side or wall.

The arm-over-the-side-of-the-face head-grip technique should only really be used on young, relatively small polled or dehorned stock. The handler, standing to one side of the animal's head, reaches over the head so that his/her arm covers the animal's far eye and cups his/her hand round the anterior part of the animal's upper lip. The handler's other hand goes to the side of the face and either grasps the lower jaw, with a finger and thumb pressing on the gum line just behind the incisor teeth, or seizes the nasal septum between the finger and thumb.

The animal is restrained by a mixture of: (i) force; (ii) calming through the use of the arm as a blind on the smallest animals; and (iii) distraction via the action of the hands/fingers inside the animal's mouth. A modified version of the arm-over-the-side-of-the-face head-grip technique can be used on animals safely held by the neck in a stall stanchion (neck yoke) or in a cattle crush head gate (see Fig. 6.6).

During all of these procedures, cattle should be kept calm. Calm cattle are easier to handle. The animal will usually remain calmer if the various holds are applied with steady force, and sudden, jerky motion should be avoided.

Fig. 6.6. Arm-over-the-side-of-the-face head grip. One advantage of this grip is that it restricts the animal's vision and has a blindfolding effect. This method is preferable to methods where the nostrils are gripped. Gripping the nostrils is highly aversive.

The use of blinds/blindfolds

Cattle can often be quietened for handling purposes by blindfolding them. The blindfold should be made of a soft, fully opaque material and be free from any foreign matter, e.g. sawdust or straw, which might enter or irritate the eyes. Blindfolding beef cattle has been shown to reduce the struggling and lower the heart rate response to routine handling (Mitchell *et al.*, 2004). See Chapters 4, 5, and 7, this volume, for more information on the effects of vision on handling.

The arm-over-the-side-of-the-face head grip (see Fig. 6.6) – as used on younger (smaller) stock – is reputed to calm by a blindfold effect, as one of the animal's eyes is covered by the handler's arm and the other is pressed against the handler's body. When polled cattle are held in a neck stanchion in a squeeze chute, they can easily be restrained for blood sampling or intravenous (IV) injection by a person leaning against their head so that a person's rear covers their eyes. The head can be gently turned sideways (the person follows this movement) and the animal will remain calm because its eyes are still covered.

The use of chemical agents for sedation or pain relief

Chemical agents are occasionally used to sedate and calm cattle before they are handled. It would be very convenient if some palatable drug could be put into the food or water of say an aggressive bull so that it was calm and cooperative by the time it was handled. However, as yet, there does not seem to be such a material. Chloral hydrate has been tried but it is not readily consumed by most cattle and, even when taken in, its action is somewhat uncertain. Many injectable agents have been used and more information on sedation and anaesthesia may be found in Thurman (1986), Houston *et al.* (2000), Hall *et al.* (2001) and Greene (2003).

The current drug of choice is still probably xylazine (Rompun, Bayer, Leverkusen, Germany) given by intramuscular injection. This agent should always be used under veterinary supervision. Low doses produce sedation; higher doses can sometimes cause the cattle to go down.

Many surgical procedures such as castration and dehorning have been traditionally done with no anaesthetic or analgesic (painkiller). Numerous studies have shown that the use of local anaesthetics and analgesics reduce both physiological and behavioural signs of stress. Stafford and Mellor (2005a, b) contains a recent review. Local anaesthesia eliminates the cortisol response after castration and reduces it after dehorning (Stafford and Mellor, 2005a, b). Anderson and Muir (2005) contains a review of pain management in ruminants. Underwood (2002) and George (2003) both present a further discussion about pain from dehorning or castration.

Practical experience has shown that the use of a local anaesthetic for dehorning small calves makes it easier for the person to hold them. Since pain is reduced, the calf struggles less. Cattle producers have also reported that the use of local anaesthetic when older calves are dehorned makes it easier to get the animals to re-enter the race in the future. Animals have an excellent memory for painful experiences.

The use of sound

Many stockmen and -women claim that the human voice can be used to calm cattle, and indeed this may be true if the animals and the humans are well acquainted and trust each other. Recorded sound (music) has sometimes been broadcast to cattle entering milking parlours in the belief that they can be more readily handled during the milking process. The benefits, if any, are probably due to the broadcast sound making the routine bangs and clatters produced by the parlour machinery less noticeable to the stock.

It should always be remembered that while cattle readily become habituated to sounds in their environment, they are easily startled by sudden, unexpected noises. See Chapter 5, this volume, for more information on the effects of sound. Research shows that loud shouting at cattle is very aversive (Pajor et al., 1999).

Common Factors

In practice, there are three key elements to most animal handling procedures which influence the chance of a successful outcome, with the minimal level of stress to the animal. First, the animals themselves: each individual animal will cope and react differently. Secondly, the facilities: the way 'hardware' such as crushes (squeeze chutes) and equipment are designed and constructed; and finally the handlers must have the knowledge, skills and the right approach. These three elements interlink and interplay with each other, and 'handling utopia' occurs only when all work and cooperate in harmony.

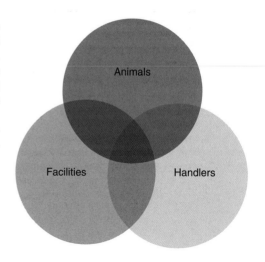

The human handlers

The persons handling the cattle should have stock skills and stock sense. Workers on the intensive, traditional family farms often have both these attributes; some of the animal attendants on the larger, more factory-like intensive units can at times be good stock system managers but poor animal handlers. The whole subject of stockmanship is complex – for details see Seabrook (1987), English et al. (1992) and Hemsworth and Coleman (1998).

Regardless of the talents (or otherwise) of the handling team, its members should have been instructed and rehearsed in the particular techniques to be used. They should also have been made aware that one person – the team leader – makes the decisions as to when each step of the technique is or is not to be carried out. The efficiency of the handling and the safety of both the humans and the animals depend on the attitudes and the skills of the handlers.

The facilities and equipment

The cattle should be handled in facilities where they are unlikely to suffer any injury or damage during the handling process or if they escape or go down. The area should, if possible, be enclosed – i.e. escape-proof, the fences/walls free of sharp projections and the floor non-slip and, if made of concrete or other hard material, well-bedded – covered with straw or other litter. Crushes must be well maintained.

The necessary equipment should be ready at hand and should have been well maintained and recently tested. Ropes, ideally should be of cotton: they should have a fairly wide diameter and be soft-surfaced and free of knots. Webbing should be clean, pliable and not frayed. All leather parts should be soft (oiled) and not perished in any way. All metal components should be free of rust and dirt and have smooth surfaces.

Whenever possible, the cattle should be kept calm and relaxed. If this can be achieved, the handling will usually be quick and safe for both humans and animals. Calmness in the animals will be encouraged if handlers understand the likely behaviour of the animals under the specific conditions and they utilize this to advantage as much as practically possible. If procedures are to be repeated periodically over an animal's lifetime, some time spent training cattle – for example through food rewards – is likely to be beneficial.

Conclusions

Cattle raised in close association with people can be restrained by a whole variety of techniques and devices, ranging from a person holding them with a halter (head collar) to complex restraining devices (see the standard texts recommended earlier). The efficiency and humaneness of the restraint depend on the stockpersons building up a routine which meets the purpose of the handling, utilizes the facilities available and is within the capacity of the handlers. An understanding of the advantages/disadvantages and the rationale of the various components common to most handling techniques is essential if success is to be assured.

Cattle are dependent on humans for the state of their health and well-being. Efficient and humane handling – a procedure very much under the control of the stockperson – can play an important part in ensuring that their welfare needs are met.

References

Albright, J.L. (1981) Training dairy cattle. In: *Dairy Science Handbook*, vol. 14. Agriservices Foundation, Clovis, California, pp. 363–370.

Albright, J.L. and Arave, C.W. (1997) *The Behaviour of Cattle*. CAB International, Wallingford, UK.

Anderson, D.E. and Muir, W.W. (2005) Pain management in ruminants. *Veterinary Clinics of North America Food Animal Practice* 21, 19–31.

Bartholomew, P.W., Khibe, T. and Ly, R. (1995) In-village studies of the use of work oxen in Central Mali. *Tropical Animal Health and Production* 27, 241–248.

Battaglia, R.A. (1998) *Handbook of Livestock Management Techniques*, 3rd edn. Upper Saddle River, New Jersey.

Bickert, W.G. (1998) Sorting, handling and restraining lactating cows for treatment and other purposes. In: *Proceedings of the 4th International Dairy Conference, St Louis, Missouri, 28–30 January 1998.* American Society of Agricultural Engineers (ASAE), St Joseph, Missouri, pp. 44–50.

Bouissou, M.F. and Boissy, A. (1988) Effects of early handling on heifers' subsequent reactivity to humans and to unfamiliar situations. In: Unshelm, J., van Putten, S., Zeeb, K. and Ekesbo, I. (eds) *Proceedings of the International Congress on Applied Ethology in Farm Animals*, Skara, Sweden, 1988. KTLB, Darmstadt, Germany, pp. 21–38.

Carli, G. (1992) Immobility responses and their behavioural and physiological aspects, including pain: a review. In: Short, C.E. and Van Poznak, A. (eds) *Animal Pain*. Churchill Livingstone, New York, pp. 543–554.

Conroy, D. (1999) *Oxen: a Teamster's Guide*. Rural Heritage Magazine, Gainesboro, Tennessee.

Creel, S.R. and Albright, J.L. (1987) Early experience. *Veterinary Clinics of North America. Food Animal Practice* 3, 251–268.

Dickfos, J.A. (1991) *Training Cattle for Scientific Experiments*. CSIRO, Rockhampton, Queensland, Australia.

English, P., Burgess, G., Segundo, R. and Dunne, J. (1992) *Stockmanship: Improving the Care of the Pig and other Livestock*. Farming Press, Ipswich, UK.

Epstein, H. and Mason, I.L. (1984) Cattle. In: Mason, I.L. (ed.) *Evolution of Domesticated Animals*. Longman, London, pp. 6–7.

Ewbank, R. (1961) The behaviour of cattle in crushes. *Veterinary Record* 73, 853–856.

Ewbank, R. (1968) The behaviour of animals in restraint. In: Fox, M.W. (ed.) *Abnormal Behavior in Animals.* Saunders, Philadelphia, pp. 159–178.

Food and Agriculture Organization (FAO) (2002) *Production Yearbook, Vol. 56,* FAO, Rome.

Fowler, M.E. (1995) *Restraint and Handling of Wild and Domestic Animals,* 2nd edn. Iowa State University Press, Ames, Iowa.

Fraser, A.F. (1960) Spontaneously occurring forms of 'tonic immobility' in farm animals. *Canadian Journal of Comparative Medicine* 24, 330–333.

Fraser, A.F. and Broom, D.M. (1990) *Farm Animal Behaviour and Welfare,* 3rd edn. Baillière Tindall, London.

George, L. (2003) Pain control in food animals. In: Steffey, E.P. (ed.) *Recent Advances in Anaesthetic Management of Large Domestic Animals.* IVIS, New York.

George, M.H., Morgan, J.B., Glock, R.D., Tatum, J.D., Schmidt, G.R., Sofos, J.N., Cowman, G.L. and Smith, G.C. (1995) Injection site lesions: incidence, tissue histology, collagen concentration, and muscle tenderness in beef rounds. *Journal of Animal Science* 73, 3510–3518.

George, M.H., Cowman, G.L., Tatum, J.D. and Smith, G.C. (1996) Incidence and sensory evaluation of injection site lesions in beef sirloin butts. *Journal of Animal Science* 74, 2095–2103.

Gilbert, J.E. (1991) *Farm Cattle Handling.* Standing Committee on Agriculture Report No. 35, CSIRO, East Melbourne, Australia.

Grandin, T. (1998a) Safe handling of large animals. In: Langley, R.I. (ed.) *Occupational Medicine: State of the Art Reviews* 14, pp. 195–212.

Grandin, T. (1998b) Handling methods and facilities to reduce stress on cattle. *Veterinary Clinics of North America. Food Animal Practice* 14, 325–341.

Grandin, T. (2000) Habitating antelope and bison to cooperate with veterinary procedures. *Journal Applied Animal Welfare Science* 3, 253–261.

Grandin, T. (2005) Behavior and handling. In: Chenowith, P.J. and Sanderson, M.W. (eds) *Beef Practice; Cow–Calf Production Medicine.* Blackwell Publishing, Ames, Iowa, pp. 109–125.

Graves, R.E. (1983) *Restraint and Treatment Facilities for Dairy Animals.* Special Circular 289, Pennsylvania State University Co-operative Extension Service, University Park, Pennsylvania.

Greene, S.A. (2003) Protocols for anesthesia of cattle. *Veterinary Clinics of North America. Food Animal Practice* 19, 679–693.

Hafez, E.S.E. and Bouissou, M.-F. (1975) The behaviour of cattle. In: Hafez, E.S.E. (ed.) *The Behaviour of Domestic Animals,* 3rd edn. Baillière Tindall, London, pp. 203–245.

Hall, L.W., Clarke, K.W. and Trim, C.M. (2001) *Veterinary Anaesthesia,* 10th edn. Saunders, Edinburgh, UK.

Hall, S.J.G. (2002) Behaviour of cattle. In: Jensen, P. (ed.) *The Ethology of Domestic Animals. An Introductory Text.* CAB International, Wallingford, UK.

Hanie, E.A. (2006) *Large Animal Clinical Procedures for Veterinary Technicians.* Elsevier Mosby, St Louis, Missouri.

Haynes, N.B. (2001) *Keeping Livestock Healthy,* 4th edn. Storey Publishing, North Adams, Massachusetts.

Hemsworth, P.H. and Coleman, G.J. (1998) *Human–Livestock Interactions. The Stockperson and the Productivity and Welfare of Intensively Farmed Animals.* CAB International, Wallingford, UK.

Hofer, M.A. (1970) Cardiac and respiratory function during prolonged immobility in wild rodents. *Psychosomatic Medicine* 32, 633–647.

Holmes, R.J. (1991) Cattle. In: Anderson, R.S. and Edney, A.T.B. (eds) *Practical Animal Handling.* Pergamon Press, Oxford, UK, pp. 15–38.

Houpt, K.A. (2005) *Domestic Animal Behavior,* 4th edn. Blackwell Publishing Professional, Ames, Iowa.

Houston, D.M., Mayhew, I.G.J. and Radostits, O.M. (2000) Handling and restraint II for clinical examination. In: Radostits, O.M., Mayhew, I.G.J. and Houston, D.M. (eds) *Veterinary Clinical Examination and Diagnosis.* W.B. Saunders, London, pp. 75–89.

Kabuga, J.D. and Appiah, P. (1992) A note on the ease of handling and the flight distance of *Bos indicus, Bos taurus* and their crossbreeds. *Animal Production* 54, 309–311.

Kennedy, G.A. (1988) Surgical restraint: cattle and sheep. In: Oehme, F.W. (ed.) *Textbook of Large Animal Surgery,* 2nd edn. Williams and Wilkins, Baltimore, Maryland, pp. 67–73.

Kilgour, R. (1987) Learning and training of farm animals. *Veterinary Clinics of North America. Food Animal Practice* 3, 323–338.

Lagerweiz, E., Nelis, P.C., Wiegart, V.M. and Van Ree, J.L. (1984) The twitch in horses: a variant of acupuncture. *Science* 225, 1172–1174.

Lawrence, A.B. (1991) Introduction: the biological basis of handling animals. In: Anderson, R.S. and Edney, A.T.B. (eds) *Practical Animal Handling.* Pergamon Press, Oxford, UK, pp. 1–13.

Leahy, J.R. (1995) *Restraint of Animals.* DVL Publishing, Liverpool, Canada.

Leigh, M.M. (1937) *Highland Homespun.* Bell, London, p. 177.

Lemenager, R.P. and Moeller, N.J. (1981) Cattle management techniques. In: Battagalia, R.A. and Mayrose, V.B. (eds) *Handbook of Livestock Management Techniques.* Burgess, Minneapolis, Minnesota, pp. 105–181.

McNitt, J.I. (1983) *Livestock Husbandry Techniques.* Granada, London.

Ministry of Agriculture, Fisheries and Food (MAFF) (1984) *Cattle Handling.* Booklet 2495, MAFF Publications, Alnwick, UK.

Mitchell, K.D., Stookey, J.M., Laternas, D.K., Watts, J.M., Haley, D.B. and Hyde, T. (2004) The effects of blindfolding on behaviour and heart rate in beef cattle during restraint. *Applied Animal Behaviour Science* 85, 233–245.

Murphey, R.M., Duarte, F.A.M. and Penedo, M.C.T. (1980) Approachability of bovine cattle in pastures: breed comparison and a breed x treatment analysis. *Behavior Genetics* 10, 171–181.

Pajor, E.A., Rushen, J. and de Passile, A.M.B. (1999) Aversion learning techniques to evaluate dairy cow handling practices. *Journal of Animal Science* 77 (Suppl. 1), 149 (abstract).

Payne, W.J.A. (1990) *An Introduction to Animal Husbandry in the Tropics,* 4th edn. Longman Scientific and Technical, Harlow, UK, pp. 197–421.

Pearson, R.A. and Vall, E. (2004) Performance and management of draught animals in agriculture in sub-Saharan Africa: a review. *Tropical Animal Health and Production* 30, 309–324.

Phillips, C.J.C. (2002) *Cattle Behaviour and Welfare,* 2nd edn. Blackwell Science, Oxford, UK.

Raussi, S. (2005) Group management of young dairy cattle in relation to animal health and welfare. Doctoral dissertation, Agrifood Research Reports 71, MIT Agrifood Research Finland, Jokioinen, Finland.

Roeber, D.L., Cannell, R.C., Wailes, W.R., Belk, K.E., Scanga, J.A., Sofos, J.N., Cowman, G.L. and Smith, G.C. (2002) Frequencies of injection site lesions in muscles from rounds of dairy and beef cow carcasses. *Journal of Dairy Science* 85, 532–536.

Rushen, J., Munksgaard, L., de Passille, A.M.B., Jensen, M.B. and Thodberg, K. (1998) Location of handling and dairy cows' responses to people. *Applied Animal Behaviour Science* 55, 259–267.

Sato, S., Shiki, H. and Yamasaki, F. (1984) The effects of early caressing on later tractability of calves. *Japanese Journal of Zootechnical Science* 55, 322–338.

Seabrook, M. (ed.) (1987) *The Role of the Stockman in Livestock Productivity and Management: Proceedings of a Seminar.* Report EUR 10982, EN Commission of the European Communities, Luxemburg.

Sheldon, C.C., Sonsthagen, T.F. and Topel, J. (2006) *Animal Restraint for Veterinary Professionals.* Mosby Publishing, Division of Elsevier, Amsterdam, Netherlands.

Shepherd, C.S. (1972) *Layout and Equipment for Handling Dairy Cattle and Suckler Cows.* Bulletin No. 157, Farm Buildings Departments, West of Scotland Agricultural College, Glasgow, UK.

Sims, B.G., Inns, F. and O'Neill, D.O. (2003) *Equipment for Working Animals, with Emphasis on Equids in Developing Countries.* World Association for Transport Animal Welfare and Studies (TAWS), Silsoe Research Institute, Swindon, UK.

Stafford, K.J. and Mellor, D.J. (2005a) Dehorning and disbudding distress and its alleviation in calves. *Veterinary Journal* 169, 337–349.

Stafford, K.J. and Mellor, D.J. (2005b) The welfare significance of the castration of cattle: a review. *New Zealand Veterinary Journal* 53, 271–278.

Steenbergen, J.M., Koolhaas, J.M. and Bohus, B. (1989) Behavioural and cardiac responses to a sudden change in environmental stimuli: effect of forced shift in food intake. *Physiology and Behaviour* 45, 729–733.

Stober, M. (1979) Handling cattle – calming by a mechanical means of restraint. In: Rosenberger, G. (ed.) *Clinical Examination of Cattle* (translated by Roy Mack from the German 2nd edn. 1977). Verlag Paul Parey, Berlin, pp. 1–28.

Taylor, A.A. and Davis, H. (1998) Individual humans as discriminative stimuli for cattle (*Bos taurus*). *Applied Animal Behaviour Science* 58, 13–21.

Thurman, J.C. (ed.) (1986) Anesthesia. *Veterinary Clinics of North America. Food Animal Practice* 2 (3), 507–785.

Underwood, W.J. (2002) Pain and distress in agricultural animals. *Journal of the American Veterinary Medical Association* 221, 208–211.

Universities Federation for Animal Welfare (eds) (1990) *Animal Training: Proceedings of a Symposium.* UFAW, Potters Bar, UK.

Wilson, R. (2002) Effects of draught animal power on crops, livestock, people, and the environment. In: *Proceedings – Responding to the Increased Global Demand.* British Society of Animal Science, p. 54, http://www.bsas.org.uk

7 Handling Facilities and Restraint of Range Cattle

Temple Grandin

Department of Animal Sciences, Colorado State University,
Fort Collins, Colorado, USA

Introduction

Even though extensively reared cattle have a large flight zone and are not completely tame, they will become calmer and easier to handle if they are trained to seeing people on foot, on horseback and in vehicles. Doing this will have the added advantage of reducing agitation and stress when the animals are transported to a feedlot or a slaughter plant. Cattle which have been handled only by people on horses may become highly agitated when they first see a person on foot.

It is important that the first experience with a mounted rider or a person on foot is a good first experience. First experiences make a big impression on animals (Grandin and Johnson, 2005; Grandin, 1997a). Handling in a new set of corrals will be easier if the animal's first experiences with the facility are positive. The first time they enter the corral, they should be walked through it and fed. Ideally, this should be done several times before any actual work is done.

Facilities that are suitable for intensively raised tame cattle are not suitable for cattle reared on large ranches or properties in the USA, Australia or South America. Facilities that take advantage of the natural behavioural characteristics of cattle will reduce stress on the animals and improve labour efficiency.

Corral Design

Round pens for gathering herding animals were invented over 11,000 years ago in the Middle East. The Syrians herded wild ungulates into a round pen to slaughter them (Legge and Conway, 1987). A round pen is efficient because there are no sharp corners in which animals can bunch up. To prevent the animals from running back out of the entrance, the pen was shaped like a heart, with the entrance between the shoulders. It is interesting that designs that are really effective keep being reinvented. The same design was used by cowboys to capture the wild horses (Ward, 1958), and fishing nets are set out in a similar manner. When designing gathering pens for cattle, sufficient space must be allocated: minimum space is 1.9 m^2 for every cow (Daly, 1970) or 3.3 m^2 for every cow–calf pair. These space recommendations are for short-term holding of less than 24 h.

In rough country, cattle can be difficult to gather, and chasing them with helicopters, horses or vehicles is stressful and labour-intensive. Cattle can easily be gathered by building corrals with trap gates around water sources (Cheffins, 1987). A trap gate acts as a valve, as the cattle can move through a closed trap gate in only one direction. Trap gate designs are described by Howard (1953), Ward (1958) and Anderson and Smith (1980). Several weeks before gathering, the previously open trap gate is closed a little more each day. On the last day, the space between the ends of the two gates is so small that a cow has to push them apart to get through. After she has gone through, she is unable to return.

Cheffins (1987) described an improved self-gathering system that has separate entrance and exit trap gates. Training the cattle is easier here, because the animals become accustomed to moving in both directions through the trap gates. To make a self-gathering system work, all water sources must be enclosed by corrals equipped with trap gates. In areas with numerous water sources, corrals baited with molasses or palatable hay can be used to trap cattle. The animals will be easier to gather if the pasture conditions are poor. When the cattle eat the feed, they trip a trigger wire that closes the corral gate (Adcock *et al.*, 1986; Webber, 1987). A more modern version of this design would be to use a radio signal or mobile telephone call to close the gate. Some newer, self-mustering systems are described in Petherick (2005) and Connelly *et al.* (2000) (see also Chapter 5, this volume).

Sorting facilities

Sorting (drafting) cattle into age, sex or condition categories is an important handling procedure. Single-file races with sorting gates are used by ranchers in the south-western USA (Ward, 1958). These systems are similar to races used for sorting sheep. Another design utilizes a triangular gate to direct the cattle (Murphy, 1987): a person standing on a platform over the race can operate the gate. The triangular design facilitates the stopping of an animal that attempts to push its way past the gate. The disadvantage of both types of single-file sorting races is that it is difficult visually to appraise cattle as they rapidly move by.

Many American ranchers hold cattle in a 3.5 m-wide alley during sorting. Individual animals are separated from the group and directed to sorting pens by a person on the ground (Grandin, 1980a). This system works well with European breeds and makes visual appraisal easier, but it may work poorly with Brahman and zebu (*Bos indicus*) cattle, due to their

greater tendency to bunch together. *Bos indicus* cattle are more reluctant to separate from a group. See Chapter 5, this volume, for more information on sorting facilities.

The Australians first developed pound yards (Daly, 1970), which are small, round pens, 3–6 m in diameter with four to eight gates around the perimeter (Powell, 1986). A single person standing on an elevated platform over the pen can operate the gates. This system works well with zebu- or Brahman-cross cattle and visual appraisal is easy because the person has more time to look at the cattle. Other sorting systems are the hub type (Oklahoma State University, 1973) and the circular alley (Arbuthnot, 1979).

Some operations use computers to operate sorting systems after cattle have been evaluated by either people or electronic devices. More information on these systems can be obtained at http://www.scaleyards.Huefner.com.au and Micro Beef Technologies (http://www.microbeef.com).

Corral layout

With the advent of truck transport, squeeze chutes (crushes for cattle restraint) and large feedlots, corral designs became more complex. Modern curved races and round crowd pens evolved independently in Australia (Daly, 1970), New Zealand (Kilgour, 1971; Diack, 1974) and the USA (Oklahoma State University, 1973). During the early to mid-1960s, the construction of large feedlots in Texas stimulated design of truly modern systems with curved single-file races, round crown pens and long, narrow diagonal pens (Paine *et al.*, undated). Prior to this time, ranchers used corrals with square pens, with little regard to behavioural principles.

Grandin (1980a) combined the best features of Texas feedlot designs with the round gathering pens from Ward (1958). Figure 7.1 illustrates a general-purpose corral. The wide, curved lane serves two functions: (i) to hold cattle going to the loading ramp or squeeze; or (ii) as a reservoir for cattle that were being sorted back into the diagonal pens by a person on the ground. The wide, curved lane facilitates the moving of cattle into the crowd pen. Each diagonal pen in Fig. 7.1 holds one truckload of 45

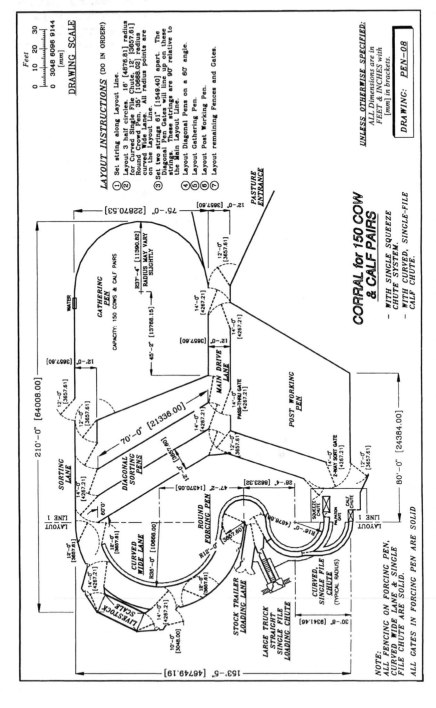

Fig. 7.1. General-purpose corral for handling, weighing, sorting and loading extensively reared cattle.

mature cattle. The curved and diagonal layout avoids sharp 90° corners for cattle in which to bunch up. For larger herds, additional diagonal pens may be added and the gathering pen enlarged. The corral is easy to build because the curved single-file race, round crowd pen and wide, curved lane consist of three half-circles with three radius points on a layout line (see Fig. 7.2).

The Weean yard is used on many ranches in Australia (Thompson, 1987). It incorporates a curved single-file race and a pound yard for sorting in an economical, easy-to-build system (Fig. 7.3; Powell, 1986; Thompson, 1987). Powell (1986) contains many corral designs that are especially adapted to Australian conditions.

Individual animal identification

Designs where cattle can be sorted after they leave the squeeze chute will become increasingly popular, as more and more people switch to individual animal identification (see Fig. 7.4). Each animal can be weighed in a squeeze chute mounted on load cells, evaluated by ultrasound or other technology and then sorted. Facilities where cattle can be easily sorted when they

leave the squeeze chute will be required when ranchers and feedlot operators sell cattle under contract, which have strict specifications for fat thickness, frame score, weight and other specifications.

This layout can also be used in electronic sorting systems; in some electronic sorting systems, two squeeze chutes are used. The animal moves into the first squeeze to have its back fat measured by ultrasound and then it moves into the next squeeze chute for vaccinations and tagging. Figure 7.4 can be modified to accommodate the additional squeeze chute. The system in Fig. 7.4 also has the advantage of training cattle to go through the squeeze chute.

Race, crowd pen and loading ramp design

Single-file races, crowd pens (forcing pens) and loading ramps should have solid sides (see Fig. 7.5; Rider et al., 1974; Grandin, 1980a, 1997b, 2004). Solid sides in these areas help to keep cattle calmer and facilitate movement, because they prevent the cattle from seeing distractions outside the fence, such as people and vehicles. Solid sides are especially important when handling wild animals that are

Fig. 7.2. Curved single-file race, round crowd pen and curved lane. The round crowd pen is a full half-circle. This is an important design feature and it works efficiently because cattle entering the single-file race think they are going back from whence they came.

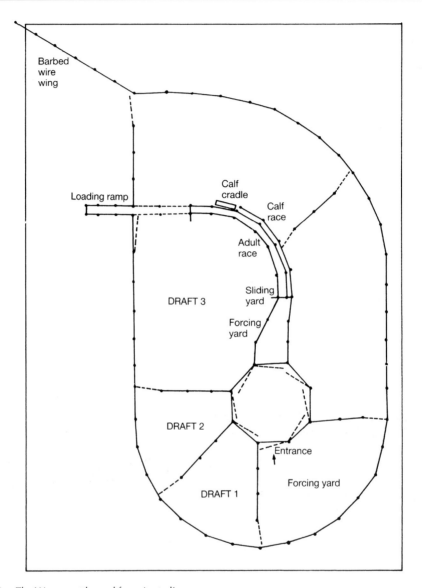

Fig. 7.3. The Weean cattle yard from Australia.

unaccustomed to close contact with people in places such as slaughter plants, feedlots and large stockyards where there are people and activity outside the race. The crowd gate that is closed behind the cattle must also be solid. Circular crowd pens and curved single-file races can reduce the time required to move cattle by up to 50% (Vowles and Hollier, 1982).

A curved single-file race is more efficient because cattle cannot see people and motion up ahead when they enter the race. Another reason why a curved race is more efficient is that the cattle think they are going back to where they have just come from. Cattle also back up less in a curved race. Observations indicate that moving people are more threatening to cattle than people who stand completely still and look away from approaching animals: staring eyes are threatening. A curved race provides the greatest advantage when cattle have to wait in line for vaccinations or other procedures, but it provides no advantage when cattle run freely

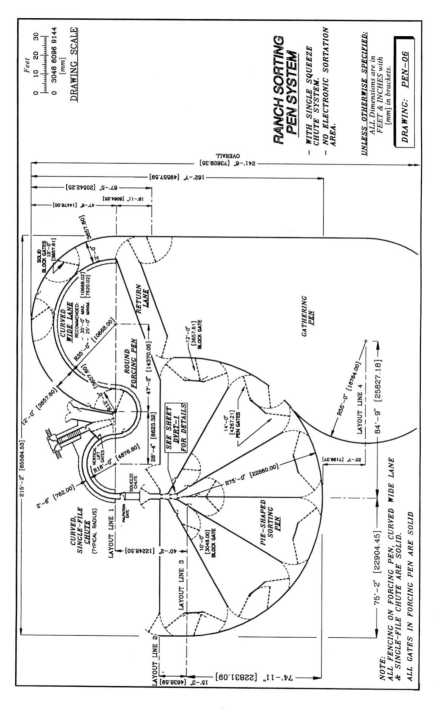

Fig. 7.4. Corral design suitable for sorting animals with individual identification through the squeeze chute.

Fig. 7.5. The round crowd pen should have a completely solid fence and catwalk as shown. Some people prefer to remove the catwalk on the single-file race and work the animal's flight zone on the ground. In these systems, the fence on the outer radius of the single-file race is completely solid and the fence on the inner radius is solid only on the bottom half.

through the race without having to be detained for handling (Vowles *et al.*, 1984). The recommended inside radius for a curved single-file chute is 3.5–5.0 m, and the radius of the crowd pen should not exceed 3.5 m.

Cattle bunch up if the crowd pen is too big. On funnel-shaped crowd pens, one side should be straight and the other should be on a 30° angle. To reduce baulking, a curved single-file race must not be bent too sharply at the junction between the single-file race and the crowd pen. A cow standing in the crowd pen must be able to see two to three body lengths up the race. If the race appears to be a dead end, the cattle will refuse to enter. Bending the single-file race too sharply where it joins the crowd pen is the most serious design mistake. Recommended dimensions for single-file chutes can be found in Grandin (1983a, 1997b, 1998) and Midwest Plan Service (1987).

If one side of the curved single-file race has to have open bars for vaccination, the outer fence should be solid and the inner fence should be solid up to the 60 cm level. This design where the catwalk is eliminated along the single-file race enables a skilled handler on the ground to work with the animal's flight zone and point of balance to move them. With less skilled

handlers, a completely solid single-file race with a catwalk is recommended (see Chapters 4 and 5, this volume). To avoid problems with cattle refusing to enter dark sheds or buildings, the wall of a building should never be located at the junction between the single-file race and the crowd pen (Grandin, 1980a, 1987).

Following behaviour can be used to facilitate cattle movement through a system. Cattle in adjacent single-file races will move when they see an animal in an adjacent race move. The outer walls are solid, but cattle can see through the bars of the inner partitions. This principle was first used at the Swift Meat Plant in Arizona in 1974 and has been used successfully by Grandin (1982, 1990a) for moving pigs and Syme (1981) for moving sheep.

John Kerston in New Zealand moves cattle into two squeeze chutes (crushes) with two parallel races. When a cow is leaving one squeeze chute, the next cow is entering the empty squeeze chute on the other side (Andre, 1991). Andre (1991) also describes a system, designed by Roy Atherton, which has three parallel, single-file races leading to a single squeeze chute. This system will work best if cattle are kept calm. Another popular variation of this design is two side-by-side single-file races leading to a single squeeze chute. If a straight single-file race has to be used, the installation of two side-by-side races is recommended.

A crowd pen should be level and never be built on a ramp. Groups of animals that are standing still on a ramp will often pile up against the crowd gate and get trampled. When animals are handled on a wide ramp, they should be kept moving. However, cattle can stand safely in a single-file race which is on a ramp. Loading ramps must not be made too steep: the recommended maximum angle is 20–25°.

On concrete ramps, stair steps are more efficient: a 10 cm rise and 30–45 cm tread length being recommended. If cleats or ridges are used, the spacing should be 20 cm between the cleats to match the stride length of cattle (Mayes, 1978). A basic principle for all species is that the animal's foot should fit easily between the cleats to provide good traction. If the cleats are too far apart, the feet will slip. Further information on loading ramps is given in Powell (1986) and Grandin (1990a, 2004).

Factors impeding movement of cattle

Corrals and races must be free of distractions that make animals baulk (see Chapter 20, this volume). Distractions such as a small piece of chain dangling in a race, reflections on vehicles or shadows will cause cattle to baulk (Grandin, 1998). In existing facilities, installing solid sides on the single-file race will often improve cattle movement and reduce baulking. Distractions and lighting problems, such as a race entrance being too dark, can ruin the efficiency of a well-designed system. The author has observed that cattle often move more easily through outdoor corrals. If a building is built over a race, it should be equipped with either skylights or plastic side-panels to let in light. White, translucent panels are best because they let in plenty of light but eliminate shadows.

The ideal lighting inside a building looks like that on a bright, cloudy day. Cattle tend to approach light. In one facility, cattle refused to enter the race unless a large door had been left open to admit light. When the door was closed, cattle baulked and refused to enter the dark race entrance. Lamps can be used to attract cattle into buildings at night, but the lighting must be indirect. The author has found that, in most facilities, cattle can be moved into squeeze chutes without electric prods. However, a lighting problem can make it almost impossible to move cattle quietly because they constantly baulk.

Restraint devices

Before the invention of squeeze chutes (crushes), range cattle were caught and restrained with a lariat (Ward, 1958). The invention of mechanized restraint devices both improved animal welfare and reduced labour requirements. These also required less skill to operate than a lariat. One of the first mechanical devices for restraining cattle was patented by Reck and Reck (1903). It has squeeze sides that pressed against the animal and a stanchion to hold the head. In the 1920s, Thompson (1931) developed a head-catching gate designed for wild, horned cattle, which was installed on the end of a single-file race.

A squeeze chute restrains the animal by two devices: a stanchion is closed around the neck and side-panels press against the animal's body to control movement (see Fig. 7.6). This device is available with either manual or hydraulic controls. The best squeeze chutes have two side-panels that close in evenly on both sides. This design enables the animal to stand in a balanced position. An animal may struggle and resist being restrained if pressure is applied to only one side of the body. Non-slip flooring in squeeze chutes is essential: animals often panic if they start slipping. Today, most adult range cattle are restrained in a squeeze chute, but lariats are still used on some large ranches for the restraint of calves for branding and vaccination. One of the first mechanical devices for holding calves was invented by Thompson and Gill (1949).

There are six different types of headgates (head bail or stanchions) for restraining cattle. These can be used either alone at the end of a single-file race or in conjunction with a complete squeeze chute. The six types are: self-catch (Pearson, 1965), positive (Thompson, 1931; Heldenbrand, 1955), scissors–stanchion, pivoting, sliding doors and rotating headgates (Moly Manufacturing, Lorraine, Kansas, USA, 1995;

Fig. 7.6. Hydraulically operated squeeze chute (crush). When the pressure is properly set, a hydraulic squeeze is safer because protruding lever arms are eliminated. The pressure should be set so that squeezing will automatically stop before cattle bellow or strain.

Cummings and Son Equipment Company, Garden City, Kansas, USA, 1997).

The advantage of pivoting or rotating headgates is that they open up to the full width of the squeeze and animals can exit more easily. These designs may also help reduce shoulder injuries. A picture of the rotating headgate may be found in Grandin (1998). The Cummings and Son (1997) headgate requires very little force to restrain the animal compared with scissors–stanchion headgates.

All types are available with either straight, vertical neck bars or curved neck bars. Stanchions with straight neck bars are recommended for general-purpose uses, as they are less likely to choke an animal. Pressure on the carotid arteries exerted by a neck stanchion will quickly kill cattle (White, 1961; Fowler, 1995). Stanchions with straight, vertical neck bars are the safest because cattle can lie down while their neck is in the stanchion: there is no pressure on the carotid arteries.

Curved-bar stanchions provide a good compromise between control of head movement and safety for the animal. Positive headgates (Thompson, 1931), which clamp tightly around the neck, provide better head control but have an increased hazard of choking the cattle (Grandin, 1975, 1980b). Both positive-type headgates and a curved-bar stanchion must be used with squeeze sides or some other apparatus to prevent the animal from lying down.

On chutes where the squeeze sides are hinged at the bottom, the width at the bottom must be narrow enough so that the V formed by the sides supports the animal's body in a standing position. A lift plate under the belly can be used to support the animal (Marshall *et al.*, 1963). The squeeze sides must be designed so that there is no tendency to throw the animal off balance.

Hydraulically activated devices

Hydraulically activated squeeze chutes have become increasingly popular. A properly adjusted hydraulic chute is safer for both people and cattle. Operator safety is improved because long, protruding, lever arms are eliminated. The pressure-relief valve must be properly set to prevent severe injury from excessive pressure: these injuries may include broken ribs and internal ruptures. Additional information on proper adjustments can be found in Grandin (1980b, 1983b, 1990b). If an animal vocalizes (moos or bellows) immediately after it is squeezed in a hydraulic chute, the pressure setting must be reduced. The valve must be set so that the squeeze sides automatically stop squeezing before excessive pressure is applied. This will prevent a careless operator from applying excessive pressure. Other indicators of excessive pressure are laboured breathing and straining.

If the chute has an additional hydraulic device attached to the headgate to hold the head still, it must have its own separate pressure-relief valve, which must be set at a pressure much lower than the pressure required to operate the squeeze. Newer models have a quiet pump and motor, in contrast to the older, noisy, hydraulic squeeze chutes. Ideally, the pump and motor should be removed and located away from the animal. Some new squeeze chutes have plastic inserts to reduce noise and prevent metal-to-metal contact when gates open and close.

Carelessness and rough handling are the major cause of injuries to cattle in squeeze chutes (Grandin, 1980a), but there is still a need to develop better restraint devices. Even under the best of conditions, bruises directly attributable to the squeeze chute occur in 2–4% of cattle. In one study, bruises occurred in five out of seven feedlots; 1.6–7.8% of the cattle had increased bruises compared with animals which had not been handled in the squeeze chute (Brown et al., 1981). Observations also indicate that cattle can be injured when the headgate is suddenly closed around the neck of a running animal. Cattle should be handled quietly so that they walk into a squeeze chute and walk out. Hitting the headgate too hard can cause haematomas and bruises. Shoulder meat may still be damaged when cattle are slaughtered.

Electronic measurement of usage of restraint devices

Progressive managers have found that quiet handling in squeeze chutes and reduction of the use of the electric prod will enable cattle to go back on feed more quickly. As stated in Chapters 1 and 20, this volume, continuous evaluation

and measurement of handling performance are essential to prevent workers from becoming rough.

The technology is now available to electronically evaluate handling of cattle in a squeeze chute. Australian researchers Burrow and Dillon (1997) used a radar unit to measure the speed at which cattle left a squeeze chute. Animals that ran out at a high speed grew more slowly. Voisinet et al. (1997) found that cattle that became agitated in a squeeze chute had lower weight gains. Canadian researchers have developed ways of measuring how hard cattle hit the headgate and to what extent animals jiggled the squeeze chute (Schwartzkopf-Genswein et al., 1998). They recorded jiggling by recording signals from the load cells in the electronic scale, which was already under the squeeze. Electronic measurements of speed and jiggling could easily be correlated with weight gain, health records and feed conversion (Grandin, 1998).

As more and more operations use electronic identification of cattle, electronic measurement of animal behaviour – both in the squeeze chute and at the exit – will be easy to carry out. Vocalization scoring, described in Chapters 1 and 20, this volume, can also be used to evaluate handling in squeeze chutes.

Methods for measuring temperament

Many breed associations have incorporated temperament (disposition) scoring into their systems for the evaluation of cattle. There are two basic methods: recording reactions in the squeeze chute and exit speed. Exit speed may be a more sensitive indicator than visually assessing struggling during restraint. A simple visual scoring system that can be used in squeeze chutes is shown below.

- standing still
- intermittent shaking
- continuous shaking
- violent struggling

A simple way to measure exit speed is to use the traditional horse gaits of walk, trot or canter (run). This method was more effective for differentiating the temperament difference between cattle breeds than restraint scoring (Baszczak et al., 2005). Exit speed measured by a laser device was highly correlated with physiological measures of stress (Curley et al., 2004, 2006; King et al., 2006). Exit speed may be most valuable for assessing older cows that have learned to stand still in the squeeze chute. Scoring reactions to restraint is usually accurate for younger animals that have had less experience with handling. Wegenhoft et al. (2005) found that temperament scores at weaning were predictive of the temperament of older cattle.

Even though exit speed may be less affected by learning than chute score, learning does have an effect on it. Scores collected when cattle first entered a feedlot had a stronger relationship with cortisol levels than scores obtained after the cattle had been in the feedlot for 70 days (King et al., 2006). Exit speed scores were significantly slower. In Nellore × Angus cattle, sire had a significant effect on temperament (Wegenhoft et al., 2005). A further study showed that cattle with the fastest exit speeds were most likely to have reduced weight gain (Müller and Von Keyserlingk, 2006).

Behavioural reactions to restraint

One of the reasons for cattle becoming agitated in a squeeze chute is due to the operator being deep inside their flight zone. They can see him/her through the open bar sides. Cattle will remain calmer in a restraining device which has solid sides and a solid barrier around the headgate to block the animal's vision (Grandin, 1992). Cattle struggled less in a restraint device if their vision was blocked until they had been completely restrained (Grandin, 1992). Cattle are less likely to attempt to lunge through the head opening if there is a solid barrier in front of the head opening preventing them from seeing a pathway of escape.

Restraint device designs that have been successfully used in slaughter plants could be adapted for handling on the ranch and feedlot (Marshall et al., 1963; Grandin, 1992; see Fig. 20.10). Most cattle stood quietly when the Marshall et al. (1963) restraint device was slowly tightened against their bodies (Grandin, 1993). Solid sides prevent the animals from seeing the operator or other people inside their flight zone. Observations also indicate that cattle unaccustomed to head restraint will remain calmer if body restraint is used in conjunction with head restraint. Head restraint without body restraint can cause stress (Ewbank et al., 1992). More information on the design and operation of restraint devices can be found in Chapter 20, this volume.

Breeders of American bison prevent injuries and agitation by covering the open-barred sides of squeeze chutes and installing a solid gate (crash barrier) about 1.0–1.5 m in front of the head gate. Covering the sides of a squeeze chute so that the animal does not see the operator standing beside it will keep animals with a large flight zone calmer. When bison are handled, the top must also be covered to prevent rearing. Commercially available squeeze chutes are now available with rubber louvres on the sides to block the animal's vision (see Fig. 7.7). The louvres are mounted on a 45° angle and the drop bars on the side of the chute can still be opened. A cow's eye view of a squeeze chute equipped with louvres may be found in Grandin (1998). Covering the open barred sides of a squeeze chute with cardboard will also result in calmer cattle. Cattle will also enter the squeeze more easily because they will not see the squeeze chute operator who is deep in their flight zone.

Cattle will remain calmer if they 'feel restrained'. Sufficient pressure must be applied to hold the animal snugly, but excessive pressure will cause struggling due to pain. There is an optimal amount of pressure. If an animal struggles due to excessive pressure, the pressure should be slowly and smoothly reduced. Many people mistakenly believe that the only way to stop animal movement is by greatly increasing the pressure. Sudden, jerky movements of the apparatus will cause agitation, and smooth, steady movements help keep the animal calm (Grandin, 1992). Fumbling a restraining procedure will also cause excitement (Ewbank, 1968). It is important to restrain the animal properly on the first attempt. See Chapter 20, this volume, for more information on restraint.

Dark-box restraint

For artificial insemination and pregnancy testing, mechanical holding devices can be eliminated by using a dark-box race (Parsons and Helphinstine,

Fig. 7.7. Rubber louvres installed on the squeeze chute to prevent incoming cattle from seeing people. This will facilitate cattle entry into the chute.

1969; Swan, 1975). This consists of a narrow stall with solid sides, a solid front and a solid top. Very wild cattle will stand still in the darkened enclosure. A cloth can be hung over the cow's rump to darken the chamber completely. Comparisons between a dark box and a regular squeeze chute with open-bar sides indicated that cows in the dark box were less stressed (Hale *et al.*, 1987). Further experiments indicated that cortisol levels were lower in the dark box, but heart rate data were highly variable due to the novelty of the box (Lay *et al.*, 1992c).

Blindfolding of both poultry and cattle reduces heart and respiration rates (Douglas *et al.*, 1984; Jones *et al.*, 1998; Don Kinsman, personal communication) (see Chapter 5, this volume). Mitchell *et al.* (2004) reported that blindfolding Hereford × Angus × Charolais heifers with several layers of opaque dark towel resulted in less struggling in the squeeze chute compared to control animals. Observations by Jennifer Lanier in our laboratory showed that blindfolds on American bison had to be opaque to give the greatest calming effect. Installation of a solid top on a squeeze chute also kept the bison calmer.

If large numbers of cattle are inseminated, two or three dark boxes can be constructed in a herringbone configuration (McFarlane, 1976; Canada Plan Service, 1984; Fig. 7.8). The outer walls are solid, with open-barred partitions between the cows. Side-by-side bodily contact helps to keep cattle calmer (Ewbank, 1968). To prevent cattle from being frightened by a novel dark box, the animals should be handled in the box prior to insemination.

The effectiveness of dark-box restraint is probably due to a combination of factors, such as blocking the view of an escape route and preventing the animal from seeing people that are inside its flight zone. Darkness, however, has a strong calming effect. Wild ungulates remain much calmer in a totally dark box. Small light leaks sometimes cause animals to become agitated. A well-designed dark box for domestic cattle has small slits in the front to admit light (see Fig. 7.9). Cattle will enter easily because they are attracted to the light. For wild ungulates, it may be desirable to block the slits after the animal is in the box.

Adaptation to restraint

Cattle remember painful or frightening experiences for many months (Pascoe, 1986), so the

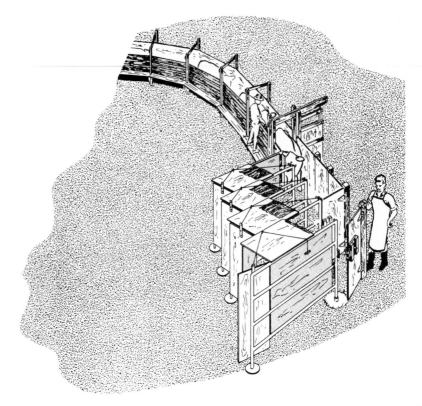

Fig. 7.8. Dark-box restraint races in a herringbone configuration. The solid sides restrict vision and keep cattle calmer for artificial insemination and pregnancy testing.

use of aversive methods of restraint should be avoided. Restraint devices should be designed so that an animal is held securely in a comfortable, upright position. If a restraint device causes pain, cattle will often become agitated and refuse to enter the device the next time it is used. For example, cattle restrained with nose tongs will toss their heads and be more difficult to restrain in the future compared with cattle restrained with a halter (Grandin, 1987, 1989a). When cattle are restrained with a halter for repeated blood sampling, they will often learn to turn their heads voluntarily. Proper use of the halter is described in Holmes (1991).

Tame animals can be trained within 1 day to voluntarily enter a relatively comfortable restraint device for a feed reward (Grandin, 1984, 1989b). The restraint device should be introduced gradually and care must be taken to avoid hurting the animal. If the animal resists restraint, it must not be released until it has stopped struggling (Grandin, 1989a).

Tame animals can be trained more quickly than wild animals. To reduce stress, a wild animal must be trained and tamed over a period of days or weeks. The 1-day method (Grandin, 1989b) is recommended for tame animals. Wilder, more excitable cattle became increasingly agitated when they were repeatedly run through the squeeze chute in one afternoon; the wilder animals need time to calm down before the next training treatment. Grandin et al. (1995) and Phillips et al. (1998) found that many weeks were required to train nyala and bongo antelope to voluntarily enter a crate for injections and blood sampling. Each new sight or sound had to be introduced very gradually in tiny increments to avoid frightening the animals. Very low, almost baseline, cortisol levels of 4.4–8.5 ng/ml were obtained (Phillips et al., 1998). Their study indicates that training an animal to voluntarily cooperate greatly reduces stress.

Extensively raised beef cattle were restrained and had nine blood samples taken from

Fig. 7.9. A cow's eye view into a well-designed dark box. Light slits attract the cow in. In some facilities the slits are covered after the cow is in the box.

the vena cava during a 16-day period. These animals had large reductions in cortisol (stress hormone) levels, and they became less excited as the days progressed (Crookshank *et al.*, 1979). It appears that four or five restraint sessions were required for the animals to become accustomed to the procedure. After a 5-day training period, which included three restraint sessions, wild cattle still reacted to 20 min of restraint with steadily increasing cortisol levels up to 30 ng/ml (Lay *et al.*, 1992a, b).

Tame animals, such as dairy cattle, which have become accustomed to restraint devices, will have a lower cortisol reaction. Possibly the wild cattle in Lay *et al.*'s (1992a) experiment had experienced some adaptation because their cortisol levels did not rise to the high of 63 ng/ml that had been recorded in a poorly designed slaughter facility (Cockram and Corley, 1991).

Stahringer *et al.* (1989) found that, in Brahman heifers, the excitable animals had higher levels of serum cortisol than calm heifers. Crookshank *et al.* (1979) also found that extensively reared calves that had been subjected to 12 h of trucking and weaning shortly before blood samping responded with increasing cortisol levels of up to 46 ng/ml for the first four samplings, and then levels dropped back down for the last five. Cattle that were not subjected to the added stress of weaning and transport had a peak level of only 24 ng/ml during the entire experiment. One can tentatively conclude that

subjecting cattle to closely spaced, multiple, stressful procedures will delay adaptation to handling.

Learning and restraint

The use of highly aversive methods of restraint, such as electro-immobilization, is not recommended (Lambooy, 1985). An electronic immobilizer restrains an animal by tetanizing the muscles with electricity: there is no analgesic or anaesthetic effect (Lambooy, 1985). Application of the immobilizer to my arm felt like a disagreeable electric shock. Cows which have been immobilized had elevated heart rates 6 months later when they approached the chute where they had received the shock (Pascoe, 1986).

A choice test in a Y-shaped race indicated that sheep preferred the tilt squeeze table to electro-immobilization. After one or two experiences, sheep avoided the race that led to the immobilizer (Grandin *et al.*, 1986). When a choice test is used to test aversiveness of restraint methods, naïve animals that have never been in the testing facility should be used. New cattle should be used for each test.

Cattle that have developed a strong preference for one of the races will often refuse to switch races to avoid mildly aversive treatment, such as being gently restrained in a squeeze chute (Grandin *et al.*, 1994). Initially, they quickly learn to avoid the aversive side, but they often refuse to switch when aversive treatment is switched to the other side. When the treatments are switched, the animal's brain registers the switch because the amount of looking back and forth at the decision point increases. However, cattle that had been accidentally struck on the head by the headgate were more likely to avoid the squeeze chute in the choice test (Grandin *et al.*, 1994).

From a species survival standpoint, it makes sense to keep going down the previously learned safe path if something mildly aversive happens – such as being restrained gently by the headgate, but when something really aversive happens, such as being struck on the head or electrically immobilized, the animal will immediately switch paths to avoid the headgate. Deer show a similar reluctance to change. After 18 training sessions with no aversive treatment, deer still quickly entered a race after the first aversive treatment (Pollard *et al.*, 1992).

Cattle will learn to differentiate between a head stanchion that strikes them on the head and a scale that causes no discomfort. Cattle that were handled five times became progressively more willing to enter a single-animal scale and somewhat less willing to enter a squeeze chute (Grandin, 1993). However, many of the animals that refused to enter the squeeze chute entered the squeeze section willingly but refused to place their head in the headgate stanchion. They had learned that pressure on the body does not cause discomfort, but the headgate stanchion hurts when it slams shut. Cattle were also more likely to become agitated in the squeeze chute: 2% of the animals became agitated on the scale but 13% became agitated in the squeeze.

Research by Virginia Littlefield in our laboratory has shown that cattle will habituate to repeated daily restraint in a squeeze chute if handled gently. The animals baulked less on each successive day and became less and less agitated in the squeeze chute. However, they will become harder and harder to drive into a squeeze chute if electric prods are used (Gooneswarden *et al.*, 1999).

The animals in our study were all restrained in a chute with a stanchion–scissors headgate, and care was taken to avoid striking them on the head. Some headgate designs are more likely to be aversive to cattle than others. Poorly adjusted, self-catching headgates may hurt the shoulders or put excessive pressure on the animal's neck. Pollard *et al.* (1992) made a similar observation in deer. Heart rate increased when the deer approached the second treatment for antler removal, but it decreased when they approached the second restraint-only treatment.

Animals make specific associations

The associations that animals make appear to be very specific. Cote (2003) reported that cattle may become afraid of specific places where they have had a frightening or painful experience. Animals create fear memories of specific objects associated with a bad experience. Grandin and Johnson (2005) described a horse that became afraid of black cowboy hats because he had been abused by a person wearing a black hat. This fear memory was very specific, as a white cowboy hat had no effect. Tame sheep approached a familiar person more quickly than did wild sheep.

Practical experience has shown that cattle can recognize a familiar person's voice. Research with pigs indicates that they can recognize people by the colour of their clothing (Koba and Tanida, 1999). Behavioural measurements of struggling indicated that taming did not generalize to other procedures such as shearing or handling in a race (Mateo *et al.*, 1991). However, an animal's learning will generalize to another similar situation. Cattle which had been handled four times in the same squeeze chute and single-animal scale were able to recognize these items when they were handled at a different location with a slightly different scale and squeeze chute.

Taming may reduce stress, even though the animal struggles during restraint. Free-ranging deer had cortisol levels that were more than double the levels in hand-reared deer during restraint (Hastings *et al.*, 1992). Both groups vocalized and actively resisted restraint. To reduce stress and improve welfare, livestock should be acclimatized to both people and specific procedures.

Dip vat design and management

Pharmaceutical products, such as ivermectin, have replaced dipping for external parasites in many places. Design information can be found in Hewes (1975), Texas Agricultural Extension Service (1979), Fairbanks *et al.*, 1980; Grandin (1980a, c, 1997b), Sweeten (1980), Kearnan *et al.* (1982), Sweeten *et al.*, 1982, Midwest Plan Service (1987), and http://www. grandin.com

Conclusions

Curved races and round crowd-pen and corral layouts that eliminate square corners facilitate efficient and humane handling of cattle under extensive conditions. These systems utilize behavioural principles, such as the natural tendency of cattle to go back from whence they came. Proper layout is essential, as layout mistakes can ruin efficiency. Wild cattle will be calmer and less stressed if their vision is blocked by solid walls on races and restraint devices that prevent them from seeing people, moving objects and other distractions outside the facility.

Stress can be reduced by gentle handling, training of animals and the use of relatively comfortable methods of restraint. Animals often panic when they slip, so non-slip flooring is essential in races and squeeze chutes. Cattle will remain calmer in a squeeze chute if the design enables them to stand and easily keep their balance. There is an optimum pressure for holding an animal of not too tight and not too loose. A common mistake is to squeeze an animal too tightly in a squeeze chute.

References

Adcock, D., Kingstone, T. and Lewis, M. (1986) Trapping cattle with hay. *Queensland Agricultural Journal* 112, 243–246.

Anderson, D.M. and Smith, J.N. (1980) A single bayonet gate for trapping range cattle. *Journal of Range Management* 33, 316–317.

Andre, P.D. (1991) Keep 'em coming. *Beef*, July, 28–35.

Arbuthnot, N.J. (1979) *Livestock Marketing*. Foundation for Technical Advancement, Shire Engineer, Shire of Mildura, Victoria, Australia.

Baszczak, J., Grandin, T., Gruber, S., Engle, T. and Tatum, J. (2005) A comparison of cattle temperament scores by breed type using different types of temperament scoring. *Journal of Animal Science* (Suppl. 11), 325 (abstract).

Brown, H., Elliston, N.G., McAskill, J.W. and Tonkinson, L.V. (1981) The effect of restraining fat cattle prior to slaughter on the incidence and severity of injuries resulting in carcass bruises. *Western Section, American Society of Animal Science* 33, 363–365.

Burrow, H.M. and Dillon, R.D. (1997) Relationships between temperament and growth in a feedlot and commercial carcass traits of *Bos indicus* crossbreds. *Australian Journal of Experimental Agriculture* 37, 407–411.

Canada Plan Service (1984) *Herringbone, A.I. Breeding Chute Plan 1819*. Agriculture Canada, Ottawa, Ontario, Canada.

Cheffins, R.I. (1987) Self-mustering cattle. *Queensland Agricultural Journal* 113, 329–336.

Cockram, M.S. and Corley, K.T.T. (1991) Effect of pre-slaughter handling on the behaviour and blood composition of beef cattle. *British Veterinary Journal* 147, 444–454.

Connelly, P., Horrocks, D., Pahl, L. and Warman, K. (2000) *Cost-effective and Multipurpose Self-mustering Enclosures for Stock*. Department of Primary Industries, Brisbane, Queensland, Australia.

Cote, S. (2003) *Stockmanship: a Powerful tool for Grazing Management*. USDA National Resources Conservation Service, Boise, Idaho.

Crookshank, H.R., Elissalde, M.H., White, R.G., Clanton, D.C. and Smolley, H.E. (1979) Effect of handling and transportation of calves upon blood serum composition. *Journal of Animal Science* 48, 430–435.

Curley, K.O., Neuendorf, D.A., Lewis, A.W., Cleere, J.J., Welsh, T.H. and Randel, R.D. (2004) Effects of temperament on stress indicators in Brahman heifers. *Journal of Animal Science* (Suppl. 1), 459 (abstract).

Curley, K.O., Pasqual, J.C., Welsh, T.H. and Randel, R.D. (2006) Technical note: exit velocity as a measure of cattle temperament is repeatable and associated with serum concentration of cortisol in Brahman balls. *Journal of Animal Science* 84, 3100–3103.

Daly, J.J. (1970) Circular cattle yards. *Queensland Agricultural Journal* 96, 290–295.

Diack, A. (1974) Design of round cattle yards. *New Zealand Farmer*, 26 September, 25–27.

Douglas, A.G., Darre, M.D. and Kinsman, D.M. (1984) Slight restriction as a means of reducing stress during slaughter. In: *Proceedings of the 30th European Meeting of Meat Research Workers*, Bristol, UK, pp. 10–11.

Ewbank, R. (1968) The behaviour of animals in restraint. In: Fox, M.W. (ed.) *Abnormal Behaviour in Animals*. Saunders, Philadelphia, Pennsylvania, pp. 159- 178.

Ewbank, R., Parker, M.J. and Mason, C.W. (1992) Reactions of cattle to head restraint at stunning: a practical dilemma. *Animal Welfare* 1, 55–63.

Fairbanks, W.C., Grandin, T., Addis, D.G. and Loomis, E.C. (1980) *Dip Vat Design and Management*. Leaflet 21190, Division of Agricultural Sciences, University of California, Riverside, California.

Fowler, M.E. (1995) *Restraint and Handling of Wild and Domestic Animals*, 2nd edn. Iowa State University Press, Ames, Iowa.

Gooneswarden, L.A., Price, M.A., Okine, E. and Berg, R.T. (1999) Behavioral responses to handling and restraint on dehorned and polled cattle. *Applied Animal Behaviour Science* 64, 159–167.

Grandin, T. (1975) Survey of behavioral and physical events which occur in restraining chutes for cattle. Master's thesis, Arizona State University, Tempe, Arizona.

Grandin, T. (1980a) Observations of cattle behaviour applied to the design of cattle handling facilities. *Applied Animal Ethology* 6, 19–31.

Grandin, T. (1980b) Good cattle restraining equipment is essential. *Veterinary Medicine and Small Animal Clinician* 75, 1291–1296.

Grandin, T. (1980c) *Safe Design and Management of Cattle Dipping Vats*. American Society of Agricultural Engineers, Paper No. 80-5518, St Joseph, Michigan.

Grandin, T. (1982) Pig behaviour students applied to slaughter plant design. *Applied Animal Ethology* 9, 141–151.

Grandin, T. (1983a) Design of ranch corrals and squeeze chute for cattle. In: *Great Plains Beef Cattle Handbook*. Cooperative Extension Service, Oklahoma State University, Stillwater, Oklahoma, pp. 5251.1–5251.6.

Grandin, T. (1983b) Handling and processing feedlot cattle. In: Thompson, G.B. and O'Mary, C.C. (eds) *The Feedlot*. Lea and Febiger, Philadelphia, Pennsylvania, pp. 213–236.

Grandin, T. (1984) Reduce stress of handling to improve productivity of livestock. *Veterinary Medicine* 79, 827–831.

Grandin, T. (1987) Animal handling. In: Price, E.O. (ed.) *Farm Animal Behaviour, Veterinary Clinics of North America, Food Animal Practice* 3 (2), 323–338.

Grandin, T. (1989a) Behavioral principles of livestock handling. *Professional Animal Sciences, American Registry of Professional Animal Scientists* 5 (2), 1–11.

Grandin, T. (1989b) Voluntary acceptance of restraint by sheep. *Applied Animal Behaviour Science* 23, 257–261.

Grandin, T. (1990a) Design of loading facilities and holding pens. *Applied Animal Behaviour Science* 28, 187–201.

Grandin, T. (1990b) Chuting to win. *Beef* 27 (4), 54–57.

Grandin, T. (1992) Observations of cattle restraint device for stunning and slaughtering. *Animal Welfare (Universities Federal for Animal Welfare, Potters Bar, UK)* 1, 85–91.

Grandin, T. (1993) Previous experiences affect behaviour during handling. *Agri-Practice* 14 (4), 15–20.

Grandin, T. (1997a) Assessment of stress during handling and transport. *Journal of Animal Science* 75, 249–257.

Grandin, T. (1997b) The design and construction of facilities for handling cattle. *Livestock Production Science* 49, 103–119.

Grandin, T. (1998) Handling methods and facilities to reduce stress on cattle. *Veterinary Clinics of North America, Food Animal Practice* 14, 325–341.

Grandin, T. (2004) Principles for the design of handling facilities and transport systems. In: Benson, G.J. and Rollin, B.E. (eds) *The Well-being of Farm Animals*. Blackwell Publishing, Ames, Iowa.

Grandin, T. and Johnson, C. (2005) *Animals in Translation*. Scribner-Simon and Schuster, New York.

Grandin, T., Curtis, S.E., Widowski, T.M. and Thurman, J.C. (1986) Electroimmobilization *versus* mechanical restraint in an avoid choice test. *Journal of Animal Science* 62, 1469–1480.

Grandin, T., Odde, K.G., Schutz, D.N. and Behrns, L.M. (1994) The reluctance of cattle to change a learned choice may confound preference tests. *Applied Animal Behaviour Science* 39, 21–28.

Grandin, T., Rooney, M.B., Phillips, M., Irlbeck, N.A. and Grafham, W. (1995) Conditioning of nyala (*Tragelaphus angasi*) to blood sampling in a crate using positive reinforcement. *Zoo Biology* 14, 261–273.

Hale, R.H., Friend, T.H. and Macaulay, A.S. (1987) Effect of method of restraint of cattle on heart rate, cortisol and thyroid hormones. *Journal of Animal Science* (Suppl. 1) (abstract).

Hastings, B.E., Abbott, D.E., George, L.M. and Staler, S.G. (1992) Stress factors influencing plasma cortisol levels and adrenal weights in Chinese Water deer. *Research in Veterinary Science* 53, 375–380.

Heldenbrand, L.E. (1955) *Cattle Headgate, U.S. Patent Nos. 2,714,872 and 2,466,102*. US Patent Office, Washington, DC.

Hewes, F.W. (1975) *Stock Dipping Apparatus, U.S. Patent No. 3,916,839*. US Patent Office, Washington, DC.

Holmes, R.J. (1991) Cattle. In: Anderson, R.S. and Edney, A.T.B. (eds) *Practical Animal Handling*. Pergamon Press, Oxford, UK, pp. 15–38.

Howard, K.F. (1953) *A Trap Gate for Cattle*. Advisory Leaflet No. 58, Queensland Department of Agriculture and Stock (reprinted from *Queensland Agricultural Journal*, July 1953).

Jones, R.B., Satterlee, D.G. and Cadd, G.G. (1998) Struggling responses of broiler chickens shackled in groups on a moving line: effects of light intensity hoods and curtains. *Applied Animal Behaviour Science* 38, 341–352.

Kearnan, A.F., McEdwin, T. and Reid, T.J. (1982) Dip vat construction and management. *Queensland Journal of Agriculture* 108, 25–47.

Kilgour, R. (1971) Animal handling in works: pertinent behaviour studies. In: *13th Meat Industry Conference, MIRINZ*, Hamilton, New Zealand.

King, D.E., Schuehle Pfeiffer Co. E., Randel, R.D., Welsh, T.H., Oliphant, R.A., Baird, B.E. *et al.* (2006) Influence of animal temperament and stress responsiveness on the carcass quality and beef tenderness of feedlot cattle. *Meat Science* 74, 546–556.

Koba, Y. and Tanida, H. (1999) How do miniature pigs discriminate between people. The effect of exchanging cues between a non-handler and their familiar handler on discrimination. *Applied Animal Behaviour*, 61, 239–252.

Lambooy, E. (1985) Electro-anesthesia or electro-immobilization of calves, sheep and pigs by Feenix Stockstill. *Veterinary Quarterly* 7, 120–126.

Lay, D.C., Friend, T.H., Randel, R.D., Bowers, C.C., Grissom, K.K. and Jenkins, O.C. (1992a) Behavioral and physiological effects of freeze and hot iron branding on crossbred cattle. *Journal of Animal Science* 70, 330–336.

Lay, D.C., Friend, T.H., Bowers, C.C., Grissom, K.K. and Jenkins, O.C. (1992b) A comparative physiological and behavioral study of freeze and hot iron branding using dairy cows. *Journal of Animal Science* 70, 1121–1125.

Lay, D.C., Friend, T.H., Grissom, K.K., Hale, R.L. and Bowers, C.C. (1992c) Novel breeding box has variable effect on heartrate and cortisol response of cattle. *Journal of Animal Science* 70, 1–10.

Legge, A.J. and Conway, R. (1987) Gazelle killing in stone age Syria. *Scientific American* 257, 88–95.

Marshall, M., Milberg, E.E. and Shultz, E.W. (1963) *Apparatus for Holding Cattle in Position for Humane Slaughtering, US Patent 3,092,871*. US Patent Office, Washington, DC.

Mateo, J.M., Estep, D.Q. and McMann, J.S. (1991) Effects of differential handling on the behaviour of domestic ewes. *Applied Animal Behaviour Science* 32, 45–54.

Mayes, H.F. (1978) *Design Criteria for Livestock Loading Chutes*. Paper No. 78- 6014, American Society of Agricultural Engineers, St Joseph, Michigan.

McFarlane, I. (1976) Rationale in the design of housing and handling facilities. In: Engminger, M.E. (ed.) *Beef Cattle Science Handbook 13*. Agriservices Foundation, Clovis, California, p. 233.

Midwest Plan Service (1987) *Beef Housing and Equipment Handbook*, 4th edn. Midwest Plan Service, Iowa State University, Ames, Iowa.

Mitchell, K.D., Stookey, J.M., Laturnas, D.K., Watts, J.M., Haley, D.B. and Huyde, T. (2004) The effects of blindfolding on behaviour and heart rate in beef cattle during restraint. *Applied Animal Behaviour Science* 85, 233–240.

Müller, R. and Von Keyserlingk, M.A.G. (2006) Consistency of flight speed and its correlation to productivity and personality in *Bos taurus* beef cattle. *Applied Animal Behaviour Science* 99, 193–204.

Murphy, K. (1987) Overhead drafting facilities. *Queensland Agricultural Journal* 113, 353–354.

Oklahoma State University (1973) *Expansible Corral with Pie Shaped Pens*. Plan No. Ex. Ok-724- 75, Cooperative Extension Service, Stillwater, Oklahoma.

Paine, M., Teter, N. and Guyer, P. (undated) Feedlot layout. In: *Great Plains Beef Cattle Handbook*. Cooperative Extension Service, Oklahoma State University, Stillwater, Oklahoma, pp. 5201.1–5201.6.

Parsons, R.A. and Helphinstine, W.M. (1969) *Rambo A.I. Breeding Chute for Beef Cattle*. One Sheet Answers, University of California Agricultural Extension Service, Davis, California.

Pascoe, P.J. (1986) Humaneness of electro-immobilization unit for cattle. *American Journal of Veterinary Research* 10, 2252–2256.

Pearson, L.B. (1965) *Automatic Livestock Headgate, US Patent No. 3,221,707*. US Patent Office, Washington, DC.

Petherick, J.C. (2005) Animal welfare issues associated with extensive livestock production. The northern Australian beef cattle industry. *Applied Animal Behaviour Science*, 92, 211–234.

Phillips, M., Grandin, T., Graffam, W., Irlbeck, N.A. and Cambre, R.C. (1998) Crate conditioning of bongo (*Tragelaphus* eurycerus) for veterinary and husbandry procedures at Denver Zoological Gardens. *Zoo Biology* 17, 25–32.

Pollard, J.C., Littlejohn, R.P., Johnstone, P., Laas, F.J., Carson, I.D. and Suttie, J.M. (1992) Behavioral and heart rate response to velvet antler removal in red deer. *New Zealand Veterinary Journal* 40, 56–61.

Powell, E. (1986) *Cattle Yards*. Information Service Q 186014, Queensland Department of Primary Industries, Brisbane, Australia.

Reck, E. and Reck, J.P. (1903) *Cattle Stanchion, US Patent No. 733.874*. US Patent Office, Washington, DC.

Rider, A., Butchbaker, A.F. and Harp, S. (1974) *Beef Working, Sorting and Loading Facilities*. Paper No. 74-4523, American Society of Agricultural Engineers, St Joseph, Michigan.

Schwartzkopf-Genswein, K.S., Stookey, J.M., Crowe, T.G. and Genswein, B.M. (1998) Comparison of image analysis, exertion force and behaviour measurements for assessment of beef cattle response to hot iron and freeze branding. *Journal of Animal Science* 76, 972–979.

Stahringer, R.C., Randel, R.D. and Nevenforff, D.A. (1989) Effect of naloxone on serum luteinizing hormone and cortisol concentration in seasonally anestrous Brahman heifers. *Journal of Animal Science* 67 (Suppl. 7), 359 (abstract).

Swan, R. (1975) About A.I. facilities. *New Mexico Stockman* 11 February, 24–25.

Sweeten, J.M. (1980) Static screening of feedlot dipping vat solution. *Transactions, American Society of Agricultural Engineers* 23 (2), 403–408.

Sweeten, J.M., Winslow, R.B. and Cochran, J.S. (1982) *Solids Removal From Cattle Dip Pesticide Solution and Sedimentation Tanks*. Paper No. 82- 4083, American Society of Agricultural Engineers, St Joseph, Michigan.

Syme, L.A. (1981) *Improved Sheep-handling Design for Self-feeding Race*. CSIRO, Division of Land Resources Management, Perth, Western Australia.

Texas Agricultural Extension Service (1979) *Portable Cattle Dipping Vat, Drawing No. 600*. Texas A & M University, College Station, Texas.

Thompson, A.C. (1931) *Cattle Holding and Dehorning Gate, U.S. Patent No. 1,799,073*. US Patent Office, Washington, DC.

Thompson, A.C. and Gill, C. (1949) *Restraining Chute for Animals, U.S. Patent No. 477,888*. US Patent Office, Washington, DC.

Thompson, R.J. (1987) Radical new yard proven popular. *Queensland Agricultural Journal* 113, 347–348.

Voisinet, B.D., Grandin, T., Tatum, J.D., O'Connor, S.F. and Struthers, J.J. (1997) Feedlot cattle with calm temperaments have higher average daily gains than cattle with excitable temperaments. *Journal of Animal Science* 75, 892–896.

Vowles, W.J. and Hollier, T.J. (1982) The influence of yard design on the movement of animals. *Proceedings of the Australian Society of Animal Production* 14, 597.

Vowles, W.J., Eldridge, C.A. and Hollier, T.J. (1984) The behaviour and movement of cattle through single file handling races. *Proceedings of the Australian Society of Animal Production* 15, 767.

Ward, F. (1958) *Cowboy at Work*. Hastings Press, New York.

Webber, R.J. (1987) Molasses as a lure for spear trap mustering. *Queensland Agricultural Journal* 113, 336–337.

Wegenhoft, M.A., Sanders, J.O., Lunt, D.K., Sawyer J.E., Herring, A.D. and Gill, C.A. (2005) *Evaluation of Four Component Traits of Disposition in Bos indicus x Bos taurus Cross Cattle*. 2005 Beef Research in Texas, Texas A & M University, College Station, Texas, pp. 5–8.

White, J.B. (1961) Letter to the editor. *Veterinary Record* 73, 935.

8 Dairy Cattle Behaviour, Facilities, Handling, Transport, Automation and Well-being

Jack L. Albright[1] and Wendy K. Fulwider[2]

[1] Purdue University, West Lafayette, Indiana, USA; [2] Colorado State University, Fort Collins, Colorado, USA

Introduction

Dairy cattle should be kept clean, dry and comfortable. Early research in Indiana, USA, showed economic and welfare advantages in providing housing for dairy cows during the cold winter months instead of leaving them outside (Plumb, 1893). To enrich their environment and to improve overall health and well-being, whenever possible cows should be moved from indoor stalls into the outdoor pen or pasture, where they can groom themselves and one another (Wood, 1977; Bolinger et al., 1997), stretch, sun themselves, exhibit oestrus behaviour and exercise.

Exercise decreases the incidence of leg problems, mastitis, bloat and calving-related disorders (Gustafson, 1993). Outdoor exercise improved bovine health and well-being regardless of tie-stall or free-stall housing (Regula et al., 2003). Free-stall housed cows with outdoor exercise had the best claw health (Bielfeldt et al., 2005), suffered less from lameness, tarsal joint lesions, teat injuries and required fewer medical treatments (Regula et al., 2004).

Providing a portion of the daily ration in an outdoor exercise yard effectively doubled the amount of time cows spent outside (Stumpf et al., 1999). After 60 days of exercise training, Davidson and Beede (2003) found cows exercised on treadmills to have improved fitness via reductions in heart rate and plasma lactate concentrations. Considerate handling of dairy cows helps to improve productivity. Other topics which will be covered in this chapter are milking centre design, cow behaviour and transport.

Housing and Facilities

Housing systems vary widely, from fenced pastures, corrals and exercise yards with shelters to insulated and ventilated barns with special equipment to restrain, isolate and treat cattle. Generally, self-locking stanchions/headlocks (one per cow), corrals and sunshades are used in warm, semi-arid regions. Free-stall housing with open sides (or no side walls) is common in hot, humid areas with rainfall > 64 cm/year or 25–30 cm in a 6-month period, e.g. San Joaquin Valley in California, USA (D.V. Armstrong, Arizona, 1999, personal communication).

The range of effective dimensions for pens and stalls for calves, heifers, dry cows, maternity or isolation, special needs, milking cows and mature bulls is constantly evolving as more information regarding the needs and behaviour of cattle is refined. Graves et al. (2006) provide design information for housing special-needs cows, which is increasingly important as herds become larger. The second author found many producers considering compost pack barns (Janni et al., 2006) for their special-needs cows.

Cows housed in compost pack barns suffered zero lesions to the hocks and stifle joints (Fulwider, unpublished data). Hughes (2001) notes that the Holstein is 15 cm taller at the shoulder and 30 cm longer nose to tail, carries 50 kg between her legs and is much more stressed than the Holstein of 30 years ago. Unfortunately, many cows are housed in barns designed for a smaller cow, resulting in more mastitis, lameness and swollen hips, hocks and stifle joints (Hughes, 2001). Anderson (2003) advises sizing stalls for the top 25% of cows in any group, while Tucker et al. (2004) suggest that wider stalls will result in longer lying times and less time spent standing with only front feet in the stall. Recommended sizes of free-stalls and tie-stalls as related to weights of Holstein female dairy cows were revised (McFarland, 2003; Table 8.1).

Maintaining high standards of hygiene may increase productivity while minimizing the incidence of mastitis, endoparasitic and foot infections. Two studies (Schreiner and Ruegg, 2003; Reneau et al., 2005) found poor hygiene of the hind legs and udder to be associated with an increase in somatic cell scores. In addition, Hughes (2001) further notes that it is unacceptable to present milking centre operators with cows requiring extensive cleaning. He also states that it would be wise for the dairy industry to remain free from reproach, since abattoirs have set cleanliness standards for animals sent to slaughter since the Escherichia coli 0157 outbreak in the mid-1990s.

Current trends and recommendations favour keeping dairy cows on unpaved dirt lots in the western USA and on concrete or pasture in northern USA throughout their reproductive lifetimes. Concrete floors should be grooved to provide good footing and to reduce injury (Albright, 1994, 1995; Jarrett, 1995). The concrete surface should be rough but not abrasive, and the micro-surface should be smooth enough to avoid abrading the feet of cattle (Telezhenko and Bergsten, 2005).

Dairy cow locomotion was studied on flooring with four different coefficients of static friction (Phillips and Morris, 2001). The optimum coefficient of static friction was found to be $\mu = 0.4\text{--}0.5$. Cows walk at a slower pace and display a different walking pattern in the presence of slurry when compared to dry or wetted concrete (Phillips and Morris, 2000).

Table 8.1. Free-stall design: recommended dimensions for cows (from McFarland, 2003).

	Weight of cow (kg)		
Dimension[1]	550	650	750
Ls = total stall length[2] (mm)	OF: 2030–2185 CF: 2335–2490	OF: 2135–2285 CF: 2440–2590	OF: 2285–2490 CF: 2590–2745
LH = head space length (mm)	430	455	480
LL = lunge space length (mm)	355	380	405
LN = length to neck rail (mm)	1575–1625	1675–1725	1775–1825
LB = length to brisket board (mm)	1575–1625	1675–1725	1775–1825
LP = stall partition length (mm)	(Ls–355) to Ls	(Ls–355) to Ls	(Ls–355) to Ls
HN = height to neck rail (mm)	1065–1170	1120–1220	1170–1270
HP = stall partition height (mm)	1065–1170	1120–1220	1170–1270
HB = brisket board height (mm)	10–15	10–15	10–15
Hc = stall kerb height (mm)	150–250	150–250	150–250
HE = stall entry height (mm)	300	300	300
HL1 = lunge clearance lower, (mm, max.)	280	280	280
HL2 = lunge clearance upper, (mm, min.)	815	815	815
Ws = stall width, centre to centre (mm)	1090–1145	1145–1220	1220–1320
SB = stall base slope (%)	1–4	1–4	1–4

[1]Outer edge of the kerb to the brisket board
[2]OF = open-front stall; CF = closed front stall

Vokey *et al.* (2003) noted that cows housed in barns with rubber alleys and sand stalls maintained balance between the lateral and medial claw, and had the lowest net growth of dorsal wall as compared to cows in other stall and alley configurations.

Data are limited on the long-term effects of intensive production systems; however, concern has been expressed about the comfort, well-being, behaviour, reproduction and udder, foot and leg health of cows kept continuously on concrete. As a safeguard, many cows are moved from concrete to dirt lots or pasture, at least during the dry period. Also, rate of detection and duration of oestrus are higher for cows on dirt lots or pastures than for cows on concrete (Britt *et al.*, 1986).

The second author visited 113 dairies during 2005–2006 and observed cows exhibiting increased activity on *dry* rubber flooring in walkways, as well as cows avoiding concrete areas when rubber flooring was an option. Dairies in north-eastern USA reported successful use of pedometers for heat detection. The link between walking activity and fertility shows the potential of the pedometer as a tool for increasing fertilization rates (López-Gatius *et al.*, 2005; Roelofs *et al.*, 2005; van Eerdenburg, 2006).

Cows seek their own level of comfort (*see* Fig. 8.1). Physical accommodations for dairy

cattle should provide a relatively clean dry area for the animals to lie down and be comfortable (Jarrett, 1995). It should be conducive to cows lying for as many hours of the day as they desire. It is also essential to provide enough stalls or space so that cows do not have to wait when they want to lie down. For every hour of resting > 7 h daily, a cow should produce an extra 1 kg of milk (Grant, 2004, 2005). Blood flow to the udder, which is related to the level of milk production, is substantially higher (28%) when a cow is lying than when standing (Metcalf *et al.*, 1992; Jarrett, 1995). Table 8.2 illustrates the daily time budget for a typical dairy cow (Grant

Table 8.2. Daily time budget for typical cow in milk (courtesy R. Grant, Miner Agricultural Research Institute, Chazy, New York).

Activity	Time spent/day (h)
Eating	5.5 (9–14 meals/day)
Resting	12–14 (including 6 of rumination)
Standing or walking in alleys (includes grooming, rumination, other)	2–3
Drinking	0.5
Total time needed	21–22

Fig. 8.1. Fair Oaks cows. Given the opportunity, cows will seek their level of comfort in well-designed stalls that have plenty of space for a cow to stretch out and relax, as on this well-managed dairy.

2004, 2005). Producers must be mindful that cows have little time to spare, and time away from the pen should be minimized.

Heat stress affects the comfort and productivity of cattle more than does cold stress (Hillman et al., 2005; Van Baale et al., 2006). Milk production can be increased during hot weather by the use of sunshades, sprinklers or other methods of cooling (Roman-Ponce et al., 1977; Armstrong et al., 1984, 1985; Schultz et al., 1985; Buchlin et al., 1991; Armstrong, 1994; Armstrong and Welchert, 1994; Gordie L. Jones, 2006, personal communication) as well as by dietary alterations. Brown Swiss tolerate heat stress better than Holsteins (Correa-Calderon et al., 2005).

In 2000, near Fair Oaks, Indiana, USA, third-generation Dutch-descent dairy families from Michigan and Western states rejuvenated dairying in Indiana by developing multiple 3000-cow units. Cow comfort, cleanliness, milk quality, nutrition and high milk production are emphasized, and they conduct public tours. Milking cows are housed in free-stall barns bedded with sand (see Fig. 8.1). When temperatures are > 21°C in the barn, the animals are cooled by sprinklers and fans. Cooled cows are found to produce 36 kg of milk as compared to 32 kg in uncooled cows (Gordie L. Jones, 2005, personal communication).

The importance of lighting is much overlooked in many of today's dairies. There may be advantages to providing good lighting, which isn't utilized despite evidence that cows may produce more milk (Phillips et al., 2000; Dahl, 2006). Bright lighting also provides a more pleasant working environment. Dahl (2006) makes a case for the incorporation of red light when cows should be experiencing night. Information with regard to proper lighting is available at http://www.traill.ucic.edu/photoperiod/

Bedding

Comfortable stalls are of the utmost importance to high-yielding dairy cattle. Of all the factors that encourage cows to use free-stalls, the condition of the bed is likely to be the most important (Bickert and Smith, 1998; Weary and Tucker, 2006). When choosing a stall bed, producers must consider climate, how management of the

bed will impact manure handling and how these decisions will affect the herd.

A Wisconsin study found that cows favour the softest available stall beds (Fulwider and Palmer, 2004b). They also favoured different beds at different times of the year due to climatic change (Fulwider and Palmer, 2004a). Cows preferred waterbeds over all other available bases during the cold of winter, probably due to their ability to retain warmth. Waterbeds must be well-bedded until the animals are acclimated to the 'wobbly' nature of this bed type. It may be advisable to acclimatize heifers to waterbeds before they are turned in with a milking group maintained on waterbeds.

Stalls should have bedding to allow for comfort. Wood shavings, straw or other fibrous material over rubber mats and mattresses help keep the base dry, minimizing the potential for bacterial infection, as well as being a 'lubricant' between the cow's skin and the mattress and to insulate the udder against cold temperatures. Some producers recess rubber crumb-filled mattresses and bed with sand (Mowbray et al., 2003), effectively providing the benefits of each bed type.

What cannot be denied are the benefits of sand regarding conformation of the cow, reduction of pressure on the joints, distribution of weight and provision of unparalleled traction. Finding a reliable, inexpensive source of high-quality sand (with no rocks or pebbles that might cause hoof damage or lameness) – and dealing with the high labour component and manure-handling complications that go with sand – deter many producers from utilizing this stall base type. Sand beds require a lot of maintenance. If the beds are not kept full (level with the kerb), the amount of time spent lying is significantly reduced (Drissler et al., 2005). Very fine sand does not stay in place, clinging to teats and udders, and needs to be removed at milking time.

Bedding should be non-abrasive, absorbent, free of toxic chemicals or residues that could injure animals or humans and of a type not readily eaten by the animals. Bedding rate should be sufficient to keep animals dry between additions or changes. Any permanent stall surface, including rubber mats, should be cushioned with dry bedding (Albright, 1983; Albright et al., 1999).

On the other hand, what is in front of the cow has as much to do with comfort as what is under her (S.D. Young, Ontario, Canada, 1999, personal communication). If neck rails are placed too low, the cow may feel cramped and be reluctant to enter or use the stall(s) (Albright and Arave, 1997). Tucker et al. (2005) suggest that producers may wish to use the neck rail to keep cows from standing in and soiling stalls and provide a comfortable flooring surface in the alley for standing on. Fulwider and Palmer (2005) reported that cows spent less time standing in stalls when rubber alley mats were installed. This may also increase the useful life of beds (avoid beds that are too hard – concrete, concrete with solid rubber mats or compacted earth).

Swollen hocks and stifle joints result from a bed that does not provide sufficient cushion. Wechsler et al. (2000) reported that cows on mats and mattresses had a higher incidence of hairless patches, scabs and wounds on the carpal and tarsal joints than those on straw beds. Findings by Sogstad et al. (2006) indicate that cows with wounds and swellings at the tarsus have more clinical mastitis and teat injuries. They further related that free-stall cows suffer from a higher prevalence of metabolic claw and infectious lesions than do tie-stall cows. Fulwider et al. (2007) found that cows maintained on waterbeds or in sand stalls suffered fewer hock lesions than cows kept on rubber-filled mattresses.

Mattresses are soft when they are new, but filling becomes compacted and the surface becomes extremely hard within a few months. Recessing mattresses several centimetres below the kerb allows for the addition of deep sand or other bedding, thus reducing tarsal joint lesions. This may, however, result in lesions at the tuber calcis when contacting the cement kerb at the rear of the stall if the depth of the additional bedding is allowed to become too low (Mowbray et al., 2003). The second author saw this successfully implemented on one dairy: PVC plastic pipe was mounted at the rear of stalls to hold sand. The PVC pipe is non-abrasive and those cows had no lesions.

Weary and Taszkun (2000) found that the number and severity of lesions increased with age and that the length of stall for cows on deep-bedded sawdust was associated with severity of lesions. Stalls that are too short are also associated with more lesions. Lameness issues were further investigated by Sogstad et al. (2006):

heifer lameness prevalence was low, but 29% had at least one lesion. Heifers in tie-stalls had fewer heel/horn erosions, sole haemorrhages and white line fissures than those in free stalls.

Recently, Weary and Tucker (2006) focused on the latest in free-stall comfort as follows: (i) neck rail: cows prefer these higher and closer to the front of the stall; (ii) brisket board: lying times are longer when these are removed; (iii) stall partitions: cows prefer these wide apart – > 123 cm improves lying time and reduces standing; (iv) stall surface: plentiful bedding prevents injuries and improves lying time; and (v) standing surface: cows prefer soft, dry surfaces (also at the feed bunk and alleyways), which can help prevent injuries and diseases. Tucker et al. (2006) suggest new approaches to dairy cow housing are needed.

Milking centre design

Until the advent of centralized milking centres, most cows were milked in their stalls. A disadvantage of this method is that it is labour intensive and hard on the knees of those milking the cows. The idea of milking cows on an elevated herringbone platform originated in Australia (O'Callaghan, 1916; Albright and Fryman, 1964). Early US designs enabled a single person to milk two cows while seated on a swivel chair or to use elevated side-opening milking centres (Albright and Fryman, 1964). Due to labour shortages and high wages, New Zealand dairy farmers were motivated to develop rotary centres (Gooding, 1971). At the time, these turnstile systems were a great innovation, but had high maintenance costs.

Simple layouts with automated gates and milking machine detachers became popular. Possibly due to shorter walking distances (Smith et al., 1998) and greater efficiency and automation at the entry and exit points, currently there is a new wave of much larger rotary milking centres available for larger herds in the USA (see Fig. 8.2). Quaife (1999) claims that today's rotary milking centres (see Fig. 8.3) will remain a viable option for some larger producers and not fade away like they did in the 1970s. Since 2000, Fair Oaks and other nearby 3000-cow multiple unit dairies are milking their cows on 72-cow rotary milking centre platforms.

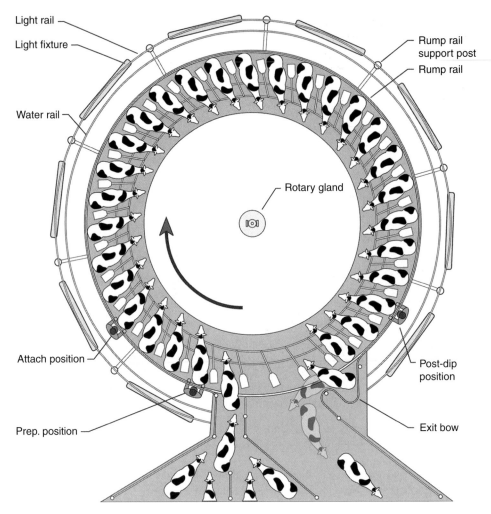

Fig. 8.2. Rotary parlour. New, large dairies prefer rotaries with 70–100 milking units. A large rotary is more efficient than older, smaller rotaries designed for 20–30 cows (diagram courtesy of DeLaval, Inc.).

Extensive time and motion studies have been conducted on different milking centre designs (Armstrong and Quick, 1986; Armstrong, 1992). The addition of automation, such as powered gates, has enabled simple designs – such as herringbones, trigons and polygons – to achieve greater labour efficiency than the early, smaller rotaries. Good reviews of these simple designs of different milking centre layouts can be found in Bickert (1977), Armstrong (1992) and Midwest Plan Service (2000).

The most common design used to be the herringbone, where two rows of cows were milked from a central pit. Currently, the parallel milking centre is the most commonly installed design in the USA. The milking machines are attached from the rear between the cow's legs instead of from the traditional side position. During milking, the cows stand at 90° relative to the pit. All the cows are released at once after milking by lifting either an entire row of stanchions or a long bar which runs in front of the animals. This design is more efficient than older-style herringbones that did not have the rapid-exit feature. Herringbone milking centres with the rapid exit design combine some of the best features of both herringbones and parallel milking centres. New heifers are easier to train to a herringbone and the cows can easily be milked from the side (Armstrong *et al.*, 1989, 1990), especially with rapid-exit stalls (see Fig. 8. 4).

Much has been made about getting cows to move more quickly into the milking centre. Cows were observed to self-load with ease in large milking centres staffed by one milker (Fulwider,

unpublished data). While one side was prepared and milkers attached, the other side loaded with no trouble. When milking centres are overstaffed, people get in the way. One producer was in the process of adapting his cows to come in without grain and it was not going well. Ceballos and Weary (2002) found that small

Fig. 8.3. Rotary milking centre, Turlock, California. These California Jerseys take a turn on the rotary. Cattle are easy to train to enter and exit the rotating platform: training often takes only 3 days.

Fig. 8.4. Herringbone rapid-exit brisket bar. The straight, rapid-exit brisket bar in a herringbone allows cows and heifers of various sizes to be milked together with ease and maximum comfort. Longer cows slide along the bar. Heifers may adapt more easily to being milked in a herringbone. Cows of varying age, size and infirmity may have difficulty making the sharp turn into a parallel milking centre (photograph courtesy of DeLaval, Inc.).

quantities of feed reduced the need to push cows or use other interventions that might negatively affect the animals. Producers with robotic milking systems effectively utilize feeding motivation to entice cows into the robot. Halachmi *et al.* (2006) suggest there is an opportunity to increase milk yields by feeding pellets rich in digestible neutral detergent fibre to selected high producers.

The use of a powered crowd gate to make the holding pen smaller induces cows to enter the milking centre voluntarily. Crowd gates should not be used to forcibly push cows or apply electric shocks. Automatic alley scrapers and crowd gates can cause injuries resulting in death or the need to dispose of cows prematurely. According to Fahey *et al.* (2002), the length of time in the holding pen indicates that exposure of cows for short (40 min) to increased (120 min) stays does not significantly affect production or stress indicators in the short term (4 weeks).

Proper training of cows and of milking centre operators will also improve the efficiency of movement through the facility. Cows should be encouraged to enter voluntarily without prodding.

The milker should avoid leaving the pit to chase the animals as this conditions them to wait for the milker to come after and chase them into the milking centre. Cows also have individual preferences for music, weather, certain people and which side of the milking centre they will enter (Albright *et al.*, 1999). Since cows are creatures of habit, it is imperative to be consistent from one milking to the next.

Automation and robotic milking

Housing and herd management developments have important effects on the well-being of dairy cattle, and the cattle enterprise is well suited to the application of electronics and automation (Albright, 1987; Smith *et al.*, 1998). Robotic milking centres have the greatest potential economic benefit for the 50–120-cow dairy (Rotz *et al.*, 2003). The second author (2005) had the opportunity to visit six dairies utilizing robotic milking centres in Canada (see Fig. 8.5). Producers appreciated that the robot gave them the opportunity to, for example, attend their

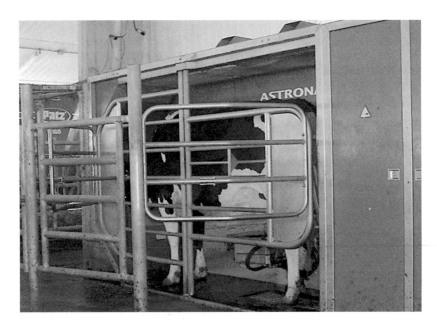

Fig. 8.5.   Robotic milking centre. Proper design of entrance and exit gates is very important in preventing dominant animals from blocking others. Robotic milking may provide the ultimate in cow choice as the cow is free to choose when to feed, rest and be milked, as well as how often. Robots are popular on smaller family dairies.

children's school events without having to plan around milking times. They also reported having more time to tend cows without the drudgery of milking.

The robotic units are helping small family dairies to stay in business and avoid the hassles of hired labour. It should also be noted that cows have more resting time in this system as the hours normally spent walking to and standing in the holding area have been eliminated. Cows have much more freedom to choose when to feed, rest, socialize or be milked, as well as how often to be milked.

Rodenburg (2004) has noted that most conventional free-stall barns can be successfully adapted for robotic milking systems. For producers who may be concerned about the adaptation of cows to the robotic milking centres, Weiss et al. (2004) have indicated that this varies widely between animals, but all adapt within days. Cows may be attracted to the milking centre by the desire to feed or to be milked, although motivation to feed is given first priority (Melin et al., 2006). When robot-milked cows were exposed to pre-recorded calf vocalizations, 24-h milk, number of milkings and milking time were not significantly different from those of cows in the control group, nor was any difference detected in behaviour after exposure to the recordings (Jones et al., 2005).

Graziers may also utilize this technology. Spörndly and Wredle (2005) recommend providing water in the pasture area for animal welfare reasons, but have found no negative effect in water provided only at the barn with walking distance up to ~300 m. Behavioural and physiological responses of robotically and conventionally milked cows have been studied (Hopster et al., 2002; Hagen et al., 2004). They reported that conventional and robotic milking were equally acceptable with regard to cow welfare. Halachmi (2004) reported that a mathematical simulation model may be useful in optimizing dairy facility efficiency and layout and should be utilized before construction begins.

Pastell et al. (2006) have assembled a system in a robotic milking centre that can weigh the cow while analysing her step and kick behaviour while in the robot, and how it may change over time. Preliminary data analysis indicate that this system is very promising for early detection of limb and hoof disorders.

Over time, capital investments for comfort and sanitary requirements have increased markedly. Labour-saving practices have been developed to reduce the drudgery of dairy farming. Many top-producing cows continue to be housed and milked in labour-intensive tie-stall barns. For these tie-stall barns with improved design (Zurbrigg et al., 2005), there are now silo unloaders, gutter cleaners, battery-operated silage carts, portable straw choppers, automatic detaching milking machines with low milk lines and mechanized manure handling.

Behaviour and Management

Few scientific data are available on cows and grooming. Cows with access to motorized brushes have a glossier hair coat than others. The second author has observed that animals will actively seek out the motorized brush and apply it to many parts of the head and body (see Fig. 8.6). Producers reported that cows without access to an automatic brush during the dry period would spend a few days at the brush upon return to the milking barn. Cows were also reported to spend time at the brush immediately after calving. The second author noted that some dairies that did not wish to invest in multiple brushes often placed a single one in a common walkway.

The dairy cow has been called 'the foster mother of the human race' (Rankin, 1925). A relationship develops between the milker and the cow which is a vital part of the milk extraction process and, as machine-milking took over from hand-milking, this relationship was considered by many to have diminished. After her calf is removed, the cow is milked with a minimum of manual stimulation in highly automated surroundings.

Individual differences were noted regarding behavioural and physiological responsiveness in primiparous cows. Van Reenen et al. (2002) evaluated primiparous heifers for stress responses at the first machine-milking on days 2, 4, and 130 of lactation. Elevated heart rate was associated with inhibition of milk ejection on days 2 and 4. A reduced level of fear may be established by 30 sessions of prepartum udder massage (Das and Das, 2004), resulting in

Fig. 8.6. Motorized brush. Cows seek out this brush to groom themselves. It is well-worn, and less than 1 year old. This motorized brush is located in the walkway and accessible to 400 cows as they enter and exit the milking centre. Producers who have provided motorized brushes for their cows report that cows spend the bulk of their free time for days at the brush if they have been previously kept in a pen without one (e.g. during the dry period). Cows that have access to a brush in the calving pen may seek comfort by the brush immediately after calving. Producers who provide a brush for cows in each pen claim that use of this device results in cleaner, glossier cows. Some producers have chosen to mount a brush in the walkway to be accessed as cows approach and return from the milking centre (from Fulwider *et al.*, unpublished).

reduced restlessness ($P < 0.05$), faster milk let-down ($P < 0.01$), higher milk flow rate ($P < 0.01$) and decreased defecation and urination rates ($P < 0.05$).

For as long as cows have been milked, there has existed the art of care that results in more milk from healthier, contented cows. It has been recognized that the dairy cow's productivity can be adversely affected by discomfort or maltreatment. Alert handlers have the perception and ability to read 'body language' in animals. For example, healthy calves and cows will exhibit a good stretch after they get up, then relax to a normal posture. Increased standing of cattle is now often taken as a sign of discomfort or discontent in studies of cow and calf confinement (Albright, 1987).

Cattle under duress show signs of distress by bellowing, butting or kicking. Behavioural indications of adjustments to the environment are always useful signs of whether the environment needs to be improved. In some cases, the way animals behave is the only clue that stress is present (Stephens, 1980; Albright, 1983). Looking-up behaviour in the holding pen has been linked to low motivation to be milked due to fear of humans (Ishiwata et al., 2005). This behaviour was most common in cows during lactation three and under, positively correlated with flight-starting distance and milking centre entrance order, and negatively correlated with productivity.

Clues to a cow's mood and condition can be obtained by observing the animal's tail. When the tail is hanging straight down, the cow is relaxed, grazing or walking, but when the tail is tucked between the cow's legs, it means the animal is cold, sick or frightened. During mating, threat or investigation, the tail hangs away from the body. When galloping, the tail is held straight out, and a kink can be observed in the tail when the animal is in a bucking, playful mood (Kiley-Worthington, 1976; Albright, 1986a; Albright and Grandin, 1993; Albright and Arave, 1997).

According to Kiley-Worthington (1976), when studying the cause and function of tail movement it is necessary to consider the whole posture of the animal as well as the contexts that give rise to it. In cattle (and horses), the immediate association one makes with lateral tail movements is with cutaneous irritation. In these species there are morphological changes of the tail that point to its use as a fly switch.

Tail-docking

Docking of tails is a controversial yet common practice performed on cows that are milked from the rear or have a filthy switch (the tail end). Tail-docking has been prohibited in the UK. Some other European countries and the Canadian and American Veterinary Medical Associations officially oppose routine tail-docking in dairy cattle (Stull et al., 2002). Under conditions of high fly numbers, tail-docked heifers' tail-flick more often with their tail stump and are forced to use alternative behaviours such as rear leg-stomps and head-turning to try to rid themselves of flies. More flies settle on tail-docked cows than on intact cows; the proportion of flies settling on the rear of the cow increases as tail length decreases. Grazing and rumination are disturbed when fly attacks are intense, and substantial losses to the US cattle industry have been attributed to flies causing interference with grazing (Albright et al., 1999).

Excellent fly control is therefore especially important for tail-docked cattle. A study of tail-docking in New Zealand (Matthews et al., 1995) found no difference in cortisol concentrations between docked and intact cows, but there were differences in milk yields, body weights, somatic cell counts, frequency of mastitis or milker comfort among the treatments studied (intact tails, trimmed tails, docked tails). Tucker et al. (2001) found little merit for tail-docking with regard to cow cleanliness, udder cleanliness and health, though they reported significant differences in cleanliness between cows.

Research on tail-docking by the USDA-ARS Livestock Behavior Research Center and Purdue University scientists from 1997 to the present demonstrated that well-being of calves (at docking) and heifers and cows (after docking) can be compromised by acute pain, increased fly numbers and irritation, and by signs of increased sensitivity or chronic pain in the stump (Eicher et al., 2000, 2001; Eicher and Dailey, 2002; Eicher, personal communication, 2006). Trimming the switch with clippers is preferred (Albright et al., 1999) as an alternative to tail-docking in dairy herds (Stull et al., 2002).

Tail stumps make it more difficult to breed cows using artificial insemination.

Stray voltage

Stray electrical voltages from malfunctioning electrical equipment can cause discomfort to dairy cows and thus lower milk production. Numerous research studies have quantified the physiological and behavioural responses of dairy cattle to electrical currents (Lefcourt, 1991; Aneshansley et al., 1992; Hannah, 2002). The electrical currents required for perception, behavioural change or physiological effects to occur are widely variable. Dairy cows can feel very low voltages of only 1.0 volt when they occur between a water bowl and the rear hooves (Gorewit et al., 1989).

Reinemann et al. (2005) stated that the current level required to produce a behavioural response was less than that required to cause a short-term reduction in feed and water intake and milk production. Some cows in the study responded by submerging the entire muzzle in the water bowl, effectively providing a larger contact surface area while reducing the maximum local current density in the muzzle. Reinemann et al. (2004) also noted that if cows have adequate time to consume water between current pulses, water was consumed at the same rate as in the absence of any current stimulus.

The sources of relatively small amounts of electrical currents passing through animals are often very difficult to locate. Stray voltage or electrical currents may arise because of poor electrical connections, corrosion of switches, frayed insulation, faulty equipment or heavily loaded power lines.

Information on how to detect and correct stray voltage problems has been available for some time (Appleman, 1991). Periodic evaluation of facilities for stray voltage is suggested. Solutions include voltage reduction, control of sources of voltage leakage, gradient control by use of equipotential planes and transition zones, and isolation of a portion of the grounding or grounded neutral system from the animals. Proper installation of electrical equipment and complete grounding of stalls and milking centre equipment should help prevent stray voltage problems. Although stray voltages and electrical currents cannot be totally eliminated, they can be reduced (Albright et al., 1991; Lefcourt, 1991; Gorewit et al., 1992). Graves (2006) concludes that stray voltage must not be used as an excuse for poor management practices that cause low milk production or high somatic cell counts.

Social Environment

Dairy cattle are social animals that function within a herd structure and follow a leader to and from pasture or milking centre. Cows exhibit wide differences in temperament, and their behaviour is determined by inheritance, instinct, physiology, hormones, prior experience and training. Cows are normally quiet (non-vocal) and thrive on gentle treatment by handlers. Handling procedures are more stressful for isolated animals; therefore, attempts should be made to keep several cows together during medical treatment, artificial insemination or when moving cows from one group to another (Whittlestone et al., 1970; Arave et al., 1974). Cattle should have visual contact with each other and with their caretakers.

Mixing multiparous and uniparous cows should be avoided, as this disrupts normal social behaviours and causes reduced milk yield. To alleviate reduced grazing time, multiparous cows compensate by becoming more dominant, with an increased rate of pasture-biting (Phillips and Rind, 2001). Evidence has been reported that cows consistently entering one side of the milking centre were more fearful in novel situations than less consistent herdmates. The consistent cows were also more successful in accessing food resources (Prelle et al., 2004). Paranhos de Costa and Broom (2001) have reported that there was no evidence of discomfort or poor welfare even when highly consistent cows were milked on their non-preferred side.

Many dairymen allow their cows to develop their own individual personalities as long as no special care or treatment is required. Mass handling of cows dictates that individual cows fit into the system rather than the system conform to the habits of the cow. The slow milker, the kicker, the boss cow, the timid cow, the explorer and the finicky eater are usually removed from larger herds, regardless of pedigree.

Although concern is expressed from time to time about temperament and behavioural problems, most attempts at reinforcing correct behaviour and disciplining improper behaviour have been successful. One dairy study showed that behaviour as a reason for disposal occurred in less than 1% of cases (Albright and Beecher, 1981; Albright, 1986b).

Although creatures of habit, gentle dairy animals may be prompted into rebellion by the use of unnecessarily severe methods of handling (e.g. shouting and shock prods) and restraint. Attempts to force an animal to do something it does not want to do often end in failure and can cause the animal to become confused, disoriented, frightened or upset. Handling livestock requires that they be 'outsmarted' rather than outfought and that they be 'outwaited' rather than hurried (Battaglia, 1998). Most tests of will between the handler and the cow are won by the cow.

Considerable self-stimulation and 'inwardness' occur in cattle due to the rumination process. During rumination, cows appear relaxed with their head down and their eyelids lowered. Resting cows prefer to lie on their chest, facing slightly uphill. Also, through cud-chewing as well as mutual and self-grooming, aggression is reduced and there is little or no boredom (Albright, 1986b).

Management developments that have improved the comfort and well-being of dairy cattle include: (i) raising calves in individual pens or hutches (Baker, 1981); (ii) providing exercise prior to calving (Lamb et al., 1979); (iii) grooving or roughening polished, smooth concrete flooring (Albright, 1983, 1994, 1995); (iv) making use of pasture or earthen exercise lots and removing slatted floors (Albright, 1983); and (v) eliminating stray voltage (Appleman, 1991).

Individual stalls (cubicles/free-stalls) have resulted in cleaner cows and fewer teat injuries than loose housing. Fregonesi and Leaver (2002) stated that providing one free-stall per cow was essential. In the low space allowance situation, high- and low-yield cows had more agonistic incidents, disturbed patterns of diurnal lying behaviour, and decreased total lying times. Dairy cattle thrive best when they are kept cool, free from flies and pests and provided with a dry, comfortable bed on which to lie down (Albright, 1986b).

Dairy cattle have traditionally been kept in groups of 40 to 100 cows. In commercial dairy herds in Arizona, New Mexico and Texas, variation in group size – small (50–99), medium (100–199) and large (> 200) – did not cause a problem per se. Large herd size, however, can affect management decisions because overcrowding with insufficient numbers of headlocks or inadequate water and feed manger space per cow, irregular or infrequent feeding and excessive walking distance to and from the milking centre have a greater impact on behaviour and well-being than does group size (Albright et al., 1999).

Self-locking mangers have become standard equipment for large dairy herd operations. In order to evaluate the effects of restraint using self-locking stanchions, 64 Holstein cows from peak to late lactation were restrained at feeding time for 4 h/day for 4 weekly periods (Bolinger et al., 1997). Milk production, somatic cell counts, mastitis or other health concerns, plasma cortisol concentrations and total daily feed intake were unaffected by restraint. For the cows locked in stanchions, their eating frequency over 24 h was significantly reduced, but dry matter intake was not affected. Total rumination frequency over 24 h was not significantly different for cows that were restrained; however, restrained cows ruminated less during the day following release.

Behaviourally, cows that were locked in the stanchions spent significantly more time lying in free stalls after release from restraint. Grooming was also one of the first behaviours performed following release. Grooming was considered to be a behavioural need and was significantly increased during all times when cows were not locked up. Acts of aggression were elevated during all periods following restraint.

The use of self-locking stanchions did not appear to affect substantially the overall well-being of the cow (Bolinger et al., 1997). In a similar lock-up trial, milk production was reduced and cortisol increased during the summer months in Utah, USA (Arave et al., 1996a, b). Advantages of headlock barriers include, in particular, reduced aggressive interactions and displacements for socially subordinate cows (Endres et al., 2005; Huzzey et al., 2006). A cow's reaction to a lock-up stanchion may be affected by how she is introduced to the stanchions. If her first experience with the stanchion is associated with painful procedures, she may be more likely to become stressed, as compared to a heifer that associates the stanchion with feeding. Grandin (1997) and Grandin and Johnson (2005) reviewed how an animal's previous experience affected reactions to handling (see Chapters 2 and 3, this volume).

Other Purdue University work with detailed observations, using intact and cannulated cows, suggests a behavioural need for the cow to rest and to ruminate on her left side (Grant et al., 1990; Albright, 1993).

Handling and Transport of Stock

Calves

Calves require special handling and care from the time they are born. The most important point to remember is to feed the newborn calf colostrum soon after birth and within the first 6 h. A calf should be given 8–10% of its body weight in fresh colostrum by bottle, bucket or tube feeder; twice within 24 h following birth. Colostrum is nutrient-rich and provides the calf with vital immunoglobulin. Good nutrition, along with proper handling, starts a calf on its way toward a healthy life. If young calves are to be marketed, the following three procedures should be used:

1. Provide individual care and colostrum for 2–3 days after birth.
2. Calves should always have a dry hair coat, a dry navel cord and walk easily before being transported either to auctions or long distances. A 1-day-old calf can stand, but it is unsteady and wobbly and is not ready for market (Albright and Grandin, 1993). In the UK and Canada, the sale of calves less than 1 week old is forbidden. Calves should not be brought to a livestock market until they are strong enough to walk without assistance. To reach adequate strength and vigour, calves need to be a minimum of 5 days old (Grandin, 1990). If the calves are going to be transported to a nearby specialized calf-rearing facility, they can be transported sooner.
3. Handle calves in transit carefully, protecting them from the sun and heat stress in the summer, and from the cold and wind chill in winter.

Bulls

The safety of humans and animals is the chief concern underlying management practices. By virtue of their size and disposition, dairy bulls may be considered as one of the most dangerous domestic animals. Management procedures should be designed to protect human safety and to provide for bull welfare.

Threat postures

There are certain major behavioural activities related to bulls. These are threat displays, challenges, territorial activities, female-seeking and directing (nudging) and female-tending. These behavioural activities tend to flow from one to another (Fraser, 1980). Threat displays in bulls and ungulates (e.g. antelope, bison) are a broadside view (see Fig. 8.7) when a person or a conspecific invades its flight zone.

The threat display of the bull puts him in a physiological state of fight-or-flight. The threat display often begins with a broadside view with back arched to show the greatest profile, followed by the head down – sometimes shaking the head rapidly from side to side, protrusion of the eyeballs and pilo-erection of the hair along the dorsal line. The direct threat is head-on, with head lowered and shoulders hunched and neck curved to the side toward the potential object of the aggression. Pawing the ground with the forefeet, sending the earth flying behind or over the back – as well as rubbing or horning the earth – are often components of the threat display. If in response to the threat display the recipient animal advances with head down in a fight mode, a short fight with butting of horns or heads ensues. If the recipient of the threat has been previously subdued by that animal, he will probably withdraw with no further interaction (Albright and Arave, 1997).

While a bull is showing a threat display, if an opponent such as another bull (or person) withdraws to about 6 m, the encounter should subside and the bull will turn away. If not, the bull will circle, drop into the cinch (flank) body position or start with a head-to-head or head-to-body pushing. At the first sign of any of the above behaviours, humans should avoid the bull and back away quickly – hopefully via a predetermined route. Do not turn your back and

Fig. 8.7. Threat display. Broadside threat display is a warning that a human has invaded his territory. A direct threat is head-on with head lowered, shoulders hunched and neck curved to the side.

run. The rapid movement of running may cause the bull to attack.

Many people lack the background, attitude and awareness of dealing with dangerous bulls and parturient cows; therefore, additional training and bull/cow behaviour information are needed. It is wise to respect and be wary of all bulls – especially dairy bulls – as they are not to be trusted. Every bull is potentially dangerous. Bull attacks are the number one cause of livestock handler fatalities. He may seem to be a tame animal, but on any given day he may turn and severely injure or perhaps kill a person, young or old, inexperienced or experienced. This is especially true when a cow is in oestrus and needs to be removed from 'his' group or the group is moved to the holding pen for milking (Albright, 2000).

Never handle the bull alone and never turn your back on him. For self-protection when moving cattle, attempt to appear larger and carry a cane, stick, handle, metal pipe or plastic pole with flap. To reduce the tendency of bulls to attack people, bull calves must be raised in a social group to prevent them from imprinting on people. Bull calves should either be raised together in a pen starting at 6 weeks of age, or on nurse cows. The most dangerous bull is one that was reared alone away from other cattle. Bull

accident reports indicate that a young bull is most likely to start attacking people when he becomes mature at 18–24 months of age (T. Grandin, personal communication, 2006). For further information about bull behaviour and handler safety refer to Albright and Arave (1997), Albright et al. (1999) and Albright (2000).

Other dairy animals

In addition to bulls, humans must be careful around certain steers, heifers and recently calved cows protecting their calves. Some animals are different and do not follow the threat display behaviour previously mentioned. Be careful of following behaviour, walking the fence, bellowing, a cow in oestrus and the bull which protects the cow, thereby attacking the handler. An animal's first attack should be its last and it should be sent to the abattoir (Wilson, 1998).

The system of management under which dairy cattle are raised and kept has a profound effect on their temperament, and this is not always taken into consideration. For example, bull calves should never be teased, played with as a calf, treated roughly or rubbed vigorously on the forehead or the area of the horns. The Fulani herdsmen stroke under the chin (rather than on top of the head) as an appeasement, taming and grooming-type behaviour. This is essentially the way cows groom each other (Hart, 1985; Albright and Grandin, 1993; Albright and Arave, 1997).

Handling and transport tips

1. When loading dairy animals for shipping, allow plenty of handling space. Cattle need ample room to turn; the leaders will then move into the chute with other animals following. This is an example of leadership–followership, as in cattle or sheep, goats and ducks.
2. Stair steps are recommended for concrete loading ramps. Each step should be 10 cm high with a 30 cm tread width.
3. Loading ramps for young stock and animals that are not completely tame should have solid sides.
4. Never attempt to transport cows which become emaciated or too weak to stand. If rehabilitation does not occur within a reasonable time, the animal should be humanely killed on the farm (Livestock Conservation Institute, 1992; National Institute for Animal Agriculture, 2004). More details may be found in Agriculture Canada (1990) and OIE (World Organization for Animal Health) (2005).
5. When transporting young dairy animals or producing cows, always handle them gently. Since cows are curious, allow them to quietly investigate their new environment and ease into it without outside distractions.
6. Try to ship dairy animals under favourable weather conditions. Avoid extremely hot or extremely cold temperatures that create undue stress and may cause sickness.

Dairy producers have much to gain when cows and young stock are properly handled and cared for (Albright and Grandin, 1993). Recently, Palmer (2005) has compiled valuable information on animal handling needs including methods, locations and possible systems.

Transport Developments

Knowledge and utilization of the flight zone (see Chapter 5, this volume) are important during the movement of dairy cattle. Cows should be moved at a slow walk, particularly if the weather is hot and humid or if the flooring is slippery.

Heart rate transmitters were implanted in lactating Holstein cows prior to travel (Ahn et al., 1992). Cows were transported 402 km in about 6 h over various road surface conditions in an 8.2 m-long livestock trailer. The two-way journeys started in the morning and ended late in the afternoon. Cows stayed overnight and were brought back in the late afternoon. This 2-day journey was repeated 1 week later. Feed and water were provided during the interim between travels, with cows receiving their normal ration for that period. Cows were milked by portable machine according to their regular schedule and confined to a fenced corral of approximately 0.4 ha.

Heart rate recorded as travel commenced averaged 89.7 bpm and differed significantly ($P < 0.01$) from all hourly readings. Average heart rate (bpm) for hours 2, 3, 4, 5 and 6 averaged 77.0, 74.8, 71.3, 74.4 and 72.9, respectively (which are all similar to a resting heart rate of 76.5). Heart rates differed significantly ($P < 0.01$) by road surface type, averaging 83.3 bpm on a dirt road, 81.2 bpm on a paved, rural three- and four-lane road, 76.1 bpm on a paved, two-lane desert road and 73.6 bpm on the paved motorway. Heart rates observed gave evidence of habituation on the day of travel and also from week one to week two.

Transport is particularly stressful for young calves, which experience mortality rates greater than 20% and bruised stifles to an incidence of 50% or more (Hemsworth et al., 1995). Research on transport of dairy calves has shown that immunological systems are affected by the age at which transport occurs within the first week of life, and cognitive changes can be detected at least 6 weeks post-transport. Additionally, adverse effects of transport of young calves can be modulated by several known modulators (yeast cell-wall products) and ascorbic acid, thereby decreasing morbidity and mortality (Eicher, et al., 2004; Eicher, 2006, personal communication). Eicher (2001) further indicates that attention must be paid to length of studies regarding young calves, as they may succumb to disease 1 month following transport.

Calves' behavioural and physiological (cortisol, heart rate) reactions to being loaded on to a truck, transported for 30 min and unloaded, were observed. It took more time and effort to load pair-raised calves than individually housed calves ($P < 0.01$) and less effort to load those that had received additional contact with people ($P < 0.01$) as compared to those who had received minimal contact. During loading, the latter group had lower heart rate ($P < 0.05$) than the former. During transport, pair-housed calves had lower heart rates ($P < 0.05$) as compared to the individually housed calves (Lensink et al., 2001). See Chapter 19 on physiology.

Human–Animal Interactions

The behaviour of the cow-handler

Studies on homogeneous dairy herds – as defined by similar feeding policy, feeding levels, breed and genetic potential, grazing management and climate – demonstrate the effect of the cow-handler's behaviour and personality on cow behaviour and productivity (Seabrook, 1972, 1977, 1991, 1994; Albright and Arave, 1997; Hemsworth and Coleman, 1998; Hemsworth et al., 2002). The highest performance cow-handlers, in terms of milk yield for a given level of input, have the following traits: self-reliant, considerate, patient, independent, persevering, difficult to get on with, forceful, confident, suspicious of change, not easy going, inadaptable, not neat, not modest, not a worrier, not talkative (quiet), uncooperative and non-social ('grumpy').

In summary, they are confident introverts. Some of these traits may seem to be socially undesirable, but it is the cows' and not another human's reaction that is critical. People with these traits were more stable and had an air of confidence, enabling them to develop a relationship with their cows that positively influenced the animal's performance. Cows under the care of such a person easily out-produced a person lacking confidence or a confident extrovert ('cheerful Charlie'), these latter tending to earn only average production achievement from their cows.

Building on this work, Reid's (1977) study on high-producing herds both in North America (Canada and USA) and the UK yielded some important results. Reid concluded that the high-production cow-handler was able to minimize output of adrenaline by the cow and obtain a higher percentage of the milk yield that her genetic capacity permitted than others would obtain from the same cow under similar conditions. The high-production cow-handler achieves this by constant attention to the behavioural patterns or performance of each individual cow in the herd.

Other interests of Reid's 'confident introverts' included vegetable growing, but the most startling fact was that they also grew either roses, gladioli or chrysanthemums, species that have different varieties requiring specific treatment and which respond to feeding at specific times of the year. The best cow-handlers were also attuned to instant recognition of each animal in the herd and the individuality of their cows, plus a close identification with the herd. In many cases it was difficult to define whether the herd was regarded as an extension of the family or the reverse.

The behaviour of the cow

Albright (1978), Seabrook (1980) and Hemsworth et al. (2000) have all shown that animal behaviour differs among dairy herds. One factor that varies both within and between groups of cows is flight distance, or how close one can approach an individual animal without it moving away. In some dairy herds this distance may be almost zero, whereas in others it may be as great as 6 m.

A more recent study indicated that the cows' flight distance was larger on dairies where there were more negative interactions such as yelling or hitting (Hemsworth et al., 2000). Their milk production was also lower. For individual animals in these herds there will be ranges of values, but they may be lower for one herd than the lowest for another herd.

Why do these differences exist, and how do they arise? Some variation could be attributed to conditional learning, e.g. the 'memory' of being struck by a handler, but there is little evidence to account for all of the differences. Seabrook (1994) has shown that animals are effective discriminators and perceive by experience and learning. Cows made the greatest number of approaches under test conditions to the familiar person and fewest to the stranger.

Cow-handler behaviour in the milking centre showed 2.1 approaches/cow/min for higher-yielding dairy units as compared to 0.5 approaches/cow/min for lower-yielding units. Likewise, cow-handler behaviour in the milking centre talking 'with' and 'to' cows were 2.1 times/min and 9.1 words/cow/min in higher-yielding dairy units. In the lower-yielding dairy units they were 0.3 times/min and 2.1 words/cow/min, respectively. Table 8.3 summarizes responses with dairy cows using pleasant or aversive handling.

Observations of identical one-person units show behaviour differences in terms of how long it takes cows to enter the milking centre. In some herds the cows are keen to enter; in others they are reluctant to do so. Studies showed the milking centres and their identically sized and shaped collecting yards to be in excellent condition. It is the relationship between the cow-handler and the cows that seems to explain differences in entry time. It is fallacious to talk about the behaviour of dairy cows in isolation: the actual pattern is a reflection of the relationship between human and cow.

This connection was realized in the 1940s by Rex Patterson, the pioneer of large-scale dairy farming in the UK, when he publicly stated that the biggest effect on herd yield and cow behaviour on his one-person dairy units was exerted by the cow-handler (Seabrook, 1972, 1977, 1980). More recently, Seabrook and Wilkinson (2000) have noted that the attitudes and behaviour of stockpersons have been little studied, in spite of being fundamental to animal well-being and performance. Verbal encouragement tends to be lacking with some managers/employers, who may be quick with criticism. They further indicate that the veterinary profession could play an important role by giving due praise to

encourage diligent fulfilment of the most disliked work, e.g. cleaning and hoof-trimming.

Research (Munksgaard et al., 1995; de Passille et al., 1996; Rushen et al., 1999; Rousing and Waiblinger, 2004) with cows and calves shows clearly that cattle learn to discriminate between humans based on their previous experience and cues based on the colour of clothing worn, approaching them positively and avoiding those who have handled them aversively. Aversive handling can result in a generalized fear of humans, making handling more difficult and increasing the chances of injury to both animal and handler. This fear can be overcome by positive handling. Discrimination was generalized to other locations, and cattle appear to be more fearful of humans in an unfamiliar location (de Passille et al., 1996).

In order to determine if an aversion corridor could be used to evaluate various handling practices, 60 cows were randomly assigned to five different treatments: electric prod, shouting, hitting, tail-twisting and control. Cows walked down a corridor and treatments were applied at the end of the corridor. Preliminary results suggest that cows found the electric prod most aversive, followed by shouting, hitting, tail-twisting and control (Pajor et al., 1998).

In a follow-up experiment, 54 cows were randomly assigned to four treatments (hit/shout, brushing, control and food). The time and force required for cows to walk down the corridor were measured. Cows on the hit/shout treatment took more time and required more force to walk through the corridor than cows on other treatments ($P < 0.001$). In addition, brushed cows took longer to move through than cows given food ($P < 0.05$) (Pajor et al., 1999). Aversion-learning methods show promise as an

Table 8.3. General response of dairy animals under different handling treatments (from Seabrook, 1991).

Action of cow	Pleasant handling	Aversive handling
Mean entry time to milking centre (s/cow)	9.9	16.1
Flight distance (nervousness) (m)	0.5	2.5
Dunging in milking centre (times/h)	3.0	18.2
Free approaches to humans (times/min)	10.2	3.0

effective method of determining which handling procedures cows find more aversive or friendly.

The implications for animal welfare

In the herd where there is a good relationship between humans and cows, production is higher, flight zones are smaller and the cows move into the milking centre more quickly. The cows also release less adrenalin to block milk let down. They are less nervous, more settled and steady in an environment created by a confident cow-handler. From an animal welfare point of view, the pertinent point is that these are not necessarily the best-equipped herds technically, e.g. in milking centre design. Cow behaviour that indicated fear of humans was moderately ($P < 0.05$) to highly ($P < 0.01$) correlated with production and composition. By regression analysis, fear of humans accounted for 19% of variation in milk yield between farms (Breuer et al., 2000).

In other words, cows may be under stress in a well-designed system if they cannot develop a good relationship with people. Similarly, they may be in a poor system technically, but may be content and under little stress if they have confidence in and a good relationship with the person who tends to them.

Efficient dairy management and animal welfare would both be served by selecting cow-handlers who have the correct traits, and then further by training them to develop a relationship with their animals, thus ensuring that the animals are able to live in an environment where stress is reduced to a minimum. Design of a system from a welfare perspective is only part of the solution. The most important factor in determining stress is the behaviour and attitude of the cow-handler (Seabrook, 1980).

There are now in place national programmes that provide animal welfare assessments or audits of dairies. An assessor will use many tools such as guidelines, tape measure, stopwatch, a body condition score card, locomotion score card and a hygiene score card (Roenfeldt, 2005).

Conclusions

Observation of dairy cattle has been going on for centuries and helps to increase knowledge and improves husbandry techniques. A logical approach to the study of cow behaviour is now advocated, linking it to dairy herd management in commercial operations. Time saved through automation should be invested in the observation of animals. Knowledge of normal behaviour patterns provides an understanding about cattle and results in improved management and handling that will achieve and maintain higher milk yields, animal comfort and well-being. Dairy cattle must fit in well with the environment – confinement or pasture – their herdmates as well as their handlers. For those who like to work with dairy cattle, proper mental attitude of handlers must blend in with skilful management and humane care in today's highly competitive, technological, urban-based and questioning society.

References

Ahn, H.M., Arave, C.W., Bunch, T.D. and Walters, J.L. (1992) Heart rate changes during transportation of lactating dairy cows. Journal of Dairy Science 75 (Suppl. 1), 166 (abstract).

Albright, J.L. (1978) The behaviour and management of high yielding dairy cows. Proceedings of the British Oil and Cake Mills–Silcock Conference, London, 31 pp.

Albright, J.L. (1983) Putting together the facility, the worker and the cow. Proceedings of the 2nd National Dairy Housing Congress, American Society of Agricultural Engineers. St Joseph, Michigan, pp. 15–22.

Albright, J.L. (1986a) Bovine body language. In: Broom, D.M., Sambraus, H.H. and Albright, J.L. (eds) Farmed Animals. Torstar Books, New York, pp. 36–37.

Albright, J.L. (1986b) Milk from contented cows – operation of a dairy farm. In: Broom, D.M., Sambraus, H.H. and Albright, J.L. (eds) The Encyclopedia of Farmed Animals. Equinox, Oxford, UK, pp. 56–57.

Albright, J.L. (1987) Dairy animal welfare: current and needed research. Journal of Dairy Science 70, 2711–2731.

Albright, J.L. (1993) Feeding behaviour of dairy cattle. Journal of Dairy Science 76, 485–498.

Albright, J.L. (1994) Behavioural considerations – animal density, concrete flooring. In: *Proceedings of the National Reproduction Workshop*, American Association of Bovine Practitioners. Rome, Georgia, pp. 171–176.

Albright, J.L. (1995) Flooring in dairy cattle facilities. *Animal Behaviour and the Design of Livestock and Poultry Systems*. NRAES-84. Northeast Regional Agricultural Engineers Service, Cornell University, Ithaca, New York, pp. 168–182.

Albright, J.L. (2000) Why and how to read a cow or bull. *Hoard's Dairyman* 145, 751.

Albright, J.L. and Arave, C.W. (1997) *The Behaviour of Cattle*. CAB International, Wallingford, UK, 306 pp.

Albright, J.L. and Beecher, S.L. (1981) Computerized dairy herd disposal study behaviour and other traits. *Journal of Dairy Science* 64 (Suppl. 1), 91 (abstract).

Albright, J.L. and Fryman, L.R. (1964) Evolution of dairying in Illinois with special emphasis on milking systems. *Journal of Dairy Science* 47, 330–335.

Albright, J.L. and Grandin, T. (1993) Understanding dairy cattle behaviour to improve handling and production (videotape). *Journal of Dairy Science* 76 (Suppl. 1), 235 (abstract).

Albright, J.L., Dillon, W.M., Sigler, M.R. and Wisker, J.E. (1991) Dairy farm analysis and solution of stray voltage problems. *Agri-Practice* 12, 23–27.

Albright, J.L., Blauwickel, R., Olson, K.E., Bickert, W.G., Morrill, J.L. and Stull, C.L. (1999) Guidelines for dairy cattle husbandry. In: Curtis, S. (ed.) *Guide for the Care and Use of Agricultural Animals in Agricultural Research and Teaching*. Federation of Animal Science Societies, Savoy, Illinois.

Anderson, N.G. (2003) Observations on dairy cow comfort: diagonal lunging, resting, standing and perching in free stalls. *Proceedings of the 5th International Dairy Housing Conference*. American Society of Agricultural Engineers, St Joseph, Michigan, pp. 026–035.

Aneshansley, D.J., Gorewit, R.G. and Price, L.R. (1992) Cow sensitivity to electricity during milking. *Journal of Dairy Science* 75, 2733–2744.

Appleman, R.D. (1991) Effects of electrical voltage/current on farm animals: how to detect and remedy problems. *United States Department of Agriculture, Agriculture Research Service*. Agriculture Handbook 696, Superintendent of Documents, Government Printing Office, Washington, DC.

Arave, C.W., Albright, J.L. and Sinclair, C.L. (1974) Behaviour, milk yield and leucocytes of dairy cows in reduced space and isolation. *Journal of Dairy Science* 59, 1497–1501.

Arave, C.W., Bolinger, D., Shipka, M.P. and Albright, J.L. (1996a) Effect of extended lock-up of lactating cows on milk production, feed intake and behaviour. *Journal of Animal Science* 74 (Suppl. 1), 43.

Arave, C.W., Sipka, M.L., Morrow-Tesch, J. and Albright, J.L. (1996b) Changes in serum cortisol following extended lock-up time of lactating cows. *Journal of Dairy Science* 79 (Suppl. 1), 191.

Armstrong, D.V. (1992) Milking systems for large dairies. *Proceedings of the Milking Centre Design National Conference*. Northeast Regional Agricultural Engineering Service, Cooperative Extension, Cornell University, Ithaca, New York.

Armstrong, D.V. (1994) Heat stress interaction with shade and cooling. *Journal of Dairy Science* 77, 2044–2050.

Armstrong, D.V. and Quick, A.J. (1986) Time and motion to measure milking parlour performance. *Journal of Dairy Science* 69, 1169–1177.

Armstrong, D.V. and Welchert, W.T. (1994) Dairy cattle housing to reduce stress in a hot climate. In: Buchlin, R. (ed.) *Dairy Systems for the 21st Century*. *Proceedings of the 3rd International Dairy Housing Conference*. American Society of Agricultural Engineers, St Joseph, Michigan, pp. 598–604.

Armstrong, D.V., Wiersma, F., Gingg, R.G. and Ammon, D.S. (1984) Effect of manger shade on feed intake and milk production in a hot arid climate. *Journal of Dairy Science* 67 (Suppl. 1), 211 (abstract).

Armstrong, D.V., Wiersma, F., Fuhrmann, T.J. and Tappan, J.M. (1985) Effect of evaporative cooling under a corral shade on reproduction and milk production in a hot, arid climate. *Journal of Dairy Science* 68 (Suppl. 1), 167 (abstract).

Armstrong, D.V., Gamroth, M.J., Welchert, W.T. and Wiersma, F. (1989) *Parallel Milking Parlour Performance and Design*. Technical Paper PR-89-1801. American Society of Agricultural Engineers, St Joseph, Michigan.

Armstrong, D.V., Gamroth, M.J., Smith, J.F., Welchert, W.T. and Wiersma, F. (1990) *Parallel Milking Parlour Performance and Design Considerations*. Technical Paper 904042, American Society of Agricultural Engineers, St Joseph, Michigan.

Baker, F.H. (ed.) (1981) *Scientific Aspects of the Welfare of Food Animals*. Report 91, Council for Agriculture Science and Technology, Ames, Iowa.

Battaglia, R.A. (1998) *Handbook of Livestock Management*, 2nd edn. Prentice Hall, Upper Saddle River, New Jersey, 589 pp.

Bickert, W.G. (1977) *Milking Parlour Types and Mechanization*. Technical Paper 77-4552, American Society of Agricultural Engineers, St Joseph, Michigan.

Bickert, W.G. and Smith, J.F. (1998) Free stall barn design and management for cow comfort. *Proceedings of the Midwest Dairy Management Conference*. Minneapolis, Minnesota, pp. 103–118.

Bielfeldt, J.C., Badertscher, R., Tölle, K.-H. and Krieter, J. (2005) Risk factors influencing lameness and claw disorders in dairy cows. *Livestock Production Science* 95, 265–271.

Bolinger, D.J., Albright, J.L., Morrow-Tesch, J., Kenyon, S.J. and Cunningham, M.D. (1997) The effects of restraint using self-locking stanchions on dairy cows in relation to behaviour, feed intake, physiological parameters, health and milk yield. *Journal of Dairy Science* 80, 2411–2417.

Breuer, K., Hemsworth, P.H., Barnett, J.L., Mathews, L.R. and Coleman, G.J. (2000) Behavioural response to humans and the productivity of commercial dairy cows. *Applied Animal Behaviour Science* 66, 273–288.

Britt, J.H., Armstrong, J.D. and Scott, R.G. (1986) Estrous behaviour in ovarectomized Holstein cows treated repeatedly to induce estrous during lactation. *Journal of Dairy Science* 69 (Suppl. 1), 91 (abstract).

Buchlin, R.A., Turner, L.W., Beede, D.K., Bray, D.R. and Hemken, R.W. (1991) Methods to relieve heat stress for dairy cows in hot, humid climates. *Applications of Engineering in Agriculture* 7, 241–247.

Ceballos, A. and Weary, D.M. (2002) Feeding small quantities of grain in the parlour facilitates pre-milking handling of dairy cows: a note. *Applied Animal Behaviour Science* 77, 249–254.

Correa-Calderon, A., Armstrong, D.V., Ray, D.E., DeNise, S.K., Enns, R.M. and Howison, C.M. (2005) Productive and reproductive response of Holstein and Brown Swiss heat stressed dairy cows to two different cooling systems. *Journal of Animal and Veterinary Advances* 4 (6), 572–578.

Dahl, G.E. (2006) Effect of photoperiod on feed intake and animal performance. *Proceedings of the Tri-State Dairy Nutrition Conference*, pp. 33–36.

Das, K.S. and Das, N. (2004) Pre-partum udder massaging as a means for reduction of fear in primiparous cows at milking. *Applied Animal Behaviour Science* 89, 17–26.

Davidson, J.A. and Beede, D.K. (2003) A system to assess fitness of dairy cows responding to exercise training. *Journal of Dairy Science* 86, 2839–2851.

de Passille, A.M., Rushen, J., Ladewig, J. and Petherick, C. (1996) Dairy calves' discrimination of people based on previous handling. *Journal of Animal Science* 74, 969–974.

Drissler, M., Gaworski, M., Tucker, C.B. and Weary, D.M. (2005) Free-stall maintenance: effects of lying behaviour of dairy cattle. *Journal of Dairy Science* 88, 2381–2387.

Eicher, S.D. (2001) Transportation of cattle in the dairy industry: current research and future directions. *Journal of Animal Science* 84 (Suppl. E), E19–E23.

Eicher, S.D. and Dailey, J.W. (2002) Indicators of acute pain and fly avoidance behaviours in Holstein calves following tail docking. *Journal of Dairy Science* 85, 2850–2858.

Eicher, S.D., Morrow-Tesch, J.L., Albright, J.L., Dailey, J.W., Young, C.W. and Stanker, L.H. (2000) Tail-docking influences on behavioural, immunological, and endocrine responses in dairy heifers. *Journal of Dairy Science* 83, 1456–1462.

Eicher, S.D., Morrow-Tesch, J.L., Albright, J.L. and Williams, R.E. (2001) Tail-docking alters fly number, fly-avoidance behaviours and cleanliness, but not physiological measures. *Journal of Dairy Science* 84, 1822–1828.

Eicher, S.D., Johnson, T.A., Cary, D.C. and Patterson, J.A. (2004) Yeast cell wall products plus ascorbic acid: potential immune modulators for neonatal calves. *Proceedings of the Conference of Research Workers in Animal Diseases*, p. 144.

Eicher, S.D., Cheng, H.W., Sorrells, A.D. and Schutz, M.M. (2006) Short communication: behavioural and physiological indicators of sensitivity or chronic pain following tail docking. *Journal of Dairy Science* 89, 3047–3051.

Endres, M.I., DeVries, T.J., von Keyserlingk, M.A.G. and Weary, D.M. (2005) Short communication: effect of feed barrier design on the behaviour of loose-housed lactating dairy cows. *Journal of Dairy Science* 88, 2377–2380.

Fahey, A.G., Schutz, M.M., Pajor, E.A., Eicher, S.D., Larkin, S.J. and Scott, K.A. (2002) The effect of holding pen time on milk production and blood components in dairy cows. *Journal of Dairy Science* 85 (Suppl. 1), 82–83 (abstract).

Fraser, A.F. (1980) *Farm Animal Behaviour*, 2nd edn. Baillière Tindall, London, 291 pp.

Fregonesi, J.A. and Leaver, D.J. (2002) Influence of space allowance and milk yield on behaviour, performance and health of dairy cows housed in strawyard and cubicle systems. *Livestock Production Science* 78, 245–257.

Fulwider, W.K. and Palmer, R.W. (2004a) Stall usage differences of thirteen different freestall base types. *Professional Animal Scientists* 20 (6), 470–482.

Fulwider, W.K. and Palmer, R.W. (2004b) Use of impact testing to predict softness, cow preference, and hardening over time of stall bases. *Journal of Dairy Science* 87, 3080–3088.

Fulwider, W.K. and Palmer, R.W. (2005) Effect of stall design and rubber alley mats on cow behaviour in freestall barns. *Professional Animal Scientists* 21 (2), 97–106.

Fulwider, W.K., Grandin, T., Garrick, D.T., Engle, T.E., Lamb, W.D. Dalsted, N.C. and Rollins, B.E. (2007) Influence of freestall base on tarsal joint lesions and hygiene in dairy cows. *Journal of Dairy Science* (in press)

Gooding, B. (1971) Farmers combine. *New Zealand Journal of Agriculture* 123 (3), 36–38.

Gorewit, R.C., Aneshansley, J.D., Ludington, R.A., Pellerin, R.A. and Shao, X. (1989) AC voltage effects in water bowls: effects on lactating Holsteins. *Journal of Dairy Science* 72, 2184–2192.

Gorewit, R.C., Aneshansley, J.D. and Price, L.R. (1992) Effects of voltages on cows over a complete lactation. *Journal of Dairy Science* 75, 2719–2725.

Grandin, T. (1990) Calves you sell should be old enough to walk. *Hoard's Dairyman* 135, 776.

Grandin, T. (1997) Assessment of stress during handling and transport. *Journal of Animal Science* 75, 249–257.

Grandin, T. and Johnson, C. (2005) *Animals in Translation*. Scribner (Simon and Schuster), New York.

Grant, R.J. (2004) Incorporating dairy cow behaviour into management tools. In: *Proceedings of the Cornell Nutrition Conference*, October 19–21, Syracuse, New York. Cornell University Agricultural Experiment Station and Department of Animal Science, pp. 65–76.

Grant, R. (2005) All cows need 'Vitamin R'. *Hoard's Dairyman* 150, 658.

Grant, R.J., Colenbrander, V.F. and Albright, J.L. (1990) Effect of particle size in forage and rumen cannulation upon chewing activity and laterality in dairy cows. *Journal of Dairy Science* 73, 3158–3164.

Graves, R.E. (2006) Is stray voltage to blame? *Hoard's Dairyman* 151, 36.

Graves, R.E., Engle, R. and Tyson, J.T. (2006) *Design Information for Housing Special Dairy Cows*. ASABE Paper No. 064034, American Society of Agricultural and Biological Engineers, St Joseph, Michigan.

Gustafson, G.M. (1993) Effects of daily exercise on the health of tied dairy cows. *Preventative Veterinary Medicine* 17, 209–223.

Hagen, K., Lexer, D., Palme, R., Troxler, J. and Waiblinger, S. (2004) Milking of Brown Swiss and Austrian Simmental cows in a herringbone parlour or an automatic milking unit. *Applied Animal Behaviour Science* 88, 209–225.

Halachmi, I. (2004) Designing the automatic milking farm in a hot climate. *Journal of Dairy Science* 87, 764–775.

Halachmi, I., Shoshani, E., Solomon, R., Maltz, E. and Miron, J. (2006) Feeding of pellets rich in digestible neutral detergent fiber to lactating cows in an automatic milking system. *Journal of Dairy Science* 89, 3241–3249.

Hannah, H.W. (2002) Stray voltage in the dairy barn. *Journal of the American Veterinary Medical Association* 220 (9), 1296.

Hart, B.L. (1985) *The Behaviour of Domestic Animals*. W.H. Freeman, New York, 390 pp.

Hemsworth, P.H. and Coleman, G.J. (1998) *Human–Livestock Interactions: the Stockperson and the Productivity and Welfare of Intensively Farmed Animals*. CAB International, Wallingford, UK, 152 pp.

Hemsworth, P.H., Barnett, J.L., Beveridge, L. and Mathews, L.R. (1995) The welfare of extensively managed dairy cattle: a review. *Applied Animal Behaviour Science* 42, 161–182.

Hemsworth, P.H., Coleman, C.J., Barnett, J.L. and Borg, S. (2000) Relationships between human–animal interactions and productivity of commercial dairy cows. *Journal of Animal Science* 78, 2821–2831.

Hemsworth, P.H., Coleman, C.J., Barnett, J.L., Borg, S. and Dowling, S. (2002) The effects of cognitive behavioural intervention of commercial dairy cows. *Journal of Animal Science* 80, 68–78.

Hillman, P.E., Lee, C.N. and Willard, S.T. (2005) Thermoregulatory responses associated with lying and standing in heat-stressed dairy cows. *Transactions of the American Society of Agricultural Engineers* 48 (2), 795–801.

Hopster, H., Bruckmaier, J.T., Van der Werff, J.T.N., Korte, S.M., Macuhova, J., Korte-Bouws, G. and van Reenen, C.G. (2002) Stress responses during milking: comparing conventional and automatic milking in primiparous dairy cows. *Journal of Dairy Science* 85, 3206–3216.

Hughes, J. (2001) A system for assessing cow cleanliness. *In Practice* 23, 517–524.

Huzzey, J.M., DeVries, T.J., Valois, P. and von Keyserlingk, M.A.G. (2006) Stocking density and feed barrier design affect the feeding and social behaviour of dairy cattle. *Journal of Dairy Science* 89, 126–133.

Ishiwata, T., Uetake, K., Kilgour, J. and Tanaka, T. (2005) 'Looking up' behaviour in the holding area of the milking parlor: its relationship with step-kick, flight responses and productivity of commercial dairy cows. *Animal Science Journal* 76, 587–593.

Janni, K.A., Endres, M.I., Reneau, J.K. and Schoper, W.W. (2006) *Compost Barns: an Alternative Dairy Housing System in Minnesota*. ASABE Paper No. 064031, American Society of Agricultural and Biological Engineers, St Joseph, Michigan.

Jarrett, J.A. (1995) Rough concrete was making cows lame. *Hoard's Dairyman* 140, 697.

Jones, C.K., Pajor, E.A., Donkin, S.S., Marchant-Forde, J. and Schutz, M.M. (2005) Effect of recorded calf vocalizations on milk production with an automatic milking system. *Journal of Dairy Science* (Suppl. 1) (abstract).

Kiley-Worthington, M. (1976) The tail movements of ungulates, canids and felids with particular reference to their causation and function as displays. *Behaviour* 56, 69–115.

Lamb, R.C., Baker, R.O., Anderson, J.J. and Walters, J.L. (1979) Effects of forced exercise on two-year-old Holstein heifers. *Journal of Dairy Science* 62, 1791–1797.

Lefcourt, A.M. (ed.) (1991) *Effects of Electrical Voltage/Current on Farm Animals: How to Detect and Remedy Problems*. Agriculture Handbook Number 696, United States Department of Agriculture, Washington, DC.

Lensink, B.J., Raussi, S., Boivin, X., Pyykkonen, M. and Vessier, I. (2001) Reactions to handling depend on housing condition and previous experience with humans. *Applied Animal Behaviour Science* 70, 187–199.

Livestock Conservation Institute/National Institute for Animal Agriculture (1992) *Proper Handling Techniques for Non-ambulatory Animals*. Downer Task Force, Madison, Wisconsin, 19 pp.

López-Gatius, F., Santolaria, P., Mundet, I. and Yániz, J.L. (2005) Walking activity at estrus and subsequent fertility in dairy cows. *Theriogenology* 63, 1419–1429.

Matthews, L.R., Phipps, A., Verkerk, G.A., Hart, D., Crockford, J.N., Carragher, J.F. and Harcourt, R.G. (1995) *The Effects of Tail Docking and Trimming on Milker Comfort and Dairy Cattle Health, Welfare and Production*. Contract Research Report to the Ministry of Agriculture and Food, Wellington, Animal Behaviour and Welfare Research Centre, Ag Research Ruakura, Hamilton, New Zealand.

McFarland, D.F. (2003) Free stall design: cow-recommended refinements. *Proceedings of the 5th International Dairy Housing Conference*, Fort Worth, Texas. American Society of Agricultural Engineers, St Joseph, Michigan, pp. 131–138.

Melin, M., Hermans, G.G.N., Pettersson, G. and Wiktorsson, H. (2006) Cow traffic in relation to social rank and motivation of cows in an automatic milking system with control gates and an open waiting area. *Applied Animal Behaviour Science* 96, 201–214.

Metcalf, J.A., Roberts, S.J. and Sutton, J.D. (1992) Variations in blood flow to and from the bovine mammary gland measured using transit time ultrasound and dye dilution. *Research in Veterinary Science* 53, 59–63.

Midwest Plan Service (2000) *Dairy Freestall Housing and Equipment Handbook*, 7th edn. Iowa State University, Ames, Iowa, 136 pp.

Mowbray, L., Vittie, T. and Weary, D.M. (2003) Hock lesions and free-stall design: effects of stall surface. *Proceedings of the 5th International Dairy Housing Conference*, Fort Worth, Texas. American Society of Agricultural Engineers, St Joseph, Michigan, pp. 288–295.

Munksgaard, L., de Passille, A.M., Rushen, J., Thoxberg, K. and Jensen, M.B. (1995) The ability of dairy cows to distinguish between people. In: Rutter, S.M., Rushen, J., Randle, H.D. and Eddison, J.C. (eds) *Proceedings of the 29th International Congress, International Society for Applied Ethology*, Exeter, UK and Universities Federation of Animal Welfare, Potters Bar, UK, p. 14.

O'Callaghan, M.A. (1916) *Farm Implement and Machinery Review*, October (Australia).

OIE (2005) *Terrestrial Animal Health Code; Guidelines for the Land Transport of Animals*. Organization Mundial de la Santa Animals (World Organization for Animal Health, Paris), http://www.oie.int/eng/en_index.htm

Pajor, E., Rushen, J. and de Passille, A.M. (1998) Development of aversion learning techniques to evaluate dairy cow handling practices. In: *Journées de Recherche et Colloque en Productions Animales*, presented at Hôtel des Gouverneurs, Sainte-Foy, Quebec, Canada, May 21–22 (abstract).

Pajor, E., Rushen, J. and de Passille, A.M. (1999) Aversion learning techniques to evaluate dairy cow handling practices. 91st Annual Meeting of the American Society of Animal Science, Indianapolis, Indiana, July 21–24. *Journal of Animal Science* 77 (abstract).

Palmer, R.W. (2005) *Animal Handling Needs*. Thomson Delmar Learning, Clifton Park, New York, pp. 51–60.

Paranhos de Costa, M.J. and Broom, D.M. (2001) Consistency of side choice in the milking parlour by Holstein-Friesian cows and its relationship with their reactivity and milk yield. *Applied Animal Behaviour Science* 70, 177–186.

Pastell, M., Takko, H., Gröhn, H., Hautala, M., Poikalainen, V., Praks, J., Veermäe, I. and Ahokas, J. (2006) Assessing cows' welfare: weighing the cow in a milking robot. *Biosystems Engineering* 93, 81–87.

Phillips, C.J.C. and Morris, I.D. (2000) The locomotion of dairy cows on concrete floors that are dry, wet, or covered with a slurry of excreta. *Journal of Dairy Science* 83, 1767–1772.

Phillips, C.J.C. and Morris, I.D. (2001) The locomotion of dairy cows on floor surfaces with different frictional properties. *Journal of Dairy Science* 84, 623–628.

Phillips, C.J.C. and Rind, M.I. (2001) The effects on production and behaviour of mixing uniparous and multiparous cows. *Journal of Dairy Science* 84, 2424–2429.

Phillips, C.J.C., Morris, I.D., Lomas, C.A. and Lockwood, S.J. (2000) The locomotion of dairy cows in passage-ways with different light intensities. *Animal Welfare* 9, 421–431.

Plumb, C.S. (1893) *Does it Pay to Shelter Milk Cows in Winter?* Bulletin 47, Purdue University Agricultural Experiment Station, West Lafayette, Indiana.

Prelle, I., Phillips, C.J.C., Paranhos de Costa, M.J., Vandenberghe, N.C. and Broom, D.M. (2004) Are cows that consistently enter the same side of a two-sided milking parlour more fearful of novel situations or more competitive? *Applied Animal Behaviour Science* 87, 193–203.

Quaife, T. (1999) Rotary parlours are here to stay. *Dairy Herd Management* 36 (3), 44–48.

Rankin, G.W. (1925) *The Life of William Dempster Hoard.* W.D. Hoard & Sons Press, Fort Atkinson, Wisconsin.

Regula, G., Danuser, J., Spycher, B. and Cagienard, A. (2003) Improvement of health and welfare of dairy cows and fattening pigs in 'animal friendly' housing systems. Proceedings of the 54th Annual Meeting of the European Association for Animal Production, Rome, Italy. *Stocarstvo* 57 (5), 345–348 (abstract).

Regula, G., Danuser, J., Spycher, B. and Wechsler, B. (2004) Health and welfare of dairy cows in different husbandry systems in Switzerland. *Preventive Veterinary Medicine* 66, 247–264.

Reid, N.S.C. (1977) *To Endeavour to Correlate any Common Factors which Exist in Herdsmen in High Yielding Dairy Herds.* Nuffield Farming Scholarship Report, London, 21 pp.

Reinemann, D.J., Stetson, L.E. and LeMire, S.D. (2004) Comparison of dairy cow aversion to continuous and intermittent current. *Transactions of the American Society of Agricultural Engineers* 47 (4), 1257–1260.

Reinemann, D.J., Stetson, L.E., Laughlin, N.E. and LeMire, S.D. (2005) Water, feed, and milk production response of dairy cattle exposed to transient currents. *Transactions of the American Society of Agricultural Engineers* 48 (1), 385–392.

Reneau, J.K., Seykora, A.J., Heins, B.J., Endres, M.I., Farnsworth, R.J. and Bey, R.F. (2005) Association between hygiene scores and somatic cell scores in dairy cattle. *Journal of the American Veterinary Medical Association* 227, 1297–1301.

Rodenburg, J. (2004) *Housing Considerations for Robotic Milking.* ASAE Paper No. 044189, American Society of Agricultural Engineers, St Joseph, Michigan.

Roelofs, J.B., van Eerdenburg, F.J.C.M., Soede, N.M. and Kemp, B. (2005) Pedometer readings for estrous detection and as predictor for time of ovulation in dairy cattle. *Theriogenology* 64, 1690–1703.

Roenfeldt, S. (2005) What do your cows say about their welfare? *Dairy Herd Management* 42 (8), 24–28.

Roman-Ponce, H.W., Thatcher, W., Buffington, D.E., Wilcox, C.J. and Van Horn, H.H. (1977) Physiological and production responses of dairy cattle to a shade structure in a subtropical environment. *Journal of Dairy Science* 60, 424–430.

Rotz, C.A., Coiner, C.U. and Soder, K.J. (2003) Automatic milking systems, farm size, and milk production. *Journal of Dairy Science* 86, 4167–4177.

Rousing, T. and Waiblinger, S. (2004) Evaluation of on-farm methods for testing the human–animal relation-ship in dairy herds with cubicle loose system-test-retest and inter-observer reliability and consistency of familiarity to test person. *Applied Animal Behaviour Science* 85, 215–231.

Rushen, J., de Passille, A.M. and Munksgaard, L. (1999) Fear of people by cows and effects on milk yield, behaviour, and heart rate at milking. *Journal of Dairy Science* 82, 720–727.

Schreiner, D.A. and Ruegg, P.L. (2003) Relationship between udder and leg hygiene scores and subclinical mastitis. *Journal of Dairy Science* 86, 3460–3465.

Schultz, T.A., Collar, L.S. and Morrison, S.R. (1985) Corral misting effects on heat stress behavioured lactating cows. *Journal of Dairy Science* 68 (Suppl.1), 239 (abstract).

Seabrook, M.F. (1972) A study to determine the influence of the herdman's personality on milk yield. *Journal of Agricultural Labour Science* 1, 45–59.

Seabrook, M.F. (1977) Cowmanship. *Farmers Weekly*, London, 24 pp.

Seabrook, M.F. (1980) The psychological relationship between dairy cows and dairy cowmen and its implica-tions for animal welfare. *International Journal for the Study of Animal Problems* 1, 295–298.

Seabrook, M.F. (1991) The human factor – the benefits of humane and skilled stockmanship. In: Carruthers, S.P. (ed.) *Farm Animals: it Pays To Be Humane*. Centre for Agricultural Strategy, University of Reading, Reading, UK, pp. 62–70.

Seabrook, M.F. (1994) Psychological interaction between the milker and the cow. In: Bucklin, R. (ed.) *Dairy Systems for the 21st Century*. American Society of Agricultural Engineers, St Joseph, Michigan, pp. 49–58.

Seabrook, M.F. and Wilkinson, J.M. (2000) Stockpersons' attitudes to the husbandry of dairy cows. *Veterinary Record* 147, 157–160.

Smith, J.F., Armstrong, D.V., Gamroth, M.J. and Harner, J. III. (1998) Factors affecting milking parlour efficiency and operator walking distance. *Applied Engineering in Agriculture* 14, 643–647.

Sogstad, Å.M., Østerås, O. and Fjeldaas, T. (2006) Bovine claw and limb disorders related to reproductive performance and production diseases. *Journal of Dairy Science* 89, 2519–2528.

Spörndly, E. and Wredle, E. (2005) Automatic milking and grazing – effects of location of drinking water and water intake, milk yield, and cow behaviour. *Journal of Dairy Science* 88, 1711–1722.

Stephens, D.B. (1980) Stress and its measurement in domestic animals: a review of behaviour and physiological studies under field and laboratory situations. *Advances in Veterinary Science and Comparative Medicine* 24, 179.

Stull, C.L., Payne, M.A., Berry, S.L. and Hullinger, P.J. (2002) Evaluation of the scientific justification for tail-docking in dairy cattle. *Journal of the American Veterinary Medical Association* 220, 1298–1303.

Stumpf, S., Wechsler, B. and Nydeggar, F. (1999) How to make the exercise yard attractive for dairy cows. *Agrarforschung* 6 (3), 95–98.

Telezhenko, E. and Bergsten, C. (2005) Influence of floor type on the locomotion of dairy cows. *Applied Animal Behaviour Science* 93, 183–187.

Tucker, C.B. and Weary, D.M. (2004) Bedding on geotextile mattresses: how much is needed to improve cow comfort? *Journal of Dairy Science* 87, 2889–2895.

Tucker, C.B., Fraser, D. and Weary, D.M. (2001) Tail docking dairy cattle: effects on cow cleanliness and udder health. *Journal of Dairy Science* 84, 84–87.

Tucker, C.B., Weary, D.M. and Fraser, D. (2004) Free-stall dimensions: effects on preference and stall usage. *Journal of Dairy Science* 87, 1208–1216.

Tucker, C.B., Weary, D.M. and Fraser, D. (2005) Influence of neck-rail placement on free-stall preference, use, and cleanliness. *Journal of Dairy Science* 88, 2730–2737.

Tucker, C.B., Zdanowicz, G. and Weary, D.M. (2006) Brisket boards reduce freestall use. *Journal of Dairy Science* 89, 2003–2007.

Van Baale, M., Smith, J., Jamison, C., Rodriguez, R. and Baumgard, L. (2006) Evaluate the efficacy 'heat stress audits' of your cooling system through core body temperature.

van Eerdenburgh, F.J.C.M. (2006) Estrous detection in dairy cattle: how to beat the bull? *Vlaams Diergeneeskundig Tijdschrift* 75 (2A), 61–69.

Van Reenen, C.G., Van der Werff, J.T., Bruckmaier, R.M., Hopster, H., Engel, B., Noordhuizen, J.P. and Blokhuis, H.J. (2002) Individual differences in behavioural and physiological responsiveness of primiparous dairy cows to machine milking. *Journal of Dairy Science* 85, 2551–2561.

Vokey, F.J., Guard, C.L., Erb, H.N. and Galton, D.M. (2003) Observations on flooring and stall surfaces for dairy cattle housed in a free-stall barn. *Fifth International Dairy Housing Proceedings*, Fort Worth, Texas, pp. 165–170.

Weary, D.M. and Taszkun, I. (2000) Hock lesions and free-stall design. *Journal of Dairy Science* 83, 697–702.

Weary, D. and Tucker, C. (2006) What's the latest in freestall comfort? *Hoard's Dairyman* 151, 357.

Wechsler, B., Schaub, J., Friedli, K. and Hauser, R. (2000) Behaviour and leg injuries in dairy cows kept in cubicle systems with straw bedding or soft lying mats. *Applied Animal Behaviour Science* 69, 189–197.

Weiss, D., Helmreich, S., Möstl, E., Dzidic, A. and Bruckmaier, R.M. (2004) Coping capacity of dairy cows during the change from conventional to automatic milking. *Journal of Animal Science* 82, 563–570.

Whittlestone, W.G., Kilgour, R., de Langen, H. and Duirs, G. (1970) Behavioural stress and the cell count of milk. *Journal of Milk and Food Technology* 33, 217–220.

Wilson, P. (1998) An animal's first attack should be its last. *Hoard's Dairyman* 143, 787.

Wood, M.T. (1977) Social grooming in two herds of monozygotic twin dairy cows. *Animal Behaviour* 25, 635–642.

Zurbrigg, K., Kelton, D., Anderson, N. and Millman, S. (2005) Stall dimensions and the prevalence of lameness, injury, and cleanliness on 317 tie-stall dairy farms in Ontario. *Canadian Veterinary Journal* 46, 902–909.

9 Cattle Transport

Temple Grandin[1] and Carmen Gallo[2]

[1] Department of Animal Sciences, Colorado State University, Fort Collins, Colorado, USA; [2] Facultad de Ciencias Veterinarias, Instituto de Ciencia Animal y Tecnologia de Carnes, Universidad Austral de Chile, Valdivia, Chile, cgallo@uach.cl

Introduction

The biggest improvements in transport since the 2000 edition have been due to: (i) better management; and (ii) holding transporters and producers accountable for losses. The first author has observed that high-value cattle in retail branded beef programmes are now being given more space on the trucks. These high-value cattle also had very low levels of bruising because both handling and transport were carried out carefully. Unfortunately, the second author reports that in South America, many trucks are still overstocked. Overstocking is most likely to occur when transporters are paid based on the number or total weight of animals delivered instead of receiving incentive pay to minimize bruising and other losses.

Some of the worst problems that occur during cattle transport are with cull cows and other animals of low economic value. Observations of hundreds of truckloads of cattle by both authors indicate that the worst cattle welfare problems occur when sick, emaciated or debilitated animals are transported. Many of these animals are simply not fit for transport. The single most important factor for maintaining an adequate level of welfare during transport is loading physically fit, healthy animals on to the vehicle (Grandin, 2003). OIE animal welfare guidelines state that animals with the following problems are unfit for travel:

- Those that are sick, injured, weak, disabled or fatigued.
- Those unable to stand unaided and bear weight on all four legs.
- Those blind in both eyes.
- Newborn with unhealed umbilical cord.
- Pregnant animals which will be in the final 10% of their gestation period at planned time of unloading.
- Females travelling without young which have given birth within the previous 48 h.
- Those whose body condition would result in poor welfare because of expected climatic conditions (OIE, 2005).

Management of Cattle Transport

Other basic factors involved in good cattle transportation are driver training, supervision and maintenance of the vehicle. In North America, Australia and South America, the first author has observed that worn-out, slippery floors are the most common equipment deficiency. Non-slip flooring prevents slips and falls.

Good driving practices are also essential: driver fatigue is a major cause of accidents. In North America, falling asleep at the wheel was a likely cause of rollovers of double-decker cattle trucks. Eighty to ninety per cent of these rollovers did not involve another vehicle and many accidents occurred between midnight and 7.00 am, when the driver would be most tired (Jennifer Woods, 2005, personal communication).

Another cause of cattle truck rollovers is going too fast around corners. Narrow roads with soft shoulders increase the risk of rollovers. In the case of South America, gravel dirt roads are very common, and this can add an extra risk, especially if these involve many bends and inclines. In some countries, trucks with trailers are used on those winding roads for transporting cattle, commonly without separating the animals into smaller groups (18–22 in one big group, for example) (Gallo et al., 2005).

The research studies reviewed in this chapter will provide information for managers, veterinarians and government officials on ways to refine and improve cattle transport. Different countries have many varied conditions but the basic principles of prevention of losses and maintenance of adequate welfare are similar.

Rest-stop requirements

Regulations governing the transport of domestic animals vary from country to country. For example, the European Union (EU) requires that journey times shall not exceed 8 h. However, this may be extended – if the transporting vehicle meets additional requirements – to 14 h of travel. Longer trips require a rest period of at least 1 h, with water. They may then be transported for a further 14 h. Two 9 h periods with 1 h rest for watering is the maximum permitted for unweaned calves. Longer times may be permitted if animals have space to lie down, bedding and access to water and feed.

In the USA and South America there are, in general, no rest-stop requirements. In Chile there is an 8 h rest-stop required for cattle after 24 h of travel. In Canada, a rest-stop is required after 48 h of travel. In North America, Australia and South America, trips of 30–40 h with no rest stops are common. In practice, unless resting facilities are adequate, the animals are unloaded with care and have sufficient time to drink water, rest-stops may be counterproductive and serve only to prolong the overall journey time.

Stressors during handling and transport

Adult cattle were able to negotiate a wide variety of ramp designs without difficulty (Eldridge et al., 1986). Difficulties at loading of commercial transport are often caused by overloading, with the last few cattle being driven forcefully on board. Poor design and/or maintenance of ramps on farms – not usually taking into consideration animal behaviour – is a common problem in South America.

For tame cattle that are trained to lead, riding in the vehicle will probably be more stressful than walking up the loading ramp. However, for wild cattle, loading and unloading may be more stressful than riding in the vehicle because the level of fear stress may be greater. Extensively raised beef cattle will have higher heart rates and cortisol levels when they are restrained and handled compared with tame dairy cows (Lay et al., 1992a, b). Numerous studies show that transport increases cortisol levels (Eicher, 2001). In wild beef cattle, handling stresses were almost as severe as hot-iron branding stress (Lay et al., 1992a, b). Stress measurements during transport for European conditions can be found in Van Hollenben et al. (2003) and Villarroel et al. (2003).

Agnes et al. (1990) reported that loading calves up a 30° angle ramp and simulated truck noise elicited cortisol levels similar to those on simulated transport. The maximum recommended angle for a cattle-loading ramp is 20–25°. For adult cattle, cleats on the ramp should have 20 cm of space between them. On concrete ramps, stair-steps are recommended, with a 10 cm rise and 30–45 cm tread length (Grandin, 1990).

Eldridge et al. (1988) concluded that, once cattle adapted to the journey, road transport was not a major physical or psychological stressor. This is in agreement with James (1997), that cattle remained calm in an aeroplane during flight. These authors recorded heart rates of beef heifers by radiotelemetry during road transport at different loading densities. The overall mean heart rates while travelling were only 15% above those recorded while animals were grazing at pasture. Similar results were obtained for bulls and steers undergoing short-haul road transport (Tennessen et al., 1984).

Honkavaara et al. (2003) also reported that heart rate was lower during longer trips compared to shorter trips. This is an indicator that the animals had time to calm down after loading. Heart rate will increase during loading and unloading (Kenny and Tarrant, 1987a; Jacobson and Cook, 1998). Stress during transport consists of both psychological stressors such as fear and physical stresses such as vibration. Fear is a strong stressor and it is reviewed in LeDoux (1998) and Grandin (1997). Both an animal's previous experiences and genetic influences on temperament will interact in complex ways to determine the relative contribution of physiological and physical stressors during transport (Grandin, 1997; Jacobson and Cook, 1998).

Transport and handling stresses affect many physiological measures (see Chapter 19, this volume). Transport may have a negative effect on both immune function and fertility. Eicher (2001) reviewed eight studies on young calves that showed transport had significant effects on the young animals' immune system. Both practical experience and scientific research indicate that young calves may be more adversely affected than older, healthy animals that are in good condition.

Restlessness

This is indicated by the frequency of changes in position in the vehicle. Restlessness increased with social regrouping on the truck, but not with motion (Kenny and Tarrant, 1987a, b). Changes in position were frequently triggered by social interactions, such as chin-resting and mounting, and also – when the truck was moving – by driving events, particularly cornering.

Standing orientation and lying down

The most common direction for cattle to face on a truck is either perpendicular or parallel to the direction of motion, with the diagonal orientations infrequently used (Kenny and Tarrant, 1987a, b; Eldridge et al., 1988; Lambooy and Hulsegge, 1988; Tarrant et al., 1988, 1992; Gallo et al., 2000). As observed by the second author, this is also the case when cattle are transported for long journeys (24 h) on roll-on, roll-off ferries (Aguayo and Gallo, 2005; Table 9.1). This may indicate that cattle have a preferred orientation to improve security of balance on a moving vehicle. Bisschop (1961) found that cattle align themselves across the direction of travel during rail transport; however, Kilgour and Mullord (1973) found no clear preference by young beef cattle during road travel.

On long journeys, the most common standing orientation was perpendicular to the direction of travel, and there was a strong bias against diagonal orientations (Tarrant et al., 1992). Cattle tend not to lie down in trucks while they are moving (Warriss et al., 1995). In 1-h and 4-h journeys to slaughter, no animals lay down in 18 loads of Friesian steers or bulls transported at relatively high stocking densities (Kenny and Tarrant, 1987a, b; Tarrant et al., 1988).

Reports on behaviour during rail transportation noted heightened activity immediately after loading, and characteristic standing and lying behaviour in moving and stationary vehicles (Bisschop, 1961). Observations by the second author indicate that during long journeys (up to 36 h), adult cattle start to lie or fall down after

Table 9.1. Orientation (position) taken by cattle (%) within trucks travelling at a density of 450 kg/m² during a 24-h maritime ferry crossing at first observation (O1, after 7 h), second observation (O2, after 14 h) and third observation (O3, after 21 h) (from Aguayo and Gallo, 2005).

	Fat cattle for slaughter (n, 1240)			Calves and cattle for fattening (n, 652)		
	O1	O2	O3	O1	O2	O3
Parallel	39	35	36	49.7	47.9	49.4
Perpendicular	40	43	47	23.9	25.9	23.3
Diagonal	21	21	17	26.4	26.2	27.3
Lying down	3	4	6	19.6	14.3	26.5
Fallen	0.2	0.2	0.4	0.2	0.2	0.6
Dead	0	0	0	0	0	0

12 h of transport (Gallo *et al.*, 2000, 2001); on long road-plus-ferry transport (up to 24 h maritime roll-on, roll-off crossing), the proportion of lying adult cattle increased with time, whereas young cattle (6–12 month) laid down earlier (Aguayo and Gallo, 2005; Table 9.1). Experiments in the USA with railcars equipped with hay and water troughs indicated that the animals would eat and drink during transport provided they had sufficient space to move to the feeders (Irwin and Gentleman, 1978).

At high stocking density, especially approaching maximum density, cattle occasionally went down, apparently involuntarily. Towards the end of long (24 h) road journeys with Friesian steers, several cattle lay down during the final 4–8 h of the journey. This was observed at all stocking densities, but only at high stocking density were animals trapped down and unable to rise. The highest stocking density is the maximum amount of cattle that can be put on a truck when it is still easy to close the rear gate. Honkavaara (1998) noted that even one restless animal was sufficient to cause continuous movements of the group; as a result, no animal could lie down during transport.

However, when cattle were transported in stock crates designed to hold two animals per pen, an animal often lay down after 2–3 h of transport (Honkavaara, 1993). This may indicate a preference for lying down when circumstances permit. Young 5–10-day-old calves will lie down during transport. Providing sufficient space to allow calves to lie down greatly reduced physiological measures of stress (Todd *et al.*, 2000). For long trips across Europe, about 20% of adult cattle lay down while the vehicle was moving (Marahrens *et al.*, 2003). These animals had been provided with enough space to lie down.

Maintenance of balance

Loss of balance on moving vehicles is a major consideration in cattle transport, in view of the hazard associated with large animals going down during transport and the risk of injury or suffocation. Observations show that minor losses of balance occur regularly and that cattle quickly respond by shifting their footing to regain their balance.

The relationship between loss of balance and driving events during 24-h journeys with Friesian steers is shown in Table 9.2. Eighty per cent of losses of balance were accounted for by braking, gear changes and cornering. Similar results were obtained for Friesian bulls and steers on shorter road journeys (Kenny and Tarrant, 1987a, b; Tarrant *et al.*, 1988).

Table 9.2 shows that cornering is the driving event that caused most losses of balance in cattle transported at high stocking density, whereas braking was a greater hazard at lower densities.

These data show that losses of balance are under the direct control of the driver. Eldridge (1988) observed that the heart rate of beef heifers was lower, indicating reduced stress, when the vehicle was traveling smoothly on highways, compared with that on rougher country roads or suburban roads with frequent intersections.

Factors likely to influence security of balance during unsteady driving are the slipperiness of the floor surface and the availability of support from adjacent structures, including vehicle sides and partitions – and other animals. It may be advisable to withhold water during the last 6 h before loading (Wythes, 1985), thus resulting in a drier truck floor, giving cattle a better footing during the journey.

Table 9.2. The association between loss of balance on a moving vehicle and driving events during 24-h journeys with 618 kg Friesian steers at three different stocking densities. Data are percentages of total losses of balance (from Tarrant *et al.*, 1992 and Tarrant and Grandin, 1993).

Driving event	Stocking density (kg/m^2)		
	448	500	585
Braking	55	58	19
Gear changing	21	17	19
Starting/ stopping	9	15	0
Cornering	5	6	50
Bumping	2	2	0
All other events	1	1	0
Uneventful	6	2	12

The major factors determining the well-being of cattle in road transport are vehicle design, stocking density, ventilation, the standard of driving and the quality of the roads. The importance of frequent inspection of the livestock and of careful driving cannot be overemphasized. In the USA, the large beef plants are collecting more and more data on the performance of trucking firms and drivers. Data collected at one large slaughter plant indicated that one trucking firm had more dark-cutters. Other unpublished industry data, from South America, show that poor driving practices such as rapid acceleration and braking increase levels of bruising, and some truck drivers are required to pay fines for bruising injuries.

Falls

The major risk in cattle transport is that of cattle going down under foot. This risk is greatly increased at the highest stocking density (Tarrant et al., 1988, 1992), especially during journeys exceeding 12 h (Gallo et al., 2000, 2001; Valdés, 2002). The normal response to a loss of balance is a change or shift of footing in order to regain balance. Shifting was inhibited at high stocking density (see Table 9.3) and there was a corresponding increase in struggling and falling at 585 kg/m^2 stocking density.

These unstable situations were caused either by driving events – typically cornering or braking – by standing on a fallen animal or by strenuous and usually unsuccessful attempts to change position in a full pen. When cattle went down at high stocking density, they were trapped on the floor by the remaining cattle 'closing over' and occupying the available standing space. Several unsuccessful attempts by fallen animals to stand up were observed. A 'domino' effect was created when standing animals lost their footing by trampling on a fallen animal. The substantial increase found in carcass bruising at the highest stocking density was explained by these observations (see Table 9.4).

Stocking density on trucks

Freedom of movement was severely restricted at 585 kg/m^2, with only 16 changes of position observed per group of cattle/1 h of transport, compared with 109 per group/1 h at 448 kg/m^2 (Tarrant et al., 1988). Similar results were obtained on long road journeys (Tarrant et al., 1992). Exploratory, sexual, aggressive behaviours were inhibited at high stocking densities, with the exception of mounting and pushing, which increased in frequency with stocking density.

Table 9.3. The effects of stocking density in the truck on loss of balance by 618 kg Friesian steers during 24-h journeys by road (from Tarrant et al., 1992 and Tarrant and Grandin, 1993).

Loss of balance	Stocking density (kg/m^2)		
	448	500	585
Shifts	153	142	26
Struggles	5	4	10
Falls	1	1	8

Table 9.4. The effect of stocking density during 24-h road journeys on plasma constituents and carcass bruising in 618 kg Friesian steers. Values for plasma cortisol, glucose and CK are the difference between the pre- and post-transport values (from Tarrant et al., 1992 and Tarrant and Grandin, 1993).

Plasma constituent	Stocking density (kg/m^2)			Level of statistical significance (P)
	448	500	585	
Plasma cortisol (ng/ml)	0.1	0.5	1.1	< 0.05
Plasma glucose (mmol/l)	0.81	0.93	1.12	< 0.15
Plasma CK (units/l)	132	234	367	< 0.001
Carcass bruise score	3.7	5.0	8.5	< 0.01

The preferred orientations adopted by animals during long-distance transportation were frustrated as the stocking density was increased. Thus, in addition to reducing mobility, an increase in the stocking density also prevented cattle from facing in the preferred direction. These effects may combine to increase the rate of loss of balance and falling, as discussed above.

In 4- and 24-h road journeys to the abattoir, the cortisol and glucose content in the plasma of Friesian steers increased with stocking density, indicating increasing stress (Tarrant et al., 1988, 1992). The activity of the muscle enzyme creatine kinase (CK) in the bloodstream also increased with stocking density, reflecting muscle damage. Carcass bruising increased with stocking density (see Table 9.4; Fig. 9.1).

Fig. 9.1.   Cattle loaded on to a truck at 377 kg/m^2 (A) and 516 kg/m^2 (B). Cattle loaded at the high density are more likely to go down during transport and have higher levels of bruising (see also Table 9.6) (Photograph credit Toby Knowles).

Dressed carcass weight was significantly reduced at high loading density (Eldridge and Winfield, 1988); this weight loss was only partly explained by the higher trimming of bruised tissue from the carcass at the highest density.

High stocking density on trucks was clearly associated with reduced welfare and carcass quality when compared with medium and low stocking densities. Attempts to reduce transport costs by overloading of trucks are offset by reduced carcass weight, downgrading of carcasses owing to bruising and increased risk of serious injury or death during travel (Eldridge and Winfield, 1988). Research in South America has shown that overloading of trucks causes more bruises on long 12–48-h trips compared to short 3-h trips.

Unfortunately, cattle transporters are more likely to overload trucks on long trips so they can make more money. Both authors have observed that on overloaded trucks there is a higher incidence of severely bruised cattle. At stocking densities of 400–500 kg/m^2, 9% of the cattle fell down during a 16-h journey, mainly at the higher density, and 0% fell after a 3-h journey (Valdés, 2002). At 585 kg/m^2, there were also more bruises and these were more severe (affecting muscle tissue, not only subcutaneous fat) and of greater extension (diameter), causing severe carcass damage and hence devaluation.

At low stocking densities, e.g. half-loads, cattle will travel very well unless subjected to poor driving techniques, such as sudden braking or swerving and emergency stops. Eldridge and Winfield (1988) transported steers of 400 kg live weight on road journeys of 6 h at high (0.89 m^2/animal), medium (1.16 m^2/animal) and low (1.39 m^2/animal) stocking densities.

Bruise scores at the high and low stocking densities were four and two times greater, respectively, than at the medium space allowance (8.2, 4.6 and 1.9 bruise scores, respectively; $P < 0.01$). These results show that the medium stocking density was superior to the low density and indicate that an optimum density may be defined by such experiments. The medium stocking densities used by Eldridge and Winfield (1988) can be found in Grandin (1981a), Animal Transportation Association (1992), Federation of Animal Science Societies (1999) and Lapworth (2004). These medium stocking densities used by Eldridge and Winfield

(1988) are very similar to a stocking density formula published by Randall (1993).

Randall recommends the use of this equation: $A = 0.01W^{0.78}$ (Randall, 1993). A = the area of the space (m^2) and W is the weight of the animal (kg). Randall's equation should be used for trips of < 5 h. Knowles (1999) contains an excellent diagram that compares stocking densities recommended by the Farm Animal Welfare Council (FAWC) and the European Union with those of Randall. The Farm Animal Welfare Council recommends 360 kg/m^2 as the maximum stocking density for cattle. Their formula is $A = 0.021W^{-0.67}$. Knowles (1999) concludes that welfare can be poor if stocking density is either too high or too low.

Carcass Bruising

Considerable financial losses are incurred by the livestock industry as a result of carcass bruising (Hails, 1978; Grandin, 1980; Wythes and Shorthose, 1984; Eldridge and Winfield, 1988). In feedlot beef, 35% of the carcasses were bruised (Smith et al., 2006). Bruising is an impact injury that can occur at any stage in the transportation chain and may be attributed to poor design of handling facilities, ignorant and abusive stockmanship or poor road driving techniques during transportation (Grandin, 1983a).

Cattle should be marketed in a manner that minimizes the number of times that they are handled or restrained immediately prior to slaughter, particularly when they are transported more than 325 km to slaughter (Hoffman et al., 1998). Cattle that were handled roughly had greatly elevated bruising compared with cattle that were handled gently (Grandin, 1981b).

The skill of the driver and the quality of the road appear to be more important than the distance travelled. Economic incentives can greatly reduce bruising. Cattle sold by live weight had twice as many bruises compared with cattle sold on a carcass basis (Grandin, 1981b). Producers selling on a carcass basis have bruising damage deducted from their payments. According to observations by the second author, efforts are being made in various countries by producing regulations, written information and graphic material to educate and train producers and animal handlers in order to improve animal welfare – and thereby meat quality – through better handling (Gallo, 2004).

Stocking density is an important consideration, and high stocking density was associated with a twofold or greater increase in carcass bruising in both short-haul (Eldridge and Winfield, 1988; Tarrant et al., 1988) and long-haul road transport (Tarrant et al., 1992) (see also Table 9.4). Barnett et al. (1984) considered that cattle with elevated blood corticosteroid concentrations as a result of chronic stress could be more susceptible to bruising damage than other cattle.

Shaw et al. (1976) and Wythes (1985) found that horned cattle had twice as much bruising. Contrary to popular belief, cutting the tips of the horns does not reduce bruising (Ramsey et al., 1976). Horn buds on calves should be removed before the animal grows horns. Cutting horns on older animals is extremely stressful and painful. Cows in late pregnancy suffered more bruising and produced tougher meat than those in early pregnancy or those that were not pregnant (Wythes, 1985).

The Dutch Road Transport Act prescribed that adult cows and heifers should be separated by a gate between every two animals when transportation was to last longer than 10 h (Lambooy and Hulsegge, 1988). However, transporters do not adhere to this rule and carry five to ten cattle per compartment. Experimentation showed that loose transport of eight heifers per pen is preferential to penning in pairs between gates, mainly because of lower risks of injury and lower frequency of lesions at contact points, e.g. hips and knees. Unfortunately, the second author has observed that separating cattle into different compartments within a truck load is not a common practice in most South American countries, because it reduces the space availability for carrying more animals, a situation which is detrimental for animal welfare and also meat quality.

Vehicle Design and Maintenance

Information on the design of loading/unloading ramps and stock crates for single- and double-deck trucks and trailers is available (Anon., 1977; Wythes, 1985; Grandin, 1991). In most developed countries, adequate loading ramps are available. However, in some developing countries, cattle are forced to jump off the truck because no ramps are available. Ramps should be used. They can be easily built from locally available materials.

Practical experience in the USA has indicated that there were fewer bruises in trailers that had doors that opened up to the full width of the trailer for unloading. Stock carried by rigid vehicles tend to experience a rougher ride than stock transported by an articulated trailer. This is mainly because rigid-body trucks, which are smaller and easier to handle, are generally driven faster than large articulated vehicles (Anon., 1977; Fig. 9.1).

Vibration stress can also be reduced by the installation of pneumatic suspension (Singh, 1991). These systems must be kept in good repair, because a damaged pneumatic suspension may produce higher vibration than a vehicle with leaf springs (Singh, 1991). Practical experience has shown that a well-maintained pneumatic suspension system will reduce stress. In the USA, a high percentage of livestock vehicles come with factory-installed pneumatic suspension.

Van de Water et al. (2003a, b) reported that calves subjected to vibrations at a frequency of 2 hz become more stressed compared with vibrations of 12 hz. These same researchers also found that calves riding in the front of the truck had higher cortisol levels (Van de Water et al., 2003a, b). Differences in vibration in the front and back of the vehicle may explain differences in cortisol levels. Over-inflated tyres will also increase vibration in a livestock truck (Stevens and Camp, 1979). Drivers over-inflate tyres to prolong the tyre life, but this practice is probably detrimental to livestock. Cattle in the USA are hauled in aluminum trailers; a lack of bedding on the aluminum floor can cause abscessed toes (Sick et al., 1982).

Ventilation

Heat builds up rapidly in a stationary vehicle. Vehicles should be kept moving and stopping should be kept to a minimum. Even during cool weather, a beef animal's temperature will rise when the vehicle is stationary (Stevens et al., 1979). In aeroplanes and ships, heat can rapidly build up to fatal levels when the vehicle is stationary (Stevens and Hahn, 1977; Grandin, 1983b). Heat stroke was the main cause of cattle death on sea transport (Norris *et al.*, 2003).

Ventilation recommendations for ships and aircraft are given in Muller (1985), Stevens (1985) and Animal Transportation Association (1992). Muirhead (1985) found that there were areas of no air movement in moving trucks. Natural ventilation in trucks through openings in the side walls results in non-uniform air circulation at animal head level in practice (Honkavaara, 1998).

Further research in Sweden indicated that during the summer, the temperature inside a truck was 6°C higher than the outside temperature (Wikner *et al.*, 2003). Honkavaara (2003–2004) stated that new designs of transport vehicles in Europe made it possible to transport cattle 8–14 h with little effect on stress levels or welfare. In the USA, trailers have adequate ventilation because the aluminum side of the trailer has numerous small holes. During cold weather, one-third to one-half of the holes are covered with plastic panels. Practical experience in very cold areas in Canada indicates that the front and rear of the trailer should be covered first because the middle of the trailer stays warmer (J. Woods, 2006, personal communication).

When cattle become wet, their ability to withstand cold greatly decreases. Wet cattle in a truck must be protected from wind chill during cold weather. A sleet storm with freezing rain can greatly increase death losses (Grandin, 1981a). Abrupt changes in the outside temperature during transport may be more detrimental than constant exposure to either high or low temperatures (Randall, 1993). In Europe, enclosed trucks with mechanical ventilation are sometimes used. Kettlewell *et al.* (2001) contains further information on this topic.

In many countries, mechanical truck ventilation would not be practical. Failure to maintain the mechanical system in a fully enclosed vehicle could result in animal death. In South America and Australia, cattle transport trucks are usually open (not roofed or enclosed) and ventilation is not required. Vehicles containing livestock that are carried in the lower, closed decks of ferries – where heat and humidity can build up quickly – must have good ventilation provided by the ship.

Meat Quality

Good-quality beef has a final pH value close to 5.5. At pH values of 5.8 and above, both the tenderness and the keeping quality of the fresh chilled meat are adversely affected. High-pH meat is unsuitable for the premium trade in vacuum-packed fresh meats and, depending on the commercial use of the product, dark-cutting meat may be discounted by 10% or more (Tarrant, 1981). In feedlot cattle in the USA, 1.9% of the steers and heifers were downgraded due to dark-cutting (Smith *et al.*, 2006). In South America, levels of dark-cutting beef in pasture-fed cattle can be as high as 5–10% (Amtmann *et al.*, 2006).

High pH in meat is caused by an abnormally low concentration of lactic acid in the meat, which in turn is a reflection of low muscle glycogen content at slaughter. Post-mortem production of lactic acid requires an adequate content of glycogen in the muscles at slaughter. Ante-mortem glycogen breakdown is triggered by increased adrenaline release in stressful situations or by strenuous muscle activity. Circumstances that trigger one or both of these glycogen breakdown mechanisms will deplete muscle glycogen, especially in the fast-twitch fibres (Lacourt and Tarrant, 1985; Shackleford *et al.*, 1994), and will result in high-pH, dark-cutting meat unless a recovery period from stress is allowed.

In practice, in many abattoirs, restful conditions with access to feed cannot be provided. Experiments in Chile have shown that there is no beneficial effect on the welfare of the animals by a long lairage time at the abattoir (Tadich *et al.*, 2005) and that increasing transport journey times (from 3 to 24 h) and lairage times (from 3 to 24 h) also increases the incidence of high-pH and dark-cutting carcasses (Gallo *et al.*, 2003; Amtmann *et al.*, 2006).

These results are in accordance with the reductions in muscle glycogen found in the same animals, as they are deprived of feed during transport and lairage. Furthermore, the rate of post-stress muscle glycogen repletion is slower in cattle than in other species (McVeigh and Tarrant, 1982), so it is better to avoid the problem than to attempt to remedy it. On average, bulls may have more dark-cutters than similar steers (Tennessen *et al.*, 1984).

The animal behaviour most closely associated with glycogen depletion and dark-cutting beef is mounting activity. Both fighting and mounting are stimulated by social regrouping and the mixing of young bulls (McVeigh and Tarrant, 1983; Warriss *et al.*, 1984; Tennessen *et al.*, 1985; Tarrant, 1988) and by heat (oestrus) in groups of females (Kenny and Tarrant, 1988). Modifications of the holding pens aimed at reducing mounting activity during penning before slaughter have been successful in preventing dark-cutting in bulls (Kenny and Tarrant, 1987c).

Social regrouping prior to transport causes a much higher incidence of dark-cutters in bulls compared with steers (Price and Tennessen, 1981; Tennessen *et al.*, 1981, 1985). Short periods of mixing greatly increase the levels of dark-cutting in bulls, but dark-cutting will increase in steers if they are mixed for more than 24 h (Grandin, 1979). Scanga *et al.* (1998) found that dark-cutting increased if there were sharp temperature fluctuations or temperature extremes 24–72 h prior to slaughter. Practical experience in large slaughter plants has also shown that feedlot cattle spending the night in the plant lairage had more dark-cutters. One of the other factors that can greatly influence the occurrence of dark-cutting beef in fed cattle is excessive use of growth promoters (Scanga *et al.*, 1998).

Short road journeys are not likely to cause dark-cutting (Eldridge and Winfield, 1988), except where trauma occurs, for example, when an animal goes down (Tarrant *et al.*, 1992). Warnock *et al.* (1978) also found much higher meat pH values in the carcasses of 'downer' cows compared with those in cows that did not go down (6.3 versus 5.7).

Long-distance road or rail transport of cattle caused a small elevation of meat pH and a corresponding increase in the incidence of dark-cutters (Wythes *et al.*, 1981; Tarrant *et al.*, 1992; Honkavaara, 1995; Gallo *et al.*, 2003). This was reversed by resting and feeding for 2 days or longer before slaughter (Shorthose *et al.*, 1972; Wythes *et al.*, 1980). The first author has observed that steers that engaged in intense fighting required up to a week on feed to recover and to have good meat quality.

Other effects of transport on meat quality include an increase in toughness (Schaefer *et al.*, 1990) and a decrease in palatability (Jeremiah *et al.*, 1992; Schaefer and Jeremiah, 1992). The sensory quality of veal was lower after long-distance transport of 20-week-old calves (Fernandez *et al.*, 1996).

Effect of Transport on Live Weight Loss

The loss of live weight and carcass yield during transportation of cattle is of both welfare and economic concern. Animals lose live weight as a consequence of excretion, evaporation and respiratory exchange (Dantzer, 1982). In cattle hauled an average of 1023 km, almost one-half of the shrinkage was due to loss of carcass weight (Self and Gay, 1972). Most of the live weight losses during transportation may be attributed to the effect of withdrawal of feed and water; the gut contents can account for 12–25% of the animal's live weight.

Fasting of 396 kg steers for 12, 24, 48 and 96 h caused live weight losses of 6, 8, 12 and 14%, respectively (Wythes, 1982). Similarly, in the USA, slaughter-weight cattle, transported for 5 and 26 h, lost 2 and 6.3% of their body weight, respectively (Mayes *et al.*, 1979). In 24-h journeys by road, the live weight losses in cattle were about 8% (Shorthose, 1965; Lambooy and Hulsegge, 1988; Tarrant *et al.*, 1992). Recovery of body weight to pre-transport values took 5 days (Warriss *et al.*, 1995).

Similar shrink losses have been observed in South America (Gallo *et al.*, 2000, 2001). In road transport of 36 h to 46 h, including roll-on, roll-off ferry crossings in Chile, the second author found weight losses of 10%. Gallo *et al.* (2003) transported cattle either 3 or 16 h and observed that carcass weights tended to be lower after the longer trips.

In 24-h road journeys under cool ambient conditions (4–16°C), there was evidence of dehydration, as shown by increases in red blood cell count, haemoglobin, total protein and packed-cell volume (Tarrant *et al.*, 1992; Warriss *et al.*, 1995). Similar results were reported in Chile (Tadich *et al.*, 2000). Moreover, Gallo *et al.* (2000) and Valdés (2002) measured the weight changes of steers during lairage, after being transported for 3, 6, 12 and 24 h, and there were increases in live weight in the steers submitted to the two longer journeys (as opposed to weight losses continuing in the steers arrived after shorter transport journeys); increases in live weight were attributed to the fact that animals transported for the longer journeys arrived thirsty and drank water during lairage.

Dehydration is, therefore, a factor in loss of live weight and also carcass weight during transportation. Lambooy and Hulsegge (1988) found slightly increased haematocrit and haemoglobin values in pregnant heifers transported by road for 24 h. The heifers had access to water and feed after 18 h of transport, and water uptake per animal ranged from 1–6 l. Shorthose (1965) calculated that the approximate rate of carcass weight loss in steers was 0.75%/day for transport and holding times lasting from 3–8 days. Providing water *ad libitum* to fasted livestock reduces shrink (Hahn *et al.*, 1978).

The effect of giving cattle access to water after a long journey in hot weather (25–36°C) was examined by Wythes (1982). Access to water for 3.5 h or longer before slaughter allowed muscle water content to increase, and this was reflected in heavier carcasses. Providing cattle with an oral electrolyte in their drinking water reduced both carcass shrink and dark-cutting (Schaefer *et al.*, 1997).

In a major study of 4685 calves and yearlings, animals subjected to the increased stress of moving through a market had greater shrink than animals purchased directly from the ranch of origin (Self and Gay, 1972). Collectively, the physiological changes observed in cattle during transport and handling – which include changes in blood cells, blood metabolites and enzymes, electrolyte balance, dehydration and increased heart rate – suggest that treatments designed to attenuate stress should be considered as a means of protecting animal welfare and benefiting carcass quality and yield (Schaefer *et al.*, 1990, 1997). The application of oral electrolyte therapy, especially if similar in constituents to interstitial fluid, seems to attenuate these physiological changes and results in less carcass shrink and reduced dark-cutting. See Chapter 19 on physiology.

South American Transport Studies

The effect of transportation on weight loss, meat quality and bruising is similar to that found in studies done in other countries. The cattle in these studies were grass-fed steers and heifers. All the animals in the studies in Chile were *Bos taurus*, mainly Friesian steers.

Compared to shorter journeys (3 and 6 h), the longer journeys (12, 16 and 24 h) were associated with higher reductions in live weight, increased bruising, higher final muscle pH and an increase in the proportion of carcasses downgraded because they were classified as dark-cutting. The space allowance in these studies was 400 kg/m² and 500 kg/m². Figure 9.1 shows cattle close to these two different stocking densities. The carcass weights also tended to be lower after the longer journeys and longer periods in lairage (Gallo *et al.*, 2000, 2001, 2003).

Regarding the effects of transport on the blood concentrations of cortisol, glucose and CK activity after transport at arrival at the slaughterhouse, it was seen that non-stop journeys lasting 24 and 36 h were detrimental to the animals' welfare (Tadich *et al.*, 2000); also, due to fatigue, animals started falling down after 12 h of transit, increasing bruising and compromising animal welfare. An 8-h rest stop with hay and water during a 36-h trip reduced severe bruising and prevented cattle from falling (see Table 9.5); the rest stop started 24 h into the journey (Gallo *et al.*, 2001).

Regarding the effects of lairage time, which is usually > 12 h in Chilean abattoirs, it was concluded that there was no beneficial effect on the welfare of cattle of these long lairage times (Gallo *et al*, 2003; Tadich *et al.*, 2005). Shorter lairage times will reduce the level of dark-cutters. The use of roofed compared with non-roofed lairage pens in rainy conditions did not affect carcass quality with 12 h lairage, but the best

Table 9.5. Effect of an 8-h rest-stop with hay and water during 36 h of truck transport on bruising and dark-cutters ($n = 20$ per group).

	Rest-stop (n)	Continuous 36-h trip (n)
Carcasses with bruises penetrating muscle	0	6
Carcasses with pH > 6.0	3	5
Fallen cattle in the truck upon arrival	0	3

carcass quality – in terms of pH and colour – was obtained when the steers were slaughtered within 2 h of arrival at the slaughterhouse (Novoa, 2003). Lairage times of 12 and 24 h significantly reduced muscle glycogen reserves in steers, increasing the risk of dark-cutting problems (Amtmann et al., 2006).

In Chile, long transport journeys (600–1000 km) without water and feed are common for cattle sent to slaughter. Surveys have shown that most cattle arriving at the slaughterhouse had a mean space allowance of 455 kg/m^2. Thirty-two per cent of the 413 loads surveyed, comprising 35.6% of the 12,931 animals transported, were carried at an estimated stocking density of 500 kg/m^2 or higher (Gallo et al., 2005).

An experimental comparison of space allowances of 400 kg/m^2 versus 500 kg/m^2 resulted in the higher density producing higher bruising scores and greater stress according to blood concentrations of cortisol and glucose after a 16-h journey. No differences were found after short journeys (3 h) (Mencarini, 2002; Valdés, 2002; Tadich et al., 2003a, b; Table 9.6).

Gallo et al. (2001) found that a rest stop during a 36-h trip reduced bruises and prevented fatigued cattle from falling down. Three cattle out of 20 with no rest fell, while none of the rested cattle fell; the truck was stocked at 500 kg/m^2. In the lighter stocking densities used by Tarrant et al. (1992), falls occurred during a 24-h non-stop trip. Both studies indicate that fatigue may be a major contributor to cattle going down during transport.

Studies in Uruguay – mainly on producing horned Hereford cattle – showed that the most common distances travelled by cattle to slaughterhouses were much shorter than in Chile (231–266 km); nevertheless, 68.8% of the cattle slaughtered between 2002 and 2003 presented bruising and there was a significant positive association of bruises with distance travelled, state of the roads and characteristics of the vehicles (Castro and Robaina, 2003; Huertas and Gil, 2003a, b). The authors indicated that the weight of condemned damaged tissue varied between 300 g and 2 kg, a situation that produces high economic losses to the Uruguayan meat industry.

In Brazil, 16,104 cattle were surveyed at one slaughter plant (R.M. Renner, 2004, personal communication); it was found that 51% of all carcasses produced had bruising, cows being the most affected, and the most common site of lesion was the leg, a region with high-value cuts. Nine per cent of the bruises observed were severe, requiring the condemnation of between 800 g and up to 55 kg of carcass weight. In this survey, 50% of the cattle had travelled only up to 100 km, and the rest up to 600 km; no clear relationship between transport distance and presence of bruises was found.

The studies in different regions of South America show that, although the risk of affecting welfare and damaging the carcass increases with distance travelled by cattle, there is no straightforward relationship, and factors such as breed, gender, age, horns, vehicle characteristics and maintenance – as well as driving skills,

Table 9.6. Effect of varying transport journey times and stocking densities (kg/m^2) on bruising in steers (from Valdés, 2002).

| Stocking density | Journey (3 h) | | | | Journey (16 h) | | | |
| | 400 | | 500 | | 400 | | 500 | |
	n	%	n	%	n	%	n	%
Total carcasses	28	100	32	100	28	100	32	100
Bruised carcasses	10	35.7	11	34.3	12	42.8	18	56.2
Grade 1 (sub-cutaneous tissue compromise only)	8	28.5	10	31.3	11	39.2	14	43.8
Grade 2 (muscle tissue compromise)	2	7.1	1	3.1	1	3.5	4	12.5

road conditions and animal handling during loading and unloading – play an important role.

Either due to the economic impact of transport in the meat industry or the increasing demands of consumers for the consideration of animal welfare in the meat chain, most countries in South America, especially those exporting meat, are implementing measures to improve live animal transport and ethical quality of the products. An important step forward has been awareness of the problem at governmental and industry level, through research results and training programmes for people handling the animals (R.M. Renner, 2004, (Auditor de Bem Estar Animal, Brazil), personal communication).

Bovine Respiratory Disease and 'Shipping Fever'

The most important disease associated with the transportation of cattle is shipping fever (bovine respiratory disease) (Fike and Spire, 2006), which is attributed to the stress caused by transporting cattle or calves from one geographical region to another. In North America, where feedlot fattening of beef cattle is common, it is estimated that 1% of cattle die as a consequence of transport stress and its aftermath (Irwin et al., 1979; Loneragen et al., 2003). Bovine respiratory disease (shipping fever) is responsible for 50% death losses in the feedlot and 75% of the sickness (Edwards, 1996; Loneragen et al., 2003; Deering, 2006). Fourteen per cent of US feedlot cattle became ill from bovine respiratory disease (shipping fever) (Snowder et al., 2006). Feedlot cattle with bovine respiratory disease gain less weight (Morck et al., 1993).

Shipping fever has also been reported in most European countries and in Asia (Hails, 1978). Differences between marketing systems in the USA and Australia that predispose cattle to shipping fever were discussed by Irwin et al. (1979). Co-mingling of weaner calves from different ranches before the journey may be more detrimental than co-mingling at their destination. To reduce losses, calves hauled long distances to feedlots – where they will receive grain – should be fed a 50% concentrate diet before shipping (Hutcheson and Cole, 1986).

Economic losses caused by death are minor compared with the cost of prophylactic treatment of affected cattle and poor growth in those that recover. The main symptoms of shipping fever are those of the bovine respiratory disease complex: this syndrome is characterized by fever, dyspnoea and fibrinous pneumonia, less often by gastroenteritis and only occasionally by internal haemorrhage. Fed cattle that had lung lesions from pneumonia (shipping fever) at slaughter had gained less weight, carcasses were downgraded for less marbling and the meat was tougher (Gardner et al., 1998). Other researchers have also found that the presence of lung lesions at slaughter was associated with reduced weight gain (Wittum et al., 1995a, b).

The pathogenesis of bovine respiratory disease involves a sequential cascade of events initiated by stress, which may have lowered the animal's resistance to infection. Very little research has been done on the detrimental effects on the immune system of heat, cold, crowding, mixing, noise and restraint (Kelley, 1980; Kelley et al., 1981). Ruminal function is impaired by transit stress. Transport imposes a greater stress on the rumen than feed and water deprivation (Galyean et al., 1981). This impairment may be explained by a decrease in rumination during transport. Kent and Ewbank (1991) reported that, during transport, rumination greatly decreased in 3-month-old calves.

In extensively reared beef cattle, the stress of transport had a greater detrimental effect on the animal's physiology than the stress of feed and water deprivation for the same length of time (Kelley et al., 1981; Blecha et al., 1984). Similar findings have been reported in Chile, when the effects of 3 or 16 h of deprivation of water and feed were compared in Friesian steers kept either confined in a pen on the farm or transported by road (Tadich et al., 2003b).

Stress-induced changes in host resistance may explain the physiological basis of shipping fever in cattle. Tarrant et al. (1992) observed an increase in total white blood cell count and neutrophil numbers and a reduction in lymphocyte and eosinophil numbers in cattle after long journeys. The reduction in lymphocytes may result in a loss of resistance to infection in cattle after long journeys.

A more recent study done with 367 kg bulls at a stocking density of 360 kg/m^2 for 18- and

24-h trips showed only minor differences in physiological measures compared with controls (Earley et al., 2006). This lack of differences may be partially explained by the fact that both the transported and control group were weaned immediately before transport (B. Earley, Grange Beef Research Centre, Ireland, 2006, personal communication). For example, Murata (1995) found that serum collected after 48 h of transportation had an immunosuppressive effect on peripheral blood neutrophils, decreasing their bactericidal activity.

The immune function of Bos indicus steers was significantly lowered after a 72-h trip (Stanger et al., 2005). Transportation of beef calves immediately after weaning can increase stress. Arthington et al. (2003) found that transport had detrimental effects on acute-phase proteins in newly weaned calves. Crookshank et al. (1979) found that calves transported immediately after weaning had higher cortisol levels compared with calves that had been weaned and placed in feedlot pens for 2 weeks prior to transport. Both weaning and transport affect the humoral immune response of calves (MacKenzie et al., 1997).

In a study of 45,000 6-month-old calves transported to feedlots, Ribble et al. (1995) found that differences between short and long hauls explained little, if any, of the variation among truckloads of calves in the risk of fatal fibrinous pneumonia. They suggested that other elements of the transportation process might be more stressful and therefore responsible for shipping fever. Longer trips have a more detrimental effect on stress physiology (Fazin et al., 2005), but other factors such as fear stress, weaning stress and mixing with strange calves are likely to be contributors to stresses which increase sickness.

Methods of preventing shipping fever

Research on 7845 calves has shown that sickness in 6-month-old calves that had been transported long distances could be greatly reduced by weaning and vaccinating of the calves 5–6 weeks prior to long-distance transport (National Cattlemen's Association, 1994; Swanson and Morrow-Tesh, 2001). Unvaccinated calves that are shipped the same day they are weaned will have more respiratory sicknesses and death losses (National Cattlemen's Association, 1994; Fike and Spire, 2006). Death losses due to respiratory disease were 0.16% in vaccinated and pre-weaned calves, 0.98% in calves which were still bawling after being removed from the cow and 2.02% in calves bought from order buyers and auctions (National Cattlemen's Association, 1994).

The best management strategy is to prevent shipping fever. Since 2000, more and more feedlot managers contract with ranchers to pre-wean and vaccinate calves 30–45 days before shipment. Feedlot managers will pay US$20 more per weaner calf to receive these preconditioned animals (Troxel et al., 2006).

Practical experience shows that cattle from pastures that were deficient in minerals had more death losses than cattle that had received mineral supplements (Peltz, 1999). Supplementation of newly arrived calves with vitamin E, chromium or an antioxidant can reduce sickness and improve performance (Barajas and Ameida, 1999; Purnell, 1999; Stovall et al., 1999). A large dose of 1600 IU/day of vitamin E in the feed was most effective. A more recent study by Carter et al. (2005) showed that 2000 IU of vitamin E reduced medical costs, but had little effect on performance. Since newly arrived feeder cattle showing signs of sickness often have reduced feed intake, they should be fed diets with increased nutrient density and be supplemented with extra vitamins and minerals to help reduce sickness (Galyean et al., 1999; Loerch and Fluharty 1999; Sowell et al., 1999). Fike and Spire (2006) go into more detail about receiving programmes for beef calves arriving at feedlots.

When wild, extensively raised calves are transported, dealers who transport thousands of calves on trips ranging from 1000–2000 km have found that the animals are less likely to get sick if they are transported within 32 h without a rest stop (Grandin, 1997). This may possibly be due to the fact that some of the calves have not been vaccinated prior to transport. Another factor is that loading and unloading may be stressful to calves that are not accustomed to handling. Factors unrelated to transport or handling may also affect susceptibility to shipping fever.

Both research and practical experiences show that cattle that eat and drink shortly after arrival are less likely to get sick (Schwartzkopf-Genswein et al., 2005). To facilitate drinking in extensively raised cattle that have watered out of ponds, the animals should be trained to drink from a water trough before the trip. The first author has observed cattle that died because they were afraid of a float-controlled water trough. Calves that received an adequate passive immunity from the mother cow's colostrum are more resistant to bovine respiratory disease (Wittum and Perino, 1995a). This implies that maternal traits and adequate milk production affect susceptibility to disease due to transport stresses later in life.

Conclusions

Some of the worst welfare problems that occur during transport are caused by putting sick, emaciated or debilitated cattle on a vehicle. Overstocking a truck with too many cattle will increase the numbers of cattle that get bruised or fall down and get trampled. To reduce the economic incentive to overload vehicles, transporters should be given incentives for the delivery of healthy cattle with a minimum of bruising. Best practices such as pre-weaning calves and vaccination 45 days before transport will help reduce sickness. Cattle transported for long distances will require more space compared to short distances, because the animals are more likely to lie down. After approximately 24 h of transport, cattle should be rested, watered, and fed.

An animal's reaction to the stress of transport may be very variable. For intensively raised cattle that are accustomed to close contact with people, riding in the vehicle may be more stressful than walking up the loading ramp. For wild, extensively raised cattle, just the opposite may be true: loading and unloading may be the most stressful part of the trip. Other essential best practices are non-slip flooring in vehicles, quiet handling and careful driving to prevent animals from being thrown off balance.

References

Agnes, F., Sartorelli, P., Abdi, B.H. and Locatelli, A. (1990) Effect of transport loading noise on biochemical variables in calves. American Journal of Veterinary Research 51, 1679–1681.

Aguayo, L. and Gallo, C. (2005) Tiempos de viaje y densidades de carga usadas para bovinos transportados vía marítima y terrestre desde la Región de Aysén a la zona centro-sur de Chile. Proceedings XII Congreso Latinoamericano de Buiatría y VII Jornadas Chilenas de Buiatría, pp. 346–347.

Amtmann, V.A., Gallo, C., van Schaik, G. and Tadich, N. (2006) Relaciones entre el manejo antemortem, variables sanguíneas indicadoras de estrés y pH de la canal en novillos. Archivos de Medicina Veterinaria 38 (3), 257–264.

Animal Transportation Association (1992) Handbook of Live Animal Transportation. Animal Transportation Association, Fort Washington, Maryland.

Anon. (1977) Livestock Building Project: Stockyard and Transport Crate Design. Report of the National Material Handling Bureau, Department of Productivity, Australia, July, 46 pp.

Arthington, J.D., Eichert, S.D., Kunkle, W.E. and Martin, F.G. (2003) Effect of transportation and co-mingling of the acute-phase protein response, growth and feed intake of newly weaned beef calves. Journal of Animal Science 81, 1120–1125.

Barajas, R. and Ameida, L. (1999) Effect of vitamin E and chromium-methionine supplementation on growth performance response of calves recently arrived to a feedlot. Journal of Animal Science 77 (Suppl. 1), 269 (abstract).

Barnett, J.L., Eldridge, G.A., McCausland, I.P., Caple, I.W., Millar, H.W.C., Truscott, T.G. and Hollier, T. (1984) Stress and bruising in cattle. Animal Production in Australia, Proceedings of the Australian Society of Animal Production 15, 653.

Bisschop, J.H.R. (1961) Transportation of animals by rail I: the behaviour of cattle during transportation by rail. Journal of the South African Veterinary Medical Association 32, 235–261.

Blecha, F., Boyles, S.L. and Riley, J.G. (1984) Shipping suppresses lymphocyte blastogenic responses in Angus and Brahman cross-feeder calves. Journal of Animal Science 59, 576–583.

Carter, J.N., Gill, D.R., Krehbiel, C.R., Confer, A.W., Smith, R.A., Lalman, D.L., Claypool, P.L. and McDowell, L.R. (2005) Vitamin E supplementation of newly arrived feedlot calves. *Journal of Animal Science* 83, 1924–1932.

Castro, I. and Robaina, R.M. (2003) *Manejo del Ganado Previo a la Faena y su Relacion con Calidad de la Carne*. INAC-Uruguay, Serie de Divulgación No. 1, September 2003.

Crookshank, H.R., Elissalde, M.H., White, R.G., Clanton, D.C. and Smalley, H.E. (1979) Effect of transportation and handling of calves upon blood serum composition. *Journal of Animal Science* 48, 430–436.

Dantzer, R. (1982) Research on farm animal transport in France: a survey. In: Moss, R. (ed.) *Transport of Animals Intended for Breeding, Production and Slaughter*. Martinus Nijhoff, The Hague, Netherlands, pp. 218–231.

Dantzer, R. and Mormede, P. (1979) *Le Stress en Élevage Intensif*. Masson, Paris, 117 pp.

Deering, M. (2006) Controlling the big problem: respiratory disease. *Western Livestock Journal*, May 15, p. 7.

Earley, B., Prendiville, D.J. and Riordan, E.G.O. (2006) Effect of transport up to 24 h followed by recovery on live weight, physiological and hematological responses of bulls. *Journal of Animal Science* (Suppl. 1), p. 306 (abstract).

Edwards, A.J. (1996) *Bovine Practice* 30, 5.

Eicher, S.D. (2001) Transportation of cattle in the dairy industry: current research and future directions. *Journal of Dairy Science* 84 (Suppl. E), E19–E23.

Eldridge, G.A. (1988) Road transport factors that may influence stress in cattle. In: Chandler, C.S. and Thornton, R.F. (eds) *Proceedings of the 34th International Congress of Meat Science and Technology*. CSIRO, Brisbane, Queensland, Australia, pp. 148–149.

Eldridge, G.A. and Winfield, C.G. (1988) The behaviour and bruising of cattle during transport at different space allowances. *Australian Journal of Experimental Agriculture* 28, 695–698.

Eldridge, G.A., Barnett, J.L., Warner, R.D., Vowles, W.J. and Winfield, C.G. (1986) The handling and transport of slaughter cattle in relation to improving efficiency, safety, meat quality and animal welfare. In: *Research Report Series No. 19*. Department of Agriculture and Rural Affairs, Victoria, Australia, pp. 95–96.

Eldridge, G.A., Winfield, C.G. and Cahill, D.J. (1988) Responses of cattle to different space allowances, pen sizes and road conditions during transport. *Australian Journal of Experimental Agriculture* 28, 155–159.

Fazin, E., Medica, P., Alberghina, D., Cavaleri, S. and Ferlazzo, A. (2005) Effect of long-distance transport on thyroid and adrenal function and haematocrit values in Limousin cattle: influence of body weight decrease. *Veterinary Research Communication* 29, 713–719.

Federation of Animal Science Societies (1999) *Guide for the Care and Use of Agricultural Animals in Agricultural Research and Teaching*. Federation of Animal Science Societies, Savoy, Illinois.

Fernandez, X., Monin, G., Culioli, J., Legrand, I. and Quilichini, Y. (1996) Effect of duration of feed withdrawal and transportation time on muscle characteristics and quality in Friesian–Holstein calves. *Journal of Animal Science* 74, 1576–1583.

Fike, K. and Spire, M.F. (2006) Transportation of cattle. *Veterinary Clinics of North America Food Animal Practice* 23, 305–320.

Gallo, C. (2004) Bienestar animal y calidad de carne durante los manejos previos al faenamiento en bovinos. *Proceedings XXXII Jornadas Uruguayas de Buiatría*, 10–12 June 2004, Paysandú, Uruguay, pp. 147–157.

Gallo, C., Perez, S., Sanhueza, C. and Gasic, J. (2000) Efectos del tiempo de transporte de novillos previo al faenamiento sobre el comportamiento, las perdidas de peso y algunas caracteristicas de la canal. *Archivos de Medicina Veterinaria* 32 (2), 157–170.

Gallo, C., Espinoza, M.A. and Gasic, J. (2001) Efectos del transporte por camión durante 36 horas con y sin período de descanso sobre el peso vivo y algunos aspectos de calidad de carne en bovinos. *Archivos de Medicina Veterinaria* 33 (1), 43–53.

Gallo, C., Lizondo, G. and Knowles, T.G. (2003) Effects of journey and lairage time on steers transported to slaughter in Chile. *Veterinary Record* 152, 361–364.

Gallo, C.P., Warriss, P., Knowles, T., Negron, R., Valdés, A. and Mencarini, I. (2005) Densidades de carga utilizadas para el transporte comercial de bovinos destinados a matadero en Chile. *Archivos de Medicina Veterinaria* 37 (2), 155–159.

Galyean, M.L., Lee, R.W. and Hubbert, M.W. (1981) Influence of fasting and transit on rumen function and blood metabolites in beef steers. *Journal of Animal Science* 53, 7–18.

Galyean, M.L., Perino, L.J. and Duff, G.C. (1999) Interaction of cattle health/immunity and nutrition. *Journal of Animal Science* 77, 1120–1134.

Gardner, B.A., Dolezal, H.G., Bryant, L.K., Owens, F.N., Nelson, J.L., Schutte, B.R. and Smith, R.A. (1998) Health of finishing steers: effects on performance carcass traits and meat tenderness. In: *1998 Animal Science Research Report P-965*. Oklahoma Agricultural Experiment Station, Stillwater, Oklahoma, pp. 37–45.

Grandin, T. (1979) The effect of pre-slaughter handling and penning procedures on meat quality. *Journal of Animal Science* 49 (Suppl. 1), 147 (abstract).

Grandin, T. (1980) Bruises and carcass damage. *International Journal of the Study of Animal Problems* 1, 121–137.

Grandin, T. (1981a) *Livestock Trucking Guide*. Livestock Conservation Institute, Madison, Wisconsin (now National Institute of Animal Agriculture, Bowling Green, Kentucky).

Grandin, T. (1981b) Bruises on southwestern feedlot cattle. *Journal of Animal Science* 53 (Suppl. 1), 213 (abstract).

Grandin, T. (1983a) Welfare requirements of handling facilities. In: Baxter, S.H., Baxter, M.R. and MacCormack, J.A.C. (eds) *Farm Animal Housing and Welfare*. Martinus Nijhoff, The Hague, Netherlands, pp. 137–149.

Grandin, T. (1983b) *A Survey of Handling Practices and Facilities Used in the Export of Australian Livestock*. Department of Primary Industry, Canberra, Australia (reprinted in *Proceedings Animal Air Transportation Association 1984*, Dallas, Texas).

Grandin, T. (1990) Design of loading facilities and holding pens. *Applied Animal Behaviour Science* 28, 187–201.

Grandin, T. (1997) Assessment of stress during handling and transport. *Journal of Animal Science* 75, 249–257.

Grandin, T. (2003) Perspectives on transportation issues. *Journal of Animal Science* 79 (Suppl. E).

Hahn, G.L., Clark, W.D., Stevens, D.G. and Shanklin, M.D. (1978) *Interaction of Temperature and Relative Humidity on Shrinkage of Fasting Sheep, Swine and Beef Cattle*. Paper No. 78.6010, American Society of Agricultural Engineers, St Joseph, Michigan.

Hails, M.R. (1978) Transport stress in animals: a review. *Animal Regulation Studies* 1, 289–343.

Hoffman, D.E., Spire, M.F., Schwenke, J.R. and Unruh, J.A. (1998) Effect of source of cattle and distance transported to a commercial slaughter facility on carcass bruises in mature beef cows. *Journal of the American Veterinary Medical Association* 212, 668–672.

Honkavaara, M. (1993) Effect of a controlled ventilation stockcrate on stress and meat quality. *Meat Focus International* 2 (12), 545–547.

Honkavaara, M. (1995) The effect of long distance transportation on live animals. In: Hinton, M.H. and Rowlings, C. (eds) *Factors Affecting the Microbial Quality of Meat 1. Disease Status, Production Methods and Transportation of the Live Animal*. University of Bristol Press, Bristol, UK, pp. 111–115.

Honkavaara, M. (1998) Animal transport. In: *Proceedings XVIII Nordic Veterinary Congress*, Helsinki, pp. 88–89.

Honkavaara, M. (2003–2004) Development phases of long distance transport vehicles for cattle in Finland. *Acta Veterinaria Scandavica* (Suppl.) 100, 27–32.

Honkavaara, M., Rintasalo, E., Ylonen, J. and Pudas, T. (2003) Meat quality and transport stress of cattle. *Deutsche Tierarztl Wochenschrift* 110, 125–128.

Huertas, S.M. and Gil, A.D. (2003a) Efecto del manejo prefaena en la calidad de las carcasas bovinas del Uruguay. *XXXI Jornadas Uruguayas de Buiatria*, Paysandú, Uruguay, 12–13 June 2003.

Huertas, S.M. and Gil, A.D. (2003b) Caraterizacion de las lesiones traumaticas y de sus posibles causas en carcasas bovinas del Uruguay. *Proceedings X Simposium Internacional de Epidemiologia y Economia Veterinaria*, Viña del Mar, Chile, 17–21 November 2003.

Hutcheson, D.P. and Cole, N.A. (1986) Management of transit-stress syndrome in cattle: nutrition and environmental effects. *Journal of Animal Science* 62, 555–560.

Irwin, M.R. and Gentleman, W.R. (1978) Transportation of cattle in a rail car containing feed and water. *Southwestern Veterinarian* 31, 205–208.

Irwin, M.R., McConnell, S., Coleman, J.D. and Wilcox, G.E. (1979) Bovine respiratory disease complex: a comparison of potential predisposing and etiologic factors in Australia and the United States. *Journal of the American Veterinary Medical Association* 175, 1095–1099.

Jacobson, L.H. and Cook. C.J. (1998) Partitioning psychological and physical sources of transport-related stress in young cattle. *Veterinary Journal* 155, 205–208.

James, M. (1997) Report on the transportation of cattle to the Isle of Man by air. *State Veterinary Journal* 7 (4), 1–2.

Jeremiah, L.E., Schaefer, A.L. and Gibson, L.L. (1992) The effects of antemortem feed and water withdrawal, antemortem electrolyte supplementation, and postmortem electrical stimulation on the palatability and consumer acceptance of bull beef after aging (6 days at 1°C). *Meat Science* 32, 149.

Kelley, K.W. (1980) Stress and immune function: a bibliographic review. *Annales de Recherches Vétérinaires* 11, 445–478.

Kelley, K.W., Osborne, C., Everman, J., Parish, S. and Hinrich, D. (1981) Whole blood leucocytes *vs.* separated mononuclear cell blastogenes in calves, time dependent changes after shipping. *Canadian Journal of Comparative Medicine* 45, 249–258.

Kenny, J.F. and Tarrant, P.V. (1987a) The physiological and behavioural responses of crossbred Friesian steers to short-haul transport by road. *Livestock Production Science* 17, 63–75.

Kenny, J.F. and Tarrant, P.V. (1987b) The reaction of young bulls to short-haul road transport. *Applied Animal Behaviour Science* 17, 209–227.

Kenny, J.F. and Tarrant, P.V. (1987c) The behaviour of young Friesian bulls during social regrouping at the abattoir. Influence of an overhead electrified wire grid. *Applied Animal Behaviour Science* 18, 233–246.

Kenny, J.F. and Tarrant, P.V. (1988) The effect of oestrus behaviour on muscle glycogen concentration and dark-cutting in beef heifers. *Meat Science* 22, 21–31.

Kent, J.E. and Ewbank, R. (1991) The behavioural response of 3-month-old calves to 18 h of road transport. *Applied Animal Behaviour Science* 26, 289 (abstract).

Kettlewell, P.J., Hoxey, R.P., Rampson, C.J., Green, N.R., Veale, B.M. and Mitchell, M.A. (2001) Design and operation of a prototype mechanical ventilation system for livestock transport vehicles. *Journal of Agricultural Engineering Research* 79, 429–439.

Kilgour, R. and Mullord (1973) Transport of calves by road. *New Zealand Veterinary Journal* 21, 7–10.

Knowles, T.G. (1999) A review of road tranport of cattle. *Veterinary Record* 144, 197–201.

Lacourt, A. and Tarrant, P.V. (1985) Glycogen depletion patterns in myofibres of cattle during stress. *Meat Science* 15, 85–100.

Lambooy, E. and Hulsegge, B. (1988) Long-distance transport of pregnant heifers by truck. *Applied Animal Behaviour Science* 20, 249–258.

Lapworth, J. (2004) *Beef Cattle Transport by Road Pre-transport Management.* Department of Primary Industries, Queensland Government, Australia.

Lay, D.C., Friend, T.H., Bowers, C.L., Grissom, K.K. and Jenkins, O.C. (1992a) Behavioral and physiological effects of freeze and hot iron branding on crossbred cattle. *Journal of Animal Science* 70, 330–336.

Lay, D.C., Friend, T.H., Bowers, C.L., Grissom, K.K. and Jenkins, O.C. (1992b) A comparative physiological and behavioral study of freeze and hot-iron branding using dairy cows. *Journal of Animal Science* 70, 1121–1125.

LeDoux, J. (1998) *The Emotional Brain.* Simon and Schuster, New York.

Loerch, S.C. and Fluharty, F.L. (1999) Physiological changes and digestive capabilities of newly received feetlot cattle. *Journal of Animal Science* 77, 1113–1119.

Loneragan, G.H., Dargatz, D.A., Morley, P.S. and Smith, M.A. (2003) Trends of mortality ratios among cattle in U.S. feedlots. *Journal of the American Veterinary Medical Association* 219, 1122–1127.

MacKenzie, A.M., Drennan, M., Rowan, T.G., Dixon, J.B. and Carter, S.D. (1997) Effect of transportation and weaning on the humeral immune response of calves. *Research in Veterinary Science* 3, 227–230.

Marahrens, M., Von Richthofer, I., Schmeiduch, S. and Hartung, J. (2003) Special problems of long-distance road transport of cattle. *Deutsche Tierarztl Wochenschrift* 110, 120–125.

Mayes, H.F., Asplund, J.M. and Anderson, M.E. (1979) Transport stress effects on shrinkage. In: *ASAE American Society of Agricultural Engineers Paper No. 79,* p. 6512.

McVeigh, J. and Tarrant, P.V. (1982) Glycogen content and repletion rates in beef muscle, effect of feeding and fasting. *Journal of Nutrition* 112, 1306–1314.

McVeigh, J.M. and Tarrant, P.V. (1983) Effect of propranolol on muscle glycogen metabolism during social regrouping of young bulls. *Journal of Animal Science* 56, 71–80.

Mencarini, I.R. (2002) Efecto de dos densidades de carga y dos tiempos de transporte sobre el contenido de glucogeno muscular y hepatico, pH y color de la carne. *Memoria de Titulo para optar al Titulo de Medico Veterinario,* Facultad de Ciencias Veterinarias, Universidad Austral de Chile, Valdivia, Chile.

Morck, D.W., Merrill, J.K. and Thorlakson, B.E. (1993) Prophylactic efficacy of tilmicosin for bovine respiratory trait disease. *Journal of the American Veterinary Medical Association* 202, 273–277.

Muirhead, V.V. (1985) *An Investigation of the Internal and External Aerodynamics of Cattle Trucks.* NASA Contract 170400, Ames Research Center, Dryden Flight Center, Edwards, California.

Muller, W. (1985) Ventilation requirements during transport of animals. *Proceedings, Animal Air Transportation Association,* Dallas, Texas.

Murata, H. (1995) Suppression of bovine neutrophil function by sera from calves transported from over a long distance by road. *Animal Science and Technology (Japan)* 66 (11), 976–978.

National Cattlemen's Association (1994) *Strategic Alliances Field Study in Coordination with Colorado State University and Texas A & M University*. National Cattlemen's Beef Association, Englewood, Colorado.

Norris, R.T., Richards, R.B., Creeper, J.H., Jubb, T.F., Madin, B. and Kerr, J.W. (2003) Cattle deaths during sea transport from Australia. *Australian Veterinary Journal* 81, 156–161.

Novoa, H. (2003) Efectos de la duracion y condiciones del reposo en ayuno previo al faenamiento de los bovinos sobre las caracteristicas de la canal. *Memoria de Titulo para Optar al Titulo de Medico Veterinario*, Facultad de Ciencias Veterinarias, Universidad Austral de Chile, Valdivia, Chile.

OIE (2005) *Guidelines for the Transport of Animals by Land*. World Organization for Animal Health, Paris.

Peltz, H. (1999) True immunity in the feedyard relies on cow–calf nutrition. *Drover's Journal* 127 (5), 26.

Price, M.A. and Tennessen, T. (1981) Pre-slaughter management and dark-cutting in the carcasses of young bulls. *Canadian Journal of Animal Science* 61, 205–208.

Purnell, R. (1999) Vitamin value: vitamin E may help incoming calves respond positively to treatment, gain faster and get sick less. *Beef* (Minneapolis, Minnesota, USA) July, 26.

Ramsey, W.R., Neischke, H.R.C. and Anderson, B. (1976) *Australian Veterinary Journal* 52, 285–286.

Randall, J.M. (1993) Environmental parameters necessary to define comfort for pigs, cattle and sheep in live-stock transporters. *Animal Production* 57, 299–307.

Ribble, C.S., Meek, A.H., Shewen, P.E., Jim, G.K. and Guichon, P.T. (1995) Effect of transportation on fatal fibrinous pneumonia and shrinkage in calves arriving at a large feedlot. *Journal of the American Veterinary Medical Association* 207, 612–615.

Scanga, J.A., Belk, K.E., Tatum, J.D., Grandin, T. and Smith, G.C. (1998) Factors contributing to the incidence of dark-cutting beef. *Journal of Animal Sciences* 76, 2040–2047.

Schaefer, A.L. and Jeremiah, L.E. (1992) Effect of diet on beef quality. In: *Proceedings of the 13th Western Nutrition Conference*. University of Saskatchewan, Saskatoon, Canada, pp. 123–128.

Schaefer, A.L., Jones, S.D.M., Tong, A.K.W. and Young, B.A. (1990) Effect of transport and electrolyte supplementation on ion concentrations, carcass yield and quality in bulls. *Canadian Journal of Animal Science* 70, 107.

Schaefer, A.L., Jones, S.D. and Stanley, R.W. (1997) The use of electrolyte solutions for reducing transport stress. *Journal of Animal Science* 75, 258–265.

Schwartzkopf-Genswein, K.S., Shah, M.A., McAllister, T.A., Genswein, B.M.A., Streeter, M., Branine, M. and Swingle, S. (2005) Relationship between feeding behaviour, morbidity, and vaccination in feedlot cattle. *Journal of Animal Science* 83 (Suppl. 1), 13 (abstract).

Self, H.L. and Gay, N. (1972) Shrink during shipment of feeder cattle. *Journal of Animal Science* 35, 489–494.

Shackleford, S.D., Koohmaraie, M., Wheeler, T.L., Cundiff, L.V. and Dikeman, M.E. (1994) Effect of biological type of cattle on the incidence of dark, firm and dry condition in the *longissimus* muscle. *Journal of Animal Science* 72, 337–343.

Shaw, F.D., Baxter, R.I. and Ramsey, W.R. (1976) The contribution of horned cattle to carcass bruising. *Veterinary Record* 98, 255–257.

Shorthose, W.R. (1965) Weight losses in cattle prior to slaugher. *CSIRO, Food Preservation Quarterly* 25, 67–73.

Shorthose, W.R., Harris, P.V. and Bouton, P.E. (1972) The effects of some properties of beef or testing and feeding cattle after a long journey to slaughter. *Proceedings of the Australia Society of Animal Production* 9, 387–391.

Sick, F.L., Bleeker, C.M., Mouw, J.K. and Thompson, W.S. (1982) Toe abscesses in recently shipped cattle. *Veterinary Medicine and Small Animal Clinician* 1385.

Singh, S.P. (1991) *Vibration Levels in Commercial Truck Shipments*. Paper No. 91–6016, American Society of Agricultural Engineers, St Joseph, Michigan.

Smith, G.C., Savell, J.W., Morgan, B. and Lawrence, T.E. (2006) National Beef Quality Audit, Colorado State University, Oklahoma State University and Texas A & M University for the National Cattlemen's Beef Association, Englewood, Colorado.

Snowder, G.D., Van Vleck, L.D., Cundiff, L.V. and Bennett, G.L. (2006) Bovine respiratory disease in feedlot cattle: environmental, genetic and economic factors. *Journal of Animal Science* 84, 1999–2008.

Sowell, B.F., Branine, M.E., Bowman, J.G.P., Hubbert, M.E., Sherwood, H.E. and Quimby, W. (1999) Feeding and watering behaviour of healthy and morbid steers in a commercial feedlot. *Journal of Animal Science* 77, 1105–1112.

Stanger, K.J., Ketheesan, N., Parker, A.J., Coleman, C.J., Lazzaroni, S.M. and Fitzpatrick, L.A. (2005) The effect of transportation on the immune status of *Bos indicus* steers. *Journal of Animal Science* 83, 2632–2636.

Stevens, D.G. (1985) Ventilation requirements for transport of livestock. In: *Proceedings Animal Air Transportation Association*, Dallas, Texas.

Stevens, D.G. and Camp, T.H. (1979) *Vibration in a Livestock Vehicle.* Paper No. 79-6511, American Society of Agricultural Engineers, St Joseph, Michigan. *Cattle Transport* 171.

Stevens, D.G. and Hahn, G.L. (1977) *Air Transport of Livestock: Environmental Needs.* Paper No. 77-4523, American Society of Agricultural Engineers, St Joseph, Michigan.

Stevens, D.G., Stermer, R.A. and Camp, T.H. (1979) *Body Temperature of Cattle During Transport.* Paper No. 79- 6511, American Society of Agricultural Engineers, St Joseph, Michigan.

Stovall, T.C., Han, H. and Gill, D.R. (1999) Effect of Agrado on the health and performance of transport-stressed heifer calves. *Journal of Animal Science* 71 (Suppl. 1), 272 (abstract).

Swanson, J.C. and Morrow-Tesh, J. (2001) Cattle transport historical, research, and future perspectives. *Journal of Animal Science* 79 (Suppl. E), E102–E109.

Tadich, N., Alvarado, M. and Gallo, C. (2000) Efectos de 36 horas de transporte terrestre con y sin descanso sobre algunas variables sanguineas indicadoras de estres en bovinos. *Archivos de Medicina Veterinaria* 32 (2), 171–183.

Tadich, N., Gallo, C., Knowles, T., Uribe, H. and Aranis, A. (2003a) Efecto de dos densidades de carga usadas para el transporte de novillos, sobre algunos indicadores sanguineos de estres. *Proceedings XXVII Reunion Anual de la Sociedad Chilena de Producción Animal*, 15–16 October, Talca, Chile, pp.15–17.

Tadich, N., Gallo, C., Echeverría, R. and van Schaik, G. (2003b) Efecto del ayuno durante dos tiempos de confinamiento y de transporte terrestre sobre algunas variables sanguíneas indicadoras de estrés en novillos. *Archivos de Medicina Veterinaria* 35 (2), 171–185.

Tadich, N., Gallo, C., Bustamante, H., Schwerter, M. and van Schaik, G. (2005) Effects of transport and lairage time on some blood constituents of Friesian-cross steers. *Livestock Production Science* 93, 222–233.

Tarrant, P.V. (1981) The occurrence, causes and economic consequences of dark-cutting in beef – a survey of current information. In: Hood, D.E. and Tarrant, P.V. (eds) *The Problem of Dark Cutting in Beef.* Martinus Nijhoff, The Hague, Netherlands, pp. 3–34.

Tarrant, P.V. (1988) Animal behaviour and environment in the dark cutting condition. In: Febiansson, S.U., Shorthose, W.R. and Warner, R.D. (eds) *Darkcutting in Cattle and Sheep. Proceedings of an Australian Workshop.* AMLRDC, Sydney, Australia, pp. 8–18.

Tarrant, P.V. and Grandin, T. (1993) Cattle Transport. In: Grandin, T. (ed.) *Livestock Handling and Transport.* CAB International, Wallingford, UK, pp. 109–126.

Tarrant, P.V., Kenny, F.J. and Harrington, D. (1988) The effect of stocking density during 4 h transport to slaughter, on behaviour, blood constituents and carcass bruising in Friesian steers. *Meat Science* 24, 209–222.

Tarrant, P.V., Kenny, F.J., Harrington, D. and Murphy, M. (1992) Long distance transportation of steers to slaughter: effect of stocking density on physiology, behaviour and carcass quality. *Livestock Production Science* 30, 223–238.

Tennessen, T., Price, M.A. and Berg, R.T. (1981) Bulls *vs.* steers: effect of time of mixing on dark cutting. In: *58th Annual Feeders Day Report.* University of Alberta, Edmonton, Canada.

Tennessen, T., Price, M.A. and Berg, R.T. (1984) Comparative responses of bulls and steers to transportation. *Canadian Journal of Animal Science* 64, 333–338.

Tennessen, T., Price, M.A. and Berg, R.T. (1985) The social interactions of young bulls and steers after regrouping. *Applied Animal Behaviour Science* 14, 37–47.

Todd, S.E., Mellor, D.J., Stafford, K.J., Gregory, N.G., Bruce, R.A. and Ward, R.N. (2000) Effect of food withdrawal and transport on 5- to 10-day-old calves. *Research in Veterinary Science* 68, 125–134.

Troxel, T.R., Barham, B.L., Cline, S., Foley, D., Hardgrave, R., Wiedower, R. and Wiedower, W. (2006) Management factors affecting selling prices of Arkansas beef calves. *Journal of Animal Science* 84 (Suppl. 1), 12 (abstract).

Valdés, A. (2002) Efectos de los densidades de carga y dos tiempo de transporte sobre el peso vivo, rendimiento de la canal y presencia de contusiones en novillos destinados al faenamiento. *Memoria de Titulo para Optar al Titulo de Medico Veterinario*, Fac. Ciencias Veterinarias, Universidad Austral de Chile, Valdivia, Chile.

Van de Water, G., Heylen, T., Swinnen, K. and Geers, R. (2003a) The impact of vertical vibrations on the welfare of calves. *Deutsche Tierarztl Wochenschrift* 110, 111–114.

Van de Water, G., Verjans, F. and Geers, R. (2003b) The effect of short distance transport under commercial conditions on the physiology of slaughter calves, pH, and colour profiles of veal. *Livestock Production Science* 82, 171–179.

Van Holleben, K., Henke, S., Schmidt, T., Bostelmann, N., Von Wenzlawowicz, and Hartung, J. (2003) Handling cattle in pre- and post-transport situations including loading and unloading on journeys up to 8 h in Germany. *Deutsche Tierarztl Wochenschrift* 110, 93–99.

Villarroel, M., Maria, G., Sanudo, C., Garcia-Belenger, S., Chacon, G. and Gebre-Senbet, G. (2003) Effect of commercial transport in Spain on cattle welfare and meat quality. *Deutsche Teirarztl Wochenschrift* 110, 105–107.

Warnock, J.P., Caple, I.W., Halpin, C.G. and McQueen, C.S. (1978) Metabolic changes associated with the 'downer' condition in dairy cows at abattoirs. *Australian Veterinary Journal* 54, 566–569.

Warriss, P.D., Kestin, S.C., Brown, S.N. and Wilkins, L.J. (1984) The time required for recovery from mixing stress in young bulls and the prevention of dark-cutting beef. *Meat Science* 10, 53–68.

Warriss, P.D., Brown, S.N., Knowles, T.G., Kestin, S.C., Edwards, J.E., Dolan, S.K. and Phillips, A.J. (1995) Effects of cattle transport by road for up to 15 h. *Veterinary Record* 136, 319–323.

Wikner, J., Gebresenbet, G. and Nilsson, C. (2003) Assessment of air quality in a commercial cattle transport vehicle in Swedish summer and winter conditions. *Deutsche Tierarztl Wochenschrift* 110, 100–104.

Wittum, T.E. and Perino, L.J. (1995a) Passive immune status at 24 h postpartum and long term health and performance of calves. *American Journal of Veterinary Research* 56, 1149–1154.

Wittum, T.E., Woollen, N.E. and Perino, L.J. (1995b) Relationship between treatment for respiratory disease and the presence of pulmonary lesions at slaughter and rate of gain in feedlot cattle. *Journal of Animal Science* 73 (Suppl. 1), 238.

Wythes, J.R. (1982) The saleyard curfew issue. *Queensland Agricultural Journal* November–December, 1–5.

Wythes, J.R. (1985) Cattle transportation strategies. In: *Proceedings of the First National Grazing Animal Welfare Symposium*. Australian Veterinary Association, Brisbane, Australia, April 1985.

Wythes, J.R. and Shorthose, W.R. (1984) *Marketing Cattle: its Effects on Live weight, Carcasses and Meat Quality*. No. 46, Australian Meat Research Committee, Sydney, Australia.

Wythes, J.R., Shorthose, W.R., Schmidt, P.J. and Davis, C.B. (1980) Effects of various rehydration procedures after a long journey on live weight, carcasses and muscle properties of cattle. *Australian Journal of Agricultural Research* 31, 849–855.

Wythes, J.R., Arthur, R.J., Thompson, P.J.M., Williams, G.E. and Bond, J.H. (1981) Effect of transporting cows varying distances on live weight, carcass traits and muscle pH. *Australian Journal of Experimental Agriculture and Animal Husbandry* 21, 557–561.

10 Behavioural Principles of Sheep-handling[1]

G.D. Hutson
Clifton Press, Kensington, Victoria, Australia

Introduction

The relationship between humans and sheep is probably more than 6000 years old (Hulet *et al.*, 1975). No doubt, the original relationship was one-sided, with humans hunting herds of wild sheep for food and clothing. But gradually, during the process of domestication, hunting changed to herding and herding changed to farming. This transition from hunter to farmer has brought about a change in behaviour and attitude towards sheep, so that humans are now responsible for the day-to-day care and well-being of sheep flocks. However, the transition has been incomplete, and many of the techniques of the hunter/herder are still used during sheep-handling.

There are about 1.2 billion sheep worldwide (Lynch *et al.*, 1992). Most of these sheep are handled at least twice a year for two essential treatments – shearing and crutching (shearing of the breech and hindlegs to prevent fouling by faeces) – and generally they are handled more often. Sheep movement is usually prompted by the use of fear-evoking stimuli, and the treatment is usually stressful and aversive (Hutson, 1982b; Hargreaves and Hutson, 1990a, 1997). The handling procedure involves mustering, often with dogs and motor bikes, movement through

yards, races and sheds and, finally, administration of a treatment, often involving isolation, manipulation and restraint. Treatments can be apparently mild, such as classing, drafting (sorting), drenching, dipping, vaccination and jetting (spraying) or more prolonged and stressful, such as foot trimming, castration, tail docking and shearing.

Behavioural Characteristics Important for Handling

Kilgour (1976) described the sheep as a:

> defenceless, vigilant, tight-flocking, visual, wool-covered ruminant, evolved within a mountain grassland habitat, displaying a follower-type dam–offspring relationship, with strong imitation between young and old in establishing range systems of tracks and best forage areas, showing seasonal breeding and a separate adult male sub-group structure.

He claimed that most behaviour seen in sheep on farms could be traced to one or other of these characteristics. I think this description of the sheep is best encapsulated in three words – flocking, following and vision; and I would add one more – intelligence. These four characteristics form the basis for all principles of sheep handling.

---

[1] This chapter is dedicated to the memory of Ron Kilgour, whose pioneering work on sheep behaviour, welfare and handling stimulated much of the research that is reviewed here.

Flocking

The ancestral sheep adapted to mountain grass-lands have evolved by natural and artificial selection into breeds occupying diverse habitats and varying climates. The social organization of sheep in the wild is probably exemplified by the feral Soay sheep on St Kilda, a small, remote island to the north-west of Scotland. Grubb and Jewel (1966) found that ewes tended to form large groups with fairly well-defined home ranges. Males associated in smaller numbers (two to three) and at mating moved from group to group. Young males left the groups of ewes at about 1 year old to form their own male–male groups.

In domestic flocks, this normal social organization is disrupted by the removal of lambs before natural weaning and by keeping sheep in flocks of uniform age and sex. Despite these disruptions, sheep still aggregate to form flocks. Crofton (1958) studied aerial photo-graphs of domestic sheep and found that the distance between individual grazing sheep varied, but they were oriented so that an angle of about 110° was subtended between the head of each sheep and the two others in front of it.

This angle corresponded to the angle between the optic axes of the eyes and implied that sheep grazed that way because that was the way their eyes were pointing. No doubt, vision is important in maintaining contact within the flock, but the 110° angle seems to be fanciful and has not been confirmed by other studies. However, Crofton's conclusions that sheep aggregate and that vision is important in social spacing remain valid.

Sheep maintain social spacing and orienta-tion, even when confined in pens. Hutson (1984) found that standing sheep oriented themselves so that they were parallel to and fac-ing in the same direction as their nearest neigh-bour. Sheep lay down parallel to and next to the sides of open-sided pens, but selected positions at random in covered pens.

Various factors will affect flock behaviour and structure while grazing, including breed, stocking rate, topography, vegetation, shelter and distance to water (Kilgour et al., 1975; Squires, 1975; Stolba et al., 1990; Lynch et al., 1992). The adaptive advantages of this behaviour are clear: it provides more efficient exploitation of seasonal food resources and protection from predators. Sheep respond to the sight of a pred-ator by flocking and flight (Kilgour, 1977; Hansen et al., 2001; Dwyer, 2004). In the wild, sheep have long flight distances, but in con-fined spaces this distance will vary according to the space available for escape. When con-fined in a 2 m-wide laneway, the flight distance of sheep to an approaching man was 5.7 m compared with 11.4 m in a 4 m-wide laneway (Hutson, 1982a). Flight distance was not affected by flock size, density or speed of approach. Individual sheep had longer flight distances than flocks.

One consequence of this social structure is that sheep have the opportunity to form stable relationships. Hinch et al. (1990) and Rowell (1991) reported that long-term social bonds may form between lambs and their mothers. Sheep may also develop a group identity. When groups of strange sheep have been mixed, they have kept to their own groups for several weeks before full integration occurred (Arnold and Pahl, 1974; Winfield et al., 1981).

The concept of a socially stable flock has important consequences for handling. Kilgour (1977) noted that, when separated from a group, an individual will run toward other sheep, irrespective of the position of the handler or dog. A sheep isolated from the flock may also show signs of tonic immobility or escape, depending on the type of restraint (Syme, 1985). Syme (1981) found that 30% of merino sheep responded either physically or vocally to less than 5 min of isolation from the mob. Kilgour (1977) suggested that four or five sheep were required in a group before the group showed signs of social cohesion.

Penning et al. (1993) found that sheep in small groups spent less time grazing and that a minimum group size of three, preferably four, was required for studies of grazing behaviour. Boissy and Dumont (2002) found that individ-ual sheep were less likely to separate from a group to graze at a preferred feeding site when they were with one companion than when they were in a group of five. Also, the frequency of vigilance behaviour increased, probably to maintain social contact with the rest of the group, when the group size was small (Dumont and Boissy, 2000).

Following

The following response of sheep is present at birth, when a newborn lamb will follow its dam during her daily activities. This response may play an important part in maintaining ewe–lamb contact, especially if the ewe has twin lambs. However, when following behaviour was tested experimentally in a circular runway, there was great variability in the response (Winfield and Kilgour, 1976). Some lambs followed the surrogate ewe and others did not follow at all. The response was strongest for lambs between 4 and 10 days of age, which suggests that the generalized following response may be replaced by a more specific response to the dam.

Nevertheless, following of conspecifics persists in the life of a sheep. Scott (1945) described 'allelomimetic behaviour' as very common in sheep. This term refers to any type of activity that involves mutual imitation. In sheep it includes walking and running together, following one another, grazing together, bedding down together and bouncing stiff-legged past an obstacle together. Behaviour within the flock also tends to be synchronized, with sheep feeding, drinking, resting and ruminating at the same time. This synchronization of behaviour by grazing herbivores might be the result of social facilitation (Rook and Penning, 1991).

Leadership also occurs in sheep flocks and may be related to independence (Arnold, 1977). Individual sheep have been identified consistently among the leaders or among the tail-enders in small sheep flocks (Squires and Daws, 1975). Hutson (1980b) identified about 10% of sheep in small flocks as leaders. It is unlikely that these sheep deliberately led the flock, but moved independently of other sheep and were then followed by them. Leadership will also depend on the setting, the size and composition of the flock and the purpose of movement (Syme, 1985). Sheep can be trained successfully to lead other sheep. Bremner (1980) used operant conditioning techniques to train sheep to walk through yards, push open and unlatch gates, accept leash restraint and lead mobs of sheep.

Vision

Sheep have excellent eyesight, as noted by Geist (1971): 'A popular myth circulates in North America that sheep vision is equal to that of a man aided by 8-power binoculars'. A wide visual field is a common characteristic of ungulates and may be an adaptation by prey animals to enable early detection of predators (Walls, 1942). In sheep, the angle between the optic axis and the midline is about 48° (Whitteridge, 1978), which indicates that the sheep should have a wide, although not panoramic, monocular field and a binocular field of about 60°.

Piggins and Phillips (1996) measured the visual field of Welsh mountain and Cambridge sheep with a retinoscope and recorded a monocular field of 185°, with binocular overlap of 61.7°. In practice, Hutson (1980a) found that the visual field of merino sheep ranged from 190° to 306°, with a mean of 270°. The main causes of obstruction to rearward vision were ears, horns and wool growth. The binocular field of sheep, where the field of vision from each eye overlapped, ranged from 4° to 77°, with a mean of 45°. The main obstruction to the binocular field was the snout.

Stereoscopic vision is the perception of three dimensions in space. For an animal to have good stereoscopic vision it is essential that it has good binocular vision but, in addition, the optic nerve fibres must decussate incompletely at the optic chiasma. Clarke and Whitteridge (1973) worked out the projections of the retina on the visual cortex of sheep. In area 18, they found an overlap of about 30° on each side of the midline, and most cells with fields up to 25° out were driven binocularly. In addition, visual acuity – as estimated from peak retinal ganglion cell density – was twice as high as that of the cat.

Clarke and Whitteridge suggested that highly developed stereoscopic vision probably forms the basis for the sure-footedness of ungulates. Tanaka et al. (1995) determined visual acuity scores in three sheep, which ranged from 0.085 to 0.19. These values indicate that sheep could resolve visual detail at about one-twelfth to one-fifth the standard threshold for humans.

Depth perception is the discrimination of a drop-off or depth downward as opposed to straight ahead. Depth can be detected using several cues: an animal may detect a difference in the density of light from similar surfaces at different depths; head movements or a change in position as the animal looks will produce motion parallax; and an animal may possess true stereoscopic vision. Walk and Gibson (1961) and Lemmon and Patterson (1964) tested sheep on the visual cliff test and found good perception of depth. One-day-old lambs placed on a sheet of glass without visual support showed an immediate protective response – freezing, stiffening the forelegs and backing off the glass.

Sheep also possess some form of colour vision, but it is not known to what extent they rely on colour for environmental discriminations. Morphological evidence suggests that the sheep is a dichromat. Jacobs *et al.* (1998) used electroretinographic techniques to study the spectral sensitivity of cones in the sheep retina. Two cone types were identified – an S cone, with a spectral peak of 445.3 nm and an M/L cone, with a spectral peak of 552.2 nm. The authors concluded that the sheep has the requisite photopigment basis to support dichromatic colour vision. A similar conclusion has been drawn for other ungulates, including cows and goats (Jacobs *et al.*, 1998) and pigs (Neitz and Jacobs, 1989).

The behavioural evidence for colour vision does not concur totally with the morphological evidence and suggests that colour sensitivity may favour the longer-wavelength end of the spectrum. Alexander and Stevens (1979) found that ewes could distinguish coloured lambs from grey-shaded lambs if the lambs were red, orange, yellow or white. However, ewes with blue, green or black lambs performed poorly.

Munkenbeck (1982) used an operant conditioning technique to demonstrate that sheep could discriminate wavelengths at 30 nm intervals in the range 520–640 nm. Tanaka *et al.* (1989a, b) also used an operant conditioning technique to demonstrate that sheep could distinguish between the three primary colours and the same shades of grey. However, 11 sessions (330 trials) were required to reach criterion on the red discrimination and 20 sessions (600 trials) for blue, and one subject failed to discriminate between green and grey after 64 sessions (1920 trials).

Thus, the limited behavioural evidence supports colour vision in the yellow–orange–red end of the spectrum, but provides conflicting evidence for the blue–green end of the spectrum. However, it should be borne in mind that in behavioural tests of colour vision it is notoriously difficult to effectively eliminate non-colour cues (Neitz and Jacobs, 1989). More recent research on dichromatic vision in grazing animals is reviewed in Chapter 4.

Kendrick and Baldwin (1987) and Kendrick (1991) have investigated visual recognition in the sheep using single-cell electrophysiological recording techniques. They investigated the responses of single neurons in the temporal cortex of sheep to various visual images. A small population of cells responded specifically to images of dog and human faces. Other cells responded to the sight of a human shape rather than the face. These cells did not distinguish between humans, their sex, what they were wearing, whether the back view or front view was presented or whether the head and shoulders were covered. Thus, decisions about an appropriate behavioural response to a potential predator could be made quickly at the level of sensory analysis.

However, there is no doubt that sheep could also learn these discriminations. Baldwin (1981) has demonstrated that sheep can perform complex visual discrimination learning tasks in the laboratory using geometrical symbols and Kendrick *et al.* (1995) have shown that sheep can discriminate in a Y-maze between the projected images of faces of different sheep and humans. Davis *et al.* (1998) used an operant conditioning technique to demonstrate that sheep can discriminate between individual humans.

Other senses

Olfaction is less important to sheep in handling situations, although sheep have well-developed olfactory apparatus and are able to make keen olfactory discriminations. Olfactory recognition of the newborn lamb may be established within 30 min of parturition (Keller *et al.*, 2003). Baldwin and Meese (1977) demonstrated that sheep were able to distinguish between conspecifics using a range of secretory and excretory products. Blissitt *et al.* (1990) found that rams could discriminate between fresh oestral and non-oestral urine odours.

Sheep will avoid grazing pasture contaminated with faeces (Cooper *et al.*, 2000), and the odour of dog faeces appears to have an innate repellent effect. Arnould and Signoret

(1993) reported that sheep refused to eat food contaminated with dog faecal odour and did not habituate to repeated exposure. Geist (1971) claimed that the sense of smell was well enough developed for sheep to be able to scent a man 350 yards away under favourable conditions.

Despite this extravagant claim, the distance over which olfactory discriminations are effective probably limits the usefulness of this sense in handling situations. Alexander (1978) found that ewes could distinguish their own lambs from aliens at close quarters, but when the lamb was more than 0.25 m away the ewes could no longer make the discrimination. Franklin and Hutson (1982a) reported that interdigital gland secretion did not influence path choice of sheep moving through a Y-maze.

Wollack (1963) investigated auditory acuity in three sheep using a conditioned leg flexion response. Auditory sensitivity increased at frequencies from 10–10,000 Hz, with a slight decrease around 1000 Hz. There was a rapid decrease in sensitivity from 10,000 to 40,000 Hz. Ames and Arehart (1972) determined auditory threshold curves by electroencephalogram (EEG) changes and behavioural responses and found maximum sensitivity at 7000 Hz. Shillito (1972) showed that the dominant frequencies in a lamb's bleat and a ewe's rumble were between 1000 and 4000 Hz, and not at the peak sensitivity of hearing. Maximum sensitivity of hearing may therefore be attuned to auditory detection of predators and danger, rather than to the calls of other sheep.

Sheep show little response towards sonic booms and jet aircraft noise. Espmark et al. (1974) found the strongest reaction to sonic booms in standing sheep, who flung up their heads and started running, forming a 'bunch' with other sheep and moving off together. However, the sheep quickly adapted so that after three exposures (five booms per day) they barely responded. Some sheep still reacted with a short and fast run, but then immediately resumed their previous activity. The sheep showed little or no response to subsonic aircraft noise ranging from 75 to 109 dB.

Ewbank and Mansbridge (1977) also found that grazing lowland sheep tended to run together in response to simulated sonic booms at their first few exposures, but quickly adapted. In contrast, hill sheep scattered, but they too

adapted. Weisenberger et al. (1996) reported that heart rates of captive mountain sheep exposed to simulated jet aircraft noise increased, but returned to resting levels in 1–3 min, and Krausman et al. (1998) reported increased heart rates in mountain sheep in only 14% of overflights by F16 aircraft and a return to preflight levels within 2 min.

Ambient noise during transport also appears to have little effect on sheep. Hall et al. (1998) reported that sheep showed no orientation away from the noise source (a generator) and no behavioural changes indicative of discomfort.

Vocal communication between sheep seems to be of relatively minor importance, since sheep do not give a vocal alarm call and vocalize only in specific situations such as: (i) isolation from the flock (Torres-Hernandez and Hohenboken, 1979); (ii) during courtship of oestral ewes, when rams utter a low-pitched rumble (Banks, 1964); and (iii) in ewe–lamb recognition (Alexander and Shillito, 1977; Dwyer et al., 1998).

Searby and Jouventin (2003) have pointed out that the vocal signature of ewes and lambs is similar, since it relies only on the mean frequency and spectral energy distribution (timbre) of the call. The simplicity of this system is probably linked to the roles played by vision and olfaction in corroborating the vocal information. Sheep vocalizations have no attractive effect on movement along races (Franklin and Hutson, 1982b).

Vocalizations have been shown to be a useful indicator of handling problems in beef slaughter plants (Grandin, 2001). However, the low frequency of vocalization in sheep and the inhibition of vocalization in the presence of predators (Dwyer, 2004) may both confound its use as a measure of distress in sheep.

Intelligence

Many farmers deride the intelligence of sheep with remarks such as 'sheep are stupid'. However, this apparent stupidity can nearly always be attributed to the overriding presence of the protective flocking instinct (Kilgour and Matthews, 1983; Hutson, 1994). There have been many studies confirming the above-average learning ability of sheep. They can be conditioned easily in classical conditioning experiments.

Liddell and Anderson (1931) were probably the first to use sheep in conditioning experiments. They conditioned sheep to make reflex leg movements in response to the beat of a metronome after eight to nine pairings of the metronome with an electrical shock.

Sheep can also be conditioned to perform operant responses, and will press panels with their muzzles to make shape discriminations (Baldwin, 1981) or push through a weighted door to obtain food (Jackson et al., 1999), operate foot treadles to obtain sodium solutions (Abraham et al., 1973) and push cards off buckets to make colour discriminations (Bazely and Ensor, 1989). They can reach high rates of response on fixed-ratio schedules for preferred foods (Hutson and van Mourik, 1981; Hutson and Wilson, 1983).

In general, sheep perform well on tasks involving discrimination between left and right turns in U- and T-mazes (Kratzer, 1971). Liddell (1925, 1954) found that individual sheep could learn to run through a simple maze in a few trials. The maze consisted of three parallel alleys, one of which was a cul-de-sac. The sheep was required to find its way up the central alley and then down one of the outer alleys to a food reward. However, when the problem was made more difficult by reversing the position of the blind alley at every trial, the sheep could not learn to run through the maze without error for four consecutive trials. This was despite continued testing of some of the sheep, three times a week for 3 years. However, Liddell reported that some of the sheep found running the maze to be a 'self-rewarding activity'. More recent studies by Hosoi et al. (1995) have shown that sheep exhibit a strong lateral preference in simple T-mazes, pay little attention to either intra- or extra-maze cues and probably do not use maze cues for decision making.

The speed and duration of learning in sheep quite clearly depend upon the nature of the task. Discrimination learning for natural objects like food and the faces of socially familiar animals is much faster than for geometrical symbols, novel objects such as bottles or socially unfamiliar animals (Kendrick, 1998). Learning to associate symbols or novel objects with food can take anything from 10 to 40 trials and learning is often only retained for a few hours or days (Kendrick et al., 1996).

In contrast, individual sheep can remember 50 other different sheep faces for over 2 years (Kendrick et al., 2001), and once lambs have learnt that wheat is a palatable food they can retain this information for up to 34 months (Green et al., 1984). Kendrick (1998) has suggested that the sheep's brain is adapted to efficiently learn associations between natural objects and reward, but not novel associations between artificial objects and reward.

Sheep appear to have an excellent spatial learning ability. Sandler et al. (1968) found that crossbred ewes learned the solution to a simple detour problem in a single trial. Lee et al. (2006) reported that sheep have the ability to learn and retain the spatial memory of a relatively complex maze after three trials. Hutson (1980b) found that sheep in a group could learn a route through yards in a relatively small number (four or five) of trials.

In contrast, Rushen and Congdon (1986a) suggested that the increased transit time taken by sheep to move along a race towards repeated aversive treatments reflected the limited learning abilities of the sheep. But this surprising conclusion is not justified by their experimental results, where one trial was sufficient to demonstrate an aversion to the most severe treatment. A more likely explanation is that repetition of the treatment itself was responsible for the increase, and that cumulative experience of aversive treatments influenced transit times. Sheep will remember an aversive experience for at least 12 weeks (Rushen, 1986a) and for up to 1 year (Hutson, 1985a).

When sheep have been tested on natural spatial memory tasks involving food-finding they have performed extremely well (Rook et al., 2005). Edwards et al. (1996) reported that sheep had the ability to retain information on the spatial distribution of a food resource after just a single exposure. Maximum efficiency was achieved in three to four trials. Sheep could learn the location of a food patch with and without cues, but learned faster when a cue was present.

Associations with cues appeared to act independently of memory of spatial locations (Edwards et al., 1997). When the location of the cue and food patches was switched randomly, sheep used spatial memory first to find the new location (Edwards et al., 1996). Sheep can use spatial memory under even more complex conditions, and learn the location of hidden food in featureless environments with only distant landmarks (Dumont and Petit, 1998).

Sheep are capable of single-trial learning, even at the cellular level. Kendrick (1990) reported that cells in the hypothalamus respond to the sight – but not the smell – of known palatable foods but not to non-food objects. Initially, the cells do not respond to the sight of an unknown food but, if a sheep eats the food just once and likes it, the cells will respond the next time it is seen, even if the sheep has not seen it for a month or more!

Similarly, Provenza and Burritt (1991) have demonstrated single-trial learning in lambs for conditioned aversions to palatable foods treated with the toxin lithium chloride. Naturally occurring plant compounds such as oxalic acid can also induce conditioned aversions after a single exposure (Kyriazakis et al., 1998). Food aversions can even be conditioned in anaesthetized sheep, which suggests that non-cognitive feedback processes are involved (Provenza et al., 1994).

In summary, sheep have excellent learning ability, can be easily conditioned, perform sensory discriminations, acquire aversions, can learn simple mazes and have a good short- and long-term memory.

Implications of Behavioural Characteristics of Sheep for Handling

These four characteristics of the sheep – its flocking behaviour, following behaviour, vision and intelligence – form the basis of all behavioural principles of sheep-handling. I will consider these principles in relation to the three key elements of an integrated sheep handling system – the design of the handling environment, the handling technique and the reason for being handled – the handling treatment.

Design

Hutson (1980c) recommended that the most crucial design criterion was to give sheep a clear, unobstructed view towards the exit, or towards where they are meant to move. This often becomes more evident by taking a sheep's eye view of the facility. Most behavioural principles

of sheep-handling are probably related to this criterion. Thus, sheep movement is generally better on the flat, rather than up- or downhill (Hitchcock and Hutson, 1979a), away from buildings and dead ends (Kilgour and Dalton, 1984), in wide, straight races (Hutson and Hitchcock, 1978) and in well-lit areas (Hitchcock and Hutson, 1979b). Sheep will stop and investigate any novel visual stimulus or change in appearance of a race.

Therefore, shadows or discontinuities on the ground (Hutson, 1980c), changes in race construction material or changes in floor type – e.g. from slats to concrete (Kilgour, 1971) – should be avoided. Handling facilities should be painted one solid colour to avoid contrasts (Grandin, 1980). Judicious use of covered and open panels can direct movement and vision. Ramps should have covered sides, and movement inside sheds should be across the direction of the grating so that sheep can obtain a better grip with their feet and cannot see through the floor or perceive heights (Hutson, 1981a).

Learning, flocking and following behaviour also affect design. Thus, sheep should always be moved through yards and sheds along the same route and in the same direction, as they will learn where they are meant to go (Hutson, 1980b). Sheep flow is better in wide races where sheep can move as a group rather than in single file (Hutson, 1980a). The sight of stationary sheep will slow down sheep movement through an adjacent race (Hutson, 1981b), but sheep will be attracted by the sight of other sheep or alternative visual stimuli, including mirrors, films, photographs and models (Franklin and Hutson, 1982c).

Dogs should be used cautiously, if at all, in confined handling situations, because sheep turn and face dogs when they cannot escape from them (Holmes, 1980). A 5-min exposure to a barking dog is used as a standard stimulus to induce stress in laboratory studies of sheep and elicits an abrupt elevation in adrenocorticotrophic hormone (ACTH) and cortisol concentrations (Cook, 1997; Komesaroff et al., 1998).

Handling technique

Humans have two conflicting roles in sheep handling: one is to act as a forcing stimulus and the other is to administer the treatment.

The findings of Whateley *et al.* (1974) suggest that sheep react to these roles. They found that the relative ease of handling of different breeds reflected that breed's tolerance of humans and dogs, but breeds that handled well in paddocks and yards resisted physical restraint. Hansen *et al.* (2001) have also reported breed differences in the response of sheep to predators, including stuffed carnivores, a man and a man with a dog.

The human as a forcing stimulus

The traditional motivation used to move sheep is the repeated application of fear-inducing stimuli. Sheep handlers use dogs – the natural predator of sheep – or auditory and visual signals, such as shouting and waving, to frighten sheep into moving. Baskin (1974) describes herders waving arms and clothing, throwing rocks and even hoisting their caps on sticks to appear unusually tall. The aim is to frighten the animals and stimulate the flight response.

However, fear-inducing stimuli do not always have the desired effect of prompting movement. Webb (1966) studied a range of stimuli and devices for driving sheep, including coloured and flashing lights, white noise, sinusoidal sound, electric shock, a mechanical sweep and air blasts. Sheep ignored the lights and sound and did not react violently to any of the shocks. Lambs quickly lost their fear of the sweep. Some lambs jumped over the pipe delivering air blasts and the stimulus could not prevent other lambs following.

McCutchan *et al.* (1992) evaluated a mild electric shock as a prompt for sheep movement in a single-file race and found that sheep responded in an unpredictable manner to the stimulus. Some sheep moved forwards, some reversed backwards and some did not respond. Vandenheede and Bouissou (1993, 1996) have shown that rams were less fearful than ewes, but that wethers were more fearful than rams in various test situations.

It appears that the effectiveness of forcing stimuli declines as the sheep approach the area where they are treated. More force is then applied, and both handler and sheep become more aroused in an escalating, vicious-circle effect. Occasionally, the sheep must be physically moved to the treatment area. The most likely explanation for this effect is the dual role of the human handler in forcing and treating sheep. The human is trying to apply more fear to make sheep move towards a fearful stimulus. In addition, increased force will result in greater arousal and less predictable and erratic responses from the sheep, including stopping, freezing, fleeing, baulking, sitting, turning, reverse movement and jumping (Holmes, 1984a; Syme, 1985; Vette, 1985).

Alternative techniques that utilize different motivations – such as the flocking/following response or positive rather than negative reinforcement – could be used. For example, Bremner *et al.* (1980) have successfully trained sheep to lead other sheep, and Hutson (1985a) has reported that food rewards can encourage voluntary movement and improve sheep-handling efficiency. Kiley-Worthington and Savage (1978) used classical conditioning to an auditory alarm to prompt movement of dairy cows. Vette (1985) has described a novel attempt to improve voluntary sheep movement into a single-file race. A rotating circular carousel holding four to six decoy sheep stimulated sheep to move into the race. Another novel attempt, using an artificial wind, was less successful (Hutson and van Mourik, 1982).

The human as a handler

Many years ago the author speculated that future improvements in sheep handling would rely on an animal perspective, which implied that temperament studies should concentrate on the handler (Hutson, 1985b). For example, Seabrook (1972) reported that in dairy herds more milk was obtained by dairymen classified as confident introverts than by non-confident extroverts (see also Chapter 8, this volume). Clear differences were noted in the willingness of cows to enter the milking parlour and return from pasture.

Thus, it is quite evident that some people have the inappropriate personality to be animal handlers, and others will need prolonged training. Hemsworth (Chapter 14, this volume) has identified similar relationships between behaviour and attitude of handlers towards pigs and subsequent production. There is no doubt that similar principles apply to sheep-handling, although handling itself is less frequent.

It is generally assumed that sheep-handling skills are acquired by experience, although there has always been debate about whether good stockhandlers are born or made (Kilgour, 1978). Ewbank (1968) has suggested that they have an understanding of animal psychology that is probably based on acute powers of observation. For example, good stockhandlers:

1. Make the minimum possible use of fearful stimuli (Rushen et al., 1999), avoid using loud noises that animals will associate with handling procedures (Waynert et al., 1999), avoid punishing animals and use positive reinforcements (Hutson, 1985b).

2. Act quickly and decisively, because if handling is fumbled animals become more difficult to restrain (Ewbank, 1968).

3. Are aware of the flight distance of animals and utilize the strategy of reverse movement, i.e. by moving towards confined animals they can prompt movement in the opposite direction more effectively than by frightening them from behind (Hutson, 1982a).

4. Are aware of the importance of arousal in animal handling and have the ability to predict animal responses in any situation (Holmes, 1984a).

Much of this knowledge is commonsense, but it is essential that these techniques are made explicit for the training of inexperienced handlers. Even relatively straightforward procedures such as ear-tagging can cause welfare problems. Studies of ear damage by Edwards and Johnston (1999) and Edwards et al. (2001) have demonstrated that care is necessary during the tag insertion procedure to avoid poor placement and unnecessary trauma. In turn, this is dependent upon appropriate instruction and training of operators. An excellent manual is available detailing sheep-handling skills for New Zealand conditions, but has universal application (Holmes, 1984a); a videotape is also available (Holmes, 1984b).

Handling treatment

Although facility design and handling technique are very important in sheep-handling, the main problem with obtaining efficient throughput of sheep is the nature of the handling treatment inflicted on the sheep. Hutson and Butler (1978) found that race efficiency fell from 93 to 73% after sheep had experienced inversion for 30 s in a handling machine. This suggested that many routine handling treatments were aversive and functioned as negative reinforcers of free movement through the handling system (Hutson, 1982b).

Research in Australia has focused on the stressfulness of various sheep-handling procedures. This research has been prompted by welfare concerns and the potential introduction of new technologies – such as robot shearing (Trevelyan, 1992). Although attempts to replace the shearer with a robot have now been abandoned, we have much more knowledge about the relative stressfulness and aversiveness of different handling treatments. Various physiological and behavioural techniques have been used.

Physiological measures of stressfulness include plasma cortisol, β-endorphin, haematocrit and heart rate. Kilgour and de Langen (1970) were the first to measure plasma cortisol concentrations in sheep for different handling procedures. They found a great deal of individual variation, but some treatments stressed sheep more than others. Dog-chasing – especially when bitten – and prolonged shearing produced the highest cortisol levels. Fulkerson and Jamieson (1982) compared patterns of cortisol release following various stressors and reported the most severe stress was associated with shearing; less stress was imposed by yarding and handling and there was no effect attributable to feeding or fasting. Fell and Shutt (1988) used salivary cortisol to assess acute stressors and ranked treatments in decreasing order of stressfulness as shearing, stop–start transport, steady transport, sham shearing, isolation, cold, jetting and yarding.

β-Endorphin has also been used to monitor the stress response of sheep to potentially painful handling or surgical procedures. Jephcott et al. (1986) found significant rises in β-endorphin after electroimmobilization in comparison with a control handling procedure, and Shutt et al. (1987) reported a threefold increase in β-endorphin concentrations 15 min after tail-docking in lambs, and a maximal eight- to tenfold increase in response to castration and/or mulesing[2] with tail-docking. Shutt and Fell (1988) found

---

[2] Mulesing is a procedure that involves removing skin folds in the tail area; it is performed to reduce the likelihood of blowfly strike.

significant increases in plasma β-endorphin in wethers at 5 min and 15 min following mulesing, and suggested that an endorphin-induced analgesic response lasted for about 1 h. However, the extent to which β-endorphin modulates pain perception is still unknown and controversial.

In another study, Fell and Shutt (1989) found elevated β-endorphin concentrations 24 h after mulesing, and Mears and Brown (1997) found that surgical castration elicited a marked and prolonged elevation of β-endorphin for up to 24 h.

Anil *et al.* (1990) reported a twofold increase in plasma β-endorphin concentrations in response to electrical stunning and a further increase after animals had regained consciousness. Fordham *et al.* (1989) reported that transport did not increase plasma β-endorphin concentrations in lambs above concentrations obtained after mustering with a dog.

Heart rate has also been used as an indicator of response to handling. Webster and Lynch (1966) found a steady increase in heart rate for the week following shearing. Syme and Elphick (1982) reported that sheep unresponsive to social isolation had a lower heart rate than responsive sheep when standing alone in a race. Also, heart rate has been reported to increase in response to visual isolation, transportation, introduction into a new flock, human approach and human approach with a dog (Baldock and Sibly, 1990).

Hargreaves and Hutson (1990a) evaluated the stress response of sheep to routine handling procedures, using plasma cortisol and haematocrit. Both parameters were significantly elevated after shearing compared with untreated sheep, but declined to basal levels within 90 min of treatment. They concluded that sheep perceived shearing, crutching and drafting as more stressful than drenching or dipping. Drenching and dipping were the only treatments in which sheep stayed together as a group.

In a further analysis of the shearing procedure, Hargreaves and Hutson (1990b, c) used several physiological measures – including haematocrit, plasma cortisol, plasma glucose and heart rate – to assess the stress response to components of the procedure. In a series of treatments, sheep were separated from other sheep, isolated, exposed to a human, blood-sampled, up-ended, exposed to shearing noise and partially shorn.

Haematocrit, plasma cortisol and glucose increased significantly after shearing but not following isolation. Shearing was the only treatment that elevated heart rate significantly above pre-treatment values. The response to noise alone was less pronounced than to actual wool removal. It was concluded that wool removal was more stressful than any of the other manipulations involved in conventional shearing.

Isolation may not be as stressful as initial investigations (Kilgour and de Langen, 1970) have suggested. Hargreaves and Hutson (1990b) did not detect a cortisol or haematocrit response to 4 min of isolation and suggested that individual handling and familiarity with the pre-treatment routine may have attenuated this response.

In contrast, Parrott *et al.* (1994) reported an increase in plasma cortisol in response to 60 min of isolation, although the magnitude of the response was less than that to standing in water or simulated transport. Parrott *et al.* (1988a) also reported an increase in plasma cortisol in response to 120 min of isolation. Coppinger *et al.* (1991) and Minton *et al.* (1995) reported that isolation, coupled with restraint (binding of the legs with adhesive tape) for 6 h, produced a robust cortisol response in lambs.

However, it is known that restraint alone will produce an acute cortisol response (Niezgoda *et al.*, 1993). Cockram *et al.* (1994) found that isolation for 24 h produced a significant increase in plasma cortisol after 1.5, 3 and 9 h, but not after 6 and 24 h. The response diminished on subsequent exposures and was not significant at the seventh and 14th repetition. The authors also noted that the response to the first period of isolation may have been affected by movement to the isolation pen and by exposure to a novel environment.

Roussel *et al.* (2004) reported that ewes repeatedly exposed to the treatment of isolation for 1 h – in the presence of a dog for half of the trials – habituated to the treatment. Integrated cortisol responses declined significantly between trials 1, 5 and 9. Degabriele and Fell (2001) reported that sheep taken from pasture and kept in isolation for 12–19 days showed a marked decline in plasma cortisol concentration from day 1 to day 3, which subsequently levelled out.

Clearly, it is hazardous comparing different experiments because of different protocols, but several points emerge from these studies of isolation. Transient isolation associated with handling treatments may not be as stressful as isolation for long periods, and sheep may habituate to long periods of isolation. Habituation may occur if sheep learn that there is an eventual escape and are able predict the frequency and duration of the stressor (Cockram et al., 1994).

The response to isolation may also be modified by the presence of various stimuli, including mirrored panels (Parrott et al., 1988b) and the sight of familiar sheep-face pictures (da Costa et al., 2004). Boivin et al. (1997) noted that lambs vocalized and moved less when in the presence of a shepherd than when isolated. These observations support the hypothesis of Price and Thos (1980), who suggested that humans can serve as an effective substitute for a conspecific and reduce the distress of sheep in isolation.

Behaviour has been used to assess relative aversiveness of handling procedures in choice tests, in aversion learning tests and in approach/avoidance conflicts in an arena test. Rushen and Congdon (1986b) showed that two forced choices were sufficient for nine out of 12 sheep to discriminate between partial shearing and electro-immobilization, although the forcing stimulus itself may have influenced the outcome of such tests. Three sheep were indifferent.

Grandin et al. (1986) reported that sheep preferred restraint on a squeeze-tilt table to electro-immobilization. Rushen (1986b) has also compared more conventional treatments using the forced choice method, and ranked the treatments in decreasing order of preference as: human presence, physical restraint, isolation, capture in isolation and inversion in isolation.

The willingness of sheep to move along a race towards a treatment area has been used to assess the aversiveness of handling treatments in aversion learning tests. Hutson and Butler (1978) found that a single experience of inversion for 30 s in a handling machine was sufficient to make sheep hesitate about moving along the race again. Hutson (1985a) reported that the time spent pushing sheep along a race and into a sheep-handling machine increased with successive trials when sheep were restrained by clamping, and increased at a greater rate when they were clamped and inverted.

Rushen (1986a) reported that the longer transit times of sheep to move to a treatment site indicated that electro-immobilization was more aversive than physical restraint – with or without electrodes attached – and that the degree of aversion decreased with experience of electro-immobilization. Rushen and Congdon (1986a) found that simulated shearing together with electro-immobilization was more aversive than either immobilization or simulated shearing alone.

Aversiveness to electro-immobilization extinguished after five non-treatment trials (J.R. Stollery, 1990, personal communication). Stafford et al. (1996) used an aversion test with rams to show that part shearing was more aversive than free movement and that electro-ejaculation was intermediate between the two.

Long-term behavioural responses to shearing have not been assessed, but it is likely that sheep develop a life-long aversion to the procedure. Sheep developed an aversion to sham shearing after four trials (Rushen and Congdon, 1986a) and cortisol and β-endorphin responses to shearing were significantly greater and longer in previously shorn sheep than in naïve sheep (Mears et al., 1999).

Fell and Shutt (1989) used an arena test to assess the aversion of sheep to the human handler after the mulesing operation. An advantage of this method is that it is not influenced by the behaviour of the experimenter, as in forced-choice tests or aversion-learning tests. Sheep were released into the midpoint of a 14×4-m arena three or four at a time. The handler who held the sheep during treatment stood quietly at one end of the arena, with the remainder of the flock behind him. On entering the arena control, sheep turned and moved towards the handler, whereas mulesed sheep turned and moved in the opposite direction. Mulesed sheep continued to show a pronounced aversion to the handler up to 36 days after the operation, but this aversion was no longer apparent after 114 days. It is not known whether the aversion is generalized to other humans. The arena test has also been used to show that a human is more aversive facing towards rather than away from sheep (Erhard, 2003) and that a dog is more aversive than a human (Beausoleil et al., 2005).

Reducing the aversiveness and stressfulness of handling

A basic dilemma plagues sheep-handling. Some of the treatments are stressful and produce pronounced aversions in sheep. For example, wool removal is the fundamental basis of sheep production and yet it is the prime contributor to the stress response to shearing. How can we improve the ease of handling and the welfare of sheep during handling if there is no prospect of changing the nature of the treatment itself? There are several possibilities:

1. Reducing the severity of the treatment, e.g. the function of the plunge or shower dip can now be performed with a pour-on chemical. Replacing the traditional method of wool severance by comb and reciprocating cutter will be more difficult. Alterations to the method of capture and restraint of sheep during shearing may be possible, and changes to the handpiece that modifies heat production, vibration and wool pull might modify the stress response to shearing. Chemical defleecing research, once a bright hope (Hudson, 1980), has encountered many difficulties, including protection of the shorn sheep from sunburn and extreme cold. An Australian company, Bioclip Pty Ltd, has commercialized the use of epidermal growth factor for wool removal and, in 2002/2003, approximately 200,000 sheep were shorn. The process is currently only available for young sheep (Australian Wool Testing Authority, 2004).

2. Clearly, it is difficult to minimize the severity of surgical treatments. Mulesing is controversial (Townend, 1985), but non-surgical methods such as topical application of a quaternary ammonium compound appear to minimize the acute stress response and prevent the development of a lasting aversion to the handler (Chapman et al., 1994). James (2006) reviews genetic alternatives to mulesing.

There is considerable debate over whether it is better to use the ring or the knife to castrate and tail-dock lambs (Shutt et al., 1987; Barnett, 1988; Shutt and Fell, 1988; Mellor and Murray, 1989; Lester et al., 1991, 1996), with the consensus view now clearly in favour of the ring (Mellor and Stafford, 2000). Even so, the ring still causes considerable distress (Grant, 2004), so that more benign methods should be actively sought (Lester et al., 1996). Spatial and temporal separation of these painful treatments from other handling procedures is advised.

Local anaesthesia can virtually abolish the cortisol response to ring castration and/or tailing. Mellor and Stafford (2000) have reviewed

the use of anaesthetics and recommended that farmers should use them if economically and practically feasible. Anaesthesia is effective for 2–3 h, but timing, injection site and mode of delivery are important variables. Oral aspirin is ineffective (Pollard et al., 2001).

Fell and Shutt (1989) recommended that mulesing of weaners should be carried out by contractors rather than by regular handlers, and that handling should be minimized for several weeks following treatment. This recommendation should also be applied to the milder surgical treatments of castration and tail-docking. These procedures should also not be performed in the usual handling area. Timing of castration can also reduce acute physiological and behavioural responses. In general, lambs castrated at 1 week had lower responses than those castrated at 4–6 weeks, regardless of method (Mellor and Stafford, 2000).

3. Training sheep by the use of food rewards. Hutson (1985a) reported that sheep trained to run through a handling system for a food reward required less labour for subsequent handling than unrewarded sheep. Grandin (1989) demonstrated that food rewards could also be used to entice previously restrained or electro-immobilized sheep to voluntarily enter and accept restraint in a squeeze-tilt table.

Thus, with time, previously acquired aversions to even severe handling treatments will diminish, and the response of sheep to a handling procedure will approach that of naïve sheep (Hutson, 1985a). Siegel and Moberg (1980) reported that neonatal stress in lambs did not influence later performance or adrenocortical response in an active avoidance task. Similarly, Cook (1996) reported a reduction in the acute cortisol response of sheep that had learned to avoid an electric shock and no change in their basal cortisol concentrations.

4. Habituation to the handling procedure. Simple repetition of a stimulus and an animal's habituation to it may be more important than whether it is perceived as being pleasant or unpleasant. Siegel and Moberg (1980) showed that the adrenocortical response of 8-month-old lambs decreased over ten sessions of avoidance conditioning to an electric shock.

However, when Hargreaves and Hutson (1990e) exposed sheep to a sham shearing procedure on four occasions at 2-week intervals, the stress response was only slightly reduced by repetition. Peak cortisol response to the procedure was not affected, but concentrations declined more rapidly to baseline levels after four exposures. Adrenal responsiveness in response to an ACTH challenge was not affected. Fordham *et al.* (1989) reported no change in cortisol concentrations after twice-daily jugular venepuncture of lambs for 5 days. Thus, repeated exposure to stressors seems unlikely to modify handling responses.

5. Hargreaves and Hutson (1990d) investigated the gentling of adult sheep as a method of reducing the aversion to subsequent handling. Gentling is often used with laboratory animals when repeated tactile, visual and auditory contact with a human makes them easier to handle. Sheep were gentled by being visually isolated, talked to and patted for 20 s/day for 35 consecutive days. Gentling reduced the flight distance and heart rate response to humans, but did not reduce the aversion to a handling procedure (sham shearing). Mateo *et al.* (1991) confirmed these results by gentling sheep for 5 min/day for 21 days. Gentling improved the approachability of sheep to humans, but did not attenuate their response to restraint or shearing.

6. Gradual conditioning of sheep to humans and handling. There is some evidence that older, multiparous ewes are less fearful of humans than young, nulliparous ewes and that the experience of motherhood may play a part (Viérin and Bouissou, 2002). Therefore, the first handling experience is critical. Markowitz *et al.* (1998) reported that 40 min of positive human contact at 1–3 days reduced the subsequent timidity of lambs towards people. Uetake *et al.* (2000) reported that gentling lambs for 10 min/day during the first 10 days after birth improved their subsequent ease of handling.

Similarly, Boivin *et al.* (2000) found that human contact – especially stroking and feeding

during the first 4 weeks after birth – strongly influenced the approach behaviour of lambs to a familiar stockperson. Tallet *et al.* (2005) demonstrated that gently handling artificially reared lambs during the days following birth increased their subsequent affinity for their stockperson. Clearly, these studies indicate that early contact with humans can reduce fear and improve ease of handling.

Miller (1960) found that rats trained to run down an alley for food could be induced to continue running, even when severely shocked in the goal box, so long as they had been exposed to a series of shocks of gradually increasing intensity. However, animals that had received an intense shock at the outset showed a complete suppression of running. Although generalizations from rats to sheep should be treated cautiously, Miller's study has important implications for training sheep to handling procedures, and indicates that training should be gradual, with exposure to innocuous treatments first before more stressful treatments such as shearing are carried out.

7. Simplifying the learning procedure, so that all handling is done in one location – preferably the shed – using a uniform method of restraint. More research on the role of conditioning in the acquisition and extinction of behavioural and physiological responses to humans and handling is required. In particular, studies of cues associated with the treatment, the place it is performed and the identity of the handler need to be carried out.

8. Finally, the conflicting roles of humans in sheep-handling need to be resolved. I suggest that only dogs, machines and electrical and mechanical devices should be used as fear-inducing forcing stimuli. This would complete the transition from hunter/herder to farmer and allow humans to concentrate on their responsibility for the care and well-being of sheep.

References

Abraham, S., Baker, R., Denton, D.A., Kraintz, F., Kraintz, L. and Purser, L. (1973) Components in the regulation of salt balance: salt appetite studied by operant behaviour. *Australian Journal of Experimental Biology and Medical Science* 51, 65–81.

Alexander, G. (1978) Odour, and the recognition of lambs by Merino ewes. *Applied Animal Ethology* 4, 153–158.

Alexander, G. and Shillito, E.E. (1977) The importance of odour, appearance and voice in maternal recognition of the young in Merino sheep (*Ovis aries*). *Applied Animal Ethology* 3, 127–135.

Alexander, G. and Stevens, D. (1979) Discrimination of colours and grey shades by Merino ewes: tests using coloured lambs. *Applied Animal Ethology* 5, 215–231.

Ames, D.R. and Arehart, L.A. (1972) Physiological response of lambs to auditory stimuli. *Journal of Animal Science* 34, 994–998.

Anil, M.H., Fordham, D.P. and Rodway, R.G. (1990) Plasma β-endorphin increase following electrical stunning in sheep. *British Veterinary Journal* 146, 476–477.

Arnold, G.W. (1977) An analysis of spatial leadership in a small field in a small flock of sheep. *Applied Animal Ethology* 3, 263–270.

Arnold, G.W. and Pahl, P.J. (1974) Some aspects of social behaviour in domestic sheep. *Animal Behaviour* 22, 592–600.

Arnould, C. and Signoret, J.P. (1993) Sheep food repellents: efficacy of various products, habituation, and social facilitation. *Journal of Chemical Ecology* 19, 225–236.

Australian Wool Testing Authority (2004) A brief history of wool harvesting research. *August Newsletter*, pp. 14–19 (http://www.awta.com.au/Publications/Newsletter/2004_August/features2.htm).

Baldock, N.M. and Sibly, R.M. (1990) Effects of handling and transportation on the heart rate and behaviour of sheep. *Applied Animal Behaviour Science* 28, 15–39.

Baldwin, B.A. (1981) Shape discrimination in sheep and calves. *Animal Behaviour* 29, 830–834.

Baldwin, B.A. and Meese, G.B. (1977) The ability of sheep to distinguish between conspecifics by means of olfaction. *Physiology and Behaviour* 18, 803–808.

Banks, E.M. (1964) Some aspects of sexual behavior in domestic sheep, *Ovis aries. Behaviour* 23, 249–279.

Barnett, J.L. (1988) Surgery may not be preferable to rubber rings for tail docking and castration of lambs. *Australian Veterinary Journal* 65, 131.

Baskin, L.M. (1974) Management of ungulate herds in relation to domestication. In: Geist, V. and Walther, F.R. (eds) *Behaviour of Ungulates and its Relation to Management*, Vol. 2. International Union for Conservation of Nature, Morges, Switzerland, pp. 530–541.

Bazely, D.R. and Ensor, C.V. (1989) Discrimination learning in sheep with cues varying in brightness and hue. *Applied Animal Behaviour Science* 23, 293–299.

Beausoleil, N.J., Stafford, K.J. and Mellor, D.J. (2005) Sheep show more aversion to a dog than to a human in an arena test. *Applied Animal Behaviour Science* 91, 219–232.

Blissitt, M.J., Bland, K.P. and Cottrell, D.F. (1990) Discrimination between the odours of fresh oestrus and non-oestrus ewe urine by rams. *Applied Animal Behaviour Science* 25, 51–59.

Boissy, A. and Dumont, B. (2002) Interactions between social and feeding motivations on the grazing behaviour of herbivores: sheep more easily split into subgroups with familiar peers. *Applied Animal Behaviour Science* 79, 233–245.

Boivin, X., Nowak, R., Desprès, G., Tournadre, H. and Le Neindre, P. (1997) Discrimination between shepherds by lambs reared under artificial conditions. *Journal of Animal Science* 75, 2892–2898.

Boivin, X., Tournadre, H. and Le Neindre, P. (2000) Hand-feeding and gentling influence early-weaned lambs' attachment responses to their stockperson. *Journal of Animal Science* 78, 879–884.

Bremner, K.J. (1980) Follow my 'leader' – techniques for training sheep. *New Zealand Journal of Agriculture* September, 25–29.

Bremner, K.J., Braggins, J.B. and Kilgour, R. (1980) Training sheep as 'leaders' in abattoirs and farm sheep yards. *Proceedings of the New Zealand Society of Animal Production* 40, 111–116.

Chapman, R.E., Fell, L.R. and Shutt, D.A. (1994) A comparison of stress in surgically and non-surgically mulesed sheep. *Australian Veterinary Journal* 71, 243–247.

Clarke, P.G.H. and Whitteridge, D. (1973) The basis of stereoscopic vision in the sheep. *Journal of Physiology* 229, 22–23.

Cockram, M.S., Ranson, M., Imlah, P., Goddard, P.J., Burrells, C. and Harkiss, G.D. (1994) The behavioural, endocrine and immune responses of sheep to isolation. *Animal Production* 58, 389–399.

Cook, C.J. (1996) Basal and stress response cortisol levels and stress avoidance learning in sheep (*Ovis ovis*). *New Zealand Veterinary Journal* 44, 162–163.

Cook, C.J. (1997) Oxytocin and prolactin suppress cortisol responses to acute stress in both lactating and non-lactating sheep. *Journal of Dairy Research* 64, 327–339.

Cooper, J., Gordon, I.J. and Pike, A.W. (2000) Strategies for the avoidance of faeces by grazing sheep. *Applied Animal Behaviour Science* 69, 15–33.

Coppinger, T.R., Minton, J.E., Reddy, P.G. and Blecha, F. (1991) Repeated restraint and isolation stress in lambs increases pituitary-adrenal secretions and reduces cell-mediated immunity. *Journal of Animal Science* 69, 2808–2814.

Crofton, H.D. (1958) Nematode parasite populations in sheep on lowland farms. VI. Sheep behaviour and nematode infections. *Parasitology* 48, 251–260.

da Costa, A.P., Leigh, A.E., Man, M.S. and Kendrick, K.M. (2004) Face pictures reduce behavioural, autonomic, endocrine and neural indices of stress and fear in sheep. *Proceedings of the Royal Society London B* 271, 2077–2084.

Davis, H., Norris, C. and Taylor, A. (1998) Wether ewe know me or not: the discrimination of individual humans by sheep. *Behavioural Processes* 43, 27–32.

Degabriele, R. and Fell, L.R. (2001) Changes in behaviour, cortisol and lymphocyte types during isolation and group confinement of sheep. *Immunology and Cell Biology* 79, 583–589.

Dumont, B. and Boissy, A. (2000) Grazing behaviour of sheep in a situation of conflict between feeding and social motivations. *Behavioural Processes* 49, 131–138.

Dumont, B. and Petit, M. (1998) Spatial memory of sheep at pasture. *Applied Animal Behaviour Science* 60, 43–53.

Dwyer, C.M. (2004) How has the risk of predation shaped the behavioural responses of sheep to fear and distress. *Animal Welfare* 13, 269–281.

Dwyer, C.M., McLean, K.A., Deans, L.A., Chirnside, J., Calvert, S.K. and Lawrence, A.B. (1998) Vocalisations between mother and young in sheep: effects of breed and maternal experience. *Applied Animal Behaviour Science* 58, 105–119.

Edwards, D.S. and Johnston, A.M. (1999) Welfare implications of ear tags. *Veterinary Record* 144, 603–606.

Edwards, D.S., Johnston, A.M. and Pfeiffer, D.U. (2001) A comparison of commonly used ear tags on the ear damage of sheep. *Animal Welfare* 10, 141–151.

Edwards, G.R., Newman, J.A., Parsons, A.J. and Krebs, J.R. (1996) The use of spatial memory by grazing animals to locate food patches in spatially heterogeneous environments: an example with sheep. *Applied Animal Behaviour Science* 50, 147–160.

Edwards, G.R., Newman, J.A., Parsons, A.J. and Krebs, J.R. (1997) Use of cues by grazing animals to locate food patches: an example with sheep. *Applied Animal Behaviour Science* 51, 59–68.

Erhard, H.W. (2003) Assessing the relative aversiveness of two stimuli: single sheep in the arena test. *Animal Welfare* 12, 349–358.

Espmark, Y., Falt, L. and Falt, B. (1974) Behavioural responses in cattle and sheep exposed to sonic booms and low-altitude subsonic flight noise. *Veterinary Record* 94, 106–113.

Ewbank, R. (1968) The behavior of animals in restraint. In: Fox, M.W. (ed.) *Abnormal Behaviour in Animals.* Saunders, London, pp. 159–178.

Ewbank, R. and Mansbridge, R.J. (1977) The effects of sonic booms on farm livestock. *Applied Animal Ethology* 3, 292.

Fell, L.R. and Shutt, D.A. (1988) Salivary cortisol and behavioural indicators of stress in sheep. *Proceedings of the Australian Society of Animal Production* 17, 186–189.

Fell, L.R. and Shutt, D.A. (1989) Behavioural and hormonal responses to acute surgical stress in sheep. *Applied Animal Behaviour Science* 22, 283–294.

Fordham, D.P., Lincoln, G.A., Ssewannyana, E. and Rodway, R.G. (1989) Plasma β-endorphin and cortisol concentrations in lambs after handling, transport and slaughter. *Animal Production* 49, 103–107.

Franklin, J.R. and Hutson, G.D. (1982a) Experiments on attracting sheep to move along a laneway. I. Olfactory stimuli. *Applied Animal Ethology* 8, 439–446.

Franklin, J.R. and Hutson, G.D. (1982b) Experiments on attracting sheep to move along a laneway. II. Auditory stimuli. *Applied Animal Ethology* 8, 447–456.

Franklin, J.R. and Hutson, G.D. (1982c) Experiments on attracting sheep to move along a laneway. III. Visual stimuli. *Applied Animal Ethology* 8, 457–478.

Fulkerson, W.J. and Jamieson, P.A. (1982) Pattern of cortisol release in sheep following administration of synthetic ACTH or imposition of various stressor agents. *Australian Journal of Biological Science* 35, 215–222.

Geist, V. (1971) *Mountain Sheep: a Study in Behaviour and Evolution.* University of Chicago Press, Chicago, Illinois.

Grandin, T. (1980) Livestock behavior as related to handling facilities design. *International Journal for the Study of Animal Problems* 1, 33–52.

Grandin, T. (1989) Voluntary acceptance of restraint by sheep. *Applied Animal Behaviour Science* 23, 257–261.

Grandin, T. (2001) Cattle vocalizations are associated with handling and equipment problems at beef slaughter plants. *Applied Animal Behaviour Science* 71, 191–201.

Grandin, T., Curtis, S.E., Widowski, T.M. and Thurmon, J.C. (1986) Electro-immobilization *versus* mechanical restraint in an avoid-avoid choice test for ewes. *Journal of Animal Science* 62, 1469–1480.

Grant, C. (2004) Behavioural responses of lambs to common painful husbandry procedures. *Applied Animal Behaviour Science* 87, 255–273.

Green, G.C., Elwin, R.L., Mottershead, B.E., Keogh, R.G. and Lynch, J.J. (1984) Long-term effects of early experience to supplementary feeding in sheep. *Proceedings of the Australian Society of Animal Production* 15, 373–375.

Grubb, P. and Jewel, P.A. (1966) Social grouping and home range in feral Soay sheep. *Symposia of the Zoological Society of London* 18, 179–210.

Hall, S.J.G., Kirkpatrick, S.M., Lloyd, D.M. and Broom, D.M. (1998) Noise and vehicular motion as potential stressors during the transport of sheep. *Animal Science* 67, 467–473.

Hansen, I., Christiansen, F., Hansen, H.S., Braastad, B. and Bakken, M. (2001) Variation in behavioural responses of ewes towards predator-related stimuli. *Applied Animal Behaviour Science* 70, 227–237.

Hargreaves, A.L. and Hutson, G.D. (1990a) The stress response in sheep during routine handling procedures. *Applied Animal Behaviour Science* 26, 83–90.

Hargreaves, A.L. and Hutson, G.D. (1990b) Changes in heart rate, plasma cortisol and haematocrit of sheep during a shearing procedure. *Applied Animal Behaviour Science* 26, 91–101.

Hargreaves, A.L. and Hutson, G.D. (1990c) An evaluation of the contribution of isolation, up-ending and wool removal to the stress response to shearing. *Applied Animal Behaviour Science* 26, 103–113.

Hargreaves, A.L. and Hutson, G.D. (1990d) The effect of gentling on heart rate, flight distance, and aversion of sheep to a handling procedure. *Applied Animal Behaviour Science* 26, 243–252.

Hargreaves, A.L. and Hutson, G.D. (1990e) Some effects of repeated handling on stress responses in sheep. *Applied Animal Behaviour Science* 26, 253–265.

Hargreaves, A.L. and Hutson, G.D. (1997) Handling systems for sheep. *Livestock Production Science* 49, 121–138.

Hinch, G.N., Lynch, J.J., Elwin, R.L. and Green, G.C. (1990) Long-term associations between Merino ewes and their offspring. *Applied Animal Behaviour Science* 27, 93–103.

Hitchcock, D.K. and Hutson, G.D. (1979a) The movement of sheep on inclines. *Australian Journal of Experimental Agriculture and Animal Husbandry* 19, 176–182.

Hitchcock, D.K. and Hutson, G.D. (1979b) The effect of variation in light intensity on sheep movement through narrow and wide races. *Australian Journal of Experimental Agriculture and Animal Husbandry* 19, 170–175.

Holmes, R.J. (1980) Dogs in works: friend or foe? *4 Quarter* 1 (3), 16–17.

Holmes, R.J. (1984a) *Sheep and Cattle Handling Skills: a Manual for New Zealand Conditions.* Accident Compensation Corporation, Wellington, New Zealand.

Holmes, R.J. (1984b) *Stockhandling Skills* (video). Accident Compensation Corporation, Wellington, New Zealand.

Hosoi, E., Swift, D.M., Rittenhouse, L.R. and Richards, R.W. (1995) Comparative foraging strategies of sheep and goats in a T-maze apparatus. *Applied Animal Behaviour Science* 44, 37–45.

Hudson, P.R.W. (1980) Technology brings home the wool harvest. *New Scientist* 1218, 768–771.

Hulet, C.V., Alexander, G. and Hafez, E.S.E. (1975) The behaviour of sheep. In: Hafez, E.S.E. (ed.) *The Behaviour of Domestic Animals.* Baillière Tindall, London, pp. 246–294.

Hutson, G.D. (1980a) Visual field, restricted vision and sheep movement in laneways. *Applied Animal Ethology* 6, 175–187.

Hutson, G.D. (1980b) The effect of previous experience on sheep movement through yards. *Applied Animal Ethology* 6, 233–240.

Hutson, G.D. (1980c) Sheep behaviour and the design of sheep yards and shearing sheds. In: Wodzicka-Tomaszewska, M., Edey, T.N. and Lynch, J.J. (eds) *Behaviour in Relation to Reproduction, Management and Welfare of Farm Animals.* University of New England Publishing Unit, Armidale, New South Wales, Australia, pp. 137–141.

Hutson, G.D. (1981a) Sheep movement on slatted floors. *Australian Journal of Experimental Agriculture and Animal Husbandry* 21, 474–479.

Hutson, G.D. (1981b) An evaluation of some traditional and contemporary views on sheep behaviour. *Wool Technology and Sheep Breeding* 29 (1), 3–6.

Hutson, G.D. (1982a) 'Flight distance' in Merino sheep. *Animal Production* 35, 231–235.

Hutson, G.D. (1982b) Sheep handling facilities. *Proceedings of the Australian Society of Animal Production* 14, 121–123.

Hutson, G.D. (1984) Spacing behaviour of sheep in pens. *Applied Animal Behaviour Science* 12, 111–119.

Hutson, G.D. (1985a) The influence of barley food rewards on sheep movement through a handling system. *Applied Animal Behaviour Science* 14, 263–273.

Hutson, G.D. (1985b) Sheep and cattle handling facilities. In: Moore, B.L. and Chenoweth, P.J. (eds) *Grazing Animal Welfare*. Australian Veterinary Association, Indooroopilly, Queensland, Australia, pp. 124–136.

Hutson, G.D. (1994) So who's being woolly minded now? *New Scientist* 1951, 52–53.

Hutson, G.D. and Butler, M.L. (1978) A self-feeding sheep race that works. *Journal of Agriculture (Victoria)* 76, 335–336.

Hutson, G.D. and Hitchcock, D.K. (1978) The movement of sheep around corners. *Applied Animal Ethology* 4, 349–355.

Hutson, G.D. and van Mourik, S.C. (1981) Food preferences of sheep. *Australian Journal of Experimental Agriculture and Animal Husbandry* 21, 575–582.

Hutson, G.D. and van Mourik, S.C. (1982) Effect of artificial wind on sheep movement along indoor races. *Australian Journal of Experimental Agriculture and Animal Husbandry* 22, 163–167.

Hutson, G.D. and Wilson, P.N (1983) A note on the preference of sheep for whole or crushed grains and seeds. *Animal Production* 38, 145–146.

Jackson, R.E., Waran, N.K. and Cockram, M.S. (1999) Methods for measuring feeding motivation in sheep. *Animal Welfare* 8, 53–63.

Jacobs, G.H., Deegan, J.F. and Neitz, J. (1998) Photopigment basis for dichromatic color vision in cows, goats, and sheep. *Visual Neuroscience* 15, 581–584.

James, P.J. (2006) Genetic alternatives to mulesing and tail docking in sheep: a review. *Australian Journal of Experimental Agriculture* 46, 1–18.

Jephcott, E.H., McMillen, I.C., Rushen, J., Hargreaves, A. and Thorburn, G.D. (1986) Effect of electro-immobilisation on ovine plasma concentrations of β-endorphin/β-lipotrophin, cortisol and prolactin. *Research in Veterinary Science* 41, 371–377.

Keller, M., Meurisse, M., Poindron, P., Nowak, R. and Ferreira, G. (2003) Maternal experience influences the establishment of visual/auditory, but not olfactory recognition of the newborn lamb by ewes at parturition. *Developmental Psychobiology* 43, 167–176.

Kendrick, K.M. (1990) Through a sheep's eye. *New Scientist* 1716, 40–43.

Kendrick, K.M. (1991) How the sheep's brain controls the visual recognition of animals and humans. *Journal of Animal Science* 69, 5008–5016.

Kendrick, K.M. (1998) Intelligent perception. *Applied Animal Behaviour Science* 57, 213–231.

Kendrick, K.M. and Baldwin, B.A. (1987) Cells in temporal cortex of conscious sheep can respond preferentially to the sight of faces. *Science* 296, 448–450.

Kendrick, K.M., Atkins, K., Hinton, M.R., Broad, K.D., Fabre-Nys, C. and Keverne, B. (1995) Facial and vocal discrimination in sheep. *Animal Behaviour* 49, 1665–1676.

Kendrick, K.M., Atkins, K., Hinton, M.R., Heavens, P. and Keverne, B. (1996) Are faces special for sheep? Evidence from facial and object discrimination learning tests showing effects of inversion and social familiarity. *Behavioural Processes* 38, 19–35.

Kendrick, K.M., da Costa, A.P., Leigh, A.E., Hinton, M.R. and Peirce, J.W. (2001) Sheep don't forget a face. *Nature* 414, 165–166.

Kiley-Worthington, M. and Savage, P. (1978) Learning in dairy cattle using a device for economical management of behaviour. *Applied Animal Ethology* 4, 119–124.

Kilgour, R. (1971) Animal handling in works – pertinent behaviour studies. In: *Proceedings 13th Meat Industry Research Conference*, Meat Industry Research Institute, Hamilton, New Zealand, pp. 9–12.

Kilgour, R. (1976) Sheep behaviour: its importance in farming systems, handling, transport and pre-slaughter treatment. In: Truscott, G.M.C. and Wroth, F.H. (eds) *Sheep Assembly and Transport Workshop*. Western Australian Department of Agriculture, Perth, Australia, pp. 64–84.

Kilgour, R. (1977) Animal behaviour and its implications: 1. Design sheepyards to suit the whims of sheep. *New Zealand Farmer* 98 (6), 29–31.

Kilgour, R. (1978) Minimising stress on animals during handling. *Proceedings 1st World Congress on Ethology Applied to Zootechnics*. International Veterinary Association for Animal Production, Madrid, Spain, E-IV-5, pp. 303–322.

Kilgour, R. and Dalton, D.C. (1984) *Livestock Behaviour*. University of New South Wales Press, Sydney, Australia.

Kilgour, R. and de Langen, H. (1970) Stress in sheep resulting from management practices. *Proceedings of the New Zealand Society of Animal Production* 30, 65–76.

Kilgour, R. and Matthews, L.R. (1983) Sheep are not stupid. *New Zealand Journal of Agriculture* 147 (4), 48–50.

Kilgour, R., Pearson, A.J. and de Langen, H. (1975) Sheep dispersal patterns on hill country: techniques for study and analysis. *Proceedings of the New Zealand Society of Animal Production* 35, 191–197.

Komesaroff, P.A., Esler, M., Clarke, I.J., Fullerton, M.J. and Funder, J.W. (1998) Effects of estrogen and estrus cycle on glucocorticoid and catecholamine responses to stress in sheep. *American Journal of Physiology* 275, E671–E678.

Kratzer, D.D. (1971) Learning in farm animals. *Journal of Animal Science* 32, 1268–1273.

Krausman, P.R., Wallace, M.C., Hayes, C.L. and DeYoung, D.W. (1998) Effects of jet aircraft on mountain sheep. *Journal of Wildlife Management* 62, 1246–1254.

Kyriazakis, I., Anderson, D.H. and Duncan, A.J. (1998) Conditioned flavour aversions in sheep: the relationship between the dose rate of a secondary plant compound and the acquisition and persistence of aversions. *British Journal of Nutrition* 79, 55–62.

Lee, C., Colegate, S. and Fisher, A.D. (2006) Development of a maze test and its application to assess spatial learning and memory in Merino sheep. *Applied Animal Behaviour Science* 96, 43–51.

Lemmon, W.B. and Patterson, G.H. (1964) Depth perception in sheep. Effects of interrupting the mother–neonate bond. *Science* 145, 835–836.

Lester, S.J., Mellor, D.J., Ward, R.N. and Holmes, R.J. (1991) Cortisol responses of young lambs to castration and tailing using different methods. *New Zealand Veterinary Journal* 39, 134–138.

Lester, S.J., Mellor, D.J., Holmes, R.J., Ward, R.N. and Stafford, K.J. (1996) Behavioural and cortisol responses of lambs to castration and tailing using different methods. *New Zealand Veterinary Journal* 44, 45–54.

Liddell, H.S. (1925) The behavior of sheep and goats in learning a simple maze. *American Journal of Psychology* 36, 544–552.

Liddell, H.S. (1954) Conditioning and emotions. *Scientific American* 190, 48–57.

Liddell, H.S. and Anderson, O.D. (1931) A comparative study of the conditioned motor reflex in the rabbit, sheep, goat and pig. *American Journal of Physiology* 97, 339–340.

Lynch, J.J., Hinch, G.N. and Adams, D.B. (1992) *The Behaviour of Sheep*. CAB International, Wallingford, UK.

Markowitz, T.M., Dally, M.R., Gursky, K. and Price, E.O. (1998) Early handling increases lamb affinity for humans. *Animal Behaviour* 55, 573–587.

Mateo, J.M., Estep, D.Q. and McCann, J.S. (1991) Effects of differential handling on the behaviour of domestic ewes (*Ovis aries*). *Applied Animal Behaviour Science* 32, 45–54.

McCutchan, J.C., Freeman, R.B. and Hutson, G.D. (1992) Failure of electrical prompting to improve sheep movement in single file races. *Proceedings of the Australian Society of Animal Production* 19, 455.

Mears, G.J. and Brown, F.A. (1997) Cortisol and β-endorphin responses to physical and psychological stressors in lambs. *Canadian Journal of Animal Science* 77, 689–694.

Mears, G.J., Brown, F.A. and Redmond L.R. (1999) Effects of handling, shearing and previous exposure to shearing on cortisol and β-endorphin responses in ewes. *Canadian Journal of Animal Science* 79, 35–38.

Mellor, D.J. and Murray, L. (1989) Changes in the cortisol responses of lambs to tail docking, castration and ACTH injection during the first seven days after birth. *Research in Veterinary Science* 46, 392–395.

Mellor, D.J. and Stafford, K.J. (2000) Acute castration and/or tailing distress and its alleviation in lambs. *New Zealand Veterinary Journal* 48, 33–43.

Miller, N.E. (1960) Learning resistance to pain and fear: effects of overlearning, exposure, and rewarded exposure in context. *Journal of Experimental Psychology* 60, 137–145.

Minton, J.E., Apple, J.K., Parsons, K.M. and Blecha, F. (1995) Stress-associated concentrations of plasma cortisol cannot account for reduced lymphocyte function and changes in serum enzymes in lambs exposed to restraint and isolation stress. *Journal of Animal Science* 73, 812–817.

Munkenbeck, N.W. (1982) Color vision in sheep. *Journal of Animal Science* (Suppl. 1) 55, 129 (abstract).

Neitz, J. and Jacobs, G.H. (1989) Spectral sensitivity of cones in an ungulate. *Visual Neuroscience* 2, 97–100.

Niezgoda, J., Bobek, S., Wronska-Fortuna, D. and Wierzchos, E. (1993) Response of sympatho-adrenal axis and adrenal cortex to short-term restraint stress in sheep. *Journal of Veterinary Medicine A* 40, 631–638.

Parrott, R.F., Thornton, S.N. and Robinson, J.E. (1988a) Endocrine responses to acute stress in castrated rams: no increase in oxytocin but evidence for an inverse relationship between cortisol and vasopressin. *Acta Endocrinologica* 117, 381–386.

Parrott, R.F., Houpt, K.A. and Misson, B.H. (1988b) Modification of the responses of sheep to isolation stress by the use of mirror panels. *Applied Animal Behaviour Science* 19, 331–338.

Parrott, R.F., Misson, B.H. and de la Riva, C.F. (1994) Differential stressor effects on the concentrations of cortisol, prolactin and catecholamines in the blood of sheep. *Research in Veterinary Science* 56, 234–239.

Penning, P.D., Parsons, A.J., Newman, J.A., Orr, R.J. and Harvey, A. (1993) The effects of group size on the grazing time in sheep. *Applied Animal Behaviour Science* 37, 101–109.

Piggins, D. and Phillips, C.J.C. (1996) The eye of the domesticated sheep with implications for vision. *Animal Science* 62, 301–308.

Pollard, J.C., Roos, V. and Littlejohn, R.P. (2001) Effects of an oral dose of acetyl salicylate at tail docking on the behaviour of lambs aged three to six weeks. *Applied Animal Behaviour Science* 71, 29–42.

Price, E.O. and Thos, J. (1980) Behavioral responses to short-term social isolation in sheep and goats. *Applied Animal Ethology* 6, 331–339.

Provenza, F.D. and Burritt, E.A. (1991) Socially induced diet preference ameliorates conditioned food aversion in lambs. *Applied Animal Behaviour Science* 31, 229–236.

Provenza, F.D., Lynch, J.J. and Nolan, J.V. (1994) Food aversion conditioned in anesthetized sheep. *Physiology and Behaviour* 55, 429–432.

Rook, A.J. and Penning, P.D. (1991) Synchronisation of eating, ruminating and idling activity by grazing sheep. *Applied Animal Behaviour Science* 32, 157–166.

Rook, A.J., Rodway-Dyer, S.J. and Cook, J.E. (2005) Effects of resource density on spatial memory and learning by foraging sheep. *Applied Animal Behaviour Science* 95, 143–151.

Roussel, S., Hemsworth, P.H., Boissy, A. and Duvaux-Ponter, C. (2004) Effects of repeated stress during pregnancy in ewes on the behavioural and physiological responses to stressful events and birth weight of their offspring. *Applied Animal Behaviour Science* 85, 259–276.

Rowell, T.E. (1991) Till death us do part: long-lasting bonds between ewes and their daughters. *Animal Behaviour* 42, 681–682.

Rushen, J. (1986a) Aversion of sheep to electro-immobilization and physical restraint. *Applied Animal Behaviour Science* 15, 315–324.

Rushen, J. (1986b) Aversion of sheep for handling treatments: paired-choice studies. *Applied Animal Behaviour Science* 16, 363–370.

Rushen, J. and Congdon, P. (1986a) Relative aversion of sheep to simulated shearing with and without electro-immobilisation. *Australian Journal of Experimental Agriculture* 26, 535–537.

Rushen, J. and Congdon, P. (1986b) Sheep may be more averse to electro-immobilisation than to shearing. *Australian Veterinary Journal* 63, 373–374.

Rushen, J., Taylor, A.A. and de Passillé, A.M. (1999) Domestic animals' fear of humans and its effect on their welfare. *Applied Animal Behaviour Science* 65, 285–303.

Sandler, B.E., van Gelder, G.A., Buck, W.B. and Karas, G.G. (1968) Effect of dieldrin exposure on detour behavior in sheep. *Psychological Reports* 23, 451–455.

Scott, J.P. (1945) Social behaviour, organization and leadership in a small flock of domestic sheep. *Comparative Psychology Monographs* 18, 1–29.

Seabrook, M.F. (1972) A study to determine the influence of the herdsman's personality on milk yield. *Journal of Agricultural Labour Science* 1, 45–59.

Searby, A. and Jouventin, P. (2003) Mother–lamb acoustic recognition in sheep: a frequency coding. *Proceedings of the Royal Society London B* 270, 1765–1771.

Shillito, E.E. (1972) Vocalization in sheep. *Journal of Physiology* 226, 45–46.

Shutt, D.A. and Fell, L.R. (1988) The role of endorphins in the response to stress in sheep and cattle. *Proceedings of the Australian Society of Animal Production* 17, 338–341.

Shutt, D.A., Fell, L.R., Connell, R., Bell, A.K., Wallace, C.A. and Smith, A.I. (1987) Stress-induced changes in plasma concentrations of immunoreactive β-endorphin and cortisol in response to routine surgical procedures in lambs. *Australian Journal of Biological Science* 40, 97–103.

Siegel, B.J. and Moberg, G.P. (1980) The influence of neonatal stress on the physiological and behavioral response of lambs during active-avoidance conditioning. *Hormones and Behaviour* 14, 136–145.

Squires, V.R. (1975) Ecology and behaviour of domestic sheep (*Ovis aries*): a review. *Mammal Review* 5, 35–57.

Squires, V.R. and Daws, G.T. (1975) Leadership and dominance relationships in Merino and Border Leicester sheep. *Applied Animal Ethology* 1, 263–274.

Stafford, K.J., Spoorenberg, J., West, D.M., Vermunt, J.J., Petrie, N. and Lawoko, C.R.O. (1996) The effect of electro-ejaculation on aversive behaviour and plasma cortisol concentration in rams. *New Zealand Veterinary Journal* 44, 95–98.

Stolba, A., Hinch, G.N., Lynch, J.J., Adams, D.B., Munro, R.K. and Davies, H.I. (1990) Social organization of Merino sheep of different ages, sex and family structure. *Applied Animal Behaviour Science* 27, 337–349.

Syme, L.A. (1981) Social disruption and forced movement orders in sheep. *Animal Behaviour* 29, 283–288.

Syme, L.A. (1985) *Intensive Sheep Management, with Particular Reference to the Live Sheep Trade.* Australian Agricultural Health and Quarantine Service, Department of Primary Industry, Canberra, Australia.

Syme, L.A. and Elphick, G.R. (1982) Heart-rate and the behaviour of sheep in yards. *Applied Animal Ethology* 9, 31–35.

Tallet, C., Veissier, I. and Boivin, X. (2005) Human contact and feeding as rewards for the lamb's affinity to their stockperson. *Applied Animal Behaviour Science* 94, 59–73.

Tanaka, T., Asakawa, K., Kawahara, Y., Tanida, H. and Yoshimoto, T. (1989a) Color discrimination in sheep. *Japanese Journal of Livestock Management* 24, 89–94.

Tanaka, T., Sekino, M., Tanida, H. and Yoshimoto, T. (1989b) Ability to discriminate between similar colors in sheep. *Japanese Journal of Zootechnical Science* 60, 880–884.

Tanaka, T., Hashimoto, A., Tanida, H. and Yoshimoto, T. (1995) Studies on the visual acuity of sheep using shape discrimination learning. *Journal of Ethology* 13, 69–75.

Torres-Hernandez, G. and Hohenboken, W. (1979) An attempt to assess traits of emotionality in crossbred ewes. *Applied Animal Ethology* 5, 71–83.

Townend, C. (1985) *Pulling the Wool.* Hale and Ironmonger, Sydney, Australia.

Trevelyan, J.P. (1992) *Robots for Shearing Sheep.* Oxford University Press, Oxford, UK.

Uetake, K., Yamaguchi, S. and Tanaka, T. (2000) Psychological effects of early gentling on the subsequent ease of handling in lambs. *Animal Science Journal* 71, 515–519.

Vandenheede, M. and Bouissou, M.F. (1993) Sex differences in fear reactions in sheep. *Applied Animal Behaviour Science* 37, 39–55.

Vandenheede, M. and Bouissou, M.F. (1996) Effects of castration on fear reactions of male sheep. *Applied Animal Behaviour Science* 47, 211–224.

Vette, M. (1985) *Sheep Behaviour: an Analysis of Yard-to-restrainer Operations in Twenty-six New Zealand Export Meat Plants.* Technical Report, Meat Industry Research Institute of New Zealand, Hamilton, New Zealand.

Viéren, M. and Bouissou, M.F. (2002) Influence of maternal experience on fear reactions in ewes. *Applied Animal Behaviour Science* 75, 307–315.

Walk, R.D. and Gibson, E.J. (1961) A comparative and analytical study of visual depth perception. *Psychological Monographs* 75, 1–44.

Walls, G.L. (1942) *The Vertebrate Eye and its Adaptive Radiation.* Cranbrook Institute of Science, Bloomfield Hills, Michigan.

Waynert, D.F., Stookey, J.M., Schwartzkopf-Genswein, K.S., Watts, J.M. and Waltz, C.S. (1999) The response of beef cattle to noise during handling. *Applied Animal Behaviour Science* 62, 27–42.

Webb, T.F. (1966) *Feasibility Tests of Selected Stimuli and Devices to Drive Livestock.* Report 52-11, Agricultural Research Service, Department of Agriculture, Beltsville, Maryland.

Webster, M.E.D. and Lynch, J.J. (1966) Some physiological and behavioural consequences of shearing. *Proceedings of the Australian Society of Animal Production* 6, 234–239.

Weisenberger, M.E., Krausman, P.R., Wallace, M.C., De Young, D.W. and Maughan, O.E. (1996) Effects of simulated jet aircraft noise on heart rate and behavior of desert ungulates. *Journal of Wildlife Management* 60, 52–61.

Whateley, J., Kilgour, R. and Dalton, D.C. (1974) Behaviour of hill country sheep breeds during farming routines. *Proceedings of the New Zealand Society of Animal Production* 34, 28–36.

Whitteridge, D. (1978) The development of the visual system in the sheep. *Archives Italiennes de Biologie* 116, 406–408.

Winfield, C.G. and Kilgour, R. (1976) A study of following behaviour in young lambs. *Applied Animal Ethology* 2, 235–243.

Winfield, C.G., Syme, G.J. and Pearson, A.J. (1981) Effect of familiarity with each other and breed on the spatial behaviour of sheep in an open field. *Applied Animal Ethology* 7, 67–75.

Wollack, C.H. (1963) The auditory activity of the sheep (*Ovis aries*). *Journal of Auditory Research* 3, 121–132.

11 Design of Sheep Yards and Shearing Sheds*

Adrian Barber[1] and Robert B. Freeman[2]

[1]*Department of Agriculture, Keith, South Australia, Australia;* [2]*Agricultural Engineering Section, University of Melbourne, Melbourne, Victoria, Australia*

Introduction

There have been significant advances in the design of sheep yards and shearing sheds. Well-designed facilities enable thousands of sheep to be handled efficiently, humanely and with a minimum of labour.

Sheep Yards

Main pens and force pens should be built of material that is clearly visible and strong. Fence height is 90 cm for fences inside the yard area and 1 m for fences at the outside edge of the yards. Sheep density in the main pens is 1.5–2 sheep/m^2 and rises to 2.5–3 sheep/m^2 in the force pens.

Where a curved force pen (a bugle) (Fig. 11.1) is used to lead sheep into a drafting (sorting) or handling race, the curved approach should be constructed of solid covering material or sheeting on the inner side for the last 6 or 8 m. Many bugle curves work well with open panels on the outer side of the draft race and force pen. However, in some cases, it may be necessary to blind off the outer draft wall and the outside fence for the last 1–3 m from the race entrance.

When the force pen leads to a handling race, it should hold a few more than one or two times the number of sheep that can fit in the race. This leaves some left over after the race is filled to act as decoys when bringing the next lot of sheep into the force pen. Generally, a force pen should not hold more than 100 sheep. Otherwise, there are too many to control and start up the race.

Triangular force pens are usually used in rectangular yards and can be built in single or double forms (Fig. 11.2). A single force has one fence as an extension of the drafting or handling race side, with the second fence set at a 30–40° angle. Double triangular forces have two-wing fences running back at similar angles and a central fence with a flip-flop gate at the race entrance to allow sheep entry from either side.

Drafting Race

An efficient drafting (sorting) system allows the operator to identify and draft the sheep he or she wishes to separate with a minimum of errors. To do this accurately requires an even flow of sheep. For small flocks, a two-way sort is satisfactory, but in large-scale sheep enterprises, particularly where crossbred sheep for prime lamb production are run, a three-way sort, using two gates, may be necessary. Four- and five-way sorts can be built for special purposes, such as on stud properties.

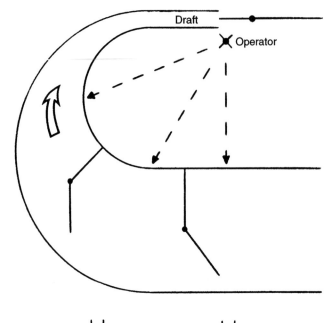

Fig. 11.1. Bugle force pen, showing operator position. The curved design provides the operator with easy access to the sheep.

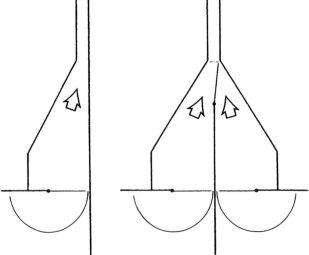

Fig. 11.2. Triangular force pens, single and double. One side is straight and the other is on a 30° angle.

The sorting race is approximately 3 m long, with the exit point showing a clear escape route for the sheep. Where a bugle force pen leads to the draft, the straight outer fence is usually 4–4.5 m long to allow for the in-flow curve.

Several points to be considered when building a drafting race are:

1. Closed-in sides;
2. Adjustable sides;
3. Tapered sides;
4. A durable floor (concrete or battens);
5. A stop gate at the outlet of the race;
6. Rubber dampers on the leading edge of the draft gate(s) to reduce noise;
7. A large-diameter vertical roller at the race entrance to prevent sheep jamming (e.g. a 200-l drum set with its surface almost flush with the angled fence);
8. A remote-control gate.

Siting the drafting race

The direction of the drafting race should minimize the effects of sunlight and shadows. Sheep appear to run better into the sun (their shadows are behind them) and the stock worker has a better chance to see ear tags or earmarks when the sun is shining on the heads of the approaching sheep. This will not work in the very early morning if the sun blinds the sheep.

The race should be directed away from, or parallel to, the shearing shed wall so that the sheep do not get an impression of approaching a dead end. The race should be on flat or slightly upwardly inclined ground, but not inclined downwards. Sheep in the drafted pens need to be clearly visible to other sheep still coming through the race, and thus act as decoys to encourage sheep flow. In some yard layouts, the drafting race will face a shearing shed wall. This can be satisfactory, if the race is set well back and there is plenty of pen area between the race and the shed.

Side panels

In most sheep yards, the drafting race has closed sheeting on both sides to prevent outward vision by the sheep. This is to direct the sheep's attention towards the exit – sheep will actually increase speed to get through the race.

However, in some yards, a race with open panels on the side opposite the stock worker has been found to work well. The sheep are encouraged to enter the race by seeing other sheep moving away from the drafting gate on the outer side.

Timber planks, steel sheeting or weatherproof plywood are suitable materials for the closed sides of the race. The sheeting must be placed on the inner faces, so as to give a completely smooth surface. Bolts should be countersunk, or round-headed coach bolts should be used with the nuts to the outside.

The draft gate

The draft gate is usually 1.3 m long, but may be in the range of 1–1.5 m. Where twin gates are used to give three-way drafting, the handles should be offset from the gate top to prevent the stock worker from jamming his or her hands between the gates. The handle should also have a looped knuckle guard, which is similar to the guard loops on handlebars of agricultural motor bikes, projecting forward to protect the stock worker's hands from charging sheep.

There is some debate on whether the draft gate should be made of open panels or closed sheeting. Reasons for using open-rail gates include:

1. The oncoming sheep can see the previous sheep moving away from the draft and are more inclined to follow;
2. Open gates are lighter, and therefore, quicker and easier to use;
3. Open gates are less affected by winds blowing across the drafting race.

Reasons for using closed sheeting gates include:

1. Such gates act as a continuation of the drafting race, thus directing the sheep into the exit pen;
2. Closed gates prevent horns or legs from getting caught.

Remote-control sort gates

There are situations where it is an advantage to be able to draft sheep from behind, for example picking daggy (dirty) sheep. This requires some type of remote-control mechanism to operate the drafting gate (Fig. 11.3).

Sheep-handling Race

A multipurpose handling race for drenching, vaccinating, branding, classing, jetting (spraying) and other agricultural activities is needed in the sheep yards. The handling race usually leads off the same force pen as the drafting race. Alternatively, it can be constructed as a forward extension of the drafting race.

Several different types of handling races can be built.

1. A single race of 52–64 cm wide where the stock worker is outside the race.
2. A single race of 70–80 cm wide where the stock worker is inside the race.

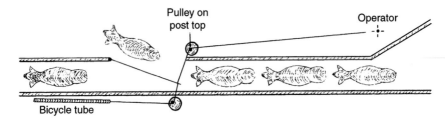

Fig. 11.3. Remote-control draft (sort) gate, with spring return.

3. An adjustable-sided race of which the width can be varied between 45 and 80 cm.

4. Double races side by side are satisfactory, but they must be worked correctly. A new batch of sheep is allowed to fill the first race while treated sheep are running out of the adjacent one. It can be difficult to fill one race if sheep are held in the second. The sheep will run about halfway along the race and then stop beside the other sheep they see through the rails.

5. Triple races are successful where the sheep fill the two outer races and the stock worker is in the centre race.

A suitable handling race is 9–15 m long with sides 85 cm high. One or two stop gates are included along its length to prevent lambs from crowding. The race should have a concrete or batten floor that extends 70–100 cm on either side to form a walkway at the same level as the race. Either steel or timber can be used to build the handling race. Steel rails are generally preferred, as they allow greater access and vision by the operator through the race sides. One section of the race wall can be removable panels or a gate section. This provides for sheep to be diverted to other sheep-handling facilities, for example, a cradle-crutching unit, an automatic jetting (spraying) race or weighing scale.

Sheep flow is often improved if the take-off point to these facilities is near the start of the race rather than at the end. If possible, a drafting gate should be positioned at the exit of the handling race to allow separation of groups of sheep after an agricultural activity, such as classing.

Bugle Yards

Bugle yards, or bugle races (force pens), work by leading the sheep around a curve while the sheep pathway narrows down to a draft or handling race (Figs 11.4 and 11.5).

Bugle yards attempt to create an improved sheep flow compared with traditional square yards. This is a debatable point that has not been proved, which indicates that sheep yard shape is not the only factor required for efficient sheep flow. Other factors, such as sunlight and shade effects, ground slope, building materials and position of the shearing shed, can also have an effect.

Bugle yards can be built with two- or three-way sorts. The main working areas of the yards are shown in the plans. Additional holding yards or mini-paddocks can be added to the sides of the plans to increase sheep-holding and working capacities to suit individual properties.

Curved-type yards can turn to either the left or right. That is, for a given plan, there will be a matching mirror image.

The dimensions of a bugle force pen leading into a race are of critical importance in obtaining a satisfactory flow of sheep. If the curves narrow down too sharply, sheep can double up and become jammed. On the other hand, if the curved entry is too wide, sheep have the opportunity to turn back away from the race.

A right-angled triangle, 4 m (4000 mm) along the base and 2 m (2000 mm) vertically, is used to establish the curved fences. A radius of 7.5 m from the toe of the triangle gives the inner curve, and a radius of 13 m from the top of the triangle gives the outer curve. A 1.5 m radius curve leads into the sort race (see Fig. 11.4).

Three-way drafting is provided. Note that the centre batch of sheep does not have to travel the length of the handling race. The sheep are diverted by a panel gate behind the main sort gates into a storage pen beside the handling race. A common drafting problem in sheep yards that have the sort and handling races in

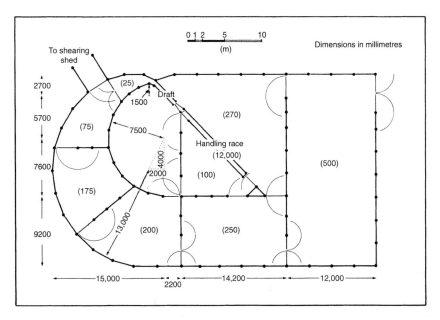

Fig. 11.4. Diagram of a bugle sheep yard.

line is that the centre batch of sheep being fed through the handling race may baulk about halfway along. Sheep then bunch up back to the sort gates. The use of a fence, diagonal to the handling race, and an extra gate to create an additional storage pen, overcomes this difficulty.

Curved-type yards can also be drafted in two ways (right or left) at the exit of the handling race.

The side and shape of the receiving pen and position of the gates may be altered to suit different sites.

Circular Yards

The circular plan has a sheep pathway 2.5–3 m wide around the circumference of the yard, with a combined draft/handling race across the diameter (Fig. 11.6). The race allows three-way drafting at the main gates and two-way drafting at the exit of the handling race. The double bugle force pen leading to the race allows for multi-mob handling for shearing or crutching. The gates can be set so that sheep can bypass the race entrance. Being symmetrical in shape, the yards can be reversed, with the race heading in the opposite direction for fitting on particular sites.

The circular design gives a high degree of flexibility of sheep movements around the yards. The plan is ideal for stud properties. The circular raceways can be readily divided into small pens for displaying rams at field days and sales.

Shearing Sheds

Older shearing sheds in Australia were designed with a centre-board layout (Fig. 11.7). Sheep are moved from the holding pens into the small catching pens. The shearers catch the sheep one at a time and pull each to the shearing board (platform the shearers work on). After each sheep is shorn, it is dispatched by sliding it down a chute to a level below the holding pens. Some sheds have races for discharging shorn sheep.

After being shorn several times, sheep become reluctant to move from the holding pens into the catching pens. The front-fill catching pen was developed to help solve this problem (Fig. 11.8). Sheep are moved from the holding pen into a filling pen and are then moved into the catching pen through a gate near the front of the pen. Some other modern designs are the parallel-flow and diagonal-flow

Fig. 11.5. Bugle sheep yards with solid sides.

methods. To reduce walking distances for wool handlers, the shearing board in a modern shed is curved (Figs 11.8 and 11.9).

The woolly sheep must walk up a ramp to get into the shed, because the floor is above ground level. The ramp should be 1.5–3 m

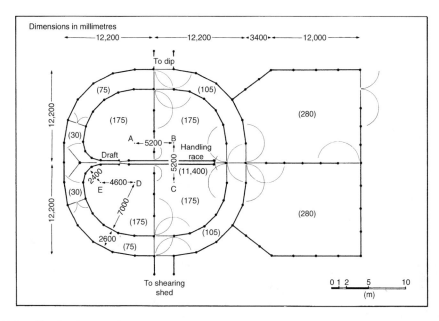

Fig. 11.6. Circular sheep yard.

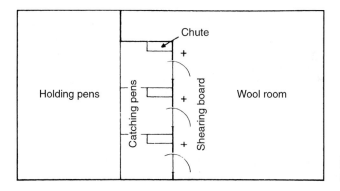

Fig. 11.7. Australian centre-board shearing shed layout.

wide, with a maximum slope of 20°. The sides of the ramp should be made of closed panels to restrict vision. Sheep will shy away from a visual cliff, such as that represented by a ramp with unprotected sides. To provide sufficient height for the discharge chutes, a minimum of 1.2 m of headroom is required under the main floor.

When wood floor gratings are used, the battens should run at right angles to the direction the sheep are required to move. When sheep walk across the battens, it restricts their vision and helps to prevent them from seeing sunlight coming up through the floor.

A shearing shed must be large enough to shelter the daily throughput of the shed from rain. Woolly sheep can be held in an area of 2.5 sheep/m². Shorn sheep can be held at 3.5 sheep/m².

Careful attention must be paid to the design of the discharge chutes. A chute should start on a 45° angle and then change to a 20° angle at the halfway point. This design will help prevent injuries.

Shearing sheds should have good lighting to facilitate sheep movement. Translucent plastic panels installed in the roof will provide evenly diffused lighting, and will help prevent the distorting influence of strong contrasts of light and dark shadows.

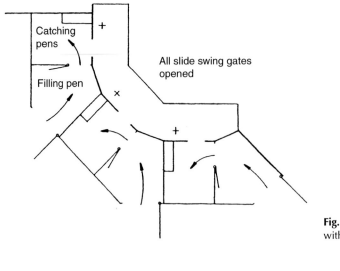

Fig. 11.8. Curved shearing board with front-fill catching pens.

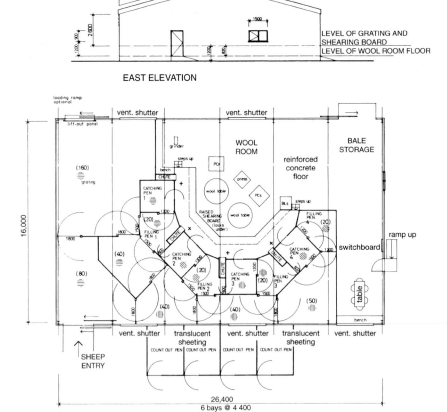

Fig. 11.9. Floor plan of three-stand (three shearers) curved-board shearing shed. (Dimensions in millimetres.)

Acknowledgements

More detailed information on sheep yards and shearing sheds can be obtained in *Design of Shearing Sheds and Sheep Yards* (1986) by A.A. Barber and R.B. Freeman, published by Inkata Press, 18 Salmon Street (PO Box 146), Port Melbourne, Victoria 3207, Australia, on which this chapter is based.

Diagrams of sheep yards and shearing sheds are reproduced with permission of Inkata Press. Sheep movement can be affected by many visual factors such as lighting, shadows, reflections or the position of people. See other chapters for more information on behavioural principles of livestock handling and how vision affects movement thoughout a handling facility.

Further Reading

Anon. (1987) Elevated race system edges towards success. *New Zealand Journal of Agriculture* 152, 20–24. *Describes elevated race system for shearing sheds.*

Belschner, H.G. (1957) *Sheep Management and Diseases*. Halstead Press, Sydney, Australia. *Sheep yard designs for drafting (sorting) many different ways.*

Burnell, R.E. (1967) Cradles make foot treatment easier. *New Zealand Journal of Agriculture* 114 (4), 50–53. *Describes sheep restraint devices.*

Grandin, T. (1984) Sheep handling and facilities. In: Baker, F.H. and Mason, M.E. (eds) *Sheep and Goat Handbook*. Westview Press, Boulder, Colorado. *Sheep handling and yard design.*

Grandin, T. (1990) Handling can improve productivity. *National Wool Grower* (American Sheep Industry Foundation, Englewood, Colorado) 79 (12), 12–14.

Jefferies, B. (1981) Sheep cradles can be used for many jobs. Fact Sheet, Agdex 430/724, Department of Agriculture, South Australia, Australia. *Information on different types of sheep restraint devices.*

McLaren, L. (1977) Sheep laneways. Fact Sheet, Agdex 430/10, Department of Agriculture and Fisheries, South Australia, Australia.

Midwest Plan Service (1982) *Sheep Housing and Equipment Handbook*. Midwest Plan Service, Iowa State University, Ames, Iowa. *Plan for easy to build tilting sheep restraint device.*

Miller, K.R. and Meade, W.J. (1984) A simple restrainer for sheep for abdominal radiography. *New Zealand Veterinary Journal* 32, 146–148. *Canvas restraint device which is easy to build.*

Pearse, E.H. (1965) *Sheep and Property Management*. The Pastoral Review Pty Ltd., Sydney, Australia. *Yard systems for drafting (sorting) in many different ways.*

Ralph, I.G. (1975) Forward planning in sheep handling systems. *Pastoral Review* 85 (10), 679–681. *Design of tilting sheep restraint devices.*

Rogers, P. and Gerhard, L. (2003) *Sheep Production Handbook*. 7th edn., American Sheep Industry Association, Centennial, Colorado.

12   Sheep Transport

Michael S. Cockram

*Division of Veterinary Clinical Sciences, Royal (Dick) School of Veterinary Studies,
University of Edinburgh, Easter Bush Veterinary Centre, Easter Bush, Roslin,
Midlothian, UK*

Introduction

Sheep are farmed throughout the world in a variety of environments for meat, wool and also milk. China has the largest sheep population (170 million), followed by Australia (106 million). India, Iran, Sudan, New Zealand, the UK, South Africa, Turkey, Pakistan, Nigeria and Spain all have sheep populations greater than 20 million (FAOSTAT data, 2005). Within each country, sheep are transported to slaughter or between farms for further fattening or breeding. As sheep are mainly kept extensively as grazing animals, they may have to be transported from one location to another due to the relative availability of pasture.

The transport of sheep for slaughter is influenced by the supply chain. For example, in some countries large supermarkets source sheep from producer marketing groups or buyers' agents direct from farms and then transport them to a limited number of large slaughterhouses that may be some distance from the farm. Traditional butchers are more likely to source their lamb from farmers that sell their sheep via livestock auction markets, dealers or direct from the farm, and the sheep may be slaughtered locally or at a larger slaughterhouse.

Farmers may also supply lamb directly to the consumer on a smaller scale after transporting sheep direct to a small or medium-sized slaughterhouse. The journey from farm to slaughterhouse might not be direct and can consist of transfer of sheep from one vehicle to another, multiple collections from a number of farms and periods of holding at either assembly points or auction markets (Murray *et al.*, 2000).

Although it is possible to slaughter sheep and then transport the meat either chilled or frozen, there is still a significant trade in live sheep (see Fig. 12.1) that are subsequently killed at their destination (often after a further, brief period of fattening) (RSPCA, 1993). The transportation of sheep is a topic that can attract controversy. For example, there is public concern over the sea transport of sheep from Australia to the Middle East (Keniry *et al.*, 2003) and the long-distance road transport of sheep within Europe. This chapter brings together the available information to clarify some of these issues. The effects of road transport on sheep sent to slaughter were reviewed by Knowles (1998). However, there are still many topics that require further investigation, and the interpretation of existing information can at times be ambiguous and subjective.

Animal Welfare

When sheep are transported they are potentially exposed to a number of factors that, either

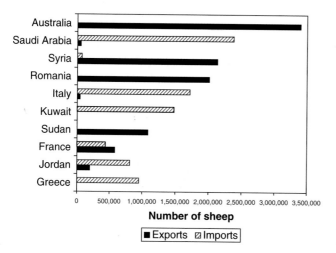

Number of sheep

■ Exports ☑ Imports

Fig. 12.1. Trade in live sheep during 2004 (FAOSTAT data, 2005).

on their own or in combination, could affect their welfare. In addition to the novelty of many aspects of transportation, sheep can be exposed to a variety of changes in their physical and social environment during each stage of transportation. Cockram and Mitchell (1999) considered the main potential effects of transport within the framework of the UK Farm Animal Welfare Council's Five Freedoms, namely:

- Freedom from thirst, hunger and malnutrition: water and feed restriction before and during a journey and changes in diet.
- Freedom from discomfort: thermal and physical discomfort as a consequence of extremes of temperature, inadequate ventilation and space, vibration, acceleration and motion.
- Freedom from pain, injury and disease: handling, impacts associated with motion, interactions between animals and infection.
- Freedom to express normal behaviour: behavioural restriction due to confinement, motion and social disruption.
- Freedom from fear and distress: handling, confinement and exposure to novel stimuli.

All of the Farm Animal Welfare Council's freedoms could be impaired during transportation. Whether this occurs will depend upon the health and fitness of the sheep, the quality of the journey and the associated handling and management of the animals. Some of the effects associated with transport can be additive, others

may interact and there are differences in how different types of sheep respond to these factors. Research that shows the manner in which sheep respond to a single factor might underestimate the net effect of that factor during transport, and studies that report the net effect of transport on sheep can be influenced by the particular circumstances of the trial.

Road Transport

Behaviour during transport

During a 12-h road journey, sheep spend most of the time standing relatively still, but may need to brace themselves and make frequent foot movements to maintain balance in response to vehicular movements (Cockram et al., 1996). They stand with their head raised for about 80% of the time and their head lowered for about 10% (Jones et al., 2002). Although it is likely to be dependent on the type of sheep and their social composition, butting can occur during transport (Cockram et al., 2004). If they are given sufficient space, sheep will lie down, but they lie down less and ruminate less than non-transported sheep kept in their home pen.

However, the reduction in lying and rumination that occurs after the confinement of sheep for 12 h on a livestock vehicle is similar regardless of whether the vehicle is driven on roads or remains stationary (Cockram et al., 1996). The amount of time spent lying and rumination is further reduced if the sheep are transported directly from pasture compared with those previously housed (Cockram et al., 2000). Almost all of the time that sheep spend lying down is with their head raised rather than with the neck relaxed and the head lowered (Cockram et al., 2004). Most rumination occurs when the sheep are lying down, but some rumination can occur while standing (Cockram et al., 1996, 2004).

As a journey progresses, the amount of lying behaviour increases (Cockram et al., 1997, 2004) and towards the end of a 24-h journey lying is the most prominent behaviour. When driven on minor or rough roads, fewer sheep lie down than when the vehicle is driven on smooth roads or motorways (Ruiz-de-la-Torre et al., 2001; Cockram et al., 2004). Driving events such as cornering or braking can disturb lying and rumination behaviour and driving style can influence the behaviour of the sheep (Cockram et al., 2004).

Lying behaviour can also be affected by floor space allowance. For example, Cockram et al. (1996) showed that after the first 3 h of a journey, a space allowance of 0.22 m²/sheep for a 35 kg sheep with a full fleece restricted the amount of lying behaviour compared with higher space allowances (0.27, 0.31 and 0.41 m²/sheep), and they showed more lying behaviour post-transport than those able to lie down during the journey.

Warriss et al. (2002) summarized information on recommended space allowances for sheep and considered that the ability of a 41 kg sheep to stand in a normal posture was impaired at a space allowance of 0.19 m²/sheep. Warriss et al. (2003) considered that the most reliable method for estimating the stocking density of sheep on a vehicle was to estimate the live weight of the sheep using girth measurements, count the number of sheep and measure the dimensions of the pen.

The vertical or deck height is also an important consideration in enabling sheep to stand in a normal posture and in allowing sufficient air space above the sheep so that air can flow into, across and out of the inlets on the side of the vehicle. Jones et al. (2002) estimated that the minimum deck height for a 65 kg sheep should be 95.5 cm.

There is no strong evidence that space allowance during road transport in Europe influences

plasma cortisol concentration, biochemical indicators of plasma dehydration, rectal temperature or the risk of injury (Jarvis and Cockram, 1994; Cockram et al., 1996; Knowles et al., 1998; Ibáñez et al., 2002). Although sheep at a low space allowance have a degree of mutual support and this can reduce the frequency of losses of balance and slips, the frequency that sheep fall over in response to vehicular movement (even when they do not have mutual support) is low (Cockram et al., 1996, 2004).

However, road type and driving style can affect the frequency of losses of balance by sheep. Cockram et al. (2004) found that the frequency of losses of balance by sheep was 21 times greater on minor roads than on a motorway. Eighty-two per cent of losses of balance were preceded by a driving event (acceleration, braking, stopping, cornering, gear changes and uneven surfaces), and about 22% of driving events were followed by a loss of balance.

Injury associated with transport

Although there is a risk that some sheep will sustain an injury associated with transport (e.g. fractures, skin injuries and bruising), studies of sheep carcasses in slaughterhouses in the UK and New Zealand show that most sheep are not bruised (Peterson, 1978; Jarvis and Cockram, 1994; Knowles et al., 1994a; Jarvis et al., 1996).

However, factors that increase the risk of bruising are: (i) the presence of a fleece (compared with sheep that had been shorn); and (ii) transport via an auction market (compared with direct from a farm). Although some studies have reported increases, most studies of sheep transported in Europe on journeys lasting from 12 to 30 h have not shown the significant increases in average values for plasma creatine kinase (CK) activity that would indicate significant muscular damage in transit (Knowles et al., 1993, 1994b, 1995, 1996; Broom et al., 1996; Cockram et al., 1996, 1997; Parrott et al., 1998a).

Loading and unloading are times when the sheep are also at risk of injury, and the angle of the ramp can affect the speed of loading and unloading. Hitchcock and Hutson (1979) showed that the time taken by sheep either to go up or down a ramp increased markedly when the angle of the ramp increased from 30 to 45°. Jarvis and Cockram (1994) studied relationships between potentially traumatic events during loading, unloading and handling of sheep transported to a slaughterhouse and the occurrence of bruising. They found potential relationships between the occurrence of wool-pulls by handlers and riding behaviour between sheep and the occurrence of bruising, but no relationship between bruising and traumatic events during loading/unloading.

Stress responses to transport

When sheep are gathered from a paddock and then loaded on to a vehicle, their plasma cortisol concentration can rise (Broom et al., 1996). However, if sheep are loaded carefully without exertion, via a loading ramp on to a single-deck vehicle, they do not show an increase in plasma cortisol concentration, whereas sheep that are loaded in this manner and subsequently transported do show an increased plasma cortisol concentration (Cockram et al., 1996).

This finding was confirmed by Parrott et al. (1998b), who loaded ewes up a ramp on to a single-deck vehicle and compared this with loading the sheep in a wooden crate on to the vehicle using a mechanical lift. They found no significant differences between the methods of loading in terms of heart rate and plasma cortisol responses to the two procedures.

Transport is stressful to sheep in that it causes a rise in the plasma cortisol concentration and heart rate during the first part of a journey (maximum after about 3 h), and then these variables decline (Broom et al., 1996; Cockram et al., 1996, 1997; Parrott et al., 1998a). Whether the reduction in plasma cortisol concentration that occurs after the initial stages of a journey is due to habituation to transport, as a result of a reduction in the perception of the severity of the stimulus or as a result of either feedback mechanisms or exhaustion of the ability to respond to the stimulus is not clear. It is possible that sheep continue to perceive transport as an aversive stimulus, but this may no longer be reflected in the peripheral plasma cortisol concentration.

Smith et al. (2003) transported ewes for 2 h and found that the concentrations of adrenocorticotrophin hormone (ACTH),

corticotrophin-releasing hormone (CTRH) and arginine vasopressin peaked 40 min into the journey, but only the concentration of ACTH and cortisol decreased during the remainder of the journey, the concentrations of (CTRH) and arginine vasopressin were elevated during the entire journey. Although the authors considered that negative feedback mechanisms may have been responsible for the hormonal changes, modelling suggested that the inputs to the hypothalamus required to stimulate pituitary secretion of CTRH and arginine vasopressin were reduced between the first and second hour of transport.

The magnitude of the plasma cortisol response to transport is similar to that of social isolation (Parrott et al., 1994), dipping, shearing or movement by use of a dog (Kilgour and de Langen, 1970). Not all ewes show a plasma cortisol response to road transport (Reid and Mills, 1962), and the cortisol response to transport is reduced for 1–2 weeks after lambing (Smart et al., 1994).

Stress responses to transport could potentially influence reproduction in breeding ewes. The release of an egg from the ovary of a ewe in oestrus is dependent on a surge in luteinizing hormone (LH) that is released from the pituitary gland. If transportation occurs 2–4 h before this event, both the amplitude of the release and its onset can be delayed which could cause fertility problems (Smith and Dobson, 2002). However, there is no evidence that short journeys of anoestrous ewes affect the onset of oestrus (Fahmy and Guilbault, 1989).

Thermoregulation during transport

If sheep have access to food and water and protection from excessive air movement, solar radiation and precipitation, they have very efficient mechanisms for responding to changes in their thermal environment (Alexander, 1974). However, not only does transport not always provide these features, but high stocking density (Fisher et al., 2002), poor ventilation (Kettlewell et al., 2001), exercise, stress (Parrott et al., 1999) and movement from their established environment could all predispose sheep to thermal environments that exceed their capacity to maintain homeostasis.

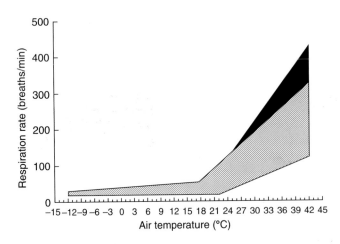

Fig. 12.2. Diagram compiled from reports in the literature of the respiration rate of non-transported sheep with a fleece in relation to air temperature. The shaded area (diagonal lines) represents the range of reported responses and the black area shows the increased responses that occur at high humidity. The responses of shorn sheep are similar, but increases in respiration rate might not occur until the air temperature reaches 25–30°C.

The thermoregulatory capacity of sheep to respond to environmental conditions varies between breeds, and also their degree of adaptation to heat or cold (Webster *et al.*, 1969; Degen and Shkolnik, 1978; Slee and Forster, 1983; Srikandakumar *et al.*, 2003). The temperature within a vehicle transporting sheep will be higher than the external temperature (due to the metabolic heat production of the sheep), especially during stationary periods such as loading or during driver breaks (when there is reduced ventilation), and the humidity will be raised if the sheep are loaded with wet fleece (Jarvis and Cockram, 1999).

Most reports of temperature and humidity during road and ferry transportation in the UK, France or New Zealand (Knowles *et al.*, 1996; Hall *et al.*, 1999; Jarvis and Cockram, 1999; Fisher *et al.*, 2002, 2004) indicate that the sheep were likely to have been within or close to a thermoneutral environment (Alexander, 1974) for most of the time. However, in other parts of the world, and during stationary periods, sheep can be exposed to more challenging thermal environments (Fisher *et al.*, 2002).

In response to a hot environment, sheep are able to maintain a body temperature of ≤ 40°C until the air temperature nears body temperature (39–40°C). This is mainly achieved by increased respiration rate to increase heat loss via respiratory evaporation of water and by allowing body temperature to fluctuate (Blaxter *et al.*, 1959). Short-term studies of non-transported sheep show that an increase in respiration rate occurs at about 20°C in sheep with a fleece

(see Fig, 12.2) and at about 25–30°C in shorn sheep.

These temperatures are only approximate and will be lowered by increased humidity, solar radiation and the restricted ability of closely packed sheep at a high stocking density to lose heat by convection and radiation. Shorn sheep can increase their rate of sweating at high temperatures, but their capacity to sweat in response to heat is less than that of cattle and man, and sweating is unlikely to contribute greatly to heat loss in sheep with a full fleece (Brook and Short, 1960).

If sheep are exposed to high temperature and high humidity, their ability to lose heat by evaporative water loss is impaired compared with conditions of lower humidity and this can result in hyperthermia (Bligh, 1963; Hales and Brown, 1974). Hyperthermia (rectal temperature > 40.5°C) has been reported in conditions of high humidity, at an air temperature of 33°C in a sheep with a fleece and at 40°C in a shorn sheep.

When a sheep becomes hyperthermic there is a change from rapid, shallow to slower, deeper, open-mouthed panting (Hales, 1969). If sheep have access to water, their intake is increased in hot conditions and this can assist cooling (Blaxter *et al.*, 1959). Water deprivation for 3 days – but not for 12 h – can reduce the ability of sheep to control their body temperature when exposed to hot and humid environments (Lowe *et al.*, 2002; Alamer and Al-hozab, 2004).

Ventilation of the vehicle is important in removing metabolic heat and water produced by the sheep, and it might also provide some direct convective cooling. Kettlewell *et al.* (2001) described the air flow around a moving livestock vehicle (with a roof, solid front and adjustable side inlets) as mainly passing over the front edge of the vehicle, then separating from the vehicle to create suction, before re-attaching along the length of the vehicle – mostly entering at the rear inlets, moving forward within the vehicle over the animals and leaving through the front grilles.

When a vehicle is stationary, air movement is dependent on the prevailing wind and thermal buoyancy. Kettlewell *et al.* (2001) suggested that the addition of mechanical fans at the front sides of the vehicle – where the air pressure is lowest during vehicle movement – could improve ventilation.

Poor ventilation can result in increased ammonia concentrations. Fisher *et al.* (2002) monitored sheep on multi-deck trailers in New Zealand while on the road and on a ferry, and recorded average ammonia concentrations ranging from 3 to 30 ppm/h, but occasionally the concentration exceeded 100 ppm/h. The concentrations were reduced if the sheep were kept at a lower stocking density.

Young lambs are very vulnerable to combinations of starvation, low temperature and air movement (Alexander, 1974), and great care should be taken if they are transported. A fit, healthy and recently fed sheep with a full fleece can withstand extremely low temperatures and might have a lower critical temperature of < –20°C, but a recently shorn sheep could have a lower critical temperature of only 28°C (Alexander, 1974). At temperatures below the lower critical temperature, a sheep would increase heat production and be able to maintain body temperature until it reached summit metabolism.

Bennett (1972) estimated that the average summit metabolism in a sheep with a 7 mm dry fleece could occur at –56°C, but this would be higher after either fasting (–45°C) or exposure to air movement (–10°C) and would be –6°C if fasted and exposed to air movement. However, a shorn sheep with a low summit metabolism, exposed to air movement, might become hypothermic at 4°C and, if wet, it might reach summit metabolism at 20°C. At temperatures < 0°C, blood flow to the extremities fluctuates to

reduce the risk of tissue freezing (Webster and Blaxter, 1966); however, contact with metal fittings on a vehicle could cause damage. Use of dry bedding, e.g. wood shavings or straw, can reduce conductive heat loss when a sheep lies down (Gatenby, 1977).

Although sheep can thermoregulate within a wide range of thermal environments, at the extremes the adjustments required, e.g. a 15-fold increase in respiration rate to 300 breaths/min in response to hot conditions (Silanikove, 2000) or shivering in a compact posture in cold conditions (Bennett, 1972) indicate that extreme conditions are likely to be aversive. Richardson (2002) provided a guidance sheet on how to avoid heat and cold stress in transported sheep that includes a table for estimation of wind-chill effects and space allowance graphs for different live weights of sheep according to fleece length.

The advice for transportation of sheep in extremely cold weather includes: (i) increasing bedding; (ii) ensuring that the sheep have sufficient room to adjust their posture; and (iii) protection of sheep from water-splashing from the road and carefully balancing the need for ventilation with the risk of wind-chill.

In hot weather, it is recommended that: (i) stocking density is reduced; (ii) the frequency and length of stops where sheep are not unloaded are kept to a minimum; (iii) parking in direct sunlight should be avoided; (iv) the sheep should be handled carefully before loading to avoid raising body temperature; and (v) sufficient ventilation should be provided by opening inlets and, whenever possible, journeys should be avoided when the temperature and humidity are high and transport should be scheduled for nighttime or early morning.

Feed and water deprivation associated with transport

When sheep are transported on long journeys there is concern about the length of time that they are without food, water and adequate rest. Some countries have introduced legislation to regulate maximum journey times, the maximum interval between staging points to provide rest, food and water and for the provision of feed and/or water on the livestock vehicle. Although the scientific basis for some of these intervals is not always apparent, sheep obviously require access to food and water to maintain bodily functions and these should at least be provided at intervals that avoid aversive effects associated with dehydration and fasting.

MacFarlane *et al.* (1961) found marked changes in the peripheral blood indicative of dehydration after sheep had been without water for 4–5 days at maximum daily air temperatures of 39–41°C and had lost 23% of their live weight. There was a 28% increase in packed cell volume (PCV), a 46% increase in total plasma protein concentration, a 34% increase in plasma osmolality and increased concentrations of blood urea and plasma potassium as urine output declined.

Several studies of sheep transport have assessed the combined effects of transportation (in conditions where the sheep would not have been expected to have markedly increased respiratory water loss) and water restriction during journeys of 12–72 h, by monitoring changes in the peripheral blood as potential indicators of dehydration and comparing these effects with non-transported sheep either with or without feed and water. In general, these studies have not identified a marked increase in the PCV (Knowles *et al.*, 1993, 1995, 1998; Broom *et al.*, 1996; Cockram *et al.*, 1996; 1997; Horton *et al.*, 1996; Parrott *et al.*, 1998a).

However, neither did Li *et al.* (2000) when they exposed non-transported sheep without access to water, for up to 3 days to raised air temperatures of 25°C, nor Lowe *et al.* (2002) when they exposed sheep without access to water for 12 h to an air temperature of 33°C and relative humidity of 85–100%.

Several transport studies have reported slight increases in the total plasma protein concentration and, in studies where the sheep have been exposed to raised air temperatures, increases in total plasma protein concentration of about 10% have been reported after a journey of 24 h (Knowles *et al.*, 1996) and after journeys in India of up to 410 km (Kumar *et al.*, 2003).

Other than the study by Knowles *et al.* (1996), where a 5% increase was reported, the plasma osmolality in sheep transported for up to 24 h in or from the UK has been found to be relatively stable (Knowles *et al.*, 1993, 1994b, 1995, 1998; Broom *et al.*, 1996; Cockram *et al.*, 1996, 1997; Parrott *et al.*, 1998a).

The values for plasma osmolality were also relatively stable when non-transported sheep were kept without food and water, for either 3 days at an air temperature of 25°C (Li *et al.*, 2000) or for 2 days at 35°C (Parrott *et al.*, 1996). However, an increased plasma osmolality was reported by Lowe *et al.* (2002) in sheep kept without feed and water for 12 h at an air temperature of 33°C and relative humidity of 85–100%.

During water restriction, water loss in the faeces and urine is reduced, the rumen acts as a water reservoir (it contains 10–25% of the total water content of a sheep) and plasma volume can be maintained by drawing water into the circulation from the rumen to the extent that the water balance in the body (excluding the contents of the gastrointestinal tract) is kept virtually unaltered during the first 2 days of water deprivation (Hecker *et al.*, 1964; Silanikove, 1994).

The rate of decrease in ruminal fluid volume is greatest during the first 2 days without food and water (6.4–3.2 l), and by the third day the rate of absorption of water from the rumen slows (Hecker *et al.*, 1964). The ability of sheep to respond to periods of water deprivation – even in hot environments – is remarkable compared with that of man, and this ability is greatest in breeds adapted to desert conditions (Meissner and Belonje, 1972).

If the sheep have an increased metabolic rate associated with either exertion, e.g. in response to vehicular movements or exposure to cold temperatures, the effects of fasting associated with long journeys can be greater than fasting alone. The fermentation of ingesta in the rumen provides dietary energy in the form of volatile fatty acids for at least 3–4 days following the previous feed. However, the metabolism of a sheep falls rapidly during the first 2 days without food and then stabilizes at a lower rate (Blaxter, 1962).

In response to reduced dietary energy, a transported sheep needs to mobilize carbohydrate reserves (in the form of glycogen from the liver, until this becomes exhausted after about 24 h) and body fat, as shown by increased plasma concentrations of free fatty acids and increased transformation of fatty acids into ketone bodies, as indicated by increased plasma concentrations of β-hydroxybutyrate (Warriss *et al.*, 1989; Knowles *et al.*, 1994b, 1995, 1996; Cockram *et al.*, 1997, 1999). Motivation to eat is high after journeys of 12 or 24 h (Cockram et al., 1996, 1997), but Horton et al. (1996) found that, after a 3-day journey, feed and water intakes in the day after transportation were lower than in controls.

Lairage

The first priority of sheep after transportation for up to 24 h is normally to eat, then to drink, lie down and ruminate, rather than immediately to lie down or drink. After a 12-h journey they eat and lie down for longer, eat more hay and drink more water than prior to transport (Cockram et al., 1996). Sheep will lie down after arrival at a slaughterhouse lairage, particularly if given adequate space (> 1 m²/sheep). However, during the day, sheep can be disturbed by human presence and by noise (Kim et al., 1994; Jarvis and Cockram, 1995).

Legislation within the European Union (EU) requires that sheep be offered water – and, if necessary, feed – during a long journey. However, under certain transport conditions, a short, mid-journey lairage period might not always offer benefits to the sheep compared with completing the journey uninterrupted.

Cockram et al. (1997) compared the effects of transporting sheep for 24 h either continuously without food, water or rest with providing a mid-journey lairage period of either 3 or 12 h. The sheep spent most of the 3-h lairage period and almost half of the 12-h lairage period standing and eating hay, but not all sheep drank during the shorter lairage period and they ruminated only during the third hour. There was no difference in any of the biochemical measurements of dehydration between sheep transported for 24 h and non-transported sheep with access to feed and water.

However, in sheep that were provided with hay and water during the 3-h mid-journey lairage period, the plasma osmolality during the second part of the journey was greater than in non-transported sheep with access to feed and water. When sheep eat, they produce a large volume of saliva and, when food enters the rumen, the increased osmolality can draw water into the rumen from the plasma. The net effect of this is a temporary decrease in plasma volume and increased plasma osmolality (Ternouth, 1967).

If sheep do not have ready access to drinking water, they can become dehydrated after eating dry food (Cockram et al., 1997). Therefore, as suggested by Parrott et al. (1996), if access to water is limited during a vehicle stop or lairage, it is advisable also to restrict the availability of dry food. However, provision of food and water when sheep have reached their destination is beneficial, as the plasma concentrations of free fatty acids and β-hydroxybutyrate that are raised during a long journey decrease during the first 3 h after sheep are given access to food.

Cockram et al. (1997) observed that sheep that had been transported continuously without food for 24 h showed an increase in eating behaviour for 5 h; in those with a 3-h, mid-journey lairage it was increased for 4 h and in those with a 12-h, mid-journey lairage it was increased for 2 h. Due to the time spent eating, the sheep lay down less during the first 12 h after transport than they did during the equivalent time on the day prior to transport. If sheep are not provided with food after 15 h of transport, they lie down more and drink less than those with access to food. This suggests that, after journeys of up to 24 h, feeding is more important than rest, that part of the increase in water consumption post-transport is associated with the intake of dry feed, but rest and water are also important for sheep after they have eaten (Cockram et al., 1999).

The priorities of sheep after transport are obviously dependent on the type of journey experienced. After rail journeys of between 2 and 4 days without food or water, sheep did not have a clear preference, but after 5 days there was a definite preference for water and they ate only after drinking copiously (Sutton and van den Heever, 1968).

Effects of transport on weight loss, meat quality and food safety

Sheep lose live weight in an approximately linear manner with increasing durations of transport from 6 to 30 h (about 8% loss of live weight after 24 h) (Knowles et al., 1995, 1996; Cockram et al., 1997; Parrot et al., 1998a). The rate of live weight loss over this period is very similar to that of sheep that have not been transported but do not have access to feed or water. However, after sheep had been transported for 72 h, there was a greater live weight loss than in non-transported sheep without access to feed and water (Horton *et al.*, 1996). Thompson *et al.* (1987) showed that the percentage of live weight loss over a 4-day period in fasted sheep with access to water was curvilinear (8% after 1 day and 16% after 4 days).

The more rapid decrease in live weight over the first 24 h in sheep without access to food is due to a decrease in gut contents, with significant decrease in carcass weight occurring after 24–48 h (Warriss *et al.*, 1987). Horton *et al.* (1996) showed that sheep transported over a 72-h period for further fattening gained weight over the subsequent month (18% increase in pre-transport live weight), compared with a weight gain of 22% in those kept without feed and water for the same duration and 29% in control sheep with continuous access to food and water.

Although Ruiz-de-la-Torre *et al.* (2001) found an effect of road type during transport on the muscle pH_{24} of young lambs, the effects of transport on meat quality are less important in sheep than in other species (Young *et al.*, 2005). Fasting of sheep for up to 72 h reduces muscle glycogen concentration, but there is little effect on the muscle pH_{24} (5.61 compared with 5.59 in non-fasted sheep) (Warriss *et al.*, 1987). Purchas *et al.* (1980) were also unable to find a correlation between the plasma cortisol concentration of sheep at the time of slaughter and the tenderness of the meat.

Sheep presented for slaughter should have a clean, dry fleece, but the fleece (especially on the abdomen and the legs) is susceptible to faecal and mud contamination. The greater the contamination, the greater the risk of bacterial contamination of the carcass. Clean, dry sheep minimize the risk of bacteriological contamination and therefore reduce the risk of meat-borne food poisoning and improve the shelf life of the meat (Hadley *et al.*, 1997).

Guidelines on the presentation of clean sheep for slaughter include provisions for transport such as: (i) adequate cleaning and disinfection of vehicles between loads; (ii) provision of clean, dry bedding; (iii) withdrawal of feed prior to transport to decrease gut-fill and reduce overall faecal contamination; (iv) avoiding transporting sheep directly from pasture (if possible, bringing them indoors on to clean, dry bedding); and (vi) keeping the sheep dry during loading (Food Standards Agency, 2005).

Health and disease

When animals are transported there is an increased risk of contact with infected animals or materials, e.g. an inadequately cleaned and disinfected vehicle. The 2001 foot-and-mouth disease epidemic in the UK illustrated how movement of sheep can rapidly spread infectious diseases (Mansley *et al.*, 2003). Biosecurity issues when sheep are held at staging points are very important, and the risk of disease transmission is a factor against unloading sheep and exposing them to potential pathogens during mid-journey lairage. Brogden *et al.* (1998) speculated that the stress, mixing and close confinement associated with the transportation of sheep could increase the risk of respiratory infection.

There is some evidence of relationships between repeated stressful situations and changes in immunity (Coppinger *et al.*, 1991), and also some case reports of increased severity of clinical conditions following transportation, e.g. Gumbrell and McGregor (1997). After transport for 72 h, Horton *et al.* (1996) did not identify any major health problems (mortality, diarrhoea or pyrexia) in the sheep, but higher nasal discharge scores were recorded than in non-transported controls (the authors attributed this to exposure of the transported sheep while at a market prior to the 72-h journey).

Knowles *et al.* (1994c) reported mortality rates of lambs either in transit to a slaughterhouse in the south of England or while in the lairage awaiting slaughter. Mortality in lambs transported directly from a farm (mean distance 62 miles) was 0.007%, and in those that had arrived via an auction market (mean distance 199 miles) it was 0.031%. Death appeared to be associated with disease present before transport that had been exacerbated by the transport, rather than as a direct consequence of transport. The highest mortality was associated with the occasions when the rate of carcass condemnations was highest. The main causes of condemnation of the carcasses as unfit for human consumption were abscess, arthritis and pleurisy.

Guidelines that include examples of the types of clinical conditions that would render a sheep unfit for transport (e.g. The Sheep Veterinary Society, 1994; Ontario Humane Transport Working Group, 2005) are extremely useful. Examples of conditions that would require a delay in transport and reassessment of the sheep are exhaustion, lambing and fever (> 39.6°C). Conditions that are so serious that the sheep should not be transported and euthanasia should be considered are: (i) inability to walk; (ii) arthritis in multiple joints; (iii) emaciation; (iv) unresponsive pneumonia with fever; (v) prolapsed uterus; and (v) nervous disorders.

Examples of conditions where it might be possible to transport the sheep on a short journey to slaughter – if special provisions such as extra bedding, loading onto the bottom deck, nearness of the ramp and separation can be provided – are abscess, prolapsed vagina or rectum and slight lameness (provided there is no evidence of pain and they can easily walk up the loading ramp). However, sheep that are so lame that they require assistance to rise – or are unable to rise or remain standing – should not be transported. However, in every case, a sheep should be transported for slaughter only if it can be loaded and transported humanely and will be fit for human consumption.

Sea Transport

Although sheep appear to cope reasonably well with road transportation, the long sea journeys from Australia to the Middle East present greater challenges and about 1 in 100 sheep fail to survive the experience. The main factors reported in the literature that affect the risk of mortality during these long sea journeys are heat distress, failure to eat sufficient food and disease (Gardiner and Craig, 1970; Black, 1993).

Prior to sea transport, large consignments of sheep are assembled in feedlots, sometimes after long road journeys with extended periods without feed. They are offered the pelleted food that is available during the sea journey, but failure to eat pellets while at the feedlot is a risk factor for death during the sea voyage (Norris et al., 1989a). The proportion of sheep that

consume pellets at the feedlot is variable and dependent on the source of the sheep. Prolonged periods of starvation before entering the feedlot increase the risk of a sheep not eating, and consumption levels are dependent upon experience of pelleted food before arrival (McDonald et al., 1990).

Examples of reported mortality and rejection rates in the 1980s were: (i) 0.005–0.05% between farms and the feedlot; (ii) 0.4–1.5% at the feedlot and before loading on to the ship; (iii) 1.5–2.7% during the sea journey; and (iv) 3% when held in a feedlot for 1 month after arrival in the Middle East (plus an additional 4% sent for emergency slaughter on arrival because of debilitation, disease or injury) (Norris et al., 1989b; Scharp, 1992). Australian Merino wethers (castrated males) are less adapted to the hot and humid conditions in the Middle East than are local sheep (Srikandakumar et al., 2003). Richards et al. (1989) reported that the most common cause of death while at the feedlot before the journey was salmonellosis (possibly related to the high stocking density and contamination at the feedlot).

Other major causes of death that were attributable to the management of the animals were trauma (attributed to injury during transport from the farm to the feedlot), inanition (marked weakness, weight loss, decreased metabolism and metabolic disorders due to starvation) and diseases associated with excessive feed intake, e.g. ruminal acidosis. During sea journeys the main causes of death were inanition (43%), salmonellosis (20%) and trauma (11%) (mainly associated with pelvic injuries thought to have been caused by slipping). Dehydration was also identified in some of the dead sheep.

The daily mortality rate (particularly that due to inanition) generally increased with the duration of the voyage; however, in some voyages there were peaks in mortality after about 1 week. Sheep that do not feed during a sea journey have raised plasma concentrations of free fatty acids, β-hydroxybutyrate and total protein (Richards et al., 1991).

Black et al. (1994) recorded a mortality rate of 2.5% (31% due to suffocation, 28% inanition, 25% pneumonia, 9% dehydration and 7% trauma and other causes) in wethers and rams kept in pens below and above deck during a 24-day sea journey from New Zealand to Saudi

Arabia. There was competition between sheep when both pellets and drinking water were offered, but the sheep gained weight during the journey. The sheep spent most of the day standing and lay down at night, but more lay down as the average daily air temperature increased during the journey (> 26°C from day 16, with relative humidity (RH) 75–85%). The percentage of sheep panting was 1–3% until the last 4 days of the journey, when it increased to 8–10%.

Bailey and Fortune (1992) recorded daily maximum temperatures of between 28 and 33°C and a relative humidity greater than 85% on a 13-day voyage during which Merino wethers lost 3% of their live weight, but those that died had lost 20% of their live weight.

In 2002, a risk assessment model for heat stress (LiveCorp, 2003) was introduced to allow exporters to enter variables such as long-range weather patterns, weather at embarkation, en route and at unloading, animal traits and the ventilation characteristics of the ships (forced-air ventilation systems are used to provide each deck with a minimum of one complete air change every 2 min). Operators can use this information to adjust stocking densities, loading configurations and animal selection to minimize the risk of problems during the journey.

Conclusions

In some countries, many of the factors discussed above that can influence the effects of a road or sea journey on the welfare of sheep, namely: (i) the fitness of the sheep; (ii) pre- and post-transport handling; (iii) the environmental conditions during transport; (iv) journey duration; (v) intervals between periods for resting, feeding and watering; and (vi) the duration and conditions provided for rest during and after a journey are regulated by legislation, codes of practice or quality assurance schemes.

Another problem is that some developing countries lack loading/unloading ramps, and sheep may be thrown or dragged. Simple, low-cost ramps can easily be built from local materials (see Fig. 12.3). Often, simple improvements in equipment and training handlers will benefit animal welfare.

Fig. 12.3.   Sheep-loading ramp that could easily be built from local materials. Cleats provide non-slip flooring.

References

Alamer, M. and Al-hozab, A. (2004) Effect of water deprivation and season on feed intake, body weight and thermoregulation in Awassi and Najdi sheep breeds in Saudi Arabia. *Journal of Arid Environments* 59, 71–84.

Alexander, G. (1974) Heat loss from sheep. In: Monteith, J.L. and Mount, L.E. (eds) *Heat Loss from Animals and Man: Assessment and Control*. Butterworths, London, pp. 173–203.

Bailey, A.N. and Fortune, J.A. (1992) The response of Merino wethers to feedlotting and subsequent sea transport. *Applied Animal Behaviour Science* 35, 167–180.

Bennett, J.W. (1972) The maximum metabolic response of sheep to cold: effects of rectal temperature, shearing, feed consumption, body posture, and body weight. *Australian Journal of Agricultural Research* 23, 1045–1058.

Black, H. (1993) The live sheep trade: diagnostic criteria for diseases of concern. *Surveillance* 20, 15–17.

Black, H., Matthews, L.R. and Bremner, K.J. (1994) The behavior of male lambs transported by sea from New Zealand to Saudi Arabia. *New Zealand Veterinary Journal* 42, 16–23.

Blaxter, K.L. (1962) The fasting metabolism of adult wether sheep. *British Journal of Nutrition* 16, 615–626.

Blaxter, K.L., Graham, N.M. and Wainman, F.W. (1959) Environmental temperature, energy metabolism and heat regulation in sheep. III. The metabolism and thermal exchanges of sheep with fleeces. *Journal of Agricultural Science* 52, 41–49.

Bligh, J. (1963) The receptors concerned in the respiratory response to humidity in sheep at high ambient temperature. *Journal of Physiology* 168, 747–763.

Brogden, K.A., Lehmkuhl, H.D. and Cutlip, R.C. (1998) *Pasteurella haemolytica*-complicated respiratory infections in sheep and goats. *Veterinary Research* 29, 233–254.

Brook, A.H. and Short, B.F. (1960) Sweating in sheep. *Australian Journal of Agricultural Research* 11, 557–569.

Broom, D.M., Goode, J.A., Hall, S.J.G., Lloyd, D.M. and Parrott, R.F. (1996) Hormonal and physiological effects of a 15 hour road journey in sheep: comparison with the responses to loading, handling and penning in the absence of transport. *British Veterinary Journal* 152, 593–604.

Cockram, M.S. and Mitchell, M.A. (1999) Rôle of research in the formulation of 'rules' to protect the welfare of farm animals during road transportation. In: Russel, A.J.F., Morgan, C.A., Savory, C.J., Appleby, M.C. and Lawrence, T.L.J. (eds) *Farm Animal Welfare – Who Writes the Rules?* Occasional Publication No. 23, British Society of Animal Science, Penicuik, Midlothian, UK, pp. 43–64.

Cockram, M.S., Kent, J.E., Goddard, P.J., Waran, N.K., McGilp, I.M., Jackson, R.E., Muwanga, G.M. and Prytherch, S. (1996) Effect of space allowance during transport on the behavioural and physiological responses of lambs during and after transport. *Animal Science* 62, 461–477.

Cockram, M.S., Kent, J.E., Jackson, R.E., Goddard, P.J., Doherty, O.M., McGilp, I.M., Fox, A., Studdert-Kennedy, T.C., McConnell, T.I. and O'Riordan, T. (1997) Effect of lairage during 24 h of transport on the behavioural and physiological responses of sheep. *Animal Science* 65, 391–402.

Cockram, M.S., Kent, J.E., Waran, N.K., McGilp, I.M., Jackson, R.E., Amory, J.R., Southall, E.L., O'Riordan, T., McConnell, T.I. and Wilkins, B.S. (1999) Effects of a 15 h journey followed by either 12 h starvation or *ad libitum* hay on the behaviour and blood chemistry of sheep. *Animal Welfare* 8, 135–148.

Cockram, M.S., Kent, J.E., Goddard, P.J., Waran, N.K., Jackson, R.E., McGilp, I.M., Southall, E.L., Amory, J.R., McConnell, T.I., O'Riordan, T. and Wilkins, B.S. (2000) Behavioural and physiological responses of sheep to 16 h transport and a novel environment post-transport. *Veterinary Journal* 159, 139–146.

Cockram, M.S., Baxter, E.M., Smith, L.A., Bell, S., Howard, C.M., Prescott, R.J. and Mitchell, M.A. (2004) Effect of driver behaviour, driving events and road type on the stability and resting behaviour of sheep in transit. *Animal Science* 79, 165–176.

Coppinger, T.R., Minton, J.E., Reddy, P.G. and Blecha, F. (1991) Repeated restraint and isolation stress in lambs increases pituitary–adrenal secretions and reduces cell-mediated immunity. *Journal of Animal Science* 69, 2808–2814.

Degen, A.A. and Shkolnik, A. (1978) Thermoregulation in fat-tailed Awassi, a desert sheep, and in German mutton Merino, a mesic sheep. *Physiological Zoology* 51, 333–339.

Fahmy, M.H. and Guilbault, L.A. (1989) Effect of transportation stress on ovarian activities and reproductive performance of ewes during the anoestrous period. *Animal Reproduction Science* 19, 229–233.

FAOSTAT data (2005) *Agriculture & Food Trade. Live Animals* (http://www.faostat.fao.org, last updated 21 December 2005). Food and Agriculture Organization of the United Nations (FAO), Rome.

Fisher, A.D., Stewart, M., Tacon, J. and Matthews, L.R. (2002) The effects of stock crate design and stocking density on environmental conditions for lambs on road transport vehicles. *New Zealand Veterinary Journal* 50, 148–153.

Fisher, A.D., Stewart, M., Duganzich, D.M., Tacon, J. and Matthews, L.R. (2004) The effects of stationary periods and external temperature and humidity on thermal stress conditions within sheep transport vehicles. *New Zealand Veterinary Journal* 53, 6–9.

Food Standards Agency (2005) *Red Meat Safety and Clean Livestock* (http://www.food.gov.uk/foodindustry/guidancenotes/meatregsguid/cleanlivestockguidance, last updated January 2005). Food Standards Agency, London.

Gardiner, M.R. and Craig, J. (1970) Factors affecting survival in the transportation of sheep by sea in the tropics and subtropics. *Australian Veterinary Journal* 46, 65–69.

Gatenby, R.M. (1977) Conduction of heat from sheep to ground. *Agricultural Meteorology* 18, 387–400.

Gumbrell, R.C. and McGregor, D.A. (1997) Outbreak of severe fatal orf in lambs. *Veterinary Record* 141, 150–151.

Hadley, P.J., Holder, J.S. and Hinton, M.H. (1997) Effects of fleece soiling and skinning method on the micro-biology of sheep carcases. *Veterinary Record* 140, 570–574.

Hales, J.R.S. (1969) Changes in respiratory activity and body temperature of the severely heat-stressed ox and sheep. *Comparative Biochemistry and Physiology* 31, 975–985.

Hales, J.R.S. and Brown, G.D. (1974) Net energetic and thermoregulatory efficiency during panting in the sheep. *Comparative Biochemistry and Physiology* 49A, 413–422.

Hall, S.J.G., Broom, D.M., Goode, J.A., Lloyd, D.M., Parrott, R.F. and Rodway, R.G. (1999) Physiological responses of sheep during long road journeys involving ferry crossings. *Animal Science* 69, 19–27.

Hecker, J.F., Budtz-Olsen, O.E. and Ostwald, M. (1964) The rumen as a water store in sheep. *Australian Journal of Agricultural Research* 15, 961–968.

Hitchcock, D.K. and Hutson, G.D. (1979) The movement of sheep on inclines. *Australian Journal of Experimental Agriculture and Animal Husbandry* 19, 176–182.

Horton, G.M.J., Baldwin, J.A., Emanuele, S.M., Wohlt, J.E. and McDowell, L.R. (1996) Performance and blood chemistry in lambs following fasting and transport. *Animal Science* 62, 49–56.

Ibáñez, M., De la Fuente, J., Thos, J. and González de Chavarri, E.G. (2002) Behavioural and physiological responses of suckling lambs to transport and lairage. *Animal Welfare* 11, 223–230.

Jarvis, A.M. and Cockram, M.S. (1994) Effects of handling and transport on bruising of sheep sent directly from farms to slaughter. *Veterinary Record* 135, 523–527.

Jarvis, A.M. and Cockram, M.S. (1995) Some factors affecting resting behaviour of sheep in slaughterhouse lairages after transport from farms. *Animal Welfare* 4, 53–60.

Jarvis, A.M. and Cockram, M.S. (1999) Environmental conditions within livestock vehicles during the commercial transport of sheep. In: Russel, A.J.F., Morgan, C.A., Savory, C.J., Appleby, M.C. and Lawrence, T.L.J. (eds) *Farm Animal Welfare – Who Writes The Rules?* Occasional Publication No.23, British Society of Animal Science, Penicuik, Midlothian, UK, pp. 129–131.

Jarvis, A.M., Cockram, M.S. and McGilp, I.M. (1996) Bruising and biochemical measures of stress, dehydration and injury determined at slaughter in sheep transported from farms or markets. *British Veterinary Journal* 152, 719–722.

Jones, T.A., Look, A., Guise, H.J. and Lomas, M.J. (2002) Head height requirements, and assessing stocking density, for sheep in transit. *Veterinary Record* 150, 49–50.

Keniry, J., Bond, M., Caple, I., Gosse, L. and Rogers, M. (2003) *The Keniry Report. Livestock Export Review.* A Report to the Minister for Agriculture, Fisheries and Forestry (http://www.affa.gov.au/content/output.cfm?ObjectID=CC9100A7-11C4-4E00-84AE69A318875267, last updated January 2006). Australian Government Department of Agriculture, Fisheries and Forestry, Canberra, Australia.

Kettlewell, P.J., Hoxey, R.P., Hampson, C.J., Green, N.R., Veale, B.M. and Mitchell, M.A. (2001) Design and operation of a prototype mechanical ventilation system for livestock transport vehicles. *Journal of Agricultural Engineering Research* 79, 429–439.

Kilgour, R. and de Langen, H. (1970) Stress in sheep resulting from management practices. *Proceedings of the New Zealand Society of Animal Production* 30, 65–76.

Kim, F.B., Jackson, R.E., Gordon, G.D.H. and Cockram, M.S. (1994) Resting behaviour of sheep in a slaughterhouse lairage. *Applied Animal Behaviour Science* 40, 45–54.

Knowles, T.G. (1998) A review of the road transport of slaughter sheep. *Veterinary Record* 143, 212–219.

Knowles, T.G., Warriss, P.D., Brown, S.N., Kestin, S.C., Rhind, S.M., Edwards, J.E., Anil, M.H. and Dolan, S.K. (1993) Long distance transport of lambs and the time needed for subsequent recovery. *Veterinary Record* 133, 286–293.

Knowles, T.G., Maunder, D.H.L. and Warriss, P.D. (1994a) Factors affecting the incidence of bruising in lambs arriving at one slaughterhouse. *Veterinary Record* 134, 44–45.

Knowles, T.G., Warriss, P.D., Brown, S.N. and Kestin, S.C. (1994b) Long distance transport of export lambs. *Veterinary Record* 134, 107–110.

Knowles, T.G., Maunder, D.H.L., Warriss, P.D. and Jones, T.W.H. (1994c) Factors affecting the mortality of lambs in transit to or in lairage at a slaughterhouse, and reasons for carcase condemnations. *Veterinary Record* 135, 109–111.

Knowles, T.G., Brown, S.N., Warriss, P.D., Phillips, A.J., Dolan, S.K., Hunt, P., Ford, J.E., Edwards, J.E. and Watkins, P.E. (1995) Effects on sheep of transport by road for up to 24 hours. *Veterinary Record* 136, 431–438.

Knowles, T.G., Warriss, P.D., Brown, S.N., Kestin, S.C., Edwards, J.E., Perry, A.M., Watkins, P.E. and Phillips, A.J. (1996) Effects of feeding, watering and resting intervals on lambs transported by road and ferry to France. *Veterinary Record* 139, 335–339.

Knowles, T.G., Warriss, P.D., Brown, S.N. and Edwards, J.E. (1998) Effects of stocking density on lambs being transported by road. *Veterinary Record* 142, 503–509.

Kumar, B.R., Muralidharan, M.R., Ramesh, V., Arunachalam, S. and Sivakumar, T. (2003) Effect of transport stress on blood profile in sheep. *Indian Veterinary Journal* 80, 511–514.

Li, B.T., Christopherson, R.J. and Cosgrove, S.J. (2000) Effect of water restriction and environmental temperatures on metabolic rate and physiological parameters in sheep. *Canadian Journal of Animal Science* 80, 97–104.

LiveCorp (2003) *Fact Sheet. Risk Management and the Heat Stress Model* (http://www.Livecorp.Com.Au/Downloads/Heat%20Stress.Pdf, accessed March 23 2006). LiveCorp, Sydney, Australia.

Lowe, T.E., Gregory, N.G., Fisher, A.D. and Payne, S.R. (2002) The effects of temperature elevation and water deprivation on lamb physiology, welfare, and meat quality. *Australian Journal of Agricultural Research* 53, 707–714.

MacFarlane, W.V., Morris, R.J.H., Howard, B., McDonald, J. and Budtz-Olsen, O.E. (1961) Water and electrolyte changes in tropical Merino sheep exposed to dehydration during summer. *Australian Journal of Agricultural Research* 12, 889–912.

Mansley, L.M., Dunlop, P.J., Whiteside, S.M. and Smith, R.G.H. (2003) Early dissemination of foot-and-mouth disease virus through sheep marketing in February 2001. *Veterinary Record* 153, 43–50.

McDonald, C.L., Norris, R.T., Speijers, E.J. and Ridings, H. (1990) Feeding behaviour of Merino wethers under conditions similar to lot-feeding before live export. *Australian Journal of Experimental Agriculture* 30, 343–348.

Meissner, H.H. and Belonje, P.C. (1972) Preliminary study on water and electrolyte metabolism during thermal and dehydrational stresss in two breeds of sheep. *South African Journal of Animal Science* 2, 97–100.

Murray, K.C., Davies, D.H., Cullinane, S.L., Eddison, J.C. and Kirk, J.A. (2000) Taking lambs to the slaughter: marketing channels, journey structures and possible consequences for welfare. *Animal Welfare* 9, 111–122.

Norris, R.T., Richards, R.B. and Dunlop, R.H. (1989a) Pre-embarkation risk factors for sheep deaths during export by sea from Western Australia. *Australian Veterinary Journal* 66, 309–314.

Norris, R.T., Richards, R.B. and Dunlop, R.H. (1989b) An epidemiological study of sheep deaths before and during export by sea from Western Australia. *Australian Veterinary Journal* 66, 276–279.

Ontario Humane Transport Working Group (2005) *Should This Animal Be Loaded? Guidelines for Transporting Cattle, Sheep & Goats* (http://www.ofac.org/pdf/Cattle-Chart_Final.pdf, last updated June 2005). Ontario Farm Animal Council, Guelph, Ontario, Canada.

Parrott, R.F., Misson, B.H. and De La Riva, C.F. (1994) Differential stressor effects on the concentrations of cortisol, prolactin and catecholamines in the blood of sheep. *Research in Veterinary Science* 56, 234–239.

Parrott, R.F., Lloyd, D.M. and Goode, J.A. (1996) Stress hormone responses of sheep to food and water deprivation at high and low ambient temperatures. *Animal Welfare* 5, 45–56.

Parrott, R.F., Hall, S.J.G., Lloyd, D.M., Goode, J.A. and Broom, D.M. (1998a) Effects of a maximum permissible journey time (31 h) on physiological responses of fleeced and shorn sheep to transport, with observations on behaviour during a short (1 h) rest-stop. *Animal Science* 66, 197–207.

Parrott, R.F., Hall, S.J.G. and Lloyd, D.M. (1998b) Heart rate and stress hormone responses of sheep to road transport following two different loading procedures. *Animal Welfare* 7, 257–267.

Parrott, R.F., Lloyd, D.M. and Brown, D. (1999) Transport stress and exercise hyperthermia recorded in sheep by radiotelemetry. *Animal Welfare* 8, 27–34.

Peterson, G.V. (1978) Factors associated with wounds and bruises in lambs. *New Zealand Veterinary Journal* 26, 6–9.

Purchas, R.W., Barton, R.A. and Kirton, A.H. (1980) Relationships of circulating cortisol levels with growth rate and meat tenderness of cattle and sheep. *Australian Journal of Agricultural Research* 31, 221–232.

Reid, R.L. and Mills, S.C. (1962) Studies on the carbohydrate metabolism of sheep. XIV. The adrenal response to psychological stress. *Australian Journal of Agricultural Research* 13, 282–295.

Richards, R.B., Norris, R.T., Dunlop, R.H. and McQuade, N.C. (1989) Causes of death in sheep exported live by sea. *Australian Veterinary Journal* 66, 33–38.

Richards, R.B., Hyder, M.W., Fry, J., Costa, N.D., Norris, R.T. and Higgs, A.R.B. (1991) Seasonal metabolic factors may be responsible for deaths in sheep exported by sea. *Australian Journal of Agricultural Research* 42, 215–226.

Richardson, C. (2002) *Avoiding Heat and Cold Stress in Transported Sheep.* Factsheet Produced by Ontario Ministry of Agriculture, Food and Rural Affairs (http://www.omafra.gov.on.ca/english/livestock/sheep/facts/02-013.htm, last updated January 2002). Ontario Ministry of Agriculture, Food and Rural Affairs. Publications, Toronto, Ontario, Canada.

RSPCA (1993) *Hook Versus Hoof – the Drive Behind the Increasing Shipment of Live Lambs to the Continent.* RSPCA, Horsham, UK.

Ruiz-de-la-Torre, J.L., Velarde, A., Diestre, A., Gispert, M., Hall, S.J.G., Broom, D.M. and Manteca, X. (2001) Effects of vehicle movements during transport on the stress responses and meat quality of sheep. *Veterinary Record* 148, 227–229.

Scharp, D.W. (1992) Performance of Australian wethers in Arabian Gulf feedlots after transport by sea. *Australian Veterinary Journal* 69, 42–43.

Sheep Veterinary Society (1994) *The Casualty Sheep.* Sheep Veterinary Society (SVS), Penicuik, Midlothian, UK.

Silanikove, N. (1994) The struggle to maintain hydration and osmoregulation in animals experiencing severe dehydration and rapid rehydration: the story of ruminants. *Experimental Physiology* 79, 281–300.

Silanikove, N. (2000) Effects of heat stress on the welfare of extensively managed domestic ruminants. *Livestock Production Science* 67, 1–18.

Slee, J. and Forster, J.E. (1983) Habituation to cold in four breeds of sheep. *Journal of Thermal Biology* 8, 343–348.

Smart, D., Forhead, A.J., Smith, R.F. and Dobson, H. (1994) Transport stress delays the estradiol-induced LH surge by a non-opioidergic mechanism in the early postpartum ewe. *Journal of Endocrinology* 142, 447–451.

Smith, R.F. and Dobson, H. (2002) Hormonal interactions within the hypothalamus and pituitary with respect to stress and reproduction in sheep. *Domestic Animal Endocrinology* 23, 75–85.

Smith, R.F., French, N.P., Saphier, P.W., Lowry, P.J., Veldhuis, J.D. and Dobson, H. (2003) Identification of stimulatory and inhibitory inputs to the hypothalamic–pituitary–adrenal axis during hypoglycaemia or transport in ewes. *Journal of Neuroendocrinology* 15, 572–585.

Srikandakumar, A., Johnson, E.H. and Mahgoub, O. (2003) Effect of heat stress on respiratory rate, rectal temperature and blood chemistry in Omani and Australian Merino sheep. *Small Ruminant Research* 49, 193–198.

Sutton, G.D. and van den Heever, L.W. (1968) The effect of prolonged rail transport on adult Merino sheep. *Journal of the South African Veterinary Medical Association* 39, 31–34.

Ternouth, J.H. (1967) Post-prandial ionic and water exchange in the rumen. *Research in Veterinary Science* 8, 283–293.

Thompson, J.M., O'Halloran, W.J., McNeill, D.M.J., Jackson-Hope, N.J. and May, T.J. (1987) The effect of fasting on liveweight and carcass characteristics in lambs. *Meat Science* 20, 293–309.

Warriss, P.D., Brown, S.N., Bevis, E.A., Kestin, S.C. and Young, C.S. (1987) Influence of food withdrawal at various times preslaughter on carcass yield and meat quality in sheep. *Journal of the Science of Food and Agriculture* 39, 325–334.

Warriss, P.D., Bevis, E.A., Brown, S.N. and Ashby, J.G. (1989) An examination of potential indices of fasting time in commercially slaughtered sheep. *British Veterinary Journal* 145, 242–248.

Warriss, P.D., Edwards, J.E., Brown, S.N. and Knowles, T.G. (2002) Survey of the stocking densities at which sheep are transported commercially in the United Kingdom. *Veterinary Record* 150, 233–236.

Warriss, P.D., Brown, S.N. and Knowles, T.G. (2003) Assessment of possible methods for estimating the stocking density of sheep being carried on commercial vehicles. *Veterinary Record* 153, 315–319.

Webster, A.J.F. and Blaxter, K.L. (1966) The thermal regulation of 2 breeds of sheep exposed to air temperatures below freezing point. *Research in Veterinary Science* 7, 466–479.

Webster, A.J.F., Hicks, A.M. and Hays, F.L. (1969) Cold climate and cold temperature induced changes in the heat production and thermal insulation of sheep. *Canadian Journal of Physiology and Pharmacology* 47, 553–562.

Young, O.A., Hopkins, D.L. and Pethick, D.W. (2005) Critical control points for meat quality in the Australian sheep meat supply chain. *Australian Journal of Experimental Agriculture* 45, 593–601.

13 Dogs for Herding and Guarding Livestock

Lorna Coppinger and Raymond Coppinger

School of Cognitive Science, Hampshire College, Amherst, Massachusetts, USA; rcoppinger@hampshire.edu

Introduction

Two completely different types of sheepdog assist livestock producers all over the world. Herding dogs are specialists at moving stock from place to place. Guarding dogs protect domestic stock from wild predators. 'The sheepdog is such a willing and uncomplaining worker and without him the farmer could not even begin to look after his sheep', wrote Lt.-Col. K.J. Price in his foreword to a popular book on the subject (Longton and Hart, 1976). Longton and Hart noted that 'There are well over one thousand million sheep in the world, and one-third of this vast total is kept in countries where the Border collie is the chief work dog'.

Guarding dogs, well known among sheep and goat producers, particularly in the high pastures of countries all the way from Portugal to Tibet, also help to manage the world's ruminants. Although these large, placid protectors were all but unknown in the USA until the mid-1970s, they have since then been widely adopted throughout the USA and Canada. As a result, many flocks that had experienced a 10% or even greater loss to coyotes have enjoyed a marked reduction in predation.

Domesticated dogs and sheep appear together in archaeological excavations dating from 3685 BP (before present) (Olsen, 1985). They become part of written history in the Old Testament ('with the dogs of my flock', Job 30, 1)

and in the writings of Cato the Elder and Marcus Terentius Varro in the two centuries before the Christian era. These treatises on Roman farm management, translated by 'a Virginia farmer' (Anon., 1913), are so full of good information that, if no other book had ever been written on flock dogs, today's farmers could learn just about all they need from Cato and Varro. 'Dogs . . . are of the greatest importance to us who feed the woolly flock, for the dog is the guardian of such cattle as lack the means to defend themselves, chiefly sheep and goats. For the wolf is wont to lie in wait for them and we oppose our dogs to him as defenders' (Anon., 1913, p. 247).

Modern stock producers still need that information on how to choose a pup, what an adult should look like, what to feed, the value of a dog, breeding, raising pups and number of dogs per flock. Much of this ancient 'manual' suggests that the first sheepdogs were primarily guardians rather than herders, although the difference is blurred.

The breeding and management of working farm dogs are not supported by an extensive academic farm dog-specific literature, but they are supported by a rich, knowledgeable and generally professional trade literature based on individual experience. In recent years, resources available via computer on the World Wide Web have greatly increased the accessibility of expert information. Herding dogs have received the most attention, with books and articles on their selection and training widely available. This popularity is driven in part by the success of organized competitions among the owners of herding dogs, especially of Border collies.

Guarding dogs have been the subject of studies by biologists in the USA since the mid-1970s, and subsequently their use has greatly increased, worldwide. Although general principles of breeding, management and health care are common to all domestic dogs, working stock dogs (and their handlers) have specific needs that benefit from technical assistance. The scientific literature about farm dogs has yet to address genetic improvement at the level so prevalent for other domestic livestock or to study their behaviour at the level enjoyed by canine pets.

Willis (1992) emphasized the lack of support for research in this field: 'Bearing in mind the enormous funding given to research into sheep breeding it is surprising that so little has been expended on understanding herding ability in dogs, for without dogs most British and Australasian sheep farmers could not function'.

Taking the need for research on dogs even further, Hahn and Schanz (1996) and Hahn and Wright (1998) emphasized the lack of studies on behavioural genetics of dogs, which limits the ability of professionals to provide useful techniques for breeders, trainers, veterinarians and end-users.

Today, scientists have sequenced the dog genome and this will help make it possible to determine how genetics and the environment interact to determine behaviour (Lindblad-Toh et al., 2005; Ostrander et al., 2005). Hopefully, these studies will provide useful tools for the selection of dogs.

The 10% loss to predators mentioned earlier is not uncommon in the USA, but businesses cannot sustain a 10% shrinkage for very many years. Research on guarding dogs needs to go beyond studies of breed differences and effect of dogs on predation. Two projects that would help the industry immediately are: (i) behavioural analyses focused on improving the success rate of guarding dogs; and (ii) closely monitored field trials of dogs learning to work in locations critical for the reintroduction and survival of endangered predatory wildlife. Subsequent transfer of information back to users – and especially to potential users – also needs improvement.

This is especially true in areas where wolves have been successfully reintroduced and predation is publicly scrutinized.

Better behavioural tests need to be developed to access the genetic component of specific herding and guarding traits. A review of the literature by Ruefenacht et al. (2002) indicated that fearfulness in dogs is highly heritable. Another study by Courreau and Langlois (2005) showed that attacking, biting and guarding behaviour is also highly heritable in Belgian shepherd dogs. Tests for finer discrimination of behaviour need to be researched.

Differentiating the Sheepdogs

It appears, from the old literature, that early 'shepherd dogs' were used both for guarding and herding. Today, distinctions between the two types are made more clearly. Herding dogs are not 24-h flock guardians, nor are guardians good at herding livestock. The reason becomes obvious when one considers morphological and behavioural differences between the two types (see Table 13.1; Fig. 13.1).

Cato and Varro divided dogs into two kinds, hunting dogs ('used against wild beasts and game') and herd dogs ('used by the shepherd'). They described 'herd' dogs in terms of their guarding abilities. Other observers, as reported by Baur (1982), may have seen what they thought was a 'driving' dog when the dog was actually just 'following' the herd: 'These dogs take the entire care of the sheep, drive them out to pasture in the morning, keep them from straying during the day, and bring them home at night. These dogs have inherited a talent for keeping sheep, but the shepherds do not depend wholly on that' (Anon., 1873, reported in Baur, 1982).

Baur commented that shepherds reinforced the dogs' talents with 'the old Spanish custom of using a foster-mother ewe to train (suckle) a puppy'. Thomas (1983) distinguished between guarding and herding dogs, noting that the lack of wolves in England at the beginning of 'the early modern period' resulted in dogs that drove sheep, while in France or Italy, where wolves still survive, sheep follow a shepherd, and a 'mastiff or wolfhound, rather than a sheepdog,

Table 13.1. Some differences between herding dogs and guarding dogs.

	Herding dogs	Guarding dogs
Examples	Australian kelpie Australian shepherd Border collie New Zealand huntaway	Anatolian shepherd Great Pyrenean Maremma Sarplaninac
Morphology	10–20 kg Ears often pricked or 'tulip'; some breeds' ears hang down Colour usually dark with white or brown markings; some are white/grey with darker spots	30–55 kg Ears hang down Colour usually white or grey, although some breeds are brown with darker markings
Behaviour	'Chase-and-bite' Active around stock Trained to respond to human commands Have motor patterns specific to job: eye, clapping, heeling, heading or voicing, which are heritable	'Never' chase or bite Passive, even lethargic, around stock Seldom trained to respond to commands Have no job-specific motor patterns; form strong social bonds with stock through early, continuous association

Fig. 13.1. Border collie (herding dog) and Maremma (guarding dog).

[goes] in front as their protector'. An early dog expert, Dr Johannes Caius, wrote about 'The Shepherd's Dogge' (1576), describing a technique still used today by shepherds directing their herding dogs to move the stock: 'The dogge, either at the hearing of his master's voice, or at the wagging and whistling of his fist, or at his shrill horse hissing, bringest the wandering weathers and straying sheepe into the self same place where his masters will and wishes is to have them'.

Kupper (1945) described dogs in the 19th-century American south-west as movers of sheep, although she did wonder about 'the wonderful sheep dogs' that would die of starvation rather than leave the flock.

In France, herding dogs are known as a fairly recent development, supplanting the guardian when large carnivores (bear, wolf) disappeared from western Europe (de Planhol, 1969; Laurans, 1975; Lory, 1989; Schmitt, 1989). Laurans advanced an explanation for herding dogs becoming more useful, for as population increased, '. . . in countries with many small parcels of land . . . the shepherd needed herding dogs in order to keep his gardens safe from damage by sheep'.

The most comprehensive modern book on farm dogs is that of Hubbard (1947), whose historical overview of working dogs differentiates between herding and guarding dogs and includes a long section on herding dogs and the competitive trials that are so popular in the UK and the USA.

Hubbard described 58 working breeds, arranged in three categories: pastoral dogs (47), draught dogs (5) and utility dogs (6).

> For the most part the pastoral dogs are Sheepdogs proper, that is, those breeds used for herding and controlling sheep only. In this group we find the well-known Collie, German Shepherd Dog and Old English Sheepdog: breeds invariably used only with sheep. Apart from these there are also Cattle Dogs, drovers' dogs or cow-herds' dogs; breeds like the Welsh Corgi, the various Bouviers of Flanders, the Australian Heeler, the Hungarian Pumi, and the Portuguese Cattle Dog. Furthermore, there are other breeds which are used in the protection, droving and controlling of other animals, such as some Russian Laiki (which herd reindeer), other Russian herding dogs (which control the dromedaries of Central Asia), and the many varieties of native races of South Sea Islands dogs (which round up the indigenous pigs of the islands). There are, of course, some Sheepdogs which work as well with cattle as with sheep, and a few, indeed, that work with goats and pigs as well as with sheep.

Today, producers in the USA and Australia report that their guarding dogs have been bonded also with llamas, ostriches, emus, turkeys and other unusual or rare species.

Discovering how modern herding dogs could be developed to such precision – and how their morphology and behaviour could be so distinctly differentiated from the guardians – is based on ethology, or the study of their behaviour. An early analysis, cut short by the Second World War, was carried out by Dawson (1965) in 1935. His goal was to study the 'inheritance of intelligence and temperament in farm animals', with dogs the experimental animal and sheep-herding one manifestation

of intelligence, 'since it was of economic importance in agriculture'.

Various cross-breedings were made between Hungarian pulis, German shepherds, Border collies, Chows and a pair of Turkish guarding dogs. Dawson reported wide variation in reactions of dogs in all tests, but could not detect differences due to sex or between the larger breed groups. He noted 'marked indications that some of the behaviour traits were inherited'.

Development of different behaviour patterns

More recently, looking specifically at differences between the two types of sheepdog, Coppinger *et al.* (1987a) wrote:

> As with juvenile wolves or coyotes, adult livestock conducting dogs displayed the first-half segment of a functional predatory system of motor patterns and did not express play or social bonding toward sheep; whereas, like wolf or coyote pups, adult livestock protecting dogs displayed sequences of mixed social, submissive, play and investigatory motor patterns and rarely expressed during ontogeny (even when fully adult) predatory behaviours. The most parsimonious explanation of our findings is that behavioural differences in the two types of livestock dogs are a case of selected differential retardation (neoteny) of ancestral motor pattern development.

In other words, herding dogs are selected to show hunting behaviours, such as eye, stalk, grip or heel. Guarding dogs are selected to show more of the wild ancestor's puppy-like or juvenile behaviour, preferring to stay in the 'litter' of livestock to which they are bonded and to react to novelty by barking an alarm. Guarding dogs are not attack dogs: they are defence dogs, although they have been aptly described as 'unbribable . . . extremely loyal, distrustful of strangers, and capable of attacking both wolves and bears' (RuHil, 1988).

Further investigations into the differences in behaviour between guarding and herding dogs were undertaken by Coppinger and Schneider (1995). They juxtaposed the behaviour of herding dogs, guarding dogs and sledge dogs to hypothesize that it is the timing of events during canine development that intensifies differences in innate motor patterns, and which in turn differentiates learning abilities of breeds. Guarding dogs, as noted above, seem to have been selected to mature at an early ontogenetic stage, before predatory sequences emerge. Otherwise they would not be trustworthy and could not be left alone with the stock. Herding dogs, represented in both studies by Border collies (see Fig. 13.2), were selected to show the predatory sequences of eye, stalk and chase, but to mature before the dangerous crush–bite–kill patterns of the true predator.

Detailed profiles of both types, based on modern understanding of dog behaviour, appear in Coppinger and Coppinger (1998). In their book, *Dogs*, Coppinger and Coppinger (2001) expand the analysis of breed differences, describing the genetic processes that can lead to the development of dogs specialized for herding, droving, guarding, racing and hunting.

Different breeds of herding dogs exhibit varying degrees of predatory behaviour. Gathering dogs such as the Border collie circle the livestock with less direct contact than heeling dogs. Breeds such as the Queensland blue heeler, kelpie and huntaway exhibit a greater degree of predatory behaviour, and their natural tendency is to chase and bite livestock instead of circling them. Breeds with grab–bite or kill–bite behaviours can be stressful to livestock

and their use should be limited to tasks such as mustering cattle in rough country.

Neurotransmitter differences

These variations in motor patterns appear to be genetically predisposed or inherited within each breed. Willis (1992) reported on the heritability of several traits, briefly mentioning the herding and guarding abilities of sheepdogs (NB: 'heritability' does not equal 'inherited'; 'heritability' includes genes plus environment as factors affecting variation). He noted three instincts of Border collies that are heritable: clapping (crouching), staring (at the sheep) and barking (not done when herding but done in other circumstances).

He also noted that studies of the effects of genetics plus environment (heritability) in guarding dogs would help breeders improve their performance. Possible chemical reasons for these behavioural differences were studied by Arons and Shoemaker (1992). They reported livestock-guarding dogs to be different from both herding dogs and sledge dogs in both the distribution and amount of neurotransmitters in various sections of the brain. The guarding dogs (several breeds) had low levels of dopamine in the basal ganglia, whereas Border collies and Siberian huskies had higher quantities of this neural transmitter. Dopamine is related to rates of neural activity.

Fig. 13.2. Border collie circling and staring at sheep. The dog is on the edge of the collective flight zone of the sheep (see Chapter 5, this volume). Some sheep are facing the dog, but others have started to turn away because the dog is entering their flight zone.

Herding Dogs

'There are expensively mechanised farms', wrote Longton and Hart (1976), 'where, to move bullocks from one field to another or bring sheep into the yards, involves turning out the entire farm staff. A good sheep dog could do the whole job more easily and economically'. Brown and Brown (1990) explain that the instinct of a gathering and herding dog such as a Border collie is to circle around the livestock. The dog will instinctively position itself opposite the handler on the other side of the livestock (see Fig. 13.3). If the handler moves to the right the dog will tend to move to the left, and vice versa. A good dog can be ruined if the handler tries to position him/herself on the same side of the livestock as the dog. The gathering instinct is part of canine predatory behaviour. This behaviour can be modified by training, to enable the handler to send the dog out and have it bring the livestock to him/her.

Training information

Modern technology has greatly enriched the methods by which livestock producers can acquire needed information about training stock dogs. The training methods may vary, but details are quickly visible when shown on videotapes or explored on the World Wide Web. Several 'standard' books, however, should never be replaced on the shelves of dog trainers. The techniques of Jones and Collins (1987) and Jones (1991) are slightly different from those of Longton and Hart (1976), but the results are similar.

The classic 'old reliable' book on training a sheepdog is *The Farmer's Dog*, by John Holmes (1976). His descriptions of training and using dogs in daily shepherding are based on a broad understanding of dog behaviour, which he applies effectively to explain the process. For example, he describes herding behaviour as inhibited (by the handler, usually) predatory behaviour by the dog.

Iley's (1978) *Sheepdogs at Work* includes chapters on the history of the working dog, the pup and its early training, trials and breeding. A right-to-the-point book is by Means (1970), who trains Border collies to work with cattle. The text includes basic training and use of a stock dog, and also what can go wrong and how to prevent or cure it, rather than assuming everything goes right the first time.

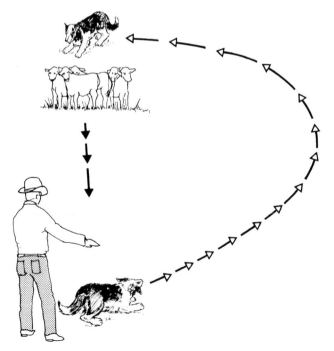

Fig. 13.3. Teaching a young Border collie to circle around the livestock and bring them to the handler (from Brown and Brown, 1990).

Other books include *Working Sheepdogs: Management and Training*, by John Templeton and Matt Mundell (1992) and *Border Collies: Everything about Purchase, Care, Nutrition, Breeding, Behavior, and Training*, by Michael Devine and David Wenzel (1997). The former is excellent, and written by two of the best working dog trainers in the UK. The second appeals more to owners of the Border collie as a pet. *Ranch and Farm Dogs: Herders and Guards*, by Elizabeth Ring (1994), is geared for younger (age 4–8 years) readers, and provides for youngsters a much-needed connection with these important tools. It has colour photographs of most of the breeds.

The above-mentioned books concentrate on Border collies, mostly with sheep, whereas Taggart (1991) discusses 21 herding breeds, including the Australian cattle dog and others bred to work more with cattle than with sheep. Training methods for all are similar, but Taggart's descriptions allow for the different styles of the different breeds. She, too, discusses common problems and suggests solutions. Diagrams and photographs of the different breeds and of dogs in action help emphasize the finer points.

The training of cattle dogs is also addressed by Little (1975), whose positive view of her job as dog trainer is evident from the title, *Happy Herding Handbook*. Little describes how to train for the various commands, and what to do about problems such as biting, grabbing or harassing stock, not obeying commands, not outrunning well and so on. Charoy's (1989) descriptions of standard selection and training methods are enhanced by the inclusion of techniques relevant to the daily needs of European shepherds, such as training a dog to conduct sheep along a road with cars or crossroads.

Among the cattle dog books is *The Complete Australian Cattle Dog* (2002) by John Holmes and Mary Holmes, long-term cattle dog trainers. Narelle Robertson's *Australian Cattle Dogs* (1996) is good for novice owners, providing a history of the breed and a discussion of the dogs' behaviour.

For people desiring to learn how to train a dog, step by step, Vergil S. Holland's *Herding Dogs: Progressive Training* (1994) is the most detailed. He writes mostly about dogs working with sheep, but his depiction of problems and solutions, and the way different dogs approach

stock, plus the amusing drawings and photographs, broaden the scope of this book to make it useful to trainers of any stock dogs.

The traits that allow herding dogs to be trained so well to their tasks are the essence of an article by Vines (1981). Fiennes and Fiennes (1968) also noted reasons for sheepdogs being so well suited for studies of how dogs process information and what they do with it. For a range of techniques and advice, the World Wide Web provides immediate access. Web sites run by experts and other practitioners contain articles, breeder lists, membership applications and links to other dedicated sites, as well as to news groups, whose participants discuss issues and give advice at all levels. On the Internet, search engines such as *Yahoo!* lead to scores of resources. Performing a search for 'stock dog' or 'herding dog' is the best way to start. One main web site is http://www.stockdog.com

Acoustic signals

One of few scientific studies on herding dogs (McConnell and Baylis, 1985) discusses: (i) locomotion and postural behaviour of Border collies when working around sheep (visual communication from dog to sheep); and (ii) acoustic signal systems used by shepherds to control their dogs. Analysis of data on 'mature trained and immature untrained Border collies shows that the [stalking] posture [of a hunting mammalian predator] was innate, but was refined by training and experience'. The signals (whistling) that stimulated the dog into activity (e.g. go fetch, go closer to the sheep, go faster) were short, rapidly repeated notes, tending to rise in frequency. Those that inhibited activity (e.g. slow down, stop) were continuous, long and descending.

McConnell (1990) subsequently looked more closely at the acoustic structure and response, and confirmed these observations. A more recent study showed that dogs respond to commands more efficiently when they came from a person instead of from a tape recorder. Dogs are highly responsive to eye and body language cues that a recording does not provide (Fukuzawa *et al.*, 2005).

The original question asked by McConnell and Baylis, 'How do individuals of two different species effectively communicate to manoeuvre a third (and usually unwilling) species', might well be asked also within the interspecific triangle of guarding dog, livestock and predator. McConnell and Baylis mention that shepherds exploit the interactions between predator and prey species (in order to move their stock), and that ethologists can exploit the shepherds' systems by studying the results of the shepherd-generated interspecific behaviour. Researchers into the behaviour of guarding dogs have begun to explore these interactions.

Guarding Dogs

The following quotes illustrate the value of guarding dogs to the stockholder:

> In the two and one half years that we had Ike we didn't have a single incidence of livestock being harmed by dogs or coyotes. Ike literally made it possible for us to continue having sheep and to begin purebred breeding with feelings of security regarding the safety of our animals.
>
> (S. Sorensen, 1990, personal communication)

> Estimates indicate coyotes . . . kill an average of 1–2.5% of the domestic adult sheep and 4–9% of the lambs in the 17 western states . . . Livestock producers reduce losses [mortality] by using various livestock management practices, frightening devices, trapping, snaring, calling and shooting, sodium cyanide guns, denning, aerial gunning, and livestock guarding dogs.
>
> (Andelt, 1992)

> Only the dogs stopped coyote predation (O'Gara et al., 1983); but they also harassed sheep'.
>
> (Linhart et al., 1979)

The ancient Romans, generations of Old World shepherds and a few 19th-century New World ranchers knew how to protect their livestock from predators with guarding dogs. But, in late 20th-century USA, guarding dogs were essentially unknown. In spite of an array of 'sophisticated' poisons and various high-tech devices, sheep and cattle producers were still losing 8–10% of their animals to predators.

In the mid-1970s, a 3-month, on-the-road survey of producers in the USA using guarding dogs with sheep and/or goats showed seven dogs out of the 12 located to be successful (Coppinger and Coppinger, 1978). Yet few producers had even heard about guarding dogs, and many were highly sceptical that a carnivore could protect a prey species from other carnivores. 'Why should I pay for my predators when I can get them for free', one of them asked.

Research began in 1976 at Hampshire College in Amherst, Massachusetts, and at the Denver Wildlife Research Center (US Fish and Wildlife Service) in Colorado. At the Denver project and its successors at the US Sheep Experiment Station in Idaho, researchers studied dog/coyote interactions under controlled conditions and ran field trials with dogs placed on western ranches.

At Hampshire, biologists and their students focused on the basic, instinctive behaviour that results in good guardians, and hypothesized that guarding dogs achieve protection by being attentive to the livestock and trustworthy with them. Both projects placed dogs on farms and ranches for on-site trials, and almost immediately reports came back of lower predation levels.

By the end of the 1980s, the data looked good: predation on dog-guarded flocks was reduced by 64–100% (Coppinger et al., 1988), dogs were an economic asset for 82% of the people surveyed (Green and Woodruff, 1988) and dogs in a Colorado survey saved an average of US$3216 worth of sheep annually per dog (Andelt, 1992). Green et al. (1984) reported that 37 ranchers spent about 9 h a month feeding and maintaining their adult guardians, corroborated by Andelt's (1992) 10 h. These authors agreed on costs of a guarding dog: first-year dollar amounts, including purchase, were US$700–900 and subsequent annual costs were US$250–290.

Use of guarding dogs may have died out among Spanish and Anglo-American ranchers in the American south-west during the late 19th century, but at least one native American group either learned from the immigrants or reinvented the system. Black (1981) and Black and Green (1985) found Navajos in Arizona using mixed-breed guarding dogs in a practice very like those still seen in Europe and Asia:

> Navajos call their dogs 'sheep dogs' but, unlike sheep dogs used by other ranchers to assist in herding and moving the flocks, Navajo dogs function primarily as guardians of sheep and goats to whom they have developed social bonds. This attraction is a result of raising dogs

essentially from birth in visual, olfactory, auditory and tactile association with sheep and goats. A minimum of handling of pups reduced the likelihood that they would bond strongly to humans. Mixed-breed dogs of the Navajo appear to exhibit all behavioral traits believed to be important in protecting flocks from predators, especially coyotes: they are attentive, defensive and trustworthy. If ranchers choose to employ dogs, the rather simple Navajo recipe for training may serve them well. Mixed-breed dogs could be quickly deployed in a variety of ranching situations to help reduce predation on livestock.

Training of guarding dogs

From the following recipe, reported in the Black and Green (1985) paper, it is obvious that guarding dogs need a different kind of training than do herding dogs:

> Raise or place mixed-breed pups in corrals with sheep, lambs, goats, and kids at 4–5 weeks of age. Feed the pups dog food and table scraps. Provide no particular shelters such as dugouts or doghouses (the pups will sleep among the sheep and will dig their own dirt beds). Minimize handling and petting. Show no overt affection. Return pups that stray to the corral (chase them, scold them, toss objects at them). Allow pups to accompany the herds onto the rangeland as age permits. Punish bad behaviour such as biting or chasing the sheep or goats, and pulling wool by scolding and spanking. Dispose of dogs that persist in chasing, biting, or killing sheep.

Within this straightforward approach is the basis of the attentive, trustworthy and protective behaviour identified as characteristic of a good guarding dog (Coppinger et al., 1988). It also contains the wisdom behind the dog's critical socialization periods identified under laboratory conditions by Scott and his colleagues (Scott and Marston, 1950; Scott, 1958). Applying the findings of Scott and other modern ethologists to the training of a livestock guarding dog results in a more formal recipe for what the Navajos (and the ancient Romans) knew all along (see Table 13.2).

Making a case against the importation and use of large, Old World guardians, Black and Green (1985) wrote that the small, mongrel-type dogs used by Navajos were cheaper, easier to get and keep, of lower liability and easier

to dispose of if they showed unacceptable behaviour.

Coppinger et al. (1985) agreed that mongrels could be an efficient dog for wider use in the USA, and better than most pure, non-guarding breeds. However, they reiterated that dogs selected specifically for the task of stock-guarding, and bred true, were likely to be more successful than non-guarding breeds. Green and Woodruff (1990) also found pure-breds to be more successful than 'others'.

This is not to imply, however, that the cross-breeding of guarding breeds is not a great idea, for many producers increase their benefits by doing so. It must be remembered that the pure breeds were selected for specific habitats and might have inappropriate conformations in a different habitat. For example, big, powerful, aggressive dogs might be suitable in remote habitats with wolves, bears and mountain lions, while such dogs might be a liability in habitats with hikers, joggers or customers picking up farm products. Small predators can be deterred by a smaller dog than can big predators. Thus, in the long term, producers should develop their own dogs, adapted to their own special needs. This is best accomplished by cross-breeding and experimenting with colour, coat or size.

The success of research in the USA has attracted the attention of wildlife managers and sheep producers in many areas of the world. A comprehensive summary of these efforts was published by Rigg (2001). In those European countries where guarding dogs were rarely, if ever, used, recent population recoveries of wolves, bears or cats have caused wildlife and agriculture agents to look at the potential for these dogs as stock protectors in their regions. A series of reports in the journal *Carnivore Damage Prevention News* has highlighted field research in Switzerland and Germany (Klaffke, 1999; Landry et al., 2005; Lüthi and Mettler, 2005; Mettler, 2005), and in Norway and Sweden (Hansen and Bakken, 1999; Smith et al., 2000; Hansen, 2005; Levin, 2005).

Other areas in Europe have traditionally used guarding dogs and have developed their own breeds. Shepherds in many of these countries had stopped using dogs when wolf or bear populations had been eradicated. The indigenous breeds existed mainly with breeders. But with depredatory carnivore populations

Table 13.2. First stages of development of livestock-guarding dogs.

Stage 1 Attentive behaviour	Stage 2 Attentive behaviour	Stage 3 Attentive behaviour	Stage 4 Trustworthy behaviour	Stage 5 Protective Behavior
Puppy, 0–8 weeks; consists of 3 phases	Early juvenile, 8–16 weeks	Late juvenile, 4–6+ months (ends at puberty)	Sub-adult, 6+–12+ months	Adult, over 12 months
Neonatal, 0–2 weeks Pup is insulated from the environment outside the litter. Reflex care-soliciting behaviour. Non-reflexive behaviour: cries, sucks, roots towards warmth, crawls *Transitional, 2–3 weeks* Eyes open, teeth appear, walks; non-reflexive learning behaviours appear; mother stops responding to pups' cries *Primary socialization, 3–8 weeks* Ears and eyes begin to function; notices other animals at a distance; begins to form primary social relationships that determine later attachments; can eat solid food; food dish dominance begins and wrestling with littermates	Secondary socialization begins; attachments made to other animals and even species. Non-reflexive care-soliciting behaviour, such as dominance–submission and food-begging, appear. These become the basis for the complex social behaviours of the adult. The target of these behaviours is determined to some degree by primary socialization. However, in guarding dogs, this is the period for bonding pups with livestock. By 16 weeks the 'critical period' – or window during which social attachments are made – is closed	Emerging social behaviours of Stage 2 must be reinforced. Pup must be kept with livestock all the time and not be allowed to play or interact extensively with other dogs or people. Exception would be if the pup is put in a pasture with another guardian dog, presumably older, which is acting as a 'teacher' dog. Any wandering or other inattentive behaviour should be stopped immediately	Onset of predatory behaviour patterns and of 'play', which includes the predatory movements of chase, grab–bite, wool-pull, ear-chew. If this behaviour is allowed to be expressed, which to the pup is a reinforcement of the behaviour, it will become common and be almost impossible to correct. If the behaviours are not reinforced, they will disappear from the pups' repertoire of behaviours. Heat cycles begin in females, sometimes resulting in unexpected behaviour such as wandering or gnawing on sheep. Males may stray if attracted by a female in heat	Care-giving and mature sexual behaviours emerge. A dog that has been properly bonded with livestock and not allowed to disrupt them should be an effective guardian at this point. First experience with serious predators must not be overwhelming; dog needs to gain confidence in its ability as it matures

recovering under protection of the law, there is renewed interest in restoring the guarding breeds to help protect the livestock. Shepherds are being urged to acquire dogs and re-learn techniques their great-grandfathers knew.

Other studies in Europe of interest to producers in the USA are those of Rigg (2005) in Slovakia, and Ribeiro and Petrucci-Fonseca (2005) in Portugal. In addition to these studies, *Carnivore Damage Prevention News* in 2005 (issues 8 and 9) included observations in Bulgaria, Poland, Romania and Spain (*CDP News* is available at http://www.kora.unibe.ch).

Further south, across the equator in the south-western African country of Namibia, guarding dogs were imported from the USA in 1994 to help protect sheep, goats and cattle from cheetah predation. These cats have been an easy target for ranchers, who have reduced the populations of cheetahs to the point where conservation experts are worried about their survival as a species (Marker *et al.*, 2005). Interestingly enough, subsequent reports noted that the dogs were effective in deterring several other predators such as leopards and baboons. Our own interviews with Herrero people in Namibia and Masai in Kenya, who regularly use guarding dogs, suggest dogs are aids against all predators including lions – at least in their ability to sound an alarm (Coppinger and Coppinger, 2000).

Are guarding dogs 100% successful at protecting livestock? Of course not, but they would not have survived as flock guardians had they not been cost-effective. Old World herders are well aware that a dog can be vulnerable to a wolf attack, and manage the dogs to minimize overt conflict. A report by Bangs *et al.* (2005) documents fatal attacks by wolves (usually a pack) on guardian dogs. Eighteen dogs were killed over a span of 10 years in three western states. No data are given on how many dogs were working during the same period. The report also makes recommendations for increasing the effectiveness of dogs to protect against wolves.

As with herding dogs, guarding dogs are well served by the World Wide Web. One main web page is at http://www.lgd.org. One of its links is to the livestock-guarding dog newsgroup (LGD-L), an active, highly informative source of advice – some of which is conflicting, but all of which is of great interest to owners of livestock-guarding dogs.

Guarding dog behaviour

To try to understand the underlying behaviours that characterize the guarding breeds, Coppinger and his colleagues at Hampshire College studied three basic guardian behavioural traits: attentive, trustworthy and protective. The attentiveness to sheep of dogs that had been used in the USA for several years was found to be similar to that of their parent and grandparent generations still working in Italy (Coppinger *et al.*, 1983b).

Looking at trustworthy and untrustworthy dogs, Coppinger *et al.* (1987a) quantified differences in motor patterns and found that untrustworthy guarding dogs shared some behaviours with herding dogs. Their protective abilities were severely tested in wild, forested wolf territory in northern Minnesota, where several guardians showed that dogs, if managed correctly, could protect stock against North American wolves (Coppinger and Coppinger, 1995).

McGrew and Blakesley (1982) found that interspecific adjustments to the dog were made by both sheep and coyotes: 'The sheep learned to run to or to stand with the dogs when attacked, and usually bedded with the dog. The coyotes learned to attack the flock when the dog was not present'. Anderson *et al.* (1988) used a trained Border collie as a mock predator in an experiment to discover the advantage of interspecific behaviour in providing protection from predators. The collie's 'aggressive, threatening' approach did result in a closer association between sheep and cows that had been bonded with each other, but non-bonded animals moved in distinctly intraspecific groups when threatened. McGrew (1982, unpublished PhD thesis) reported that early exposure of pups to sheep (i.e. bonding) was important, and that a dog's value was determined by what breed it was and by its own personality.

Guard dogs on the open range

Early doubts that dogs could protect sheep on the huge, unfenced ranges of the western USA were diminished by reports by Coppinger *et al.* (1983a) and Green and Woodruff (1983). Methods of increasing the dogs' effectiveness were studied: by looking at causes of pre-senescent mortality (Lorenz et al., 1986) and by transferring inattentive or untrustworthy dogs to new ranches (Coppinger et al., 1987b). That guarding dogs are a positive presence in managing livestock is evident in reports from other researchers. Pfeifer and Goos (1982) found them to be effective in North Dakota, and Hagstad et al. (1987) noted that the presence of a dog with dairy goats reduced predation in Louisiana.

Conclusions

Sheepdogs – both herding and guarding types – are highly efficient for moving or protecting livestock. Within the herding types, different breeds tend to specialize in moving different stock, although most breeds are versatile and adaptable to sheep, cattle, llamas – whatever needs moving. Livestock are safer when moved by herding dogs than by people, for dogs are fast at heading off a stampede and deft at aiming reluctant beasts into the pen or changing a maverick's mind.

Sheep are safer, too, when protected by guarding dogs, whose senses match those of most predators and which live with the stock day and night. Sheepdogs extend a herder's control over the stock, saving time, energy and animals. Many livestock growers who use dogs – both herding and guarding dogs – confirm, in chorus, 'Without our dogs, we would be out of business'.

References

Andelt, W.F. (1992) Effectiveness of livestock guarding dogs for reducing predation on domestic sheep. *Wildlife Society Bulletin* 20, 55–62.

Anderson, D.M., Hulet, C.V., Shupe, W.L., Smith, J.N. and Murray, L.W. (1988) Response of bonded and non-bonded sheep to the approach of a trained Border Collie. *Applied Animal Behaviour Science* 21, 251–257.

Anon. (1873) How they train sheep dogs in California. *Forest and Stream* (New York) 1 (11 December), 279.

Anon. (1913) *Roman Farm Management: the Treatises of Cato and Varro* (circa *150 BC*) a Virginia Farmer, translated by Macmillan, New York.

Arons, C. and Shoemaker, W.J. (1992) The distribution of catecholamines and beta-endorphin in the brains of three behaviorally distinct breeds of dogs and their F_1 hybrids. *Brain Research* 594, 31–39.

Bangs, E., Jimenez, M., Niemeyer, C., Meier, T., Asher, V., Fontaine, J., Collinge, M., Handegard, L., Krischke, R., Smith, D. and Mack, C. (2005) Livestock guarding dogs and wolves in the northern Rocky Mountains of the United States. *Carnivore Damage Prevention News* 8, 32–39.

Baur, J.E. (1982) *Dogs on the Frontier*. Denlinger's, Fairfax, Virginia.

Black, H.L. (1981) Navajo sheep and goat guarding dogs. *Rangelands* 3, 235–237.

Black, H.L. and Green, J.S. (1985) Navajo use of mixed-breed dogs for management of predators. *Journal of Range Management* 38, 11–15.

Brown, L. and Brown, M. (1990) *Stock Dog Training Manual*. Eagle Publishing Company, Ekalaka, Montana.

Caius, J. (1576) *Of English Dogges*. Rychard Johnes, London.

Charoy, G. (1989) L'éducation du chien. *Ethnozootechnie* 43, 35–50.

Coppinger, R. and Coppinger, L. (1978) *Livestock Guarding Dogs for US Agriculture*. Hampshire College, Amherst, Massachusetts.

Coppinger, R. and Coppinger, L. (1995) Interactions between livestock guarding dogs and wolves. In: Carbyn, L.N., Fritts, S.H. and Seip, D.R. (eds) *Ecology and Conservation of Wolves in a Changing World*. Occasional Publication number 35, Canadian Circumpolar Institute, University of Alberta, Edmonton, pp. 523–526.

Coppinger, R. and Coppinger, L. (1998) Differences in the behavior of dog breeds. In: Grandin, T. (ed.) *Genetics and the Behavior of Domestic Animals*. Academic Press, San Diego, California, pp.167–202.

Coppinger R. and L. Coppinger (2001) *Dogs: a Startling New Understanding of Canine Origin, Behavior and Evolution*. Scribner, New York.

Coppinger, R. and Schneider, R. (1995) The evolution of working dog behavior. In: Serpell, J. (ed.) *The Domestic Dog: its Evolution, Behavior and Interactions with People*. Cambridge University Press, Cambridge, UK, pp. 21–47.

Coppinger, R., Lorenz, J. and Coppinger, L. (1983a) Introducing livestock guarding dogs to sheep and goat producers. In: Decker, D.J. (ed.) *Proceedings of the First Eastern Wildlife Damage Control Conference*, Cornell University, Ithaca, New York, pp. 129–132.

Coppinger, R., Lorenz, J., Glendinning, J. and Pinardi, P. (1983b) Attentiveness of guarding dogs for reducing predation on domestic sheep. *Journal of Range Management* 36, 275–279.

Coppinger, R.P., Smith, C.K. and Miller, L. (1985) Observations on why mongrels may make effective livestock protecting dogs. *Journal of Range Management* 38, 560–561.

Coppinger, R., Glendinning, J., Torop, E., Matthay, C., Sutherland, M. and Smith, C. (1987a) Degree of behavioral neoteny differentiates canid polymorphs. *Ethology* 75, 89–108.

Coppinger, R., Lorenz, J. and Coppinger, L. (1987b) New uses of livestock guarding dogs to reduce agriculture/wildlife conflicts. In: *Proceedings of the Third Eastern Wildlife Damage Control Conference*, Gulf Shores, Alabama, pp. 253–259.

Coppinger, R., Coppinger, L., Langeloh, G., Gettler, L. and Lorenz, J. (1988) A decade of use of livestock guarding dogs. In: Crabb, A.C. and Marsh, R.E. (eds) *Proceedings of the Thirteenth Vertebrate Pest Conference*, University of California, Davis, California, pp. 209–214.

Courreau, J.F. and Langlois, B. (2005) Genetic parameters and environmental effects which characterize the defense ability of the Belgian Shepherd dog. *Applied Animal Behaviour Science* 91, 233–245.

Dawson, W.M. (1965) *Studies of Inheritance of Intelligence and Temperament in Dogs*. ARS 44–163, Animal Husbandry Research Division, US Department of Agriculture, Beltsville, Maryland.

de Planhol, X. (1969) Le chien de berger: développement et signification géographique d'une technique pastorale. *Bulletin de l'Association des Géographes Français* 370, 351.

Devine, M. and Wenzel, D. (1997) *Border Collies*. Barrons Educational Series, Hauppauge, New York.

Fakazawa, M., Mills, D.S and Cooper, J.J. (2005) More than just a word: non-semantic command variables affecting obedience in the domestic dog (*Canis familiaris*). *Applied Animal Behaviour Science* 91, 129–141.

Fiennes, R. and Fiennes, A. (1968) *The Natural History of Dogs*. Bonanza Books, New York.

Green, J.S. and Woodruff, R.A. (1983) The use of three breeds of dog to protect rangeland sheep from predators. *Applied Animal Ethology* 11, 141–161.

Green, J.S. and Woodruff, R.A. (1988) Breed comparisons and characteristics of use of livestock guarding dogs. *Journal of Range Management* 41, 249–251.

Green, J.S. and Woodruff, R.A. (1990) *Livestock Guarding Dogs: Protecting Sheep from Predators*. Agriculture Information Bulletin No. 588, US Department of Agriculture, Beltsville, Maryland.

Green, J.S., Woodruff, R.A. and Tueller, T.T. (1984) Livestock-guarding dogs for predator control: costs, benefits, and practicality. *Wildlife Society Bulletin* 12, 44–49.

Hagstad, H.V., Hubbert, W.T. and Stagg, L.M. (1987) A descriptive study of dairy goat predation in Louisiana. *Canadian Journal of Veterinary Research* 5, 152–155.

Hahn, M.E. and Schanz, N. (1996) Issues in the genetics of social behavior revisited. *Behavior Genetics* 26, 417–422.

Hahn, M.E. and Wright, J.C. (1998) The influence of genes on social behavior of dogs. In: Grandin, T. (ed.) *Genetics and the Behavior of Domestic Animals*. Academic Press, San Diego, California, pp. 299–318.

Hansen, I. (2005) Use of livestock guarding dogs in Norway – a review of the effectiveness of different methods. *Carnivore Damage Prevention News* 8, 2–8.

Hansen, I. and Bakken, M. (1999) Livestock-guarding dogs in Norway: Part I. Interactions. *Journal of Range Management* 52, 2–6.

Holland, V.S. (1994) *Herding Dogs: Progressive Training*. Howell Book House, New York.

Holmes, J. (1976) *The Farmer's Dog*. Popular Dogs, London.

Holmes, J. and Holmes, M. (2002) *The Complete Australian Cattle Dog*. Interpet Publishing, Dorking, UK.

Hubbard, C.L.B. (1947) *Working Dogs of the World*. Sidgwick and Jackson, London.

Iley, T. (1978) *Sheepdogs at Work*. Dalesman Books, Clapham, UK.

Jones, H.G. (1991) *Come Bye! and Away!* Farming Press Videos, Ipswich, UK.

Jones, H.G. and Collins, B.C. (1987) *A Way of Life: Sheepdog Training, Handling and Trialling*. Farming Press Books, Ipswich, UK.

Klaffke, O. (1999) The company of wolves. *New Scientist* (http://www.newscientist.com, accessed 6 February 1999).

Kupper, W. (1945) *The Golden Hoof*. Alfred A. Knopf, New York.

Landry, J.-M., Burri, A., Torriani, D. and Angst, C. (2005) Livestock guarding dogs: a new experience for Switzerland. *Carnivore Damage Prevention News* 8, 40–48.

Laurans, R. (1975) Chiens de garde et chiens de conduite des moutons. *Ethnozootechnie* 12, 15–18.

Levin, M. (2005) Livestock guarding dogs in Sweden: a preliminary report. *Carnivore Damage Prevention News* 8, 8–9.

Lindblad-Toh, K. *et al.* (2005) Genome sequences, comparative analysis and haplotype structure of the domestic dog. *Nature* 438, 803–819.

Linhart, S.B., Sterner, R.T., Carrigan, T.C. and Henne, D.R. (1979) Komondor guard dogs reduce sheep losses to coyotes: a preliminary evaluation. *Journal of Range Management* 43, 238–241.

Little, M.E. (1975) *Happy Herding Handbook*. Wheel-A-Way Ranch, Riverside, California.

Longton, T. and Hart, E. (1976) *The Sheep Dog: its Work and Training*. David and Charles, North Pomfret, Vermont.

Lorenz, J., Coppinger, R. and Sutherland, M. (1986) Causes and economic effects of mortality in livestock guarding dogs. *Journal of Range Management* 39, 293–295.

Lory, J. (1989) Le chien de berger, son utilisation. *Ethnozootechnie* 43, 27–34.

Lüthi, R. and Mettler, D. (2005) Experiences with the Maremmano-Abruzzese as a livestock guarding dog in Switzerland. *Carnivore Damage Prevention News* 9, 39–44.

Marker, L., Dickman, A. and Schumann, M. (2005) Using livestock guarding dogs as a conflict resolution strategy on Namibian farms. *Carnivore Damage Prevention News* 5, 28–32.

McConnell, P.B. (1990) Acoustic structure and receiver response in domestic dogs, *Canis familiaris*. *Animal Behaviour* 39, 897–904.

McConnell, P.B. and Baylis, J.R. (1985) Interspecific communication in cooperative herding: acoustic and visual signals from human shepherds and herding dogs. *Zeitschrift für Tierpsychologie* 67, 302–328.

McGrew, J.C. and Blakesley, C.S. (1982) How Komondor dogs reduce sheep losses to coyotes. *Journal of Range Management* 35, 693–696.

Means, B. (1970) *The Perfect Stock Dog*. Ben Means, Walnut Grove, Missouri.

Mettler, D. (2005) The institutionalisation of livestock protection in the Alps with respect to the small-scale agriculture of Switzerland. *Carnivore Damage Prevention News* 9, 36–38.

O'Gara, B.W., Brawley, K.C., Munoz, J.R. and Henne, D.R. (1983) Predation on domestic sheep on a western Montana ranch. *Wildlife Society Bulletin* 11, 253–264.

Olsen, J.W. (1985) Prehistoric dogs in mainland East Asia. In: Olsen, S.J. (ed.) *Origins of the Domestic Dog: the Fossil Record*. University of Arizona Press, Tucson, Arizona, pp. 47–70.

Ostrander, E.A. and Lindblad-Toh, K. (2005) *The Dog and its Genome*. Cold Spring Harbor University Press, Cold Spring Harbor, New York.

Pfeifer, W.K. and Goos, M.W. (1982) Guard dogs and gas exploders as coyote depredation control tools in North Dakota. *Proceedings of the Vertebrate Pest Conference* 10, 55–61.

Ribeiro, S. and Petrucci-Fonseca, F. (2005) The use of livestock guarding dogs in Portugal. *Carnivore Damage Prevention News* 9, 27–33.

Rigg, R. (2001) *Livestock Guarding Dogs: their Current Use Worldwide*. IUCN/SSC Canid Specialist Group occasional paper No. 1, Gland, Switzerland.

Rigg, R. (2005) Livestock depredation and livestock guarding dogs in Slovakia. *Carnivore Damage Prevention News* 8, 17–27.

Ring, E. (1994) *Ranch and Farm Dogs*. Millbrook Press, Brookfield, Connecticut.

Robertson, N. (1996) *Australian Cattle Dogs*. TFH Publications, Neptune City, New Jersey.

Ruefenacht, S., Gebberdt-Henrich, S., Miyake, T. and Gaillard, C. (2002) A behavior test on German Shepherd dogs: heritability of seven different traits. *Applied Animal Behavior Science* 79, 113–132.

RuHil, R. (1988) *The Yugoslav Sheepdog – Tarplaninac*. Jugoslovenska Revija, Belgrade.

Schmitt, R. (1989) Chiens de protection des troupeaux. *Ethnozootechnie* 43, 51–58.

Scott, J.P. (1958) *Animal Behavior*. University of Chicago Press, Chicago, Illinois.

Scott, J.P. and Marston, M. (1950) Critical periods affecting the development of normal and maladjustive social behavior of puppies. *Journal of Genetic Psychology* 77, 25–60.

Smith, M.E., Linnell, J.D.C., Odden, J. and Swenson, J.E. (2000) Review of methods to reduce livestock depredation: I. Guardian animals. *Acta Agriculturæ Scandinavica Section A, Animal Science* 50, 279–290.

Taggart, M. (1991) *Sheepdog Training: an All-breed Approach*. Alpine Publications, Loveland, Colorado.

Templeton, J. and Mundell, M. (1992) *Working Sheepdogs: Management and Training*. Crowood Press, Ramsbury, UK.

Thomas, K. (1983) *Man and the Natural World*. Pantheon, New York.

Vines, G. (1981) Wolves in dogs' clothing. *New Scientist* 91, 648–652.

Willis, M.B. (1992) *Practical Genetics for Dog Breeders*. H.F. & G. Witherby, London.

Further Reading

Adams, H. (1980) *Diary of Maggie (the Komondor)*. Middle Atlantic States Komondor Club, Princeton, New Jersey.

Aurigi, M. (1983) Migration of sheep dogs of Abruzzo (Il cane da pastore abruzzese emigra). *Informatore Zootecnico* (Bologna) 30, 30–33.

Austin, P. (1984) *Working Sheep with Dogs*. Department of Primary Industry, Canberra, Australia.

Dalton, C. (1983) Training working dogs: short lessons and simple commands are the key. *New Zealand Journal of Agriculture* 146, 42.

Ellegren, H. (2005) The dog has its day. *Nature* 438, 745–746.

Fytche, E. (1998) *May Safely Graze: Protecting Livestock Against Predators*. Creative Bound, Toronto, Canada.

Gordon, J. (1985) More from the fool at the foot of the hill. *New Zealand Journal of Agriculture* 150, 47–49.

Hartley, S.W.G. (1967/1981) *The Shepherd's Dogs: a Practical Book on the Training and Management of Sheepdogs*. Whitcoulls, Sydney, Australia.

Henderson, F.R. and Spaeth, C.W. (1980) *Managing Predator Problems: Practices and Procedures for Preventing and Reducing Livestock Losses*. Bulletin No. C-620, Cooperative Extension Service, Kansas State University, Manhattan, Kansas.

Howe, C.E. (1983) Training the farm dog. In: Vidler, P. (ed.) *The Border Collie in Australasia*. Gotrah Enterprises, Kellyville, New South Wales, Australia.

Jones, A.C. and Gosling, S.D. (2005) Temperament and personality in the dog (*Canis familiaris*). A review and evaluation of past research. *Applied Animal Behavior Science* 95, 1–53.

King, T. (2003) Fear of novel and startling stimuli in domestic dogs. *Applied Animal Behavior Science* 82, 45–64.

Lorenz, J.R. (1989) *Introducing Livestock-guarding Dogs*. Extension Circular 1224 (revd.), Oregon State University, Corvallis, Oregon.

Lorenz, J.R. and Coppinger, L. (1989) *Raising and Training a Livestock-guarding Dog*. Extension Circular 1238 (rev.), Oregon State University, Corvallis, Oregon.

Parsons, A.D. (1986) *The Working Kelpie: the Origins and Breeding of a Fair Dinkum Australian*. Nelson, Melbourne, Victoria.

Ravinsky, A. and Sampson, J. (2001) *The Genetics of the Dog*. CAB International, Wallingford, UK.

Sherrow, H.M. and Marker, L. (1988) *Livestock Guarding Dogs in Namibia*. Cheetah Conservation Fund, Otjiwarongo, Namibia.

Sims, D.E. and Dawydiak, O. (1990) *Livestock Protection Dogs: Selection, Care and Training*. OTR Publications, Fort Payne, Alabama.

Tatarskii, A.A. (1960) *Herd Dogs in Sheep Farming. (Pastush'i sobaki v ovtsevodstve)*. Sel'khozgiz, Moscow.

Von Thüngen, J. and Vogel, K. (1991) *Como Trabajar con Perros Pastores*. Instituto Nacional de Tecnologia Agropecuaria, Bariloche, Rio Negro, Argentina.

Wick, P. (1998) *Le Chien de Protection sur Troupeau Ovin: Utilisation et Méthode de Mise en Place*. Editions Artus, St Jean de Braye, France.

Zernova, M.V. (1979a) Working dogs in agriculture: training recommendations. *Tvarynnytstvo-Ukrainy* (Ministerstvo sil's'koho hospodarstva, Kyiv, URSR) 9, 54–55.

Zernova, M.V. (1979b) General training of herd dogs. *Tvarynnytstvo-Ukrainy* (Ministerstvo sil's'koho hospodarstva, Kyiv, URSR) 10, 54–55.

Zernova, M.V. (1980) Rex! To me! Recommendations for training dogs for herding animals. *Tvarynnytstvo-Ukrainy* (Ministerstvo sil's'koho hospodarstva, Kyiv, URSR) 2, 50–51.

Zernova, M.V. (1981) Restraining and keeping in line of a herd: recommendations for training sheep dogs. *Tvarynnytstvo-Ukrainy* ('Urozhai', Kyiv, URSR) 7, 52–53.

Useful Web Sites (see also list of useful web sites at the end of this volume, p. 371)

Bud Williams (stockmanship specialist: http://www.stockmanship.com).

Stock Dog Server (information on stock dogs and herding: http://www.stockdog.com).

Stock Dogs Journal (http://www.stockdogsmagazine.com).

14 Behavioural Principles of Pig Handling

P.H. Hemsworth

Animal Welfare Science Centre, The University of Melbourne and the Department of Primary Industries (Victoria), Parkville, Victoria, Australia; phh@unimelb.edu.au

Introduction

The ability of farm animals to be handled easily may be affected by fear and social behaviour, with fear particularly affected by the animals' contact with humans as well as the animals' sensory characteristics and physical environment (Hemsworth and Coleman, 1998). Learning, including experiences during rearing, can have long-term effects on fear (Schaefer, 1968; McFarland, 1981). Thus, understanding the animal's behavioural and sensory characteristics is important in handling animals efficiently as well as ensuring high animal productivity and welfare.

Fear-provoking stimuli within the animal's physical environment are likely to elicit overt behavioural responses that may be inconsistent with ease of handling. Furthermore, research – particularly in the intensive livestock industries – has shown that the interactions between stockpeople and their animals can limit animal productivity and welfare (Hemsworth and Coleman, 1998).

While many of these interactions are routinely and, at times, habitually used by stockpeople and may appear harmless to the animals, this research has shown that the frequent use of some of these routine behaviours by stockpeople can result in farm animals becoming highly fearful of humans. It is these high fear levels, through stress, that appear to limit animal productivity and welfare. For instance, fear-provoking stimuli, by leading to stress, may interfere with animal productivity.

One key function of the stress response is to divert food and substrates – such as acetates, glucose and amino acids – away from normal day-to-day functions such as growth and reproduction (Sapolsky, 1992) and thus, during stress – particularly chronic stress – the substrates essential for growth and reproduction may be diverted elsewhere, thereby interfering with growth and reproduction.

Fear is an undesirable emotional state of suffering (Jones and Waddington, 1992), and the implications of fear of humans on the welfare of livestock are highlighted by the substantial between-farm variation in the avoidance response of commercial dairy cows, pigs and poultry to humans (Hemsworth and Coleman, 1998).

An additional outcome of these behavioural and stress responses to fear-provoking stimuli within the animal's environment is reduced stockperson or operator comfort (including perceived increases in workload and lower job satisfaction). The stockperson's frustration with difficulties in handling animals can lead to increased negative interaction with the animals by the stockperson, which in turn may further exacerbate animal fear responses within this environment. Figure 14.1 depicts these series of interactions for dairy cows in the milking environment (adapted from Arnold, 2005), but the general principles are applicable to animal handling in

©CAB International 2007. *Livestock Handling and Transport,*
3rd edn. (ed. Grandin, T.)

Milking Environment

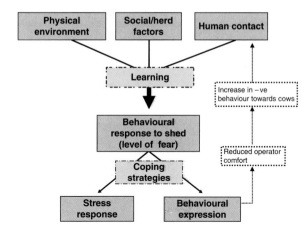

Fig. 14.1 Diagram showing relationships between various factors influencing dairy cow behaviour within milk harvesting systems (from Arnold, 2005).

most livestock production systems, particularly in intensive systems such as the pig industry.

Thus, understanding the animal's behavioural and sensory characteristics is important in efficiently handling animals as well as in ensuring high productivity and welfare of these animals. This chapter will review some of the main principles of pig handling based on current knowledge of the fear and explorative responses, learning capability, sensory characteristics and social behaviour of commercial pigs. The second part of this chapter briefly considers opportunities for improved handling of pigs.

Principles of Pig Handling

There has been considerable debate over the concept of fear (see Hinde, 1970; Murphy, 1978). Gray (1987) defines fear as a hypothetical state of the brain, or the neuro-endocrine system, arising from certain conditions and eventuating in certain forms of behaviour. Fear is usually listed among the emotions and, as such, fear can be viewed as a form of emotional reaction to the threat of punishment, where punishment refers to a stimulus that the animal works to terminate, escape from or avoid (see Gray, 1987).

For the purpose of this chapter, fear will be considered as a state of motivation, and fear will be viewed as eliciting escape or avoidance responses. Furthermore, fear-provoking stimuli will include those to which the animal may not have had previous exposure (e.g. novel and so-called 'sign stimuli') and those to which a fear response has become attached through the process of conditioning (McFarland, 1981).

Exploratory behaviour has been defined as 'behaviour which serves to acquaint the animal with the topography of the surroundings included in the range' (Shillito, 1963), and the amount of exploration of an object will depend on characteristics of the object (i.e. its stimulus properties) such as: (i) its novelty (the time since it was last encountered or its degree of resemblance to other encountered situations); (ii) its complexity; (iii) its intensity and contrast; and (iv) the poverty of the preceding environment (Berlyne, 1960). In this chapter, exploration (or curiosity, Toates, 1980) will be considered as a state of motivation, and exploration will be considered to involve behaviours resulting in close contact with a novel stimulus.

A sudden environmental change will usually elicit the movement of turning towards the source, a response called the orientation response, which may be followed by startling responses and defensive or flight reactions by the animal (Hemsworth and Barnett, 1987). As fear responses wane, the animal will also approach and examine the stimulus (Hinde, 1970). Exploration will be terminated once the animal is somewhat acquainted with the stimulus.

Therefore, the responses of animals to novel stimuli can be considered to contain elements of both fear and exploratory responses. Furthermore, in response to a novel stimulus or a stimulus perceived as aversive (with or without previous experience), the initial avoidance and subsequent exploration can be viewed as a consequence of the conflicting motivations of fear and exploration and the waning of fear responses. Thus fear and exploration are important states of motivation that affect the ease of handling animals, and these two emotions will be considered in this section.

Fear of humans

Fear and ease of handling

Most of the limited research on fear of humans and ease of handling pigs and other farm animals indicates that animals that are highly fearful of humans are generally the most difficult to handle. As shown in Table 14.1, moderate to large correlations have been found between the behavioural response of pigs to humans and their ease of handling; pigs that show low fear responses to humans are easy to handle (see Fig. 14.2).

These correlations indicate that pigs showing high levels of fear of humans, based on their avoidance behaviour of, or lack of approach to, an experimenter in a standard approach test, were the most difficult pigs to move along an unfamiliar route: fearful pigs took longer to move, displayed more baulking and were subjectively scored as the most difficult to move by the handler.

While it may intuitively appear that fearful animals may be easy to move, animals that are fearful of humans are likely to be the most difficult to handle for several reasons. When alarmed and stressed, pigs will often pack together, and sorting and handling may become chaotic. As discussed later, stressed pigs may release pheromones in their urine and saliva that communicate to other pigs that they have encountered a difficult situation. Other pigs following will respond fearfully to these signals, making these pigs also difficult to handle.

Furthermore, pigs are generally wary of moving towards an unfamiliar or unpredictable situation and, if they are fearful of both this environment and the handler, they may show exaggerated behavioural responses to handling, such as baulking or fleeing back past the handler.

Other authors have also reported that high levels of fear of humans will decrease the ease of handling pigs (Gonyou et al., 1986; Grandin et al., 1987). In contrast, Hill et al. (1998) found no effect of fear of humans on the time taken to regularly move pigs to and from a weighing area. With a routine handling task, fear of humans may not affect ease of handling since the predominant response of pigs that are highly fearful of humans may be avoidance of the human rather than of the location. In contrast, as suggested previously, if pigs are fearful of humans and if they are being moved to an unfamiliar location, the conflicting responses to the handler and the

Table 14.1. Correlations between the behavioural response of pigs to an experimenter in a standard approach test to assess fear of humans and the ease of movement of 24 pigs along an unfamiliar route by an unfamiliar handler (Hemsworth *et al.*, 1994b).

	Variables recorded in ease of movement test		
	Time to move	Baulks	Score[a]
Variables recorded in test to assess fear of humans			
Time to approach experimenter	0.34	0.44*	−0.63**
Number of interactions with experimenter	−0.42**	−0.42**	0.51*

Correlation coefficients with * = $P < 0.05$ and ** = $P < 0.01$.
[a] Score was given based on ease of movement, with 0 reflecting substantial difficulty and 4 reflecting little or no difficulty in moving the pig.

Fig. 14.2. Pigs showing low fear responses to humans are generally easy to handle.

location may lead to exaggerated behavioural responses to handling, such as baulking or fleeing back past the handler.

A number of studies have reported that pigs regularly moved out of their pens prior to slaughter were quicker to move during the early stages of transport, such as moving out of their home pen and into a transport crate or box (Abbott *et al.*, 1997; Geverink *et al.*, 1998). While increased familiarity with locations early in the transport process may be responsible for these effects, increased human contact may also be implicated since Eldridge and Knowles (1994) reported that commercial grower pigs that were regularly handled and moved out of their pens to a range of locations were easier to move in an unfamiliar environment.

Handling effects on subsequent ease of handling have been found in other livestock. Studies on cattle, sheep and horses have shown that positive handling, generally involving speaking and touching the animals – particularly during infancy – improved their subsequent ease of

handling (for example, Waring, 1983; Boissy and Bouissou, 1988; Lyons, 1989; Hargreaves and Hutson, 1990; Mateo *et al.*, 1991; Boivin *et al.*, 1992; Hemsworth *et al.*, 1996b).

A number of authors have commented that genetics may affect ease of handling pigs (Grandin, 1991), but there is little evidence of such effects apart from an experiment by Hill *et al.* (1998). As reviewed by Price (2000), although there is little evidence that domestication has resulted in the loss of behaviours from the species repertoire or that the basic structure of the motor patterns for such behaviours has been changed, in nearly all cases, behavioural differences between wild and domestic stocks are quantitative in character and best explained by differences in response thresholds.

An excellent example of this is fear of unfamiliar or novel stimuli (i.e. general fear), and studies on wild and domestic Norway rats indicate that the former are more reactive or fearful of novelty (see Price, 2000). In farm animals, therefore, any breed differences in fear of novelty are likely to affect ease of handling.

Factors affecting the pig's fear of humans

The pig's response to a stockperson in an intensive farming system may have components of both stimulus-specific fear and general fear. While the initial response of a naïve pig to humans may involve a response to novelty or unfamiliarity (i.e. general fearfulness), with subsequent experience of humans and as a consequence of learning there is the development of a specific response to humans, that is a learned response to humans (Hemsworth and Coleman, 1998). The initial response of a naïve farm animal to humans may be similar to the animal's response to an unfamiliar object or to unfamiliar animals of another species. Furthermore, Suarez and Gallup (1982) have suggested that the predominant response of naïve animals to humans may be a response to a predator.

As a consequence of the amount and nature (e.g. positive, neutral or negative in nature for the animal) of interactions with humans, commercial pigs will develop a stimulus-specific response to humans. Therefore, although there will be some components of novelty in the response of experienced animals to humans, which will occur with changes in the stimulus property of humans (e.g. changes in behaviour, clothing, location of interaction, etc.), a major component of this response will be experientially determined.

There is some evidence that the behavioural response of relatively naïve pigs to humans, which may be predominantly a result of general fearfulness, may be moderately heritable; however, subsequent experience with humans will increasingly affect the response to humans (Hemsworth et al., 1990): the behavioural response of experienced pigs to humans accounted for less than one-quarter of the variance of their behavioural response to humans earlier in life, when they were relatively inexperienced with humans.

There is considerable support for this view of the development of a stimulus-specific response of farm animals to humans. For example, numerous handling studies have shown that handling treatments varying in the nature of human contact – but not in the amount of human contact – resulted in rapid changes in the level of fear of humans by pigs (Hemsworth et al., 1981, 1986a, 1987; Gonyou et al., 1986; Hemsworth and Barnett, 1991).

Considerable research has been conducted in the livestock industries over the past two decades on the sequential relationships between stockperson attitude, stockperson behaviour and animal fear. These human–animal relationships were studied in the pig industry by measuring stockperson beliefs about handling and working with pigs (attitudes), conducting direct behavioural observations on stockpeople and assessing fear of humans by measuring the time taken by pigs to approach and interact with a stationary experimenter in a standard approach test. These studies have shown significant sequential relationships between the stockperson's attitudes and behaviour towards animals and the fear of humans by pigs (Hemsworth et al., 1989; Coleman et al., 1998).

These results on commercial farms generally confirm the predictions of handling studies that have been conducted under experimental conditions (Hemsworth et al., 1981, 1986a, 1987; Gonyou et al., 1986; Paterson and Pearce, 1989; Pearce et al., 1989; Hemsworth and Barnett, 1991) and indicate that conditioned approach-avoidance responses develop as a consequence of associations between the stockperson and aversive and rewarding elements of the handling bouts.

The main aversive properties of humans for pigs include hitting, slapping and kicking by the stockperson, while the rewarding properties include patting, stroking and the hand of the stockperson resting on the back of the animal. This research has shown that it is the percentage of these negative tactile interactions of the total tactile interactions that appear to determine the commercial pig's fear of humans.

While auditory interactions by stockpeople may not be highly important in the regulation of these fear response (Hemsworth et al., 1986b), visual interactions by the stockperson such as speed of movement and unexpected movement may also affect fear of humans. Furthermore, pigs with limited experience with humans may habituate with repeated exposure to humans in a neutral context.

Evidence from a number of handling studies on pigs supports the view that the animal's response to a single human might extend to include all humans, through the process of stimulus generalization (Hemsworth and Coleman, 1998). For example, pigs which had previously been briefly, but regularly, handled by either a handler in a predominantly negative manner or by two handlers who differed markedly in the nature of their behaviour towards pigs, showed similar behavioural responses to familiar and unfamiliar handlers (Hemsworth et al., 1994b). Similar evidence is also available from studies in poultry and sheep (Barnett et al., 1993; Jones, 1993; Bouissou and Vandenheede, 1995).

Such results suggest that in commercial situations, the behavioural response of pigs to one handler may extend to other humans. However, it is possible that there are handling situations in which pigs may discriminate between particular people. In situations in which there is intense handling, animals may learn to discriminate between one handler and other handlers to which the animals may be subsequently exposed.

Following an extensive period of intense human contact, Tanida et al. (1995) found that young pigs showed greater approach to the familiar handler than to an unfamiliar handler, even though both handlers wore similar clothing. Furthermore, in situations in which the physical characteristics of the handlers might differ markedly, farm animals may learn to discriminate between the handlers.

For example, in a series of experiments, de Passille *et al.* (1996) found that dairy calves exhibited clear avoidance of a handler who had previously handled them in a negative manner in comparison with handlers wearing different-coloured clothing who were either unfamiliar to the calves or had previously handled them in a positive manner. Initially, there was a generalization of the aversive handling, with calves showing increasing avoidance of all handlers but, with repeated treatment, calves discriminated between handlers and, in particular, between the 'negative' and 'positive' handlers. It is of interest that discrimination was greatest when tested in the area in which handling had previously occurred rather than in a novel location.

These data indicate that discrimination between people by farm animals will be easier if the animals have some distinct cues on which they can discriminate, such as colour of clothing or location of handling.

Although several species of farm animals – including pigs – are capable of discriminating between stockpeople, they do not appear to do so under normal commercial circumstances. Nevertheless, even when farm animals learn to discriminate between humans, fear responses to humans in general are likely to increase in response to the most aversive handler (Hemsworth *et al.*, 1994b; de Passille *et al.*, 1996). Such a finding has important implications in situations in which several stockpeople may interact with pigs.

The development of fear responses in commercial pigs is not surprising considering that stockperson interactions may be biased towards negative ones, given that opportunities for positive human contact are probably reduced in modern pig units and that many routine husbandry tasks undertaken by stockpeople may contain aversive elements. In situations in which contact may be markedly reduced, such as in the grower and finisher sections of pig units, the importance of regular contact and the nature of this contact is particularly important for subsequent ease of handling.

For example, studies by Eldridge and Knowles (1994), Abbott *et al.* (1997) and Geverink *et al.* (1998) suggest that regular handling and moving may facilitate subsequent ease of handling from farm to slaughter. This bias towards negative interactions, together with the sensitivity of pigs to even moderate negative interactions, highlights the challenge confronting stockpeople in improving their interactions with pigs and this aspect will be considered in more detail later in this chapter.

Fear of novelty or unfamiliarity

Pigs, like other animals, are initially fearful of strange objects and locations and will generally baulk. Therefore, features such as floor surfaces, floor levels and wall types should be as consistent as possible throughout a race or corridor to reduce baulking. It is probably particularly important to minimize such changes at critical points in the route, such as pen exits, corners and entrances to corridors or races, where unfamiliarity is likely to have a greater effect on ease of handling. Novel objects in the race or moving and flapping objects will also cause baulking.

If pigs become fearful in an unfamiliar location, it is preferable to allow them some time to familiarize themselves with the environment. Trying to move pigs quickly in this situation may be costly in terms of time and effort, as well as risking injury to both pigs and handlers. Habituation to novel stimuli of moderate intensity will occur over time, as the animal's fear is gradually reduced by continuous exposure in a neutral context.

Grandin (1982–1983) reported that a smooth, concrete floor with a wet, slippery surface inhibited pig movement. Furthermore, Grandin (1988) observed that fattening pigs reared on metal mesh floors were difficult to move on concrete floors. Lack of confidence in gaining a firm footing or an unfamiliar surface may be responsible for these effects. The novelty of a concrete floor for these fattening pigs may also have adversely affected movement.

There are conflicting data in the literature on the behaviour and welfare of livestock provided with additional complexity in their environments (Jongman, 2000). Such manipulations, which are often called 'environmental enrichment', may be similar to the phenomenon of 'infantile stimulation' seen in laboratory animals. Infantile stimulation, which involves handling of young animals, has at times been shown to advance behavioural and physiological maturation (Schaefer, 1968; Hinde, 1970). One notable effect of infantile stimulation is decreased general fearfulness or fear of novelty.

It has been proposed that these subsequent effects on maturation act via an acute stress response early in life (Schaefer, 1968). Grandin *et al.* (1987) found that fattening pigs given novel objects to manipulate, such as rubber hoses, were easier to handle and Pearce *et al.* (1989) found that young pigs housed in pens with novel objects, such as chains and tyres, showed less avoidance of humans than those reared in barren pens. In contrast, Hill *et al.* (1998) found no effect of the provision of such novel objects on the behavioural response of pigs to humans in their pens nor in their ease of movement.

It has been reported that pigs are subsequently more easily startled and more difficult to handle when reared in darkness (Grandin, 1991) or semi-darkness (Warriss *et al.*, 1983). In the latter study, control pigs were reared outdoors, while in the former study the control pigs were presumably reared indoors with greater illumination. As discussed earlier, a number of studies have reported that pigs regularly moved out of their pens during rearing were quicker to move during the early stages of transport (Abbott *et al.*, 1997; Geverink *et al.*, 1998).

In a study examining the effects of environmental enrichment in which toys were provided, Moore *et al.* (1994) found that the provision of toys for early-weaned piglets reduced their fear of humans. It is possible that these effects observed on general fearfulness and ease of handling may have been a consequence of increased environmental stimulation.

Learning ability

As reviewed by Houpt and Wolski (1982) and Kilgour (1987), pigs easily learn a variety of tasks. For example, they readily learn multiple-choice spatial tasks. Pigs are generally considered easy to classically condition: they are rapidly capable of learning to show a range of conditioned or associative responses, such as salivary and cardiac responses to a range of stimuli – including auditory stimuli (Houpt and Wolski, 1982). Pigs can be trained quite easily to perform operant responses and will respond, for example, by pushing or manipulating levers with their snout to receive sensory rewards such as lighting or temperature (Houpt and Wolski, 1982) or jumping across an obstacle to avoid

the punishment of an electric shock (see Craig, 1981). They perform well in maze learning tests, but often perform poorly in visual discrimination tests (Kilgour, 1987).

Pigs have good short-term and long-term memory (Houpt and Wolski, 1982), and this can be used to develop handling routines. With breeding pigs, there is an excellent opportunity for pigs to learn to move easily to and from commonly used locations by regular exposure. By allowing the pigs initially to move at their own pace and thus by minimizing aversive experiences, pigs are likely to move more easily on subsequent introductions.

Rewards for pigs can be used to mask or minimize aversive husbandry experiences or even to train them. Positive rewards such as: (i) handlers patting and talking; or (ii) opportunities for pigs to explore novel settings – since mild novelty is rewarding – or a palatable feed reward, can be used to alleviate moderately aversive husbandry procedures such as restraint and vaccination, or to promote movement to or from a location.

For example, Hemsworth *et al.* (1996a), in examining the aversiveness of daily injections, surprisingly found that daily intramuscular injections, using a low-penetration gas injection gun, were not highly aversive to pigs. The authors suggested that there may have been some rewarding elements for the pigs in these handling bouts such as the presence of and opportunity to closely approach and interact with the handler before and after injection. Indeed, Jones (1993) has suggested that since animals find increased complexity attractive, human contact may be rewarding for farm animals in environments that lack complexity.

Sensory characteristics

Pigs have good colour vision (see Grandin, 1987), and thus may respond to the novelty of a change in the colour of the routine clothing of handlers. Pigs have a wide angle of vision (310°, Prince, 1977), and therefore walls of corridors, pen fronts and gates should be solid (at least up to pig height) to prevent the pigs that are being moved becoming distracted by what they see, such as other people or pigs (Grandin, 1980).

Stockpeople should follow behind, and slightly to one side, and use a solid board to prevent the pigs turning back. Livestock have sensitive hearing and thus may avoid excessive and unfamiliar noise (Grandin, 1990; Waynert *et al.*, 1999). While livestock appear to move more easily on a level surface, excessively steep ramps were avoided by pigs in a preference test: 20–24° ramps were preferred to 28–32° ramps (see Grandin, 1990).

Pigs, like other animals, have a tendency to move towards a more brightly lit area (Van Putten and Elshof, 1978). Tanida *et al.* (1996) found that piglets preferred to move from dark to lighter areas and were encouraged to move to darker areas with the provision of lighting. Experiments indicate that the light should be even and diffuse (see Grandin, 1980). Thus lights can be used to encourage movement into poorly illuminated areas, such as races and dark corridors. Although Grandin (1987) suggests that shadows will cause baulking, Tanida *et al.* (1996) found no effect on the movement of piglets of either shadows or lines on the floor. While the light intensity of a flashlight (160 lux) did not affect pig movement (Tanida *et al.*, 1996), excessive light (1200 lux) caused avoidance (see Grandin, 1990).

Furthermore, pigs avoid black and white patterns on the floor (Tanida *et al.*, 1996) and, since pigs are only moderate judges of distance (Grandin, 1980), they are reluctant to cross changing light patterns, drain grates, steps,

puddles of water, gutters and other high-contrast objects. Batching gates and pig boards should be solid to block the vision of pigs and thereby to encourage their movement away from the gates or boards. Figure 14.3 shows a well-designed loading ramp with solid outer walls to block distractions outside the ramp that may make pigs baulk.

Pigs will avoid sites containing urine from stressed pigs (Vielle-Thomas and Signoret, 1992). Therefore, handling of fearful pigs or handling that creates stress in pigs may cause other following animals to respond fearfully to these olfactory signals, making these pigs also difficult to handle.

Social behaviour

Several aspects of the pigs' social behaviour will affect ease of handling. Herding or flocking, in which social spacing and orientation are maintained, is most pronounced in sheep but is also evident in other livestock, including pigs. Following behaviour, in which there is synchrony of behaviour such as walking, running, feeding and lying, is commonly seen in pigs and other livestock. Pigs show pronounced herding and following behaviour (Van Putten and Elshof, 1978).

This motivation of pigs to follow other pigs and maintain body and visual contact with other

Fig. 14.3. A 'see-through' partition promotes following in this loading ramp. The outer fences are solid.

pigs obviously can and should be utilized in the moving of pigs. For example, the walls of corridors, pen fronts and gates should be solid (at least up to pig height) to prevent the pigs being moved becoming distracted by adjacent animals; however, race design should utilize the attraction of pigs to the sight of others moving ahead.

Thus, corridors or races should be wide enough to provide the animals with a clear view ahead and of other animals moving ahead: Grandin (1990) suggests that races for pigs should be 1 m wide. If one animal needs to be isolated from the group, it may be preferable to move it within a small group to a location where the animal can easily be drafted from the rest or to use a large pig board in the pen to direct the pig out of the pen or away from the group.

Main Recommendations on Handling Pigs Arising from an Understanding of Human Behaviour, Pig Behaviour and Sensory Capacity

Achieving desirable human contact

Modern pig production involves several levels of interaction between stockpeople and their animals. Many interactions are associated with regular observation of the animals and their conditions, and thus this type of interaction often involves only visual contact between the stockperson and the pigs, perhaps without the stockperson entering the animals' pen. Pigs in most production systems have to be moved by the stockperson, and this often also involves tactile and auditory interaction with the animals. Growing pigs are occasionally moved from pen to pen, in order to provide accommodation suitable to their stage of growth, and breeding pigs are regularly moved according to their stage of the breeding cycle. It is during these situations that human–animal interactions have considerable potential to influence animal behaviour.

Considerable between-farm variation exists in the pig industry in both the behaviour of stockpeople and the fear response of pigs to humans (see Hemsworth and Coleman, 1998), highlighting the opportunities to reduce fear responses in pigs and improve the ease of handling, welfare and productivity of pigs. As discussed earlier, studies in the pig industry have also

shown sequential relationships between the stockperson's attitudes and behaviour towards their animals and the fear of humans and productivity of farm animals (Hemsworth et al., 1994a; Coleman et al., 2000).

This research indicates that the best way to predict how a stockperson will interact with their animals is by knowing what their attitude is toward the activity itself. Attitudes of an individual are generally considered by psychologists to be determined by a combination of the individual's beliefs about the outcomes that are likely to occur following a particular behaviour and the individual's evaluation of those outcomes, and thus the idea that attitudes best predict how stockpeople behave is critical in achieving behavioural change.

Studies in the livestock industries – including the pig industry – using cognitive behavioural training have successfully reduced fear of humans in commercial farm animals (Hemsworth et al., 1994a, 2002). This approach in improving the attitudes and behaviour of stockpeople has been described in detail by Hemsworth and Coleman (1998).

Basically, cognitive–behavioural training techniques involve retraining people in terms of their behaviour by firstly targeting both the beliefs that underlie the behaviour (attitude) and the behaviour in question and, secondly, maintaining these changed beliefs and behaviour. This process of inducing behavioural change is really a comprehensive procedure in which all of the personal and external factors that are relevant to the behavioural situation are explicitly targeted.

Hemsworth et al. (1994a) found that targeting those attitudes and behaviour of the stockperson that are correlated with level of fear of humans in pigs resulted in stockpeople having a more positive attitude towards their pigs, with subsequent reductions in the proportion of negative interactions towards their animals and reductions in the animals' fear of humans (see Table 14.2). Furthermore, there was a marked tendency for an improvement in the reproductive performance of the pigs at the farms in which this training programme was introduced.

Thus there is a strong case for utilizing stockperson training courses in the livestock industries that target the attitudes and behaviour of the stockperson. Commercial multimedia training programmes called 'ProHand Pigs' and 'ProHand

Table 14.2. Summary of the effects of a cognitive–behavioural training, applied at control and modification farms, targeting the attitudes and behaviour of stockpeople in the pig industry (from Hemsworth et al., 1994a).

Variables	Control farms	Modification farms
Human attitudes[a]		
Beliefs about petting and talking	89.2	102.9
Beliefs about effort needed to handle	89.8	92.2
Human behaviour		
Negative tactile behaviours (%)	55.8	38.6
Fear levels		
Time spent near experimenter(s)	15.6	21.9
Number of interactions with experimenter	1.3	2.0
Reproductive performance		
Piglets born/sow/year	22.2	23.8

[a] High score indicates positive beliefs.

Abattoir' have been developed to train stockpeople in handling pigs in commercial piggeries and abattoirs, respectively, and these are currently being used in Australia, New Zealand and the USA. A similar training programme for the dairy industry ('ProHand Dairy Cows') has been developed, and details of these training programmes are available at http://www.animalwelfare.net.au

While it is clear that appropriate training that targets attitudes and behaviour can improve stockperson behaviour and, subsequently, pig behaviour (Hemsworth and Coleman, 1998), it is also important to identify those characteristics that identify potentially good stockpeople when they have little experience working with pigs. As discussed by Coleman (2004), attitudes are unreliable predictors of behaviour when they are not associated with relevant experience (Ajzen, 1988).

Therefore, it is important to persist with investigations into dispositional factors that are not dependent upon specific animal care experience. A recent study by Coleman (2001) gives some indication that personal attributes such as empathy and attitudes towards both animals and aspects of work may well be useful in identifying those from amongst inexperienced people who are likely to be good stockpeople.

As Hemsworth and Coleman (1998) have discussed, the stockperson also requires a basic knowledge of both the behaviour and requirements of farm animals, together with a range of

well-developed husbandry and management skills to effectively care for and manage farm animals. These are skilled tasks, and stockpeople are therefore required to be competent in many of these tasks. Clearly, the conventional training of stockpeople to develop such competencies should be a systematically and soundly implemented process in which the requirements of both the stockperson and the industry are addressed.

Utilization of the characteristics of pigs for their handling and control

An understanding of the learning and sensory ability and social behaviour of pigs can be effectively utilized in handling pigs and designing features of handling facilities. The scientific and popular literature on livestock handling describes many of the features of livestock that are relevant to livestock handling, and the main features that should be considered in handling and controlling pigs include:

- Familiarity with direction of flow and route will assist pigs to learn where to go.
- The sight of stationary pigs adjacent to the race will slow the movement of pigs, and thus race walls adjacent to other animals should be covered.
- Wide, clear, well-lit areas will promote movement.

- Lighting can be used to promote movement. For example, at the time of handling, darkening the area in which the pigs are held and brightening the area in which they are to move will promote movement.
- Races with a clear, unobstructed view towards the exit or where the pigs are meant to move will help promote movement.
- Pigs will be attracted by the sight of other pigs moving ahead, and thus visual contact with these animals needs to be maintained and not obstructed.
- Changes in race construction material or changes in floor type (e.g. slats to concrete) will inhibit flow.
- Walls painted one colour to avoid contrasts will promote flow.
- Use of covered and open panels can direct movement and vision.
- Ramps with covered sides will not allow pigs to see elevation.
- It is easier to move pigs as a group, rather than individually, to a holding facility where individual animals can be separated and treated.
- Direction of movement inside a shed should be across the direction of the grating or slats to improve footing and to reduce the opportunities for pigs to see through the floor or perceive heights.
- Reduction of excessive noise – such as banging gates, shouting and engines – will facilitate movement.
- Reduction of the aversiveness of handling and treatment in a location will promote subsequent entry and movement in the location.

Thus, it is useful to consider the following: (i) minimize or reduce the duration of restraint and severity of handling and treatment, provide rewards after any treatment (opportunity for exploration, feed, interaction with handler, etc.); (ii) allow habituation to location and handler if possible (through repeated exposure); and (iii) apply all aversive husbandry treatments in a location other than those in which pigs are routinely introduced or handled.

For detailed information on the designs of races and loading and unloading facilities for livestock, including pigs, readers are referred to the review by Grandin (1990).

Conclusion

Human–animal interactions are a key feature of modern pig production. Research has shown that the quality of the relationship that is developed between stockpeople and their animals can have surprising effects on both the animals and the stockpeople. Handling studies on pigs and observations in the pig industry show that human–animal interactions may markedly affect the behaviour, productivity and welfare of pigs. Furthermore, by influencing the behavioural response of the pigs to humans, these interactions can affect the ease with which pigs can be observed, handled and managed by the stockperson. In addition to human contact, physical features of the environment will also influence animal movement and thus animal handling. Therefore an understanding of the behavioural and sensory characteristics of pigs is also important in effectively handling and controlling pigs.

There is a clear, ongoing need for the pig industry to both train their personnel to effectively handle and move their stock as well as to ensure that current knowledge on the characteristics of pigs is utilized in the design of handling facilities. The role and impact of the stockperson should not be underestimated: to do so will seriously risk the welfare and productivity of pigs. Indeed it is possible that the stockperson may be the most influential factor affecting pig-handling and animal welfare (Hemsworth and Coleman, 1998).

Furthermore, is likely that in the near future both the livestock industries and the general community will place an increasing emphasis on ensuring the competency of stockpeople in managing the welfare of livestock.

References

Abbott, T.A., Hunter, E.J., Guise, H.J. and Penny, R.H.C. (1997) The effect of experience of handling pigs on willingness to move. *Applied Animal Behaviour Science* 54, 371–375.

Ajzen, I. (1988) *Attitudes, Personality and Behaviour.* Open University Press, Milton Keynes, UK.

Arnold, N.A. (2005) Assessment of the behavioural responses of the dairy cow (*Bos taurus*) to environmental features of the milking shed. PhD thesis, University of Melbourne, Australia.

Barnett, J.L., Hemsworth, P.H. and Jones, R.B. (1993) Behavioural responses of commercial farmed laying hens to humans: evidence of stimulus generalization. *Applied Animal Behaviour Science* 37, 139–146.

Berlyne, D.E. (1960) *Conflict, Arousal and Curiosity*. McGraw-Hill, New York.

Boissy, A. and Bouissou, M.F. (1988) Effects of early handling on heifers' subsequent reactivity to humans and to unfamiliar situations. *Applied Animal Behaviour Science* 20, 259–273.

Boivin, X., Le Neindre, P. and Chupin, J.M. (1992) Establishment of cattle–human relationships. *Applied Animal Behaviour Science* 32, 325–335.

Bouissou, M.F. and Vandenheede, M. (1995) Fear reactions of domestic sheep confronted with either a human or a human-like model. *Behavioural Processes* 43, 81–92.

Coleman, G.J. (2001) Selection of stockpeople to improve productivity. Paper presented at the *4th Industrial and Organisational Psychology Conference*, Sydney, Australia.

Coleman, G.J. (2004) Personnel Management in Agricultural Systems. In: Benson, G.J. and Rollin, B.E. (eds) *The Well-Being of Farm Animals: Challenges and Solutions*. Blackwell Publishing Professional, Ames, Iowa, pp.167–181.

Coleman, G.C., Hemsworth, P.H., Hay, M. and Cox, M. (1998) Predicting stockperson behaviour towards pigs from attitudinal and job-related variables and empathy. *Applied Animal Behaviour Science* 58, 63–75.

Coleman, G.J., Hemsworth, P.H., Hay, M. and Cox, M. (2000) Modifying stockperson attitudes and behaviour towards pigs at a large commercial farm. *Applied Animal Behaviour Science* 66, 11–20.

Craig, V.C. (1981) *Domestic Animal Behaviour: Causes and Implications for Animal Care and Management*. Prentice-Hall, Englewood Cliffs, New Jersey.

de Passille, A.M., Rushen, J., Ladewig, J. and Petherick, C. (1996) Dairy calves' discrimination of people based on previous handling. *Journal of Animal Science* 74, 969–974.

Eldridge, G.A. and Knowles, H.M. (1994) *Improving Meat Quality through Improved Pre-slaughter Management*. Pig Research & Development Corporation Research Report (DAV91P), Victorian Institute of Animal Science, Werribee, Australia.

Geverink, N.A., Kappers, A., van de Burgwal, J.A., Lambooij, E., Blokhuis, H.J. and Wiegant, V.M. (1998) Effects of regular moving and handling on the behavioural and physiological responses of pigs to preslaughter treatment and consequences for subsequent meat quality. *Journal of Animal Science* 76, 2080–2085.

Gonyou, H.W., Hemsworth, P.H. and Barnett, J.L. (1986) Effects of frequent interactions with humans on growing pigs. *Applied Animal Behaviour Science* 16, 269–278.

Grandin, T. (1980) Livestock behaviour as related to handling facilities design. *International Journal for the Study of Animal Problems* 1, 33–52.

Grandin, T. (1982–1983) Pig behaviour studies applied to slaughter-plant design. *Applied Animal Ethology* 9, 141–151.

Grandin, T. (1987) Animal handling. In: Price, E.O. (ed.) *The Veterinary Clinics of North America, Food Animal Practice*, Vol. 3. W.B. Saunders, Philadelphia, Pennsylvania, pp. 323–338.

Grandin, T. (1988) Livestock handling pre-slaughter. *Proceedings of the 34th International Congress of Meat Science and Technology*, Copenhagen, Denmark, pp. 41–45.

Grandin, T. (1990) Design of loading facilities and holding pens. *Applied Animal Behaviour Science* 28, 187–201.

Grandin, T. (1991) Handling problems caused by excitable pigs. *Proceedings of the 37th International Congress of Meat Science and Technology*, Kulmbach, Germany, Vol. 1, pp. 249–251.

Grandin, T., Curtis, S.E. and Taylor, I.A. (1987) Toys, mingling and driving reduce excitability in pigs. *Journal of Animal Science* (Suppl. 1) 65, 230 (abstract).

Gray J.A. (1987) *The Psychology of Fear and Stress*, 2nd edn. Cambridge University Press, Cambridge, UK.

Hargreaves, A.L. and Hutson, G.D. (1990) The effect of gentling on heart rate, flight distance and aversion of sheep to a handling procedure. *Applied Animal Behaviour Science* 26, 243–252.

Hemsworth, P.H. and Barnett, J.L. (1987) Human–animal interactions. In: Price, E.O. (ed.) *The Veterinary Clinics of North America, Food Animal Practice*, Vol. 3. W.B. Saunders, Philadelphia, Pennsylvania, pp. 339–356.

Hemsworth, P.H. and Barnett, J.L. (1991) The effects of aversively handling pigs either individually or in groups on their behaviour, growth and corticosteroids. *Applied Animal Behaviour Science* 30, 61–72.

Hemsworth, P.H. and Coleman, G.J. (1998) *Human–Livestock Interactions: the Stockperson and the Productivity and Welfare of Intensively-farmed Animals*. CAB International, Wallingford, UK.

Hemsworth, P.H., Barnett, J.L. and Hansen, C. (1981) The influence of handling by humans on the behaviour, growth and corticosteroids in the juvenile female pig. *Hormones and Behavior* 15, 396–403.

Hemsworth, P.H., Barnett, J.L. and Hansen, C. (1986a) The influence of handling by humans on the behaviour, reproduction and corticosteroids of male and female pigs. *Applied Animal Behaviour Science* 15, 303–314.

Hemsworth, P.H., Gonyou, H.W. and Dzuik, P.J. (1986b) Human communication with pigs: the behavioural response of pigs to specific human signals. *Applied Animal Behaviour Science* 15, 45–54.

Hemsworth, P.H., Barnett, J.L. and Hansen, C. (1987) The influence of inconsistent handling on the behaviour, growth and corticosteroids of young pigs. *Applied Animal Behaviour Science* 17, 245–252.

Hemsworth, P.H., Barnett, J.L., Coleman, G.J. and Hansen, C. (1989) A study of the relationships between the attitudinal and behavioural profiles of stockpersons and the level of fear of humans and reproductive performance of commercial pigs. *Applied Animal Behaviour Science* 23, 310–314.

Hemsworth, P.H., Barnett, J.L., Treacy, D. and Madgwick, P. (1990) The heritability of the trait fear of humans and the association between this trait and the subsequent reproductive performance of gilts. *Applied Animal Behavioural Science* 25, 85–95.

Hemsworth, P.H., Coleman, G.J. and Barnett, J.L. (1994a) Improving the attitude and behaviour of stockpeople towards pigs and the consequences on the behaviour and reproductive performance of commercial pigs. *Applied Animal Behaviour Science* 39, 349–362.

Hemsworth, P.H., Coleman, G.J., Cox, M. and Barnett, J.L. (1994b) Stimulus generalisation: the inability of pigs to discriminate between humans on the basis of their previous handling experience. *Applied Animal Behaviour Science* 40, 129–142.

Hemsworth, P.H., Barnett, J.L. and Campbell, R.G. (1996a) A study of the relative aversiveness of a new daily injection procedure for pigs. *Applied Animal Behaviour Science* 49, 389–401.

Hemsworth, P.H., Price, E.O. and Bogward, R. (1996b) Behavioural responses of domestic pigs and cattle to humans and novel stimuli. *Applied Animal Behavioural Science* 50, 43–56.

Hemsworth, P.H., Barnett, J.L., Hofmeyr, C., Coleman, G.J., Dowling, S. and Boyce, J. (2002) The effects of fear of humans and pre-slaughter handling on the meat quality of pigs. *Australian Journal of Agricultural Research* 53(4), 493–501

Hill, J.D., McGlone, J.J., Fullwood, S.D. and Miller, M.F. (1998) Environmental enrichment influences on pig behaviour, performance and meat quality. *Applied Animal Behaviour Science* 57, 51–68.

Hinde, R.A. (1970) *Behaviour: a Synthesis of Ethology and Comparative Psychology.* McGraw-Hill, Kogakusha, Japan.

Houpt, K.A. and Wolski, T.R. (1982) *Domestic Animal Behavior for Veterinarians and Animal Scientists.* Iowa State University Press, Ames, Iowa.

Jones, R.B. (1993) Reduction of the domestic chick's fear of humans by regular handling and related treatments. *Animal Behaviour* 46, 991–998.

Jones, R.B. and Waddington, D. (1992) Modification of fear in domestic chicks, *Gallus gallus domesticus*, via regular handling and early environmental enrichment. *Animal Behaviour* 43, 1021–1033.

Jongman, E.C. (2000) Environment enrichment for experimental farm animals: science and myths. Proceeding of the ANZCCART Conference 'Farm Animals in Research – can We Meet the Demands of Ethics, Welfare, Science and Industry?, 30 November–1 December 2000, South Australia, pp. 64–66.

Kilgour, R. (1987) Learning and Training of Farm Animals. In: Price, E.O. (ed.) *The Veterinary Clinics of North America, Food Animal Practice*, Vol. 3. W.B. Saunders, Philadelphia, Pennsylvania, pp. 269–284.

Lyons, D.M. (1989) Individual differences in temperament of dairy goats and the inhibition of milk ejection. *Applied Animal Behaviour Science* 22, 269–282.

Mateo, J.M., Estep, D.Q. and McCann, J.S. (1991) Effects of differential handling on the behaviour of domestic ewes (*Ovis aries*). *Applied Animal Behaviour Science* 32, 45–54.

McFarland, D. (1981) *The Oxford Companion to Animal Behaviour.* Oxford University Press, Oxford, UK.

Moore, E.A., Broom, D.M. and Simmins, P.H. (1994) Environmental enrichment in flatdeck accommodation for exploratory behaviour in early-weaned piglets. *Applied Animal Behaviour Science* 41, 269–279.

Murphy, L.B. (1976) A study of the behavioural expression of fear and exploration in two stocks of domestic fowl. PhD dissertation, Edinburgh University, UK.

Murphy, L.B. (1978) The practical problems of recognizing and measuring fear and exploration behaviour in the domestic fowl. *Animal Behaviour* 26, 422–431.

Paterson, A.M. and Pearce, G.P. (1989) Boar-induced puberty in gilts handled pleasantly or unpleasantly during rearing. *Applied Animal Behaviour Science* 22, 225–233.

Pearce, G.P., Paterson, A.M. and Pearce, A.N. (1989) The influence of pleasant and unpleasant handling and the provision of toys on the growth and behaviour of male pigs. *Applied Animal Behaviour Science* 23, 27–37.

Price, E.O. (2002) *Animal Domestication and Behaviour*. CAB International, Wallingford, UK.

Prince, J.H. (1977) The eye and vision. In: Swenson, M.J. (ed.) *Dukes Physiology of Domestic Animals*. Cornell University Press, Ithaca, New York, pp. 696–712.

Sapolsky, R.M. (1992) Neuroendocrinology of the stress-response. In: Becker, J.B., Breedlove, S.M. and Crews, D. (eds) *Behavioral Endocrinology*. MIT Press, Cambridge, UK, pp. 287–324.

Schaefer, T. (1968) Some methodological implications of the research on early handling in the rat. In: Newton, G. and Levein, S. (eds) *Psychobiology of Development*. Charles Thomas, Springfield, Illinois, pp. 102–141.

Shillito, E.E. (1963) Exploratory behaviour in the short-tailed vole, *Microtus agrestis. Behaviour* 21, 145–154.

Suarez, S.D. and Gallup, G.G., Jr (1982) Open-field behaviour in chickens: the experimenter is a predator. *Journal of Comparative Physiology and Psychology* 96, 432–439.

Tanida, H., Miura, A., Tanaka, T. and Yoshimoto, T. (1995) Behavioural response to humans in individually handled weaning pigs. *Applied Animal Behaviour Science* 42, 249–259.

Tanida, H., Miura, A., Tanaka, T. and Yoshimoto, T. (1996) Behavioural responses of piglets to darkness and shadows. *Applied Animal Behaviour Science* 49, 173–183.

Toates, F.M. (1980) *Animal Behaviour – a Systems Approach*. John Wiley and Sons, Chichester, UK.

Van Putten, G. and Elshof, W.J. (1978) Observations on the effect of transport on the well-being and lean quality of slaughter pigs. *Animal Regulation Studies* 1, 247–271.

Vielle-Thomas, R.K. and Signoret, J.P. (1992) Pheromonal transmission of aversive experiences in domestic pigs. *Journal of Chemical Endocrinology* 18, 1551 (abstract).

Waring, G.H. (1983) *Horse Behaviour. The Behavioural Traits and Adaptations of Domestic and Wild Horses, Including Ponies*. Noyes Publications, Park Ridge, New Jersey.

Warriss, P.D., Kestin, S.C. and Robinson, J.M. (1983) A note on the influence of rearing environment on meat quality in pigs. *Meat Science* 9, 271–279.

Waynert, D.F., Stookey, J.M., Schwartzkopf, K.S., Watts, J.M. and Waltz, C.S. (1999) The response of beef cattle to noise during handling. *Applied Animal Behaviour Science* 62, 27–42.

15   Transport of Pigs

E. Lambooij

Animal Sciences Group of Wageningen UR, Animal Production Division,
Lelystad, The Netherlands; bert.lambooij@wur.nl

Introduction

Europe has a long history of transporting live-stock over long distances. By the beginning of the 17th century, tens of thousands of oxen were travelling every year by road and by sea from Denmark and Schleswig-Holstein to the Netherlands. Around 1750, no less than one quarter of a million oxen were traded on the continent each year. By way of comparison in 2000, 11.9 million pigs and 3 million cattle were transported within the EU, or 12 million and 3.8 million, respectively, if one includes import and export (Gijsberts and Lambooij, 2005).

Cowboys or shepherds drove large numbers of animals to the towns on the prairies of the USA. This ranged from large numbers of cattle and pigs transported by truck or train in the West to – in the far East – single pigs tied and transported by bicycle or trailer to the slaughterhouse. At present, transport by train is uncommon, because the animals have to be transported to a station and reloaded, thus increasing the adverse effects of loading and lengthening of some journeys. However, conditions by train can be very good.

The use of aircraft is limited to transport of breeding animals and day-old-chicks, because it is expensive. A large number of sheep are shipped from Australia to the Middle East by sea; cattle are also sometimes transported by ship.

The most common means of transport for all farm animals today, however, is the road vehicle, even though it is generally found that truck transport is worse for the animals – in welfare terms – than rail, sea or air transport. Pigs are usually transported in large trucks that may hold over 200 animals; in the EU, most of those trucks are equipped with a loading lift.

Nowadays, transport distances of farm animals by road to another farm or to the slaughterhouse are expanding because of the economic consequences of greater opportunities for long-distance and international trade, improved infrastructure and increased demand for live animals for fattening and slaughtering. Within the EU, free movement of animals from one member state to another and more uniformity have resulted in more long-distance travel to slaughter. Regulations for protection of animals during transport are laid down in an EC Council directive (1991). Different EU member states have legislation from the viewpoints of: (i) health and welfare of animals; (ii) ethical considerations; and (iii) protection and safety of man and animal to prescribe rules for animals and their handling. When animals are transported, housed, restrained, stunned, slaughtered or killed, one should spare any avoidable excitement, pain or suffering.

Quality schemes are developed or are in development in several countries:

- The UK (RSPCA Standards for Pigs, 1998; Lindsay, 2000)
- New Zealand (Code of Animal Welfare Advisory Committee, 1994)
- Australia (Standing Committee on Agriculture and Resource Management, 1999)
- Denmark (Barton-Gade, 2002)
- France (Chupin *et al.*, 2003)
- The Netherlands (Lambooij, 2004).

Welfare

Responsibilities

The farmer or the owner is responsible for the careful selection of livestock for the loading on to the road vehicle and for the presentation only of pigs fit for travel. The nature and duration of the proposed journey should be considered when determining the degree of fitness required. Animals must be able to stand and bear weight on all four limbs and be fit in order to withstand the journey without suffering unnecessary pain or distress. Where emergency transport is considered necessary, special conditions apply. The farmer, owner or agent is responsible for the provision of well-maintained loading facilities.

Transport operators have the right to refuse pigs for transport. They must refuse animals when they consider them to be unfit and should ask for a veterinary opinion if they have any doubts. The driver of a road vehicle is responsible for the care and welfare of animals during transport. They must stop and assist a distressed or injured animal as soon as practical after they become aware of the problem. Drivers must be trained in animal handling to ensure the welfare of pigs in their charge and be familiar with the contents of the appropriate welfare code. Learner drivers are not permitted to transport pigs without supervision. Inspectors and authorities appointed under the various national acts can obtain possession of an animal where the inspector believes that an offence is being or has been committed in respect of that animal.

Stress

Farm animals are kept under specific housing conditions for several months, the exact time depending on the species and production system. After that period, the animals have to be transported. Transport causes physical and behavioural problems, because the animals are not accustomed to transport conditions and procedures. Figure 15.1 illustrates all the interacting factors that affect stress and animal welfare.

The loading procedures, design and the other transported animals are unfamiliar to the animals and will frighten them. Drivers may not treat the animals in a proper way to minimize stress and sometimes animals are mixed or regrouped, thus increasing stress and resulting in fighting for determination of social order

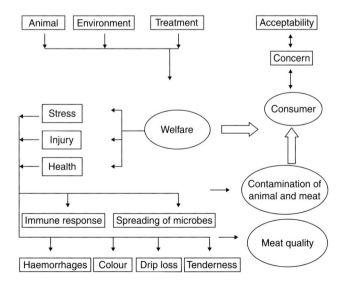

Fig. 15. 1. Factors that affect stress, meat quality and contamination.

(Geverink et al., 1998). During loading and unloading, transport injuries and bruising may occur in all animal species. These defects occur by forceful contacts in passageways, in compartments and in containers, by fighting between animals and by mounting. Skin blemish is a serious commercial problem. The skin blemish score reflects the amount of fighting in which pigs have indulged pre-slaughter. Lesions from fighting can be easily quantified and may be used to determine which genetic line of pigs are most aggressive (Turner et al., 2006).

A well-known disease related to transport is the porcine stress syndrome (PSS) (Tarrant, 1989). This syndrome is the acute reaction to stress, mediated by the sympathetic nervous system, which can cause severe distress and even death. The affected animals show severe signs of dyspnoea, cyanosis and hyperthermia, and may develop rigor in the muscles before death occurs. These characteristics are supported by the fact that the pig has a relatively small heart mass, with bradycardia (low heart rate), osteochondrosis and white muscle types.

The heart mass of present crossbreds is approximately 0.003% of the body weight – small compared with that of the wild pig at 0.007% (Yang and Lin, 1997). On the electrocardiogram, a lower heart rate and ST-deviation in the QRS complex is observed, especially in heavyweight pigs. At the same time, a higher percentage of white muscle is found. These muscles have a lower O_2 consumption at the cellular level (Geers et al., 1990). Osteochondrosis is a general enchondrial dysformation process. Many names have been introduced: arthrosis deformans, epiphysiolysis capitis femoris, apophysiolysis tuberis ischiadic, osteoarthrosis and polyarthrosis.

Environment

During road transport, weather conditions (temperature, air velocity, humidity), loading density and duration of the trip are important factors, all capable of influencing the condition of the animals (Augustini, 1976; Hails, 1978, Lewis et al., 2005). On long trips of 2–3 days,, pigs may be exposed to wide variations in weather conditions. In general, live weight losses during transport of 1–2 days are 40–60 g/kg, whereas the mortality is 0.1–0.4% (Hails, 1978; Holloway, 1980; Grandin,

1981; Markov, 1981; Lambooij, 1983, 1988). Piglets may lose close to 7% of their body weight after weaning if not transported or transported up to 24 h (Lewis et al., 2005).

For pigs being transported from the farm to a nearby slaughterhouse, death loss has ranged from 0.1 to 1.0% (Fabianson et al., 1979; Allen et al., 1980; Warriss 1998). Losses increase during hot weather conditions (Smith and Allen, 1976; van Logtestijn et al., 1982). Clark (1979) found that 70% of the Canadian losses occurred on the truck.

Vehicle motion and vibration are known to have effects on humans: comfort, postural stability and ability to perform a task can be severely compromised, as well as the generation of motion sickness. It is likely that a similar response occurs in pigs; however, the relevant ranges of frequencies could be different (Randall, 1992). Vibration magnitudes are highest in a small, towed, twin-axle trailer as used by farmers for transport of up to ten pigs. A small, fixed-body truck provides conditions that could be classified as somewhat uncomfortable on the best roads to fairly uncomfortable on minor roads. A large, fixed-body truck with air suspension provides a very smooth ride, classified as not uncomfortable to a little uncomfortable (Randall et al., 1996). In piglets, the following have been observed: a decreasing effect with increasing frequency for a given acceleration, an independence of acceleration at higher frequencies and a large between-animal variability (Perremans et al., 1996, 2001).

Coping mechanisms

Individual pigs respond in different ways to stress factors. The response is dependent on the genotype, coping style, treatment and experience of the animal. In general, barren, housed pigs respond more actively to a novel environment in the home pen (Beattie et al., 2000) and they tended to be more active than enriched, housed pigs at the lorry and in lairage (De Jong et al., 2000).

Coping mechanisms may correlate with response to transport and associated conditions. Transport conditions involve exposure to social stress (e.g. mixing with unfamiliar pigs) and non-social stress (rough handling). Individual differences in behaviour in home pen conditions and during mild challenge tests may be related to subsequent reaction during transport, driving and mixing (Bradshaw et al., 1996a; Geverink, 1998; Geverink et al., 1998).

During fattening – when the pigs are aged between 14 and 20 weeks – piglets can be tested for social status by scoring the agonistic interactions or by using a 'food competition test'. In the first test, social ranking of individuals may be determined by allowing a total of 4 h of focally sampled data on agonistic interactions (biting, head knocking, threatening, displacement, avoidance). In the second test, the pigs have no access to food for 29 h prior to the test.

At the start of the test, a fixed amount of food is placed in the trough. For 15 min, interactions and frequency and duration of eating are scored. Finally, an 'open-door test' may be used to test individual activity and exploration. In this test, the door of a pen is opened and the reaction of pigs is scored. Pigs that were regularly given the opportunity to leave their pen and, in addition, were accustomed to transport showed increased willingness to move during preslaughter treatment. Pigs that are easier to move are less likely to be subject to rough handling, which implies improved welfare, while the workload for stockmen is reduced (Geverink, 1998).

The most detrimental effect of coping is death, which normally follows a period of very poor welfare. Stress activates hormones via the pituitary–adrenal (glucocorticoids) and sympathetic–adrenal medullary systems (cathecholamines), resulting in behavioural (overreaction to normal stimuli) and clinical (increase in heart and respiration rate) deviations from normal functioning, followed by exhaustion. The death rate following transport varies between 0.1 and 1% (Warriss, 1998). See Chapter 19 on physiology.

Thermoregulation

The variation in temperatures encountered by the pigs during transport may increase by up to approximately 20°C. This variation in temperature within the vehicle is related to variation in temperature outside (Lambooij, 1988; Lewis et al., 2005). Therefore, ventilation rate during transport should be adapted to the inside

temperature, which is the result of both heat flowing from outside to inside and heat produced by the animals. Data of heat production at climatic conditions occurring during transport of slaughter pigs are not known. Little quantitative information is available on thermal thresholds during transport (Schrama et al., 1996).

Pigs are homeothermic animals, maintaining a constant body temperature by balancing heat loss and heat production, as presented in Fig. 15.2. Animals may maintain a constant body temperature in zone AC. Heat loss is kept constant by regulation of both sensible and evaporative heat losses in the thermoneutral zone BC. Factors such as feeding level, physical activity and stress determine the level of heat production. Mechanisms for reduction and control of heat loss are depleted in zone AB. In order to maintain homeothermia, the animal has to increase its heat production. Climatic conditions in the thermoneutral zone are optimal for the animal (Fig. 15.2; Table 15.1; Schrama et al., 1996).

Heat production at maintenance can be assumed normally to be around 420 kJ/kg $^{0.75}$/day (Holmes and Close, 1977). At feeding time, heat production will increase by about 30% (van der Hel et al., 1986). Lambooij et al. (1987) simulated a 48-h transport of slaughter pigs that were non-fed and held at a loading density of 225 kg/m^2 and at environmental temperatures of 8, 16 and 24°C. It was calculated that the animals lost 824–944 g body fat during exposure, the loss at 16°C being the lowest. The metabolic

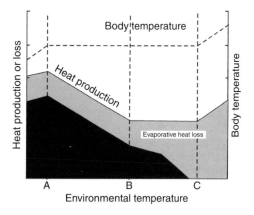

Fig. 15.2. Relationships between heat production, (non-)evaporative heat loss, body temperature and environmental temperature (from Mount, 1974).

rate was, on average, above the maintenance requirement, as expected. The mean heat production at 16°C was 551 kJ/kg$^{0.75}$/day. The animals had produced this heat as a result of their maintenance and response to their environment. The heat production increased during light and decreased during darkness, with a minimum early in the morning (see Fig. 15.3).

Heat production values during environmental temperatures of 8 and 24°C tended to be higher than during 16°C. This extra heat production may be due to some extra activity (see Fig. 15.3). It appeared that 8°C is below thermoneutrality, which agrees with data derived from the literature (Holmes and Close, 1977). A higher weight loss during an air velocity of 0.8 m/s compared to 0.2 m/s may be related to a lower water consumption and a higher heat production. It is assumed that 16°C and an air velocity of 0.2 m/s are at thermoneutrality level (Lambooij et al., 1987).

Table 15.1. Calculated effect of age on the thermoneutral zone of individually housed pigs fed at maintenance and under standard environmental conditions (from Verstegen, 1987).

	Live weight (kg)	Thermoneutral zone (°C)
Piglets	2	31–33
Growing piglets	20	26–33
Feeder pigs	60	24–32
Slaughter pigs	100	23–31

Temperature differences in the vehicle followed the ambient seasonal temperature. The compartment temperatures are reflected in ear and rectal temperature and in behaviour in piglets. During summer, the temperature meets the comfort zone (24–34°C), whereas they do not during autumn and winter (Lewis et al., 2005; Lewis and Berry, 2006). A decrease in temperature in the compartment was followed by a decrease in both ear and rectal temperature during the last 12 h of transport. However, both temperatures remained within normal limits (Lewis et al., 2005). In autumn and winter, resting frequency is observed to be low initially and to increase substantially after 12 h of transport. The resting frequency is wider for summer transport (Lewis and Berry, 2006).

During transport, the ventilation rate cannot be altered, in general. Therefore, a solution might be the use of adjusted vents or an artificial ventilation system (Lambooij, 1988). At the moment, different ventilation systems are under development and are used more and more. See Chapter 19 on thermoregulation.

Product Quality

Meat

During loading and unloading, transport injuries and bruising occur commonly in all animal species (Grandin, 1990). These defects occur by forceful contacts in passageways, in compartments and in containers, through fighting between

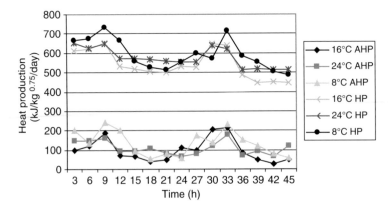

Fig. 15.3. Heat production (HP) and activity-related heat production (AHP) of non-fed slaughter pigs in a calorimeter at environmental temperatures of 8, 16 and 24°C (from Lambooij et al., 1987).

animals and during mounting (Connell, 1984). Skin blemish is a serious commercial problem. The skin blemish score reflects the amount of fighting in which pigs have indulged pre-slaughter (Barton-Gade et al., 1996). It was observed that 63% of 5484 carcasses from pigs in the EU had incurred some damage; however, only in about 10% of carcasses was this moderate to severe. Based on the association between the level of skin blemish and increased muscle ultimate pH value, a probable factor contributing to this is fighting between mixed groups of unfamiliar animals (Warriss et al., 1998).

Loss of live and carcass weight are the result of excretion, evaporation and respiratory exchange (Dantzer, 1982; Warriss, 1993), which is a normal physiological reaction. However, when food and water are withdrawn for a longer time, dehydration and mobilization of fat and muscle glycogen may occur (Tarrant, 1989; Warriss et al., 1989).

Transport conditions may affect post-mortem meat quality, either via adrenal or other stress responses or by fatigue of the animals. Important meat quality parameters related to stress and exhaustion before slaughter are pH, rigor mortis, temperature, colour and water-binding capacity. Acidity of the meat after slaughter is caused by a breakdown of glycogen to lactate. The rigor mortis value is indirectly caused by a decrease in the energy store. Temperature is increased by chemical metabolism (Sybesma and van Logtestijn, 1967). Colour and water-binding capacity are determined by protein denaturation, caused by a rapid acidification after death (Tarrant, 1989). The rate of acidification after death is controlled by the degree of hormonal and contractile stimulation of muscle immediately before and during slaughter, whereas muscle temperature at death and rate of cooling are also important (Warriss, 1987; Tarrant, 1989; Monin and Ouali, 1992).

It is assumed that stress before slaughter leads to an increased breakdown of glycogen and a greater decrease in the energy store, and thus a rapid acidification, an earlier and increased rigor mortis value and an increased post-mortem muscle temperature. In pigs, this results in pale, soft and exudative (PSE) meat (Tarrant, 1989; Gispert et al., 2000). This explanation is too simply postulated, because the physiological response to stress factors from the environment is partly influenced by the genotype of the animal (see Fig. 15.1).

Slaughter pigs of different genotypes from the same production unit subjected to identical pre-slaughter handling may show different values of meat quality parameters (Tarrant, 1989; Klont et al., 1993; Klont and Lambooij, 1995a, b); stress-resistant Hampshire pigs can have a low water-binding capacity (Monin and Sellier, 1985). Information from about 5500 pigs killed in the EU has shown large differences in the prevalence of potentially PSE meat.

No apparent relationships between indices of stress and characteristics associated with PSE meat were observed. Nevertheless, carcasses with high temperature and low pH values early at post-mortem result in inferior meat quality (Hambrecht et al., 2003).

In contrast, greater stress tended to be reflected in more dark, firm and dry (DFD) meat (Warriss et al., 1998). The occurrence of DFD meat is more readily attributable to effects of the transport environment and is less variable amongst genetic lines. It occurs when the animals are fatigued. In this case, glycogen energy store is exhausted at slaughter, resulting in no acidification, an increase in rigor mortis value and dark-coloured meat. In well-fed, rested animals, meat pH falls to about 6.0–6.5, the rigor mortis value is < 10 and the temperature is < 40°C about 45 min post-mortem (Sybesma and van Logtestijn, 1967; Klont et al., 1993).

Meat quality is influenced by increased muscle temperature (Klont and Lambooij, 1995b), the normal range in resting pigs varying between 37.0 and 39.6°C (Hannon et al., 1990). An increase to the upper level of the normal range may already cause an increased incidence of PSE meat, especially in stress-susceptible pigs (Klont and Lambooij, 1995b). Pre-slaughter stress factors such as exercise and high ambient temperatures in summer might easily elevate the body temperature up to values between 39.0 and 41.0°C, with the highest increases in stress-susceptible pigs compared with stress-resistant pigs (Gariepy et al., 1989; Geers et al., 1992; Gispert et al., 2000).

Higher post-mortem muscle temperatures, in combination with increased lactate formation after normal slaughter conditions, will lead to a greater incidence of PSE meat. Showering and resting pigs for 2–4 h will help in reducing the incidence of PSE (Malmfors, 1982; Smulders et al., 1983). When the period is longer, the percentage of carcasses with DFD may increase (Verdijk, 1974; Culau et al., 1991; Warriss, 1993). Stress before slaughter also affects the microbiological contamination in the live animal by influencing the meat quality, which may result in a more heavily contaminated carcass. In PSE and DFD carcasses, growth of microorganisms is enhanced.

Contamination

Animals require proper preparation before transport. This means that, for pigs, feed should be withheld for 16–24 h (Eikelenboom et al., 1990; Warriss, 1993). Other advantages of feed withdrawal are less labour at slaughter, less contamination of the carcass and a lower percentage of PSE meat (Eikelenboom et al., 1990, Warriss, 1993). It is thought that after great physical and psychological labour in clinically healthy animals carrying Salmonella and other pathogenic microorganisms, the excretion pattern from the intestinal tract may be changed from intermittent to constant shedding. This disturbance may also lower the immunological response and facilitate the spreading of intestinal bacteria. Feeding, environmental conditions during transport and lairage – including the total time involved and the mixing of animals from several herds – have been shown to be the main factors.

Experiments have shown that in pigs delivered to the slaughterhouse no Salmonellae were isolated: after delivery to the abattoir 0.1% of samples were positive, while after slaughter this percentage had increased to 0.7. It was concluded that stress factors had been responsible for this increase in the carrier percentage (Slavkov et al., 1974). When pigs stayed in lairage for a longer time, in larger pens and in poorer hygienic conditions (cross-)contamination also increased.

Carcass contamination was caused by Salmonellae of intestinal origin, as demonstrated by the Salmonella recovery rate and the Salmonella serotypes from caecal contents and the carcass surface (Morgan et al., 1987). A high level of contamination with Salmonella of the lairage can have a significant effect on the number of Salmonella-infected pigs at slaughter (Swanenburg et al., 2001) (see Chapter 21).

Since exploration of the environment is normal porcine behaviour, they can self-infect, whereafter the infection can spread around the whole body (Loynachan and Harris, 2005). Conflicting interests concerning animal welfare (rough floors) and hygiene (smooth floors) are a basic dilemma when trying to improve the situation (Swanenburg et al., 2001).

Transit Stations

Conditions

Weather conditions are very dependent on the location, the time of the day and the season. Any impairment of airflow will result in the accumulation of heat and moisture, which will impose heat stress. The application of adjusted vents, an artificial ventilation system or air conditioning system is necessary. After arrival at the slaughterhouse or transit station, the animals need to be unloaded carefully and as soon as possible, because ventilation in stationary vehicles is often not good. Driving of animals should be performed quietly. Electric goads should not be used, thereby reducing the incidence of petechial haemorrhages (Grandin, 1988).

In transit stations, pigs require to be showered and to be allowed a few hours of rest. Before long journeys they can be fed with a thin porridge consisting of one part feed (high sugar content) and three parts water, in order to reduce in-transit weight loss. A resting period of 2–4 h before slaughter of pigs is recommended. This may result in a lower percentage of carcasses with PSE meat. When this period is longer, the percentage of carcasses with DFD may increase (Verdijk, 1974; Culau et al., 1991; Warriss, 1993).

During transit or lairage, pigs should be showered intermittently. Showering has the following advantages: (i) the stress reaction as a result of the transport is reduced, due to the calming and cooling effect; (ii) fighting will be reduced, resulting in less skin damage; and (iii) the animals are cleaner (van Putten et al., 1983; Tarrant, 1989; Warriss, 1993). Practical experience and experiments have demonstrated that showering is most effective during hot weather conditions and that it should not be used when the temperature drops below 11°C (Santos *et al.*, 1997; Grandin, personal communication). A loading density of 2 pigs/m² is recommended, because higher densities diminish meat quality (van Putten *et al.*, 1983).

Design and construction

Below are summarized the main recommendations for holding yards, ramps and vehicles (see Fig. 15.4):

Holding yards

- The waiting area and parking duration of loaded vehicles need to be logistically adjusted to prevent temperature increase in the compartments.
- The floors should be clean and dry and have no changes in colour. At the intersection of ramp and floor, straw may be used.
- The walls need to be smooth to prevent skin damage.
- Misting or showering should be included for arriving pigs to help in calming down.
- A special reception area for injured and sick animals should be available.
- A facility for emergency slaughter should be available.

- Boars and sows must be penned separately.

Ramps

- Loading ramps should be at deck height because all species have problems with descending loading ramps. An alternative is the deck lift, in which the whole deck moves hydraulically upwards or downwards.
- The passageways need to be solid, while projections and channels should be avoided – pigs walk more easily from a dark to a lighter place.
- The floors must not be slippery.
- The width of the passageway is sufficient for four to five animals to walk side by side. Groups of approximately 15 or more animals should be driven.
- The use of goads should be banned; unwanted reversal of animals can be reduced by the use of gates.
- The angle of the loading ramp should not be greater than 15–20°.

Vehicles

- The design and construction should be in compliance with legislation.

Fig. 15.4. Modern pig lairage in a large slaughter plant (photograph by Temple Grandin).

- The outside of the vehicle should be light-coloured.
- The compartments should be free from obstructions, projecting objects and hazards that could cause skin damage.
- Ventilation of compartments and containers should be by vents positioned at the upper part of the left and right sides. The vents should be opened according the legal prescriptions.
- Materials used in the construction must be capable of being cleaned effectively.
- Drinking nipples should be available for journeys over 8 h and must be equipped with a low-level signal system.

Treatment during Transport

Procedures at loading and unloading

Farm animals are kept under specific housing conditions for several months, the exact time depending on the species and production system. After that period, the animals have to be transported to a slaughterhouse. Transport causes physical and behavioural problems, because the animals are not accustomed to transport conditions and procedures. The loading procedures, the design and the other in-transit animals are all unfamiliar to these animals and will frighten them. Drivers may not treat the animals in a proper way to minimize stress, and sometimes animals are mixed or regrouped, thus increasing stress and resulting in fighting amongst the animals to determine social order (Connell, 1984).

Climbing a loading ramp is easy for horses, cattle and sheep, when ramp design and handling procedures are good. For pigs, climbing a loading ramp is difficult since the situation is often psychologically disturbing. The animals may simply refuse to try and even turn their sides towards the ramps. As a result the heart rate may increase to a level where the heart starts to loose synchronization. The angle of the loading ramp should not be greater than 15–20° (van Putten and Elshof, 1978; Phillips et al., 1988; Fraser and Broom, 1990). Descending a loading ramp steeper than 20° is difficult for all animals and should be avoided (Grandin, 1981).

Driving

The driver of the vehicle is responsible for ensuring that animals are provided with reasonably comfortable and secure accommodation when pigs are carried on a light vehicle or a heavy trailer. Health, comfort, postural stability and ability can be severely compromised, as well as generating motion sickness. Vehicles conveying animals must be driven steadily, avoiding rapid acceleration and braking as far as possible. Drivers must be aware that sudden braking can subject animals to horizontal loads as high as 33% of their own weight. Sudden acceleration and rapid cornering can cause horizontal forces of up to 20% of the animals' weight. Such loads will cause stress and may result in falls and injuries.

Vibration magnitudes are highest on a small, towed, twin-axle trailer – as used by farmers to transport up to ten pigs. A small and fixed-body truck provides conditions that could be classified as slightly uncomfortable (on the best roads) to fairly uncomfortable (on minor roads). A large, fixed-body truck with air suspension provides a very smooth ride, classified as not uncomfortable to a little uncomfortable.

Feeding and watering

Animals should be fit for the intended journey, irrespective of the purpose of the journey. Pigs are likely to suffer motion sickness during road transport, which may result in vomiting after having eaten 4 h prior to transport (Bradshaw et al., 1996a, b). For this reason, pigs require careful preparation before transport, and comprehensive plans for the journey should be made.

This means that feed should be withheld for 16–24 h before slaughter (Warriss and Brown, 1983; Eikelenboom et al., 1990). Depending on the distance to the slaughterhouse or transit station, feeding should be stopped the night before transport, but water should be available. It is suggested that pigs suffer from motion sickness as a result of low-frequency vibrations (Randall, 1992) and may vomit when the stomach is full. Pigs may die by inhalation of their own vomit (Guise, 1987). Other advantages of feed withdrawal are less labour at slaughter, less contamination of the carcass, a decreased weight loss of the carcass during chilling and a lower percentage of PSE meat (Eikelenboom *et al.*, 1990; Warriss, 1993).

During stops in transport of long duration, piglets were observed to drink from bite nipples (Barton-Gade and Vorup, 1991). Observations of slaughter pigs during such journeys showed that pigs drank only 0.65 and 1.6 l water per pig when available via bite nipples during motion or in a trough after unloading, respectively (Lambooij, 1983, 1984). If feed was supplied, vomiting was observed. There were no physiological differences between watered and non-watered animals, due to mobilization of water within the fat cells (Lambooij *et al.*, 1985). However, an increased blood protein concentration was found in pigs after 6 h of transport (Warriss *et al.*, 1983).

The stress of handling, transport and fasting lowered blood glucose levels; however, additional energy was obtained from fat breakdown. Most liver glycogen was utilized in the first 18 h of transport and fat was broken down 9 h after feed withdrawal (Warriss and Brown, 1983). It is recommended that food and water should be available at appropriate times (Lambooij, 2004). If piglets are weaned and held for a period prior to transportation, they will be unlikely to consume enough feed and water to prevent weight loss; this weight loss will be in addition to transport losses. Therefore, it would benefit the producer to wean piglets as near to the onset of the journey as is feasible (Berry and Lewis, 2001).

Loading density

Animals must be able to stand in their natural position and all must be able to lie down at the same time. For animals, which may stand during the journey, the roof must be well above the heads of all animals when they are standing with their heads up in a natural position. This height will ensure adequate freedom of movement and ventilation and will depend on the species and breed concerned (see welfare codes, above).

Loading density has a major effect on animal welfare and, hence, post-mortem meat quality. At loading densities of > 200 kg/m^2, pigs showed increased body temperature, heart rate and breathing frequency after a short journey and a high frequency of PSE meat post-mortem (Heuking, 1988; von Mickwitz, 1989). When the loading density is 275 kg/m^2, not all pigs are able to lie down, hence there is a continual changing of positions and the pigs cannot rest (see Fig. 15.5). The consequences are more skin blemishes, rectal prolapses and poor meat quality (Lambooij *et al.*, 1985; Guise and Penny, 1989; Lambooij and Engel, 1991).

A loading density for slaughter pigs of 235 kg/m^2 is suggested as being acceptable as a compromise between animal welfare, meat quality and economics of transport (see Fig. 15.6). The recommended loading densities for different

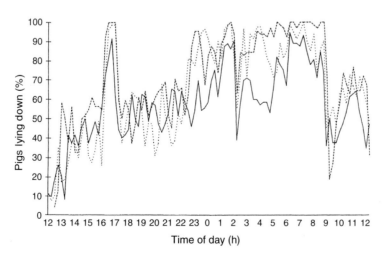

Fig. 15.5. Percentages of slaughter pigs lying down during transport. The loading densities were 186 (---), 232 (···) and 278 kg/m^2(—). Stops in transport were between 16.00 and 17.00 h, 1.00 and 2.00 h and 6.00 and 9.00 h, while the driver was changed at 21.00 h (from Lambooij and Engel, 1991).

Fig. 15.6. During long-distance transport, all pigs should have enough space to lie down at the same time without lying on top of each other.

weight groups are presented in Table 15.2 (Lambooij, 2004) (see Chapter 1).

Microclimate

The effects of climatic conditions are difficult to measure. The weather conditions are dependent of the location, the time of the day and the season. Especially during summer, heat stress may become significant, while chilly temperatures in winter also impose stress (Gispert *et al.*, 2000). Following winter and summer transport, piglets rest for 3 days, which is higher than the average of 2 days following transport, indicating that fatigue had been present during both seasons (Lewis and Berry, 2006). Experiments in climate-controlled calorimeters showed that the lowest heat production ante-mortem and best meat quality post-mortem in slaughter pigs occurred at an environmental temperature of 16°C and an air velocity of 0.2 m/s (Lambooij *et al.*, 1987).

The most common method of ventilating compartments and containers is via vents positioned in the upper part of the left and right sides. Figure 15.7 shows a truck with removable panels for adjustable ventilation. During the

Table 15.2. Recommended loading densities for pigs during transport.

	Live weight (kg)	Loading density (m^2/animal)
Piglets	< 25	0.15
Feeder pigs	60	0.35
Slaughter pigs	100–120	0.42
Slaughter pigs	120–140	0.45
Heavy pigs	> 140	0.71

journey the correlation between the outside and inside temperature was positive and significant, but for humidity this was not the case. The temperature during stops increased by 1–4°C in the compartments (Lambooij, 1988).

Thus, the microclimate depends on the ambient weather conditions. Placing covers at the holes that can be opened or closed allows adjustment of ventilation. In pigs, this variable ventilation improved meat quality when it was combined with showering during the journey (Lambooij and Engel, 1991).

Artificial ventilation in compartments and containers may improve conditions during transport. When pigs were ventilated artificially during

Fig. 15.7. Truck trailer for transporting pigs, with removable panels for adjustment of ventilation. More panels are added during cold weather but removed in summer. The panels are constructed from thin plastic and they slide into slots on the trailer.

transport, the rigor mortis value post-mortem was decreased, which pointed to a decreased energy loss in the muscles (Lambooij, 1988). For breeding pigs, air conditioning during transport has been developed and applied during short and long journeys by road. The air velocity is low and the temperature is held at 16°C.

Air, rail and sea transport

Specially designed containers are used to ship young breeding pigs by air. Each container holds 25 to 30 piglets. The floor of the containers is covered with a thick layer of sawdust, approximately 10 cm in depth. Each container should have at least two drinking nipples with sufficient water supply. It is common to tranquillize piglets during transport to calm them down; however, this is not recommended. Pigs in stressful conditions are calmed by the presence of their penmates. Another comforting factor is a familiar, well-lit environment. To prevent fighting during air transport, pigs should be divided into groups that will fly together in a container at the farm of origin. Dust from dry sawdust might cause frequent sneezing. Heat stress occurs when the aircraft is on the ground.

To prevent death losses at the destination, prompt unloading of the containers is essential. Upon arrival at the destination airport, the pigs have to be transported to the farm of destination by truck. Piglets may suffer from jet lag: their diurnal rhythm is completely out of phase, but they gradually become accustomed to a new rhythm. To reduce death losses during air transport, shippers should: (i) avoid use of tranquillizers; (ii) pre-mix pig groups; (iii) provide adequate water supply; (iv) keep lighting dim during the flight; (v) control dust levels; and (vi) avoid heat stress (van Putten, personal communication).

Transport by rail is often condemned, because it takes much more time than transport by road. However, rail transport is considered to have positive effects on animal welfare. Since the trip is longer, there has to be an accompanying attendant to look after the animals and provide them with water and feed. Jackson (1973) reported that pigs eat and drink during transport and they gain weight instead of losing it. Railway wagons have to be provided with artificial ventilation, because they may stand in sunshine for hours. In the USA, there is one particular slaughterhouse to which slaughter pigs are transported almost 2000 km by rail. These pigs are fed and watered from large troughs in the wagon, and generally arrive at their destination in good condition (Grandin, personal communication).

Transport by ship is not feasible, because it takes too long and it is difficult to deliver the pigs to inland destinations. It is sometimes used in a roll-on, roll-off situation, where the truck with pigs is placed on a ferry to transport them across a short stretch of water. The truck is placed on the upper deck to guarantee ventilation.

Summary

1. Nowadays, transport distances of farm animals by road to another farm or to the slaughterhouse are expanding because of the economic consequences of greater opportunities for long-distance and international trade, improved infrastructure and increased demand for live animals for fattening and slaughtering. The quality awareness of the consumer is not limited only to intrinsic characteristics, but often involves extrinsic characteristics such as environment and animal welfare aspects related to production.

2. Quality schemes are developed or are under development in several countries. The owner is responsible for the careful selection of livestock for the loading on to the road vehicle and the presentation of only pigs fit for travel. The driver of a road vehicle is responsible for the care and welfare of animals during transport. Inspectors appointed under national acts can obtain possession of an animal where an offence is being or has been committed in respect of that animal.

3. Transport conditions involve exposure to social stress (e.g. mixing with unfamiliar pigs) and non-social stress (rough handling). A well-known disease related to transport is the porcine stress syndrome. This syndrome is the acute reaction to stress, mediated by the sympathetic nervous system, and can cause severe distress and even death.

4. Transport conditions may affect post-mortem meat quality, either via adrenal or other stress responses or by fatigue of the animals. Important meat quality parameters related to stress and exhaustion before slaughter are pH, rigor mortis, temperature, colour and water-binding capacity. Since exploration of the environment is normal porcine behaviour, they can self-infect, whereafter the infection can spread throughout the entire pig. Conflicting interests concerning animal welfare (rough floors) and hygiene (smooth floors) are a basic dilemma in improvement of the situation.

5. Transport causes physical and behavioural problems, because animals are not accustomed to transport conditions and procedures. The loading procedures, the design and the other animals are unfamiliar to these animals and will frighten them. Drivers have to treat the animals in a proper way to minimize stress. Depending on the distance to the slaughterhouse or transit station, feeding should be stopped the night before transport, but water should be available during the whole transport period. Animals must be able to stand in their natural position and all must be able to lie down at the same time in the compartment. The microclimate depends on the ambient weather conditions. Especially in summer, heat stress may become significant, while chilly temperatures in winter also impose stress.

References

Allen, W.M., Herbert, C.N. and Smith, L.P. (1980) Death during and after transportation of pigs in Great Britain. *Veterinary Record* 94, 212–214.

Animal Welfare Advisory Committee (1994) Code of recommendations and minimum standards for the welfare of animals transported within New Zealand. *Code of Animal Welfare No. 15*, Ministry of Agriculture and Fisheries, Wellington, New Zealand.

Augustini, C. (1976) ECG – und körper temperature messungen an schweinen während der mast und auf dem transport. *Fleischwirtschaft* 56, 1133–1137.

Barton-Gade, P. (2002) Welfare of animal production in intensive and organic systems with special reference to Danish organic pig production. *Proceedings 48th International Congress of Meat Science and Technology*, Rome, pp 79–83.

Barton-Gade, P. and Vorup, P. (1991) Investigation into various aspects of long-distance transport in pigs. *Danish Meat Research Institute Report* 02.487/93, Roskilde, Denmark.

Barton-Gade, P., Blaajberg, L. and Christensen, I. (1992) New lairage system for slaughter pigs: effect on behaviour and quality characteristics. *Proceedings 38th International Congress of Meat Science and Technology*, Clermont-Ferrand, France, pp. 161–164.

Barton-Gade, P., Warriss, P.D., Brown, S.N. and Lambooij, B. (1996) Methods of improving pig welfare and meat quality by reducing stress and discomfort before slaughter – methods of assessing meat quality. *Proceedings EU Seminar 'New Information on Welfare and Meat Quality of Pigs as Related to Handling, Transport and Lairage Conditions'*. FAL-Sonderheft 166, Völkenrode, Germany, pp. 23–34.

Beattie, V.E., O'Connell, N.E. and Moss, B.W. (2000) Influence of environmental enrichment on the behaviour, performance and meat quality of domestic pigs. *Livestock Production Science* 65, 71–79.

Berry, R.J. and Lewis, N.J. (2001) The effect of duration and temperature of simulated transport on the performance of early-weaned piglets. *Canadian Journal Animal Science* 81, 199–204.

Bradshaw, R.H., Parrott, R.F., Goode, J.A., Lloyd, D.M., Rodway, R.G. and Broom, D.M. (1996a) Behavioural and hormonal responses of pigs during transport: effect of mixing and duration of journey. *Animal Science* 62, 547–554.

Bradshaw, R.H., Parrott, R.F., Forsling, M.L., Goode, J.A., Lloyd, D.M., Rodway, R.G. and Broom, D.M. (1996b) Stress and travel sickness in pigs: effects of road transport on plasma concentrations of cortisol, beta-endorphin and lysine vasopressin. *Animal Science* 63, 507–516.

Chupin, J.M., Lucbert, J. and Carrotte, G. (2003) *Ensuring the welfare of cattle and the safety of men; good practice guide to live cattle transport*. Institut de L'elevage Interbev, Paris.

Clark, E.G. (1979) A post-mortem survey of transport deaths in Saskatchewan market hogs. *Western Hog* 1, 34–36.

Connell, J. (1984) *International transport of farm animals intended for slaughter*. Commission of the European Community B-1040, Brussels.

Culau, P.O.V., Ourique, J.M. and Nicolaiewsky, S. (1991) The effect of transportation distance and pre-slaughter lairage time on the pig meat quality. *Proceedings 37th International Congress of Meat Science Technology*, Kulmbach, Germany, pp. 224–228.

Dantzer, R. (1982) Research on farm animal transport in France: a survey. In: Moss, R. (ed.) *Transport of Animals Intended for Breeding, Production and Slaughter*. Current Topics Veterinary Medicine and Animal Science Vol. 18, Martinus Nijhoff, The Hague, Netherlands, pp. 218–231.

De Jong, I.C., Prelle, I.T., nan de Burgwal, J.A., Lambooij, E., Korte, S.M., Blokhuis, H.J. and Koolhaas, J.M. (2000) Effects of rearing conditions on behavioural and physiological responses of pigs to preslaughter handling and mixing at transport. *Canadian Journal of Animal Science* 80, 451–458.

EC Council Directive (1991) On the protection of animals during transport. *Official Journal of the European Community* 1, 340/17.

Eikelenboom, G., Bolink, A.H. and Sybesma, W. (1990) Effects of feed withdrawal before delivery on pork quality and carcass yield. *Meat Science* 29, 25–30.

Fabianson, S., Lundström, K. and Hansson, I. (1979) Mortality among pigs during transport and waiting time before slaughter in Sweden. *Swedish Journal of Agricultural Research* 9, 25–28.

Fraser, A.F. and Broom, D.M. (1990) *Farm Animal Behaviour and Welfare*. Baillière Tindall, London.

Gariepy, C., Amiot, J. and Nadai, S. (1989) Ante-mortem detection of PSE and DFD by infrared thermography of pigs before stunning. *Meat Science* 25, 37–41.

Geers, R., Parduyns, G., Goedseels, V., Bosschaerts, L. and De Ley, J. (1990) Skeletal muscularity and heart function in growing piglets. *Annales de Recherche Vétérinaire* 21, 231–236.

Geers, R., Ville, H., Janssens, S., Goedseels, V., Goosens, K., Parduyns, G., van Bael, J., Bosschaerts, L. and Heylen, L. (1992) Changes of body temperatures of piglets as related to halothane sensitivity and treadmill exercise. *Journal of Thermal Biology* 17, 125–130.

Geverink, N.A. (1998) Preslaughter treatment of pigs: consequences for welfare and meat quality. PhD thesis, University of Wageningen, Netherlands.

Geverink, N.A., Bradshaw, R.H., Lambooij, E., Wiegant, V.M. and Broom, D.M. (1998) Effects of simulated lairage conditions on the physiology and behaviour of pigs. *Veterinary Record* 143, 241–244.

Gijsberts, W. and Lambooij, E. (2005) Oxen for the axe. A contemporary view on historical long-distance livestock transport. In: *By, Marsk og Geest 17*. Kulturhistorisk aarbog for Ribe-egnen, Ribe Byhistoriske Arkiv and Den antikvariske Samling I Ribe Forlaget Liljeberget, Sweden, pp. 58–78.

Gispert, M., Faucitano, L., Oliver, M.A., Guardia, M.D., Coll, C., Siggins, K., Harvey, K. and Diestre, A. (2000) A survey of pre-slaughter conditions, halothane gene frequency, and carcass and meat quality in five

Spanish pig commercial abattoirs. *Meat Science* 55, 97–106.

Grandin, T. (1981) *Livestock Trucking Guide*. Livestock Conservation Institute, Madison, Wisconsin.

Grandin, T. (1988) Effect of temperature fluctuation on petechial haemorrhages in pigs. *Proceedings Workshop on Stunning of Livestock, 39th International Congress of Meat Science and Technology*, Brisbane, Australia, pp. 23–26.

Grandin, T. (1990) Design of loading facilities and holding pens. *Applied Animal Behaviour Science* 28, 187–201.

Guise, J. (1987) Moving pigs from farm to factory. *Pig International* December, 8–12.

Guise, H.J. and Penny, R.H. (1989) Factors influencing the welfare and carcass meat quality of pigs. *Animal Production* 49, 511–515.

Hails, M.H. (1978) Transport stress in animals: a review. *Animal Regulation Studies* 1, 289–343.

Hambrecht, E., Eissen, J.J. and Verstegen, M.W.A. (2003) Effect of processing plant on pork quality. *Meat Science* 64, 125–131.

Hannon, J.P., Bossone, C.A. and Wade, C.E. (1990) Normal physiological values for conscious pigs used in biomedical research. *Laboratory Animal Science* 40, 293–298.

Heuking, L. (1988) Die beurteilung des verhaltens von schlachtschweinen bei LKW-transporte in abhängigkeit von der ladedichte mit berücksichtigung des blutbildes, der herzfrequenz und der körpertemperatur zur erfassung tierschutzwidriger transportbedingungen. PhD thesis, Freien University, Berlin.

Holloway, L. (1980) The Alberta Pork Producers Marketing Board Transit Indemnity Fund. *Proceedings Livestock Conservation Institute*, Madison, Wisconsin.

Holmes, C.W. and Close, W.H. (1977) The influence of climatic variables on energy metabolism and associated aspects of productivity in the pig. In: Haresign, W., Swan, H. and Lewis, D. (eds) *Nutrition and the Climatic Environment*. Butterworth, London, pp. 51–74.

Jackson, W.T. (1973) To Italy – with 120 pigs. *Veterinary Record* 92, 121–122.

Klont, R.E. and Lambooij, E. (1995a) Effects of preslaughter muscle exercise on muscle metabolism and meat quality studied in anesthetized pigs of different halothane genotypes. *Journal of Animal Science* 73 (1), 108–117.

Klont, R.E. and Lambooij, E. (1995b) Influence of preslaughter muscle temperature on muscle metabolism and meat quality in anesthetized pigs of different halothane genotypes. *Journal of Animal Science* 73 (1), 96–107.

Klont, R.E., Lambooij, E. and Logtestijn, J.G. van (1993) Effect of pre-slaughter anesthesia on muscle metabolism and meat quality of pigs of different halothane genotypes. *Journal of Animal Science* 71, 1477–1485.

Lambooij, E. (1983) Watering pigs during 30 hours road transport through Europe. *Fleischwirtschaft* 63, 1456–1458.

Lambooij, E. (1984) Watering and feeding pigs during road transport for 24 hours. *Proceedings 30th European Conference of Meat and Meat Research Workers*, Bristol, UK, pp. 6–7.

Lambooij, E. (1988) Road transport of pigs over a long distance: some aspects of behaviour, temperature and humidity during transport and some effects of the last two factors. *Animal Production* 46, 257–263.

Lambooij, E. (2004) Welfare and hygiene code for pigs during transport: implementation of a security scheme. *ASG Report 04/0010107*, Lelystad, Netherlands.

Lambooij, E. and Engel, B. (1991) Transport of slaughter pigs by truck over a long distance: some aspects of loading density and ventilation. *Livestock Production Science* 28, 163–174.

Lambooij, E., Garssen, G.J., Walstra, P., Mateman, G. and Merkus, G.S.M. (1985) Transport by car for two days: some aspects of watering and loading density. *Livestock Production Science* 13, 289–299.

Lambooij, E., van der Hel, W., Hulsegge, B. and Brandsma, H.A. (1987) Effect of temperature on air velocity two days pre-slaughtering on heat production, weight loss and meat quality in non-fed pigs. In: Verstegen, M.W.A. and Henken, A.M. (eds) *Energy Metabolism in Farm Animals: Effect of Housing, Stress and Disease*. Martinus Nijhoff, The Hague, Netherlands, pp. 57–71.

Lewis, N.J. and Berry, R.J. (2006) Effect of season on the behaviour of early-weaned piglets during and immediately following transport. *Applied Animal Behaviour Science* 100, 182–192.

Lewis, N.J., Berry, R.J., Crow, G. and Wamnes, S. (2005) Assessment of the effects of the season of transport on the performance of early-weaned piglets. *Canadian Journal of Animal Science* 85, 449–454.

Lindsay, L. (2000) *Assured British Pigs*. Assured British Meat, Milton Keynes, UK.

Loynachan, A.T. and Harris, D.L. (2005) Dose determination for acute *Salmonella* infection in pigs. *Applied and Environmental Microbiology* 71, 2753–2755.

Malmfors, E. (1982) Studies on some factors affecting pig meat quality. *Proceedings 28th Conference of European Meat and Meat Research Workers*, Madrid, Spain, pp. 21–23.

Markov, E. (1981) Studies on weight losses and death rate in pigs transported over a long distance. *Meat Industry Bulletin* 14, 5.

Monin, G. and Ouali, A. (1992) Muscle differentiation and meat quality. In: Lawrie, R.A. (ed.) *Developments in Meat Science*. Elsevier Applied Science, London, pp. 89–159.

Monin, G. and Sellier, P. (1985) Pork of low technological quality with a normal rate of muscle pH fall in the intermediate post-mortem period. The case for the Hampshire breed. *Meat Science* 13, 49–63.

Morgan, J.R., Krautil, F.L. and Craven, J.A. (1987) Effect of time and lairage on caecal and carcass Salmonella contamination of slaughter pigs. *Epidemiology and Infections* 98, 323–330.

Mount, L.E. (1974) The concept of thermal neutrality. In: Monteith, J.L. and Mount, L.E. (eds) *Heat Loss from Animals and Man*. Butterworths, London, pp. 425–439.

Perremans, S., Randall, J., Ville, H., Stiles, M., Duchateau, W. and Geers, R. (1996) Quantification of pig response to vibration during vertical motion. *Proceedings EU Seminar 'New Information on Welfare and Meat Quality of Pigs as Related to Handling, Transport and Lairage Conditions'*. FAL-Sonderheft 166, Völkenrode, Germany, pp. 135–141.

Perremans, S., Randall, J.M., Rombouts, G., Decuypere, E. and Geers, R. (2001) Effect of whole-body vibration in the vertical axis on cortisol and adrenocorticotropic hormone levels. *Journal of Animal Science* 79, 975–981.

Phillips, R.A., Thompson, B.K. and Fraser, D. (1988) Preference tests of ramp designs for young pigs. *Canadian Journal of Animal Science* 68, 41–48.

Randall, J.M. (1992) Human subjective response to lorry vibration: implications for farm animal transport. *Journal of Agricultural Engineering Research* 52, 295–307.

Randall, J.M., Stiles, M.A., Geers, R., Schütte, A., Christensen, L. and Bradshaw, R.H. (1996) Vibrations on pig transporters: implications for reducing stress. *Proceedings EU Seminar 'New Information on Welfare and Meat Quality of Pigs as Related to Handling, Transport and Lairage Conditions*. FAL-Sonderheft 166, Völkenrode, Germany, pp. 143–159.

RSPCA (1998) *Welfare Standards for Pigs*. RSPCA, Horsham, UK.

Santos, C., Almeida, J.M., Matias, E.C., Fraqueza, M.J., Roseiro, C. and Sardina, L. (1997) Influence of lairage environmental conditions and resting time on meat quality in pigs. *Meat Science* 45 (2), 253–262.

Schrama, J.W., van der Hel, W., Gorssen, J., Henken, A.M., Verstegen, M.W. and Noordhuizen, J.P. (1996) Required thermal thresholds during transport of animals. *Veterinary Quarterly* 18 (3), 90–95.

Slavkov, I., Iodanov, I., Milev, M. and Danov, V. (1974) Study of factors capable of increasing the number of Salmonella carriers in clinically normal pigs before slaughter. *Veterinary Medicine Nauki, Bulgaria* 11, 88–91.

Smith, L.P. and Allen, W.M. (1976) A study of the weather conditions related to the death of pigs during and after their transportation in England. *Agricultural Meteorology* 16, 115–124.

Smulders, F.J.M., Romme, A.M.T.S., Woolthuis, C.H.J., de Kruijf, J.M., Eikelenboom, G. and Corstiaensen, G.P. (1983) Pre-stunning treatment during lairage and pork quality. In: Eikelenboom, G. (ed.) *Stunning of Animals for Slaughter*. Martinus Nijhoff, The Hague, Netherlands, pp. 90–95.

Standing Committee on Agriculture and Resource Management (1999) *Land Transport of Cattle*. SCARM Report No 77, CSIRO Publishing, Collingwood, Victoria, Australia.

Sybesma, W. and van Logtestijn, J.G. (1967) Rigor mortis and fleischqualität. *Fleischwirtschaft* 4, 408–410.

Swanenburg, M., Urlings, H.A.P., Keuzenkamp, D.A. and Snijders, J.M.A. (2001) *Salmonella* in the lairage of pig slaughterhouses. *Journal of Food Protection* 64, 12–16.

Tarrant, P.V. (1989) The effects of handling, transport, slaughter and chilling on meat quality and yield in pigs – a review. *Irish Journal of Food Science and Technology* 13, 79–107.

Turner, S.P., Farnsworth, M.J., White, I.M.S., Brotherstone, S., Mendi, M., Knap, P., Penny, P. and Lawrence, A.B. (2006) The accumulation of skin lesions and their use as a predictor of aggressiveness in pigs. *Applied Animal Behaviour Science* 96, 245–259.

van der Hel, W., Duijghuisen, R. and Verstegen, M.W.A. (1986) The effect of ambient temperature and activity on the daily variation in heat production of growing pigs kept in groups. *Netherlands Journal of Agricultural Science* 34, 173–184.

van Logtestijn, J.G., Romme, A.M.T.C. and Eikelenboom, G. (1982) Losses caused by transport of slaughter pigs in the Netherlands. In: Moss, R. (ed.) *Transport of Animals Intended for Breeding, Production and Slaughter*. Martinus Nijhoff, The Hague, Netherlands, pp. 105–114.

van Putten, G. and Elshof, W.J. (1978) Observations on the effect of transport on the well-being and lean quality of slaughter pigs. *Animal Regulation Studies* 1, 247–271.

van Putten, G., Corstiaensen, G.P., van Logtestijn, J.G. and Zuidhof, S. (1983) Het douchen van slachtvarkens. *Tijdschrift voor Diergeneeskunde* 108, 645–652.

Verdijk, A.Th.M. (1974) Oorzaken van afwijkende vleeskwaliteit bij stress-gevoelige varkens. PhD thesis, University of Utrecht, Netherlands.

Verstegen, M.W.A. (1987) Swine. In: Johnson, H.D. (ed.) *World Animal Science, B5: Bioclimatology and the Adaptation of Livestock.* Elsevier, Amsterdam, pp. 245–258.

von Mickwitz, G. (1989) Lowest possible requirements for dealing with slaughter pigs beginning with the loading–transport–resting time, and ending with stunning seen from the animal welfare and meat quality point of view. *Proceedings European Association for Animal Production,* Dublin.

Warriss, P.D. (1987) The effect of time and conditions of transport and lairage on pig meat quality. In: Tarrant, P.B., Eikelenboom, G. and Monin, G. (eds) *Evaluation and Control of Meat Quality in Pigs.* Martinus Nijhoff, Dordrecht, Netherlands, pp. 245–264.

Warriss, P.D. (1993) Ante-mortem factors which influence carcass shrinkage and meat quality. *Proceedings 39th International Congress on Meat Science and Technology,* Calgary, Canada, pp. 51–56.

Warriss, P.D. (1998) The welfare of slaughter pigs during transport. *Animal Welfare* 7, 365–381.

Warriss, P.D. and Bevis, E.A. (1987) Liver glycogen in slaughtered pigs and estimated time of fasting before slaughter. *British Veterinary Journal* 143, 254–360.

Warriss P.D. and Brown, S.N. (1983) The influence of pre-slaughter fasting on carcass and liver yield in pigs. *Livestock Production Science* 10, 273–282.

Warriss, P.D., Dudley, C.P. and Brown, S.N. (1983) Reduction in carcass yield in transported pigs. *Journal of the Science of Food Agriculture* 34, 351–356.

Warriss, P.D., Bevis, E.A. and Ekins, P.J. (1989) The relationships between glycogen stores and muscle ultimate pH in commercially slaughtered pigs. *British Veterinary Journal* 145, 378–383.

Warriss, P.D., Brown, S.N., Barton-Gade, P., Santos, C., Nanni Costa, L., Lambooij, E. and Geers, R. (1998) An analysis of data relating to pig carcass quality and indices of stress collected in the European Union. *Meat Science* 49 (2), 137–144.

Yang, T.S. and Lin, J.H. (1997) Variation of heart size and its correlation with growth performance and vascular space in domestic pigs. *Animal Science* 64, 523–528.

16 Handling and Transport of Horses

K.A. Houpt

Department of Clinical Sciences, College of Veterinary Medicine, Cornell University, Ithaca, New York, USA; kah3@cornell.edu

Handling of Horses

Livestock species, with the exception of dairy cows, are handled rarely. Most livestock neither are trained nor is their value dependent on their trainability. In contrast, most horses are usually handled daily as individuals. Their ability to be handled safely and trained to respond to physical and verbal cues is the reason for their existence and the basis of their value. For these reasons, most of the handling methods described will be for individual animals.

Handling methods to be covered include 'imprint training' of the newborn foal, training the young horse to lead, round pen or tackless training, methods of catching, trailering and methods of restraint. Retraining of problem horses will also be addressed.

Early handling

The easiest way to ensure that a horse will be tractable is to handle the foal from birth. One method of handling foals is termed 'imprint training' (Miller, 1991). The term itself is probably something of a misnomer because the foal does not become imprinted on to humans as would a duckling. The foal does not follow people nor does it show sexual behaviour toward people. Imprint training involves handling the foal as soon after birth as possible. The goal is to accustom the foal to manipulations that will be done to it in the course of saddling, shoeing and medicating.

For that reason, the handler puts his/her arms around the foal's chest and squeezes as if a girth were tightening. The hoofs are each tapped with the hand 50 times or until it does not respond, to simulate shoeing and hoof care. Foals seem to resist manipulation of their hooves more than the other forms of handling. The foal is rubbed all over. Fingers are inserted into its mouth, ears and (while wearing a rubber glove) the rectum so that the foal will not be frightened by bits, thermometers or otoscopes. Clippers are held, while running, against its head, neck and body so it will not be afraid of their sound or vibration.

There is some controversy about how soon after birth foals should be handled. If the foal is manipulated before it can stand, the phenomenon of 'learned helplessness' is probably involved. Learned helplessness is the phenomenon whereby an animal that cannot escape a situation will not even try to escape or avoid later when escape or avoidance are possible: the animal simply gives up (Maier and Seligman, 1976). The neonatal foal cannot escape and so, in later life, may not try to escape from a farrier or a pair of clippers. Handling the foal a few hours or days later may not be as effective because the foal can move vigorously. This would indicate that early handling is desirable.

However, care must be taken to avoid disturbing the mare. Some mares will become so agitated at the presence of a person that they will either attack the person or redirect their aggression toward the foal. The odour of the handler may interfere with the mare's recognition of the foal, but in some cases the smell of the handler may reassure the mare. More importantly, the process of imprinting – especially applying stimuli until the foal fails to respond – may take so long that antibodies in the colostrum will no longer be absorbed (gut closure), resulting in failure of passive immunity.

Handling extending over the first few weeks of life is important, as found in the study of Mal and McCall (1996), in which foals were handled for 10 min per day, 5 days per week for either the first or the second 7 weeks of their lives. The foals handled for the first 42 days were easier to halter-break than those handled later. The handling involved rubbing the foals all over their bodies. In an earlier study, Mal et al. (1994) had found that foals handled for 10 min twice a day for the first week were no easier to manage at 120 days than were unhandled foals.

Jezierski et al. (1999) tested the effects of environment and handling on Konik (Polish primitive) horses. There were four groups: SIH (stable-reared, intensively handled), SNH (stable-reared, not handled), RIH (reserve-reared, intensively handled) and RNH (reserve-reared, not handled). Stable-reared foals were raised under typical domestic conditions in a stable. Reserve-raised foals (R) were born and lived in a harem group within a semi-reserve and were not handled at all until they were 10 months old.

At that time, they were divided into intensively handled (RIH) and non-handled groups (RNH). Handling occurred 10 min per day, 5 days per week from 2 weeks of age until 2 years for SIH foals and from 10 months until 2 years for RIH. Handling consisted of haltering, rubbing the foal all over its body and picking up the hooves. The two non-handled groups (SNH and RNH) were not handled except for routine management and veterinary care.

The test, given at 6 months (for the SIH and SNH foals only) and at 12, 18 and 24 months for all groups, consisted of three people catching the horse in a paddock, leading it to and from the stable, picking up its feet and holding it while a stranger approached. The horses were given a score for each test, with high being easy to handle and low being difficult. The intensively handled foals were easier to manage than the non-handled foals and the stable-reared horses were easier to manage than the reserve horses at 12 months but, by 18 months, handling had proved more important than early rearing environment. Fillies had higher heart rates than colts, but their test scores were similar.

There have been three experiments using Miller's technique and it is interesting that Spier et al. (2004) obtained the best results with only two sessions, at birth and 24 h, whereas Simpson (2002) showed only minimal differences between imprint-trained and non-trained foals and Williams et al. (2002) found no difference at all. See Table 16.1 for a summary of researchers' methods.

Perhaps the best and least invasive way to obtain socialized, tractable foals is to handle their dams (Henry et al., 2005). If mares were softly brushed for 15 min/day and hand-fed pellets daily for the first 5 days of their foals' lives, the foals were friendlier, less frightened and more tolerant of handling (saddle pad placement) – not only at 1 month, but also at 1 year of age. A passive human standing in the stall did not have any effect. Another worthwhile technique is to handle the foal daily for the first 2 weeks after weaning/haltering, petting the entire body, picking up the feet and leading, resulting in foals that are easier to restrain and less reactive than foals only handled later (Lansade et al., 2004).

Handling later in the foal's life may be as important or more important than handling at birth. The Japan Racing Association (Ryo Kusunose, personal communication) tested the behavioural characteristics of 270 foals on 25 thoroughbred breeding farms and compared the rearing methods used on each farm. This is the most extensive test of foal behaviour in relation to handling method.

The following test was for the foal's response to six manipulations. The foal's head was touched and the circumference of the front and hind cannon bones were measured. The height at the withers, the heart girth and the hip width were also measured. Foals were tested at 50, 90 and 140 days of age and the scores were added for each foal. The reaction of each foal while a handler touched its head and took each measurement was scored. If the foal stood still during measuring, its score was four; if the foal moved a little, the score was three points. If the handler could not measure the foal without applying a twitch or having someone help hold the foal, or if the foal reared, its score was only one point. The maximum, if the foal always stood still – motionless whatever the handler did to the foal – was 24 points. The minimum was six points if the foal rejected all of the handlers' manipulations.

The average score increased with the foal's age, indicating that they became calmer as they matured, but differed between farms. On one stud farm (A), all of the foals scored 24 or 23 points. These foals were very calm and gentle. However, on farm Y, there were far fewer gentle foals than at A. The behavioural characteristics of the foals reared on the same farm resembled each other. The differences between the farms were not great when the foals were 50 days old but, by 90 days, the behavioural characteristics of foals reared on the same farm were found to resemble each other. There was no further change at 140 days.

The farm's size was not related to the behaviour of foals, but the number of mares per worker was inversely related to the mean scores of foals. The top three rankings of farms had 3.7, 5.3 and 3.1 mares/handler. On the contrary, the bottom three farms had 7.0, 7.5 and 7.5 mares/handler. The fewer mares and foals each handler was responsible for, the more obedient the foals were. There were marked differences in foal handling at these farms. The handlers on the high-scoring farms groomed their foals twice a day, every morning and evening.

One of the most important differences was in the manner of moving the foals from stable to pasture: the two methods used were leading and driving. When leading a mare and a foal to pasture, the handler walked along between the mare and the foal with a lead on each animal. There was a lot of contact between the human and the foal, the leading method apparently allowing the foal to develop confidence in humans. The further the mare and foal were led per day, the quieter the foal was during testing. On other farms, where the mares and foals were driven to pasture, the scores were lower.

On some farms, the halter was taken off at the gate of the pasture every day for security because the foal might catch its halter on a tree or fence and injure itself. Foals from farms on which the halters were taken off at the pasture gate every day and replaced in the evening were easier to handle.

Table 16.1. Summary of imprint training methods.

Imprint age	Repetitions	Test	Age at test	Author
14 days	Until 24 months	Hoof, lead approach	6, 12, 18, 24 months	Jezierski et al., 1999
10 min	at 24 h	Hoof, restraint[a] halter; worming	3 months	Spier et al., 2004
Birth	Twice a day for 7 days	Approach responses to stimuli	4 months	Mal et al., 1994
24 h	Until 42 days	Halter training[a]	84 days	Mal and McCall, 1996
43 days	Until 4 days	Halter training	84 days	Mal and McCall, 1996
2–8 h	5 days	Approach[a] hoof; responses to stimuli	4 months	Simpson, 2002
2 h	12, 24, 28 h	Approach; haltering stimuli	1, 2, 3 months	Williams et al., 2002

[a] Imprinted foals significantly better.

Stroking and massaging

One method of handling horses that has gained considerable popularity is that of Linda Tellington-Jones (Tellington-Jones and Bruns, 1985). The basis of this method is no different from most training in that the horse is rewarded for correct or desirable behaviour and reprimanded for incorrect or undesirable behaviour, but emphasis is on positive methods, particularly pleasurable tactile stimuli.

The equipment needed is a 4-foot (120 cm)-long wand or dressage whip without a lash and lead shank with a long (30 inch, 76 cm) chain. The chain is threaded through the lower nearside (left) and right halter ring and fastened to the offside (right) upper halter ring. This serves to control the horse if it should try to evade or react aggressively to the handler. The wand is used for two purposes: (i) the thickened base is used to rap the horse on the nose if it plunges ahead when led; and (ii) the tapered end is used as an extension of the handler's arm to caress the horse or, more rarely, to block the horse from moving forward.

The first step is usually to stroke the horse with the wand on the back and withers, then down the rump and the hind limbs. The wand is also used to stroke the forelegs and belly. This approach – stroking the body without even attempting to touch the horse directly – is an excellent way of overcoming resistance to handling of the tail, the feet the flanks, etc., and accustoms the horse to touches that can later be used as cues. When the horse relaxes and accepts the wand strokes without moving away or flinching, it is ready for direct manual contact.

Another use of the wand is to reprimand the horse for invading the handler's personal space, by knocking it on the forehead or nose. The horse can be taught to halt by applying slight pressure on the lead line and tapping the line with the whip while standing directly in front of the horse.

The manual contact is in the form of light digital pressure. The digital pressure is applied in a circular motion. A small (3 cm) circle is made, usually in the clockwise direction from 6 o'clock, 360 degrees and beyond to 9 o'clock. The amount of pressure is no more than that which is comfortable when applied to a person's eyelid, i.e. a light touch. The massage can be done anywhere on the body.

A good place to start is on the shoulders, because this is where horses most frequently mutually groom one another. When horses groom, they pick up a fold of skin with their incisors and pull back, allowing the skin to slip out of their teeth. Mutual grooming is believed to serve two purposes: (i) to groom areas such as the back and withers that the horse cannot reach itself; and (ii) to strengthen bonding between horses. Fillies are more likely than colts to mutually groom and, as adults, mares groom their preferred associates and those close in social rank (Waring, 1983). Feh and de Mazières (1993) found that scratching a horse on the withers, but not on the chest, reduces heart rate, i.e. is calming.

The massage method, commercially termed the TTouch®, is good for calming an anxious horse and is a useful adjunct to veterinary examination. Massaging the area around the tail can make a horse much more relaxed and willing to accept a thermometer. Head shyness can be improved by massaging the head closer and closer to the ears or other sensitive areas. The circular motion should not be applied again and again at the same spot during one session or the horse will be irritated. Instead, work across a given area. Pre-treatment with massage can relax a horse sufficiently that it will accept a thorough veterinary examination without becoming tense.

Other exercises consist of manipulating and rotating the horse's ears and massaging its gums. One can tell whether the technique is relaxing the horse by its posture. The horse's head will drop and its upper lip lengthen when it is relaxed, especially when pleasurable sensations are being perceived. This technique is also useful for reassuring a horse on first acquaintance.

Greeting

Blowing in a horse's nostrils has been recommended (Woodhouse, 1984) as a technique for greeting a strange horse. That is, in fact, the usual behaviour when two horses meet (Houpt, 2005). Usually, however, one or both horses squeal; and they may also strike. Therefore, when a person blows in a horse's nostril, it may threaten in a similar manner. Rubbing the horse is a much safer approach.

Physical and Chemical Restraint

Tying

Horses are restrained by tying or cross-tying. Descriptions of the methods used to train a horse to be tied can be found in Wright (1973) and in Miller (1975). Ideally, a horse should be taught to lead and be fully accepting of the pressure of a halter before tying is attempted. Then, the first tie session should be accomplished in a safe, non-confining place with halter and lead (including snaps) that will not break. Few horses that respect the halter will do more than 'test' the tie and the few that do usually give up quickly if they don't 'escape' the tie when they pull hard the first few times.

There are several important points to remember. Quick-release knots or safety fasteners should always be used so the horse can be freed if it becomes entangled. Ideally, the point to which the horse is fastened should be higher than its eyes. Although well-trained horses may be tied to hitching rails, fences, etc., this is not recommended because a downward pull exerts pressure on the poll and is more frightening to the horse than an upward pull. These tying situations are *not* recommended by 4-H horse manuals, even for quiet horses.

Two methods of restraining horses that attempt to break ropes have been suggested (McBane, 1987). Both methods involve use of ropes around the horse's body and both have the rope passing through – but not attached – to the halter. In the first method, a long rope is attached around the heart girth using a bowline knot (non-slip) and the free end of the rope is run between its forelegs, through the centre halter ring and then to a post or wall. If the horse pulls back, the rope presses down on the withers and pulls up on the chest. It should be tied at the horse's eye level or above.

The second method is similar but involves an even longer rope that is looped around the horse from in front, passing under its tail. The two lengths of rope are knotted twice at the croup and at the withers. The loose ends pass through the lower halter rings on each side and are attached to the post or wall. When the horse pulls back, pressure will be exerted under its tail. For some horses, a bungee tether can solve tying problems. When the horse pulls back, this gives and doesn't exert pressure on the poll; however, if the tether breaks, the snap may fly back and injure the horse or its handler.

Hand restraint

Most movement of horses is achieved by means of a halter and lead. Descriptions of training to lead can be found in Wright (1973) and Miller (1975). A trained gelding or mare can usually be handled by means of a rope and snap fastened to the ring in the noseband of the halter. For stallions and fractious horses, a little more control can be exerted using a lead shank with a chain. There are various techniques of restraint using a chain shank, the effectiveness of which depends on the amount of leverage afforded the handler and the sensitivity of the area contacted by the chain.

In all methods using a chain on the face, the chain should be threaded through the rings, with the snap opening facing down toward the horse's face. For minor restraint by this method, the chain is threaded from one side-ring to the other across the horse's nose. The reason for placing the chain across the nose is that pressure applied to the nose is painful and will cause the horse to stop or back up. A properly fitted and used chain should release quickly and not apply painful pressure for more than a second or two. The principle is that punishment must be applied instantly, preferably as the horse misbehaves, and should be removed as soon as the misbehaviour ceases.

This is the same principle as that used in dog-training, in which leash corrections are applied with a choke-chain collar. In both cases, the quick application and release is much more effective than a steady pull, because the animal quickly habituates to a steady pull or begins to fight it if too painful.

Another method has the chain running through the lower right ring behind the horse's jaw to fasten on the upper left ring. This gives leverage without the danger of twisting the halter. The chain can be threaded through the lower left halter ring and fastened to the upper right ring. The chain can then be tightened to pass through the horse's mouth. The chain applies pressure to the gums and the commissure of the lip, causing the horse to stop and raise its head. Placing the chain behind the chin applies pressure to the sensitive soft tissue there. It causes the horse to raise its head and move forward rather than stop.

The use of a chain to treat head-shy horses has been demonstrated in a video, *Influencing the Horse's Mind*, by Robert Miller, DVM (B&B Equestrian Films, 1325 Thousand Oaks Blvd, #102, Thousand Oaks, CA 91366). The principle is that the horse is rewarded for desirable behaviour and punished for undesirable behaviour. The reward is scratching the horse with flexed fingers while clucking (cha chah cha chah). The scratching is directed to the horse's face if head shyness is the problem, but can just as easily be applied to any part of the horse.

The head-shy horse is stroked on the head, but each time it throws its head the lead shank is jerked. The scratching and clucking must stop as soon as the head goes up and resume as soon as it comes down, but there are no verbal reprimands. If the goal is to bridle the horse, the scratching should progress up toward the horse's ears and any head-tossing should be punished. If the goal is to place a crupper on a horse that resents handling of its tail, the scratching should be applied to the neck, then the withers, then the back, then the rump, and finally the area under the tail should be rubbed. If the horse picks up a hind foot or moves away, the lead shank should be jerked.

These activities can be tried first with a chain just over the nose, and the lip chain – which is very severe – should be a last resort. To use the severe method, the chain is run through the left lower ring and attached to the right lower or upper ring, and is worked under the upper lip with the left hand. It should only be used to make the horse stand still.

The twitch

Another method of restraint used almost exclusively on horses is the twitch. The twitch is made of a loop of rope or chain attached to a pole. The horse's upper lip is grasped and the loop slipped over the lip and twisted. This applies pressure and, presumably, produces pain. Most horses will stand quietly when the twitch is applied. It had been assumed that this was to avoid more pain that would ensue if the horse moved against the restraint. Lagerweij *et al.* (1984) found that administration of the opiate blocker, naloxone, greatly reduced the effectiveness of twitching, indicating that horses are being sedated by endogenous opiates. Apparently, the pain sensed when the twitch is first applied leads to release of opiates, which sedate the horse. The calm facial expression and hanging head of many horses when twitched indicates that they are somewhat sedated.

A bridle, The Stableizer®, has been developed that makes use of a cord beneath the lip and over the poll that can be tightened. This is supposed to calm the horse and may operate on the same principle as the twitch. A halter, the Dually®, is equipped with a second strap over the nose, to which a lead rope may be attached for extra control.

Stocks and chutes

Although most handling of horses can be done by tying or holding the horse, there are occasions when it is necessary to further restrain the animal. These occasions are usually for veterinary care or any painful procedure. Stocks consisting of pipes at the level of the horse's chest serve the purpose well when the horse is fairly tractable. The vertical pipes should be 80–90 cm apart, attached to the ceiling and be provided with cleats so that ropes can be attached to cross-tie the horse. The horizontal pipe should be approximately 120 cm from the floor. It must have a non-slip floor. Animals panic when they start to slip.

The purpose is to prevent lateral movement. If more restraint is needed, solid panels can form the front and back of the stocks. This type of arrangement is ideal for rectal palpation, ultrasonic examination and other reproductive procedures. If an unhandled horse must be restrained, solid sides are necessary. This is the manner in which halters can be placed on feral horses that have never before encountered humans. Typical livestock chutes with metal panels separated from the next panel by spaces are very dangerous for horses. A horse can easily thrust a limb through the space between the panels and injure itself.

Chemical restraint

The methods discussed above are physical, but is also possible to use chemical restraint. The most commonly used drugs are the tranquillizer acepromazine (0.04–0.10 mg/kg intravenously (IV)) or the alpha-adrenergic agonist xylazine (0.4–2.2 mg/kg intramuscularly (IM) or 0.2–1.1 mg/kg IV). A more potent alpha-adrenergic agonist, detomidine, has been developed and is useful for restraint and pain control: the dose is 20–40 µg/kg (0.02–0.04 mg/kg IV or IM (Robinson, 1992). These should be used only under veterinary supervision.

Social Facilitation

When horses have to be moved, one can take advantage of the phenomenon of social facilitation. Social facilitation is the technical term for a behaviour that is influenced by the behaviour of other animals. Because horses are herd animals, there is considerable social facilitation of their behaviour. They graze and rest at the same time (Tyler, 1972; Arnold, 1984/1985) and they usually move together to water and to escape from danger. This behaviour that motivates one horse to do what the other horses are doing can be a nuisance if one is trying to take one horse away from a group, but the astute handler can use it to advantage. If a horse will not enter a barn, cross a creek or walk into a trailer alone, it may be perfectly willing to follow another horse into those areas.

Catching

Horses are prey animals that have evolved flight as the best means of defence against predators. Therefore, it is not surprising that horses quickly learn to escape, especially if they are not too eager to be ridden.

One of the easiest ways to catch a horse is to teach it to come. In general, horses will not usually come when called simply for attention. The exception is the single horse – especially a young horse – that may want companionship. Food is the best reward for most horses, although even grain may not be more attractive than a field of young grass: (i) approach the fence and call the horse; (ii) give it grain, apples or carrots to entice the approach; and (iii) try to accustom the horse to being caught by repeating the feed-and-catch routine many times. The ratio of catches to catches followed by something unpleasant should be low, so don't catch the horse only to work it. Catch the horse and let it go or catch the horse and take it out of the pasture only to graze or to be groomed and massaged.

The problem with using food is that some horses at all times and most horses at some time are more motivated to avoid capture than to eat. In that case, another technique must be used, but even then it is best to have some food to reward the horse when it is caught. To catch the horse, avoid direct eye contact because that is threatening to the horse, and avoid running at it from behind because that will stimulate the animal to flee. Move with the horse at the level of its shoulder, but stop when it stops. Leaning away from the horse will encourage it to approach, whereas leaning forward or crouching down will encourage it to move away. Decrease the distance by walking at an angle.

When a difficult horse is to be caught, plan to take several hours. Horses would rather not be in continual motion (the basis of round-pen training), so most will eventually allow themselves to be approached unless the field is big enough that they can run off and rest while the human has to walk a considerable distance to approach them. When the horse is caught, stroke and reward it. No matter how frustrating the process may have been, do not reprimand or yell at the horse.

Herd Behaviour and Fear

When handling horses, one must always bear in mind that they are herd animals. This is reflected in the behavioural (Houpt and Houpt, 1988) and physiological (Mal et al., 1991) response to isolation. The handler should take advantage of this to move horses and to keep them calm in a strange environment. Conversely, extra care must be taken to prevent escape or injury of a horse that is isolated.

When moving horses, whether as a group or while riding, driving or leading a single horse, one should be aware of those things that frighten horses. Horses are reluctant to enter a dark area from a brightly lit one; they are also reluctant to step on anything that sounds or feels unusual. They often shy when crossing a drain, presumably because of the hollow sound made by their hooves. Many horses are reluctant to cross a stream or even a puddle. The unfamiliar sound of rushing water probably frightens them and then the wet feel of water only serves to increase the fear factor. One can desensitize a horse to water by starting with a trickle of water from a hose in a familiar situation. When the horse will cross the trickle, the trickle can be increased to a stream or to create a puddle.

Windy days are particularly apt to result in problems. Some horses seem frightened by the feel of the wind. In addition, many horses will shy from bits of paper or light-coloured leaves blowing near them. Pennants or flags can also frighten them. Horses may be reluctant to pass through gates with flapping flags. This can be especially dangerous if the horse is pulling a wagon that can catch on the gate as the horse swerves to avoid the flag.

Learning Ability of Horses

There have been many studies of learning in horses; this subject has been reviewed by McCall (1990) and Nicol (2002). Unfortunately, few of these studies dealt with learning tasks of interest to riders or drivers. The tasks most frequently studied were visual discrimination and maze learning. Some information which is relevant to handling may be found in the study by Heird et al. (1981), who found that horses handled moderately (once a week) learned more quickly than either those handled extensively (daily) or which had not been handled at all. Trainers could predict the trainability of a horse after working with it for 10 days. In a later study, Heird et al. (1986) found that unhandled horses learned more slowly than handled horses.

Mader and Price (1980) found that Quarter Horses learned visual discrimination more rapidly than did thoroughbreds; they concluded that the thoroughbreds were more distractable. Others working with one breed (Quarter Horses) have also found that the more emotional the horse, the slower it is to learn (Fiske

and Potter, 1979; Heird et al., 1986). Haag et al. (1980) found that there was no correlation between position in the equine dominance hierarchy and learning, but there was a correlation between avoidance and maze-learning – that is, horses that learn one task quickly learn another.

Horses can form concepts. One experiment involved an operant-conditioning task: the horse simply had to push one of two hinged panels. The correct panel was unlocked, allowing the horse access to a bowl of grain. The incorrect panel was locked, but there was a bowl of grain behind it to ensure the horse did not choose this panel because it could smell the grain. The correct choice was not always on the same side. The first problem was a simple discrimination between a black panel and a white panel. Next, it had to discriminate between a cross and a circle. The third problem was to distinguish a triangle from a rectangle, and the next to discriminate triangles from half-circles and various other patterns. In the true test of conceptual learning, the horse had to choose between two shapes it had never seen before, one triangular and the other non-triangular (Sappington and Goldman, 1994).

Other tests of conceptual learning demonstrated that horses could learn to judge relative size (Hanggi, 2003) and to categorize open and filled symbols (Hanggi, 1999). It is surprising that horses can do all those higher order tasks, when McLean (2004) found that they could not remember a location for even 10 s.

Horses have also been shown to be capable of conditional discrimination – in this case identity matching to sample. The horse had to touch the two identical cards (marked with X's or with circles) rather than a third, distractor card that was marked with a square or star. The horses could learn with 73–83% accuracy (Flannery, 1997).

The ability of horses to perceive objects is influenced by their size and by environmental conditions that affect contrast, such as overcast skies. Horses responded by hesitating at a distance of 15.2 cm from a narrow stripe (1.27 cm) when contrast was poor, or by hesitating at a distance of 2.3 m for a wide stripe (10.2 cm) when contrast was good (Saslow, 1999). Horses are more likely to hesitate when crossing a yellow or white mat compared with a grey or black one (Hall and Cassaday, 2006). The animals crossed the mats more easily during the second trial. Horses are able to discriminate colours (Hall *et al.*, 2006).

There have been four experiments attempting to prove that horses can learn by observation (Baer *et al.*, 1983; Baker and Crawford, 1986; Clarke *et al.*, 1996; Lindberg *et al.*, 1999). None proved that they could, but the tasks – visual discriminations – are probably not as relevant to equine ecology as learning which food is safe to eat or which place to avoid. It is probably not a good idea to let a young horse watch another horse misbehave, despite the lack of scientific evidence for observation learning in horses.

Ability to make a visual discrimination (a black from a white bucket) and then a reversal (the formerly incorrect choice is now correct) is not correlated with ability to learn how to jump a hurdle or to cross a wooden bridge, despite the fact that food was the reward in all three tasks. Furthermore, those horses that learned to jump quickly were not those that learned to cross the bridge quickly (Sappington *et al.*, 1997).

Speed of learning to avoid a shock or a puff of air is not correlated with learning for a positive reinforcer – food (Haag *et al.*, 1980; Visser *et al.*, 2003). Nor is learning a detour correlated with ability to open a chest, both for food rewards (Wolff and Hausberger, 1996). Performance can't be predicted from a discrimination learning task, nor from a different type of performance.

Emotionality or nervousness has been shown to interfere with learning. Furthermore, injuries to itself and to handlers are often the result of the horse's response to a stimulus it perceives as frightening. Therefore, it is useful to be able to predict which horses are likely to shy, bolt or baulk in fear. McCann *et al.* (1988a, b) have developed a temperament or emotionality test for horses based on their response to being herded, isolated and approached by people. The subjective rating of the observers correlated well with the heart rate response. Handling or reserpine treatment did not improve the latter response of the horses to frightening situations.

One of the most difficult training concepts is distinguishing negative reinforcement from punishment. Negative reinforcement is *not* punishment. Punishment follows behaviour or misbehaviour. The behaviour will be performed less often when the animal learns that the punishment is contingent on the behaviour. Negative reinforcement proceeds until the horse performs the desired behaviour. The reason it is so important to make the technical distinction is that most horse-training uses pressure cues that are negative reinforcement.

For example, the untrained horse is 'nudged' with the rider's heels or tapped with a whip until it moves. Gradually, subtle cues such as a slight pressure by the rider's legs can replace the more forceful ones (see Fig. 16.1). When the rider pulls on the reins, he or she is applying negative reinforcement – pressure or pain on the gums or lips until the horse slows down (McLean and McGreevy, 2004).

Clicker training, a method in which sound is used as a secondary reinforcer, is very

Fig. 16.1 Pat Parelli demonstrates riding a horse with no bridle. The horse responds to subtle body movements, such as leaning forward to induce the horse to move forward and leaning back to stop. Turning is accomplished by the rider turning their head and shoulders in the desired direction. Gentle methods of training reduce stress, and colts and fillies that are treated gently will usually respond well to these methods, and can be ridden with no pressure applied to the mouth.

popular, but apparently does not speed learning to touch a target in comparison to using a primary reinforcer – food – alone (Williams et al., 2004).

Positive reinforcement is a reward when the desired behaviour is performed. The horse is rubbed when it walks up to the trailer or given a piece of sugar for coming when called. Positive reinforcement is probably not used enough in horse-training. The tackless training and free-lunging methods do use subtle forms of reinforcement – the negative reinforcement is chasing the horse by approaching it from behind and removing that stimulus by backing off when the horse speeds up. Circus and trick horse-trainers make much more use of positive reinforcement: the horse is rewarded with a food treat for bowing, for example.

Dougherty and Lewis (1993) studied stimulus generalization. They taught horses that they would receive a reward if they pressed a panel after having felt a touch in a particular spot on their back; the touch was then moved away from the original location. The horse's response decreased with distance from the original spot. This is certainly applicable to horse training. If a horse is taught to move sideways to a touch on one location on his flank, but the next rider cues it in a slightly different location, it may not respond.

The secret of a good trainer is excellent timing. The negative reinforcement should be removed as soon as the horse begins to perform the desired behaviour, and the positive reward also should be given as soon as the horse performs the desired behaviour (both within a few seconds' time).

Most important is timing of punishment. Punishment is a punishment only if it reduces the frequency of a behaviour. The frequency will be reduced only if the horse realizes which behaviour is being punished. The punishment must occur within 1 s of the misbehaviour. If the horse kicks and is immediately struck with a whip, it will be less inclined to kick, but if it kicks and the handler must walk across the barn to get a whip and then strikes the horse, the horse will not be punished for kicking: the horse will be punished for allowing someone to approach with a whip. The other important factors are to give only one signal for each behaviour to avoid confusing the horse and to avoid signalling two behaviours simultaneously (McLean and McGreevy, 2004).

There has been a change in equine training methods over the past 50 years. Gentling has replaced 'breaking', so that inductive methods rather than intimidation and physical force are used. Perhaps the most popular of the spokespersons for these methods is Monty Roberts, whose book, Listening to Horses, achieved renown within a readership much larger than that of equestrians. Roberts was by no means the first of these people. The Jeffrey method of horse training (Wright, 1973), first used in Australia on their feral horses (brumbies), uses an approach and withdrawal technique with a horse in a paddock wearing a halter and lead rope. The trainer approaches when the horse is calm, but withdraws if the horse is even slightly evasive. Using that method, a horse can be saddled, bridled and ridden within a few hours.

In the USA, round-pen training is favoured. The first trainer to use this method was Archibald – and later, Ray Hunt, Pat Parelli and John Lyons toured the country giving demonstrations and clinics and writing in the popular equine press.

The best description is given in Mackenzie's book, Fundamentals of Free Lungeing (1994). The main thrust is to move the horse around the pen – at first, simply in a circle, but later inducing the horse to turn toward the outside of the pen and reverse, and to turn toward the inside of the pen and reverse according to the body movements (mostly head and hip) of the person in the centre of the ring.

Many trainers use a whip or a rope to keep the horse at the perimeter of the circle and to keep it moving fast. At least part of the technique is simply to tire the horse, but what may be more important is that it teaches the horse to watch the human and to pay attention to his or her movements. The theory is that the human is assuming a dominant or leadership role over the horse. There may also be an element of predator avoidance because, to push the horse forward, the person leans forward assuming an almost quadrupedal stand and 'nips' at the horse with rope or whip. To encourage the horse to approach him or her, the person stands up straight and leans back, avoiding eye contact – which is threatening to the horse, just as sheep avoid Border collies that give them 'the eye,' i.e. stare at them.

One of the most interesting aspects of round-pen training is that horses will make a mouth movement, opening their mouth enough to run their tongue in and out. Although Roberts calls this a 'signal meaning I am a herbivore', the same mouth movement is used by horses when anticipating food or when giving in to a farrier's demands. It may be a care-soliciting expression or simply subordination. It is not the 'snapping' behaviour of immature horses that open and close their mouths, but also show their teeth. Snapping is shown in situations where a young horse is ambivalent, wishing to stay, but frightened enough to flee. It is seen when a colt approaches a stallion: it need not approach, but the positive aspects of approach outweigh the negative ones. It signals that it is young by snapping, perhaps to discourage aggression. The mouthing movements seem more subordinate without the ambivalence of snapping.

The principle of round-pen training seems to be that horses would rather not run, and they are rewarded by being allowed to stop: they appear to be attracted to the trainer when he allows them to stop. The horse is then rewarded further by stroking. Some trainers stroke the horse before they begin the circling manoeuvre. The next steps are bridling and saddling the horse. Once the saddle is in place, the horse is again induced to circle until it stops bucking or, if it does not buck, until it has become accustomed to the feel of the saddle and of the stirrups flapping against its sides.

The skilled trainers like Lyon and Roberts can induce a horse to approach them – 'join up' – within a few minutes. The less skilled rely on exhaustion to control a horse. When a horse has been run for hours around a pen, the owners are usually not happy. The difference in those who master the technique is the ability to observe subtle changes in the horse's demeanour – ears, tail and muscle tension – and an excellent sense of timing, so the horse can be rewarded instantly for obeying or for approaching.

There is one impediment to application of round-pen training or free lungeing: it requires a round pen. If a conventional square or rectangular enclosure is used, the horse will run into a corner. Good practitioners of free lungeing can use a cornered area, but it is much easier –

especially for a neophyte – to use a round pen. The best round pens have solid walls, so the horse is not distracted by anything outside the pen. A solid-walled round pen is also much safer than either a rectangular pen or one with rails, because the horse does not have so many opportunities to injure itself.

Handling of horses requires both an understanding of horse behaviour and a willingness to patiently apply the principles of learning.

Transportation Needs of the Horse Industry

Horses were probably transported first across seas, as the earliest reports have warhorses disembarking near Carthage in the 9th century BC, fit enough to proceed immediately to battle (Cregier, 1989).

The use to which a horse is put – work, meat or recreation – has a large influence on the frequency, distance and method of transport to which the horse is subjected. Horses used for livestock ranch work, farm work and hauling of produce and passengers in extensive agricultural enterprises or developing countries are unlikely to travel very far from the area where they were born and used. Horses intended for slaughter are usually cull animals from the work and recreational groups and are purchased from widely dispersed areas and are generally loaded in loose groups on to large livestock transports for movement to distant processing plants.

There is increasing public concern about the methods used and care given to horses during transit for slaughter and, within the last couple of years, a few states and the US federal government have initiated research into controls over the transport of these animals. In 1998, the State of California passed a proposition to prevent the sale of horses for the purpose of slaughter for human consumption.

The last use group, recreational horses, includes a very large, diverse group of individuals that vary greatly in the frequency, distance and method of movement. The simple pleasure mount is likely to be transported less distance and less frequently than a competitive show or racehorse. Most transport of recreational horses is accomplished using a private towed horse trailer or large commercial-type van; and, in nearly all cases, these horses have individual stalls and care while in transit. An increasing number of the international – and some national level – competitors and imported breeding stock are transported by air, a method that significantly reduces travel time for long distances.

Transport via sea shipping may be used and, as with air shipment, the horses are normally placed in individual stalls in customized, containerized, freight-type crates (Doyle, 1988). Some welfare groups have mentioned inhumane conditions in the shipment of slaughter horses via sea.

It is difficult to establish the number of times or distance that an average horse may be moved in its lifetime. Certainly, this would be much greater than for other large livestock, such as cattle or pigs, simply because of the use and longer lifespan. Casual observation of today's equine competitions indicates that during the most active portion of the competition season, a horse might easily be transported every 2–3 weeks to some type of competition within an 80 km radius of its home. Racehorses, especially harness horses in the north-east and Midwest portions of the USA, are vanned every week to a different track during their summer race circuits, while in Europe, the prominent thoroughbred racehorses may be flown from one country to another to compete in stake races.

Summer and winter horse show circuits for a variety of breeds have been in existence for decades in the USA. Currently, hundreds of hunter-jumper and dressage horses travel from as far as California and Canada to Florida for 8–12 weeks of showing in the winter. In addition to a possible 5000 km+ two-way round trip from home (requiring several days), these horses compete in shows several times per week and may be moved 50–150 km to a different show facility each week during their stay.

Also, many brood mares are annual commuters from their home farms to the stallion's location, covering distances of a few to over 1500 km, while heavily pregnant or with foal at side. Acceptance and use of shipped semen by the breed registries will greatly reduce the necessity of transport for breeding animals.

Recreational horses are often handled by non-professionals who have minimal experience. This can contribute to behavioural problems

and injuries acquired both during loading/ unloading and transit.

Some of the horses subjected to extremely long transport, especially when combined with pre- or post-stresses from infectious diseases, contract serious illnesses, such as pleuropneumonia.

Requirements for Transport

Reasonably safe, low-stress and humane transport of horses from point A to point B can be accomplished if the following points are addressed.

- Use of a transport vehicle suitable to the type of horse(s) being moved and its proper maintenance and operation.
- Preconditioning of the animals to be transported, both behaviourally and medically.
- Completion of required governmental medical vaccinations, tests and quarantine specifications necessary for leaving and entering controlled areas (country/state).
- Careful loading, movement and offloading that avoids traumatic injury to the animals.
- Proper care while in transit to assure that the horses arrive in healthy condition, free of long-term stress and ready to perform.
- Monitoring of the stress accumulation of horses subjected to repeated and/or prolonged transportation situations, especially when additional stressful factors precede or follow.
- Trained handling and medical personnel to properly accomplish the above.

Transport Vehicles and Stalls

Most of the currently used methods and practices for the transportation of horses have been established over a period of time by the demands of the industry, with few governmental or industry standards applicable. This has resulted in a wide variety of road transport vehicles. Conveyances with individual stalls (boxes) where the head and rearquarters of the horse are restrained have been the norm for many years. This prevents interaction and possible injury due to kicking and biting between horses, but the space given to each horse, convenience and safety of the stall for horse and handler – and the loading designs – vary greatly. A discussion of necessities for transport stalls and some advantages/disadvantages to other design features follow.

Styles and sizes of transport vehicle

Vehicles, usually called vans, are motorized trucks where the bed is enclosed and fitted with individual stalls to accommodate two or more horses. Trailers (also called boxes or floats) are beds made to be separable from the motorized portion, which are enclosed and may or may not be fitted with individual stalls. Trailers that are 'loose' or not fitted with individual stalls are called stock trailers. Trailers that connect over the pulling vehicle's rear axles (goose-neck and tractor-trailer) are more stable for towing than trailers connected to the rear of the vehicle (bumper-pull or tag-along).

Space and engineered weight restrictions (axle-carrying weight, etc.) determine the number of horses that may be carried, and may range from one to over ten animals for individual-stall vehicles and trailers. Large, stock-type trailers may carry more. Double-deck or pot-belly livestock trailers designed to transport cattle are not well suited to the transport of horses because they lack sufficient height and their internal ramps are difficult for horses to negotiate (Grandin et al., 1999).

For example, the upper compartment in a two-level cattle trailer may be only 164 cm (64.5 inches) in height (Wilson Trailer Company, Sioux City, Iowa). Abrasion- and laceration-type injuries (mostly to the head and face) were found by Stull (1998) to occur in 29.2% of the horses transported in double-deck (pot-bellied) trailers compared to only 8% of horses hauled in straight-deck livestock trailers.

Grandin et al. (1999) observed that taller horses (those over 162 cm height at withers) were especially prone to injuries to the head and topline (withers, back and croup), and recommended that they not be transported in such trailers.

Individual stall construction

An average size horse (height 158 cm, weight 550 kg) should have approximately 90 cm width and 2.4 m in both length and height of standing space. This allows the horse to use its head and legs to balance during the motion of the transport vehicle. The standing space should be void of projections, edges and protrusions that could cause injury or be contacted by the horse's head or legs. Padding can be added around the walls to prevent bruising when the horse bumps the walls.

The construction of the box should be exceedingly strong and include the use of durable material capable of withstanding the forces of weight, pressure, striking, chewing and excrement of the horse. This usually means strong timber (> 5 cm thick) and heavy-gauge metal, often in combination with each other. Construction of the sides, floor and ceiling should be mostly solid. Openings that could trap head or limbs should be avoided.

Other important factors that need to be considered include adequate ventilation and light. The horse in one stall should be able to see other horses nearby to facilitate ease of loading and to calm behaviour. Adaptation of the stall to prevent extremes of hot (> 25°C) or cold (< 10°C) temperatures is critical to the horse's life and general health (Leadon et al., 1989).

Construction of rear-facing (opposite to the direction of travel) transport stalls should be considered, as there is evidence here that horses maintain balance better and show less muscle fatigue (Cregier, 1982; Clark et al., 1988; Kusunose and Torikai, 1996), have a lower heart rate and a lower-stress postural stance (Waran et al., 1996) and have a preference for this position when transported untethered in large boxes (Smith et al., 1994; Kusunose and Torikai, 1996).

Slant-stalled trailers are now popular (see Fig. 16.2) because they provide for more horses and storage space in an overall shorter length of trailer and may increase ease of loading for some horses. Also, designing the front of the stall to allow the horse to lower its head to shoulder height has recently been shown to be important to normal respiratory function during transits of more than a few hours (Racklyeft and Love, 1990). Most transport stalls have head ties and chest bars that require the horse to maintain its head at 0.25 m or more above the shoulder. Researchers recommended that feed, including hay, be maintained at shoulder level or below.

Fig. 16.2   Slant-load three-horse trailer.

Floors need special consideration in both construction and maintenance. Wooden boards of 5×15 cm or larger are usually used in smaller trailers, but heavy-gauge metal can also be used. Rubber mats and/or bedding are almost a necessity to prevent slipping on most floors once they get wet. Bedding also encourages the horse to eliminate wastes during long transits, which is desirable. Cleaning, maintenance and inspection for wear, rot, rust and weakness are a constant requirement. Unfortunately, horses can and do take fatal falls through improperly maintained floors while in transit.

Stall and vehicle entrance

A horse's heart rate increases dramatically during the loading and unloading process (Waran et al., 1996); therefore, its behavioural requirements should be considered in the construction of both the transport stall and the entrance to the trailer or van. Because the horse is, by instinct, afraid of confinement and its eyes neither adapt rapidly to light changes nor see items clearly at close range, the ideal entrance should be wide, well lit and uncomplicated (no steps or ramps) to traverse. A level, funnelled walk into the transport vehicle – such as that provided by the special ground loading ramps seen at most large

horse facilities – is the most ideal. However, loading in less than ideal situations is often necessary.

The two most common types of entrances are: (i) the step-up, where the horse steps upward directly from ground level into the trailer (see Fig. 16.3); and (ii) the ramp type (see Fig. 16.4), where the horse walks up a sloped platform into the trailer. In both cases, the horse should be allowed to go slowly and look down at its footing during loading.

The step-up design, which should have a rubber bumper guard on the leading edge of the floor to prevent injury to the horse's shins, usually allows the horse to get closer to scrutinize the inside of the stall before entering. This design requires the horse to learn to pick its feet up to enter, whereas the platform of the ramp type, which solves the stepping up problem, may be yielding, slippery and hollow-sounding to the horse's weight and step, all of which may frighten the horse.

The disadvantages of both designs need to be minimized and combined with careful preconditioning and patience to teach the horse to accept. Doors that swing out and with side ramps create a funnel effect to help the horse gradually enter the more confining area of the stall. Also, stall construction that allows one of the side walls to be swung wider for entry and then back into place after the horse enters is very advantageous for young and large horses. Entrances with low step-ups or very short, sturdy ramps are adapted to by most horses very quickly. The longer, steeper loading ramps required by high-bedded vans should always have sides for guiding and safety.

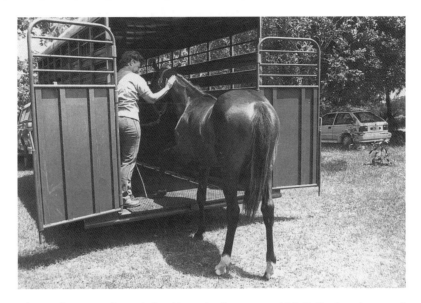

Fig. 16.3   This yearling is standing quietly with one leg in a step-up 16-ft (0.48 m) stock-type trailer and being rewarded with a pat on the neck.

Fig. 16.4   Two-horse trailer with ramp entrance.

Horse restraint in the transport stall

As the horse enters the stall, a prompt, quiet and easy method must be available to solidly close off the rear (or front) of the stall. Be sure that open latch handles do not stick out where the horse can be impaled on them and injured during the loading process. Many trailers and vans have very cumbersome or dangerous closures, including having to raise a tail gate from the ground. A simple-to-latch but strong, side-swinging door that closes off one stall at a time is probably the best. The latching mechanism should be quick and positive to avoid injury to handlers should the horse bump into the door if it comes out prematurely. Where butt or chest bars are used, they must be adjustable to the height of the horse to prevent the horse slipping under them. Head restraint must always be accomplished *after* exit from the stall is closed off by putting in place the chest or rump bar.

When unloading the horse, this sequence should be reversed; or, the horse's head restraint should be removed *before* the chest or butt bars are released. Head ties may be installed, but should be adjustable and fitted with quick-release snaps. Total stall size and restraints should allow 0.3 m or so of movement forward and back, and the head tie should remain slightly slack when the horse's rear is touching the butt bar or back wall. The purpose of the head tie(s) is to prevent the horse from turning around in an individual stall. It is questionable whether a horse's head should be restrained under conditions where there is no rearquarters restraint.

The horse inexperienced in tying or one that becomes afraid during transit is likely to pull against a head tie and be injured when it is the only restraint used. If a horse is not accustomed to restraint by the halter, as is often the case with youngsters, then it should not be tied, but preferably be transported in a large loose box of approximately 1.5×2.0 m. Obviously, vehicles hauling loose horses should not have openings in any cross-gates or an entrance door large enough to encourage the horse to attempt escape.

All transport stalls should allow safe and easy access to the horse's head for tying and care purposes. Shipping crates for air and sea transport may additionally require access to the horse's rearquarters, since it would not be possible to remove the horse in an emergency. In the case of road trailers, extreme care must be taken by both manufacturers and horse owners to properly construct and use escape doors meant for humans only, so that horses do not attempt to use them for exit and, thereby, become trapped or injured.

Transport equipment choice and operation

All transports used for livestock, especially those used on the roadways, should be properly engineered for stability and weight-carrying requirements. Transport equipment soundness is paramount to safety on the road. There are few governmental standards regarding transport equipment, in particular, the pull-type trailers currently being used to transport the majority of recreational horses.

Bumper-pull or tag-along single-axle or tandem-axle trailers can be difficult to attach to the pulling vehicle and tow safely because they are more predisposed to attachment problems, load imbalance, weaving while being towed and jack-knifing during the braking phase of driving. The ball and hitch arrangement and braking system must be heavy-duty enough for the weight of the trailer plus load to be pulled. For balance reasons, it is important that the trailer is level or slightly uphill at its front end when hooked to the pulling vehicle.

When a single horse is loaded into a two-horse or larger trailer, the horse should be placed with most of its weight towards the front (over or in front of axles) and driver's side of the trailer, because weaving problems may arise when most of the horse's weight is behind the trailer's axles. In addition, putting the horse on the non-driver's side places it on the low part of the road crown, causing the trailer to drift to the outer road edge.

Low or uneven tyre pressure on either the pulling vehicle or trailer can cause swaying and weaving of a moving trailer. Various trailer equipment attachments are available to help adjust the tongue weight of bumper-pull loads and to help reduce swaying, thereby aiding the stability of the rig. Most states in the

USA require a safety chain near where the trailer is attached to the pulling vehicle and some, in addition, require a breakaway system that automatically sets the brakes on the trailer should the trailer become disconnected during travel. It is highly recommended that the driver make a last, careful inspection of the entire rig just prior to the start of a journey to catch hook-up and other safety problems.

Little research in trailer design has been done, but recently Smith *et al.* (1996a, b) found that a two-horse bumper-pull trailer having leaf-spring suspension with bias-ply tires produced a smoother ride than the same trailer having torsion bar suspension with normal-pressure radial tyres. No difference in the horses' well-being was found for each type after a 24-h haul (Smith *et al.*, 1996a, b).

It is up to the purchaser to check for sound engineering design and purchase only equipment made for the purpose intended. The hitch, axle and brake capacity for trailers and pulling capacity and weight ratings of pulling vehicles and vans should be acquired from the vehicle and trailer manufacturers to determine their suitability for the purpose intended (i.e. total weight). Try to include a safety margin in the equipment.

Horse owners must consider that they are transporting an animal that is large, top-heavy and, at inopportune times, fractious. These factors readily contribute to sway and weaving of the vehicle/trailer at high speeds. Also, uniform tyre pressure, load weight distribution and sound undercarriage structure are other factors under the control of the driver that contribute to trailer and vehicular stability on the road. Poor driving technique, neglect of maintenance, improperly used equipment and transport of valuable horses do not mix and, more often than necessary, result in a tragic accident.

Careful maintenance of the running soundness of the motor and wearable parts will reduce the chances of being stranded on the roadside with animals. Extreme cold or hot weather conditions and difficult horses may contribute to the seriousness of a road breakdown. Horse owners should choose the transportation method, design and size best suited to their horse's behaviour and size.

Training the Horse to Load, Haul and Unload

Horses to be transported can be separated into two groups, halter-trained and untrained (or insufficiently trained). Recreational horse-users and trainers have been dealing with the situation by a great variety of methods. What follows are some commonsensical suggestions based upon the horse's normal behavioural patterns.

Horses that have been well trained to be individually handled by halter should be taught to properly load well before the day of an antici-pated transport, in order to reduce the stress of the first haul. Loading training may take as little as a couple of days to as long as several weeks. Generally, a longer training period is required for teaching horses to enter and stand in two-horse trailers than in more spacious-type trans-ports (stock trailer or vans). Loading training should be slow and deliberate and without scar-ing the horse or the use of excess force. If the animal panics and fights or falls, the procedures are in error and need rethinking.

Frequently, loading a horse that is already a good loader first and in the presence of the horse learning and/or careful use of food rewards – along with asking for only one step at a time – will load most horses. Mares and foals are best loaded by holding the mare at the trailer entrance and loading the foal using two people, one on each side, locking arms behind the foal and nudging it gently forward straight into one side of the trailer. One handler restrains the foal in the trailer while the mare is loaded. This method has an added advantage of entic-ing a poor-loading mare to load in order to stay with her foal. Loading the mare first usually causes the mare to lose sight of her foal, which may cause her to struggle in the trailer. If there is insufficient time for the loading training, then these horses should be treated like untrained horses when loaded and transported.

Tranquillization sometimes helps with problem or inexperienced loaders when there is insufficient time for proper training. Horses – such as suckling foals and youngsters that may lead using the halter but do not tie, back up or stand well – need special treatment when load-ing and hauling in individual stalls.

A survey of owners of horses with trailer problems revealed no breed differences in prevalence of trailer problems (Lee et al., 2001). Loading was the most common problem, closely followed by problems in transit. A smaller number had trouble unloading. Many horses misbehaved both when loaded and while travelling: presumably the aversiveness of travel accounted for their reluctance to load. Changing the trailer type, especially from two-horse to livestock, was most likely to result in an improvement. Horses that had once travelled well were more likely to improve than those that had always had problems.

An excellent method of improving trailer-loading behaviour while simultaneously reducing stress is that of Shanahan (2003). Horses were trained during six 30-min sessions to walk on lead using light taps on the croup, to lower the head in response to a downward pull, to stop and to back up. They were taught to walk up to and on to tarpaulins, to cross poles on the ground, to step on planks on the ground, to step on to a raised platform, to walk through a narrow space and to walk under a tarpaulin held overhead. The Tellington Touch Equine Awareness Movements – see Stroking and Massaging, above – were used to relax the horse. Cortisol levels, heart rate and time to load were all reduced by this regime.

Medical Preconditioning, Care in Transit and Stress Monitoring

Several weeks prior to moving a horse, a licensed veterinarian should be consulted to determine the proper disease testing, vaccinations and paperwork required for its intrastate, interstate or international transit. Requirements vary from state to state and country to country, and two or more weeks may be needed to accomplish the vaccinations and required tests. The veterinary-issued interstate health certificate is usually acquired a few hours or days before transit.

Transits of 4–12 h or longer tend to be measurably stressful, and horses will reduce their water and feed intakes significantly while in a moving vehicle or trailer (Anderson et al., 1985; Traub-Dargatz et al., 1988; Mars et al.,

1992; Smith et al., 1996b; Friend et al., 1998). Therefore, consideration should be given to body temperature monitoring (pre-, post- and during transit) and the giving of immunostimulants, prophylactic antibiotics and fluids (pre- and during) (Nestved, 1996). These precautions aid the immune system and functioning of the renal and gastrointestinal systems.

The transport personnel should stop for brief rest periods and offer water every 3–6 h of transit to encourage the consumption of water and hay; and, after every 16–24 h of transit, a complete offloading of the horse for an extended 12-h rest period is recommended. Research has shown that even normal, healthy horses offered water en route will dehydrate, and that the consumption of food and water is greatly increased when the van/trailer is standing rather than moving (Kusunose and Torikai, 1996).

Transits of up to 24 h were tolerated by healthy horses that had been rested in a stopped transport and offered water for 15 min out of every 4 h en route. They made an uneventful recovery from mild dehydration (Smith et al., 1996b). However, some healthy horses, after 24 h of transit without water and in hot environments, became severely dehydrated, fatigued and unsuited to continue travel. Some horses refuse all food and drink while being transported and may need special care during long trips (Friend et al., 1998).

Very cold weather conditions may require that horses, especially in more open trailers, be blanketed while en route; conversely, in hot environments, care should be taken not to leave a horse in a parked trailer/van in the sun, as heat stroke can occur from extreme temperature rises within it. The use of leg wraps to protect the lower limbs and head bumper guards is not recommended unless: (i) the horse is completely accustomed to them; (ii) they are properly applied and checked periodically en route; and (iii) they are not kept on for overly long periods of time.

Horses that have an elevated temperature and show signs of infectious disease should not be transported, especially long distances. A horse that begins to show a fever while in transit should be offloaded at the earliest opportunity and receive veterinary supervision and rest until healthy enough to resume travel.

Transport Stress and Post-transport Performance

Limited research indicates that short (≤ 1 h) transits just prior to a submaximal exercise performance are not detrimental to a horse (Beaunoyer and Chapman, 1987; Russoniello et al., 1991; Covalesky et al., 1992). For normal, healthy horses, longer transits (4–24 h) – even though they produce some measurable changes in weight loss, heart rates, dehydration and some metabolites, hormones and other blood factors, including cortisol (Clark et al., 1988, 1993, White et al., 1991; Smith et al., 1996b; Friend et al., 1998) – do not appear to be injurious to a horse's general well-being or health, and require only 1–2 days of recovery for most horses.

Inexperienced show horses, young horses being hauled for the first time and horses experiencing very rough transport conditions where they had been exposed to quick starts and abrupt stops were found to be moderately to highly stressed (Russoniello et al., 1991; Covalesky et al., 1992; Kusunose and Torikai, 1996). Continuous transport of breeding mares for 9–12 h, which produced measurable stress, did not interfere with the normal reproductive functions of oestrus cycle and early gestation (Baucus et al., 1990a, b).

Mild to severe respiratory dysfunction and disease changes – even pneumonia – can occur during or following transportation by ground and air and tend to remain for several days to weeks after clinical signs of disease disappear (Anderson et al., 1985; Traub-Dargatz et al., 1988). These respiratory problems may be related to high head position restraint during transit (McClintock et al., 1986; Racklyeft and Love, 1990). However, Smith et al. (1996b), who studied horses during and after 24 h of transit, speculated that exposure to pathogenic agents that initiate injury to the respiratory epithelium before or during transport, trailers that are not well ventilated and horses that have a greater individual stress response may be the factors responsible for the development of respiratory disease post-transport.

Transport of Injured and Rescued Horses

Many racetracks and large horse facilities now have specially designed horse trailers for the transport of severely injured, non-ambulatory horses. After sedation and stabilization of its injury, the horse is loaded into the special trailer either by a very low mobile floor on to which the horse is initially slid or rolled, or the horse is raised and moved in a harness using an overhead hoist system suspended from a beam that extends from inside the trailer. Horses have even been rescue airlifted out of inaccessible areas via helicopter using a special overhead support device and body sling developed by C.D. Anderson (Madigan, 1993).

Research Related to Transportation

There is still little research addressing horse trailer and van design in the scientific literature. However, Smith et al. (1996a) evaluated several combinations of vehicle suspension (lead-spring and torsion bar), tyre types and inflation rate and shock absorber use for smoothness of ride and common frequencies of vibration using a two-horse, bumper-pull, tandem-axle, forward-facing trailer. The leaf-spring suspension with low-pressure radial tyres (or bias-ply tyres) and without shock absorbers provided the smoothest ride. The torsion-bar suspension combined with normal-pressure radial tyres was the roughest. Shock absorbers did not improve the ride quality.

Horses travelling on the right side of the trailer experienced more vibration than horses on the left side, which the researchers thought might be caused by the poor conditions of asphalt roads near the shoulders. The International Air Transport Association has standards for horse stalls carried on aircraft in their Live Animals Regulations (Doyle, 1988).

Several researchers have looked at the relationship of the horse's position in the trailer and others have worked at evaluating well-being during longer transits and transport conditions for slaughter horses. The expectation of moving a number of the world's leading performance horses to Sydney, Australia for the 2000 Olympics, brought researchers together in early 1999 for a workshop on equine transportation stress (Miguarese, 1999). This may now result in the initiation of more research and new guidelines for equine transport.

The need for such research was brought to the fore when several horses transported from the USA to Dubai in the Middle East for the 1998 World Endurance Championships experienced mild tying-up to full-blown, massive myositis (muscle damage) when exercise was initiated post-shipping. The horses had been confined to their shipping boxes for up to 58 h during their flight and layover in Europe (Teeter, 1999). The latest advancement in horse transport appears to be a report that one can now Federal Express one's horse across the USA via their air transport system (Bryant, 1999).

Further evidence is accumulating to indicate that horses do have a preference for certain body positions when in transit. Waran et al. (1996) found lower heart rates in horses facing rearwards versus forward to the direction of travel. Horses facing backwards tended to rest on their rump more. When facing forward, horses tended to move more frequently and hold their neck higher than normal and to vocalize more frequently. The authors postulated that rearward-facing horses could use their forelimbs more effectively than their hind limbs to balance for lateral trailer movements and placed more weight on the forelimbs and protected their head better.

Even loading methods and entrances may make a difference to the horse, as walking into a van from a platform at the same level and backing into a stall produced lower peak heart rates than walking up a sloped ramp directly into the stall.

Kusunose and Torikai (1996) observed the behaviour of six pairs of untethered yearling thoroughbred horses (which had been transported only once previously) in a horse-carrying vehicle; they found the time spent feeding was only 10.5% of the total behaviour activity when the vehicle was moving compared to 67 and 64%, respectively, when it was parked or parked idling. Half of their driven pairs were exposed to five repeated abrupt stops during each driving trial and the other half to normal, slow acceleration and deceleration driving conditions without the abrupt stops.

The normal-driving pairs increased their amount of backward facing behaviour (to the direction of travel) and decreased the number of body position changes with each succeeding driving trial, while the abrupt-stop pairs maintained a high incidence of changing position and could not seem to settle into a favoured

standing position when driven. The researchers felt that, since the number of body direction changes did not decrease with the number of trials, either in the parking or idling periods, it appeared that the abrupt stops had created constant and tremendous stress for the horses.

Smith et al. (1994) found mature horses spent more time facing backward to the direction of travel when the trailer was in motion, but not when it was parked. Several horses displayed strong individual preferences for the direction they faced during road transport. They also found higher heart rates when the trailer was moving, but no difference in heart rate when the horse was tethered facing forward or backward versus untethered.

In partial disagreement with past research, Smith et al. (1996b) studied four horses during transport for 24 h in a two-horse trailer and could detect no decrease in pulmonary (respiratory) function using aerosol clearance rates; however, changes were observed in the red blood cell count, packed cell volume and levels of haemoglobin, plasma protein and cortisol. The horses lost weight and were slightly dehydrated, and water and hay intake rates were lower during transport than pre-transport. Heart rates were higher during only the first 120 min of travel.

Smith et al. (1996b) also looked at trailer environment during transit and found that ammonia and carbon monoxide concentrations in the trailer during transport were within acceptable limits for human exposure; however, respirable particulates (dust particles probably from the hay) in the atmosphere were too high. The authors did not address head restraint position–respiratory interactions found by other researchers, but instead suggested that exposure to pathogens either pre- or during shipping and poorly ventilated transports might be the cause of respiratory illnesses.

Nestved (1996) has further substantiated the seriousness of respiratory illnesses during transit by finding a 60.9% illness rate (upper respiratory diseases requiring treatment) in non-treated controls (142/233 horses); however, he managed to reduce this to a 18.4% incidence of disease by pre-shipment administration of the immunostimulant, Propionibacterium acnes (EQSTIM™ Immunostimulate) (40/217 horses) in horses being transported from 390–2300 miles (625–3680 km) with transit times of 8–50 h.

A few recent, well carried out studies looking at long-distance shipping stress and slaughter transport have been conducted. Friend et al. (1998) made extensive physiological measures of 30 untethered horses on four treatment groups: penned with water, penned without water, transported with water and transported without water. Twice as many horses were assigned to the transported as the penned groups.

The transported groups contained in one 16-m long open-top livestock trailer and tractor rig were driven for 4 h and offloaded for 1 h out of each 5-h period continuously, until after five trips (24 h) it was determined that three horses (all in transport without water group) were not fit enough to continue. Two of these three horses had an elevated body temperature (39.6 and 40.6°C or 104 and 105°F), and the third was classified as too weak to continue. The 'transported with water' group drank less water (offered in last 10 min of each 1-h break) than did the 'penned with water' group (20.9 versus 38.2 l, respectively).

Physical measures such as capillary refill time, mucous membrane score and skin turgor were not useful predictors of dehydration or welfare assessment. Body temperature, serum sodium and total serum protein concentrations were more useful but had great variation. These researchers did not recommend any character (factor) that they would use to predict which horses were approaching a critical condition.

Blood values indicated that horses had hypertonic dehydration, so rehydration with plain water was required rather than the addition of electrolytes, which would have exacerbated the loss of water. Although the transported horses drank less than the penned horses, they appeared to consume enough to delay severe dehydration. All horses starting the experiment were in good, fit condition and calm, easy to handle individuals accustomed to extensive handling, and the researchers were concerned that less tame and more stressed horses like those often found in slaughter transport might be more reluctant to drink and that under hot environmental conditions, 24 h is probably the longest period horses should be transported without access to water.

Friend and co-workers further stated that neither of these standards was based on research that has been published in scientific journals and indicated that the European Council of the Animal Transportation Association had published recommendations for animals transported by road in which all horses had to be fed and offered water at least every 15 h – and preferably every 6, with recommendations that horses not travel in groups larger than four or five and that foals and young horses be able to lie down when journeys exceed 24 h. In addition, some commercial trailers being used to transport slaughter horses in Europe are now being fitted with fold-out water troughs.

Stull (1998) reported research to help formulate USDA (US Department of Agriculture) regulations for the 1995 Senate Bill (S.B. 2522) entitled 'The Humane and Safe Commercial Transportation of Horses for Slaughter Act' (Anon., 1996). Nine loads (five straight-deck trailers plus four potbelly/double-deck trailers) totalling 306 horses were studied on trips to packing plants with distances of 230–963 miles (370–1550 km) at durations of 5.75–30.00 h.

The differences in some pre- and post-transit factors were measured. The straight-deck trailer had greater changes in white blood count (WBC), neutrophils (N), lymphocytes (L), N:L ratio, lactate, body weight and rectal temperature. This makes the straight-deck trailer appear more stressful for transport than the potbelly trailer, but this could have been caused by poor ventilation in the particular straight-deck trailer used in the study, since the climate was described as hot and humid.

Stocking density (1.24–1.54 m^2/horse) affected changes in body weight and WBC, N, L and N:L ratio, while duration/distance of travel affected rectal temperature, N and lactate. Only 8% of the straight-deck trailer horses were injured, while 29.2% were injured in the potbelly trailer, the most prevalent location of injuries (abrasion/lacerations) being the head and face. No horses died in transit.

A survey of trucking practices and injury to slaughter horses was conducted by Grandin et al. (1999), consisting of 63 trailer loads (1008 horses) arriving at two slaughter plants in Texas in July and August of 1998. Nearly half (49%) of the horses were transported on gooseneck trailers while 42% were on double-deck (potbelly) and 9% on straight, single-deck semi-trailers. The authors found that 92% of the horses arrived in good shape while 7.7% had severe welfare problems and 1.5% were not fit for travel. Most of the severe welfare problem horses were in conditions caused by owner neglect or abuse, and these would have been present before transport; only 1.8% (18 horses) had transport and marketing injuries severe enough to be rated a severe welfare problem.

Examples given of origin of welfare problems were: loaded with a broken leg, emaciated, foundered, racehorses with bowed tendons and horses that were too weak to be transported. Most of the injuries acquired during transit were to the head, face, withers, back, croup and tailhead. Some moderately severe back injuries were attributed to the double-deck trailers being too low for tall horses and to internal trailer ramps that are not well suited for horses.

Horses in double-deck trailers appeared to unload better at night, where the authors postulated that they were attracted to the lighted barn. The authors felt that fighting (biting and kicking) due to dominance expression in loose groups was a major problem, because 13% of the carcasses had bruises caused by bites or kicks and 55% of all carcass bruises had been caused by bites or kicks.

Also, loads from dealers that had picked up horses from more than one auction had more external injuries and carcass bruises than direct loads, probably from repeated mixing of strange horses. These authors made additional observations during one day of a large horse sale (New Holland Sale in Pennsylvania, USA) in July 1998, when 168 horses were sold. At this sale, they found that all horses were individually handled with a halter or were tied up (in contrast to group slaughter shipment, which involves loose handling). Most of the horses arrived either on gooseneck (90%) or horse trailers.

Fresh abrasions were found on only five horses (4%) and all were minor, and 11.6% were classified as welfare problem animals for various reasons (skinny 3.5%, behavioural problems 7%, physically abused 1.1%). All horses were fit for travel. Prices ranged from US$200 to over US$1,000, and the sale company did not accept horses that were severely lame or in very poor body condition.

At this sale, most of the injured horses were located in the 'drop-off' pens where dealers can unload horses bought at a previous sale for temporary rest, feeding and watering. These horses

are loose in their pens and tended to fight with the dominant individuals, causing injury to the less dominant individuals.

Overall recommendations of the authors for slaughter horse handling and transport include:

- Educate horse owners that they are responsible for horse welfare.
- Horse associations should all have animal care guidelines.
- Station USDA/APHIS trained welfare inspectors in slaughter plants (US Department of Agriculture/Animal and Plant Health Insection Service).
- Fine those individuals who transport horses unfit for travel.
- Segregate aggressive mares and geldings in the same manner as stallions.
- Improve horse identification.
- Implement procedures to immediately euthanize horses with severe injuries – such as fractures – when they arrive outwith normal slaughter plant hours.
- Inspect horse transport vehicles at truck-weighing stations and at auctions.
- To prevent transport of slaughter horses to Mexico or underground markets, the four currently existing horse slaughter plants should be encouraged to remain open. A lack of slaughter facilities will increase the number of horses that will die from neglect.
- Double-deck trailers should not be used to transport tall horses.
- Educate horse owners to improve training methods to prevent behavioural problems that can cause a horse to be sold for slaughter.

Conclusions

Limited research to date indicates that, while horses do show an acute physiological stress response to transportation, with proper care this response is neither long-lasting, nor does it appear to interfere with either exercise or reproduction. More research is needed on the accumulated stress from extended transit, with emphasis on prevention of transit dehydration and post-transit performance horse metabolism and muscle function, and respiratory and immune factors. Stress on young and naïve horses needs more study. The behavioural aspects of transportation have been little studied and could be closely related to traumatic injuries acquired during loading or in transit. Experiments directed at improved trailer construction for the horse – which should include facilitating the lowering of the head – are required, as well as those aimed at general engineering to improve road stability.

In the USA, USDA slaughter horse transport standards are in the development stages, but there are still none for the regular horse industry. A greater coordination of efforts between caretakers, regulatory agents, shippers and airport management will reduce the stress of international horse shipments of long duration. With the increased movement of horses and welfare concerns, research should proceed and improvements in transportation standards – especially those concerning the movement of slaughter horses and horses using road transit – should be made, either by the horse industry or national government.

References

Anderson, N.V., DeBowes, R.M., Nyrop, K.A. and Dayton, A.D. (1985) Mononuclear phagocytes of transport-stressed horses with viral respiratory tract infection. *American Journal of Veterinary Research* 46(11), 2272–2277.

Anon. (1996) The Humane and Safe Commercial Transportation of Horses for Slaughter Act of 1995, S 1283, Pub L. 104-127, Title IXDC, Subtitle A, SSU901 to 905, 4 April 1996; 110 Stat 1184. USDA, Washington, DC.

Arnold, G.W. (1984/1985) Comparison of the time budgets and circadian patterns of maintenance activities in sheep, cattle and horses grouped together. *Applied Animal Behaviour Science* 13, 19–30.

Baer, K.L., Potter, G.D., Friend, T.H. and Beaver, B.V. (1983) Observational effects on learning in horses. *Applied Animal Ethology* 11, 123–129.

Baker, A.E.M. and Crawford, B.H. (1986) Observational learning in horses. *Applied Animal Behaviour Science* 15, 7–13.

Baucus, K.L., Squires, E.L., Ralston, S.L., McKinnon, A.O. and Nett, T.M. (1990a) Effect of transportation on the estrous cycle and concentrations of hormones in mares. *Journal of Animal Science* 68, 419–426.

Baucus, K.L., Ralston, S.L., Nockels, C.F., McKinnon, A.O. and Squires, E.L. (1990b) Effects of transportation on early embryonic death in mares. *Journal of Animal Science* 68, 345–351.

Beaunoyer, D.E. and Chapman, J.D. (1987) Trailering stress on subsequent submaximal exercise performance. *Proceedings of 11th Equine Nutrition and Physiology Symposium*, Oklahoma State University, Stillwater, Oklahoma, pp. 379–384.

Bryant, J.O. (1999) Dressage competition nurtures tomorrow's talent. *Horse Show* 62 (3), 34–37.

Clark, D.K., Dellmeier, G.R. and Friend, T.H. (1988) Effect of the orientation of horses during transportation on behaviour and physiology. *Journal of Animal Science* 66 (suppl. 1), 239 (abstract).

Clark, D.K., Friend, T.H. and Dellmeier, G.R. (1993) The effect of orientation during trailer transport on heart rate, cortisol and balance in horses. *Applied Animal Behaviour Science* 38, 179–189.

Clarke, J.V., Nicol, C.J., Jones, R. and McGreevy, P.D. (1996) Effects of observational learning on food selection in horses. *Applied Animal Behaviour Science* 50, 177–184.

Covalesky, M., Russoniello, C. and Malinowski, K. (1992) Effects of show-jumping performance stress on plasma cortisol and lactate concentrations and heart rate and behaviour in horses. *Journal of Equine Veterinary Science* 12 (4), 244–251.

Cregier, S.E. (1982) Reducing equine hauling stress: a review. *Journal of Equine Veterinary Science* 2 (6), 186–198.

Cregier, S.E. (1989) Transporting the horse: from BC to AD. *Live Animal Trade and Transport Magazine* April, 39.

Dougherty, D.M. and Lewis, P. (1993) Generalization of a tactile stimulus in horses. *Journal of the Experimental Analysis of Behaviour* 59, 521–528.

Doyle, K.A. (1988) The horse in international commerce – regulatory considerations in an Australian perspective. *Journal of Equine Veterinary Science* 8 (3), 227–232.

Feh, C. and de Mazières, J. (1993) Grooming at a preferred site reduces heart rate in horses. *Animal Behaviour* 46, 1191–1194.

Fiske, J.C. and Potter, G.D. (1979) Discrimination reversal learning in yearling horses. *Journal of Animal Science* 49, 583–588.

Flannery, B. (1997) Relational discrimination learning in horses. *Applied Animal Behaviour Science* 54, 267–280.

Friend, T.H., Martin, M.T., Householder, D.D. and Bushong, D.M. (1998) Stress responses of horses during a long period of transport in a commercial truck. *Journal of the American Medical Association* 212 (6), 838–844.

Grandin, T., McGee, K. and Lanier, J.L. (1999) Prevalence of severe welfare problems in horses that arrive at slaughter plants. *Journal of the American Veterinary Medical Association* 214 (10), 1531–1533.

Haag, E.L., Rudman, R. and Houpt, K.A. (1980) Avoidance, maze learning and social dominance in ponies. *Journal of Animal Science* 50, 329–335.

Hanggi, E.B. (1999) Categorization learning in horses (*Equus caballus*). *Journal of Comparative Psychology* 113 (3), 243–252.

Hanggi, E.B. (2003) Discrimination learning based on relative size concepts in horses (*Equus caballus*). *Applied Animal Behaviour Science* 83 (3), 201–213.

Hall, C.A. and Cassaday, H.J. (2006) An investigation into the effects of floor colour on the behaviour of the horse. *Applied Animal Behaviour Science* 99, 301, 314.

Hall, C.A., Cassaday, H.J., Vincent, C.J. and Derrington, A.M. (2006) Cone excitation ratio correlate with colour discrimination performance in the horse (*Equus Caballus*). *Journal of Comparative Psychology* 120, 438–448.

Heird, J.C., Lennon, A.M. and Bell, R.W. (1981) Effects of early experience on the learning ability of yearling horses. *Journal of Animal Science* 53 (5), 1204–1209.

Heird, J.C., Whitaker, D.D., Bell, R.W., Ramsey, C.B. and Lokey, C.E. (1986) The effects of handling at different ages on the subsequent learning ability of 2-year-old horses. *Applied Animal Behaviour Science* 15, 15–25.

Henry, S., Hemery, D., Richard M.-A. and Hausberger, M. (2005) Human–mare relationships and behaviour of foals toward humans. *Applied Animal Behaviour Science* 93 (3/4), 341–362.

Houpt, K.A. (2005) *Domestic Animal Behaviour*, 4th edn. Blackwell Publishing, Ames, Iowa.

Houpt, K.A. and Houpt, T.R. (1988) Social and illumination preferences of mares. *Journal of Animal Science* 66, 2159–2164.

Jezierski, T., Jaworski, Z. and Górecka, A. (1999) Effects of handling on behaviour and heart rate in Konik horses: comparison of stable- and forest-reared youngstock. *Applied Animal Behaviour Science* 62, 1–11.

Kusunose, R. and Torikai, K. (1996) Behaviour of untethered horses during vehicle transport. *Journal of Equine Science* 7 (2), 21–26.

Lagerweij, E., Nelis, P.C., Weigant, V.M. and van Ree, J.M. (1984) The twitch in horses: a variant of acupuncture. *Science* 225 (4667), 1172–1174.

Lansade, L., Bertrand, M., Boivin, X. and Bouissou, M.-F. (2004) Effects of handling at weaning on manageability and reactivity of foals. *Applied Animal Behaviour Science* 87, 131–149.

Leadon, D., Frank, C. and Backhouse, W. (1989) A preliminary report on studies on equine transit stress. *Journal of Equine Veterinary Science* 9 (4), 200–202.

Lee, J., Houpt, K. and Doherty, O. (2001) A survey of trailering problems in horses. *Journal of Equine Veterinary Science* 21 (5), 237–241.

Lindberg, A.C., Kelland, A. and Nicol, C.J. (1999) Effects of observational learning on acquisition of an operant response in horses. *Applied Animal Behaviour Science* 61, 187–199.

Mackenzie, S.A. (1994) *Fundamentals of Free Lungeing: an Introduction to Tackless Training*. Half Halt Press, Boonsboro, Maryland, 97 pp.

Mader, D.R. and Price, E.O. (1980) Discrimination learning in horses: effects of breed, age and social dominance. *Journal of Animal Science* 50, 962–965.

Madigan, J. (1993) Evaluation of a new sling support device for horses. *Journal of Equine Veterinary Science* 13 (5), 260–261.

Maier, S.F. and Seligman, M.E.P. (1976) Learned helplessness: theory and evidence. *Journal of Experimental Psychology* (General) 105, 3–46.

Mal, M.E. and McCall, C.A. (1996) The influence of handling during different ages on a halter training test in foals. *Applied Animal Behaviour Science* 50, 115–120.

Mal, M.E., Friend, T.H., Lay, D.C., Vogelsang, S.G. and Jenkins, O.C. (1991) Physiological responses of mares to short-term confinement and social isolation. *Equine Veterinary Science* 11 (2), 96–102.

Mal, M.E., McCall, C.A., Cummins, K.A. and Newland, M.C. (1994) Influence of preweaning handling methods on post-weaning learning ability and manageability of foals. *Applied Animal Behaviour Science* 40, 187–195.

Mars, L.A., Kiesling, H.E., Ross, T.T., Armstrong, J.B. and Murray, L. (1992) Water acceptance and intake in horses under shipping stress. *Journal of Equine Veterinary Science* 12 (1), 17–20.

McBane, S. (1987) *Behaviour Problems in Horses*. David & Charles, 1 North Pomfret, Vermont, 304 pp.

McCall, C.A. (1990) A review of learning behaviour in horses and its application in horse training. *Journal of Animal Science* 68, 75–81.

McCann, J.S., Heird, J.C., Bell, R.W. and Lutherer, L.O. (1988a) Normal and more highly reactive horses. I. Heart rate, respiration rate and behavioural observations. *Applied Animal Behaviour Science* 19, 201–214.

McCann, J.S., Bell, R.W. and Lutherer, L.O. (1988b) Normal and more highly reactive horses. II. The effect of handling and reserpine on the cardiac response to stimuli. *Applied Animal Behaviour Science* 19, 215–226.

McClintock, S.A., Hutchins, D.R., Laing, E.A. and Brownlow, M.A. (1986) Pulmonary changes associated with flotation techniques in the treatment of skeletal injuries in the horse. *Equine Veterinary Journal* 18 (6), 462–466.

McLean, A.N. (2004) Short-term spatial memory in the domestic horse. *Applied Animal Behaviour Science* 85 (1/2), 93–105.

McLean, A. and McGreevy, P. (2004) Training. In: McGreevy, P. (ed.) *Equine Behaviour: a Guide for Veterinarians and Equine Scientists*. W.B. Saunders, Philadelphia, Pennsylvania and Elsevier, New York, pp. 291–311.

Miguarese, N. (1999) A smoother road to Sydney. *Horse Show* 62 (4), 23.

Miller, R.M. (1991) *Imprint Training of the Newborn Foal*. Western Horsemen, Colorado Springs, Colorado, 143 pp.

Miller, R.W. (1975) *Western Horse Behaviour and Training*. Dophin Books of Doubleday & Company, Garden City, New York, 305 pp.

Nestved, A. (1996) Evaluation of an immunostimulant in preventing shipping stress-related respiratory disease. *Journal of Equine Veterinary Science* 16 (2), 78–82.

Nicol, C.J. (2002) Equine learning: progress and suggestions for future research. *Applied Animal Behaviour Science* 78, 193–208.

Racklyeft, D.J. and Love, D.N. (1990) Influence of head posture on the respiratory tract of healthy horses. *Australian Veterinary Journal* 67 (11), 402–405.

Robinson, N.E. (1992) *Current Therapy in Equine Medicine*, 3rd edn. W.B. Saunders, Philadelphia, Pennsylvania, 847 pp.

Russoniello, C., Racis, S.P., Ralston, S.L. and Malinowski, K. (1991) Effects of show-jumping performance stress on hematological parameters and cell-mediated immunity in horses. *Proceedings of 12th Equine Nutrition and Physiology Symposium*, University of Calgary, Calgary, Canada, pp. 145–147.

Sappington, B.F. and Goldman, L. (1994) Discrimination learning and concept formation in the Arabian horse. *Journal of Animal Science* 72, 3080–3087.

Sappington, B.K.F., McCall, C.A., Coleman, D.A., Kuhlers, D.L. and Lishak, R.S. (1997) A preliminary study of the relationship between discrimination reversal learning and performance tasks in yearling and 2-year-old horses. *Applied Animal Behaviour Science* 53, 157–166.

Saslow, C.A. (1999) Factors affecting stimulus visibility for horses. *Applied Animal Behaviour Science* 61, 273–284.

Shanahan, S. (2003) Trailer loading stress in horses: behavioral and physiological effects of nonaversive training (TTEAM). *Journal of Applied Animal Welfare Science* 6 (4), 263–274.

Simpson, B.S. (2002) Neonatal foal handling. *Applied Animal Behaviour Science* 78, 303–317.

Smith, B.L., Jones, J.H., Carlson, G.P. and Pascoe, J.R. (1994) Body position and direction preferences in horses during road transport. *Equine Veterinary Journal* 26 (5), 374–377.

Smith, B.L., Miles, J.A., Jones, J.H. and Willits, N.H. (1996a) Influence of suspension, tires, and shock absorbers on vibration in a two-horse trailer. *Transactions of the American Society of Agricultural Engineers* 39 (3), 1083–1092.

Smith, B.L., Jones, J.H., Hornof, W.J., Miles, J.A., Longworth, K.E. and Willits, N.H. (1996b) Effects of road transport on indices of stress in horses. *Equine Veterinary Journal* 28 (6), 446–454.

Spier, S.J., Pusterla, J.B., Villarroel, A. and Pusterla, N. (2004) Outcome of tactile conditioning of neonates, or 'imprint training' on selected handling measures in foals. *The Veterinary Journal* 168, 252–258.

Stull, C.L. (1998) Health and welfare parameters of horses commercially transported to slaughter. *Journal of Animal Science* 76 (Suppl. 1), 88.

Teeter, S. (1999) Valerie Kanavy, new world endurance champion. *Horse Show* 62 (2), 32–35.

Tellington-Jones, L. and Bruns, U. (1985) *An Introduction to the Tellington-Jones Equine Awareness Method.* Breakthrough Publications, Millwood, New York, 180 pp.

Traub-Dargatz, J.L., McKinnon, A.O., Bruyninckx, W.J., Thrall, M.A., Jones, R.L. and Blancquaert, A.-M.B. (1988) Effect of transportation stress on bronchoalveolar lavage fluid analysis in female horses. *American Journal of Veterinary Research* 49 (7), 1026–1029.

Tyler, S.J. (1972) The behaviour and social organization of the New Forest ponies. *Animal Behaviour Monographs* 5, 85–196.

Visser, E.K., van Reenen, C.G., Schilder, M.B.H., Barneveld, A. and Blokhuis, H.J. (2003) Learning performances in young horses using two different learning tests. *Applied Animal Behaviour Science* 80, 311–326.

Waran, N.K., Robertson, V., Cuddeford, D., Kokoszko, A. and Marlin, D.J. (1996) Effects of transporting horses facing either forwards or backwards on their behaviour and heart rate. *Veterinary Record* 139, 7–11.

Waring, G.H. (1983) *Horse Behaviour: the Behavioural Traits and Adaptations of Domestic and Wild Horses, Including Ponies.* Noyes Publications, Park Ridge, New Jersey, 292 pp.

White, A., Reves, A., Godoy, A. and Martinez, R. (1991) Effects of transport and racing on ionic changes in thoroughbred race-horses. *Comparative Biochemistry and Physiology, A. Comparative Physiology* 99, 343–346.

Williams, J.L., Friend, T.H., Toscano, M.J., Collins, M.N., Sisto-Burt, A. and Nevill, C.H. (2002) The effects of early training sessions on the reactions of foals at 1, 2, and 3 months of age. *Applied Animal Behaviour Science* 77, 105–114.

Williams, J.L., Friend, T.H., Nevill, C.H. and Archer, G. (2004) The efficacy of a secondary reinforcer (clicker) during acquisition and extinction of an operant task in horses. *Applied Animal Behaviour Science* 88, 331–341.

Wolff, A. and Hausberger, M. (1996) Learning and memorisation of two different tasks in horses: the effects of age, sex and sire. *Applied Animal Behaviour Science* 45, 137–143.

Woodhouse, B. (1984) *Barbara's World Of Horses And Ponies.* Summit Books, New York, 127 pp.

Wright, M. (1973) *The Jeffery Method of Horse Handling.* Griffin Press, Netley, South Australia, 92 pp.

17 Deer Handling and Transport

L.R. Matthews

Animal Behaviour and Welfare Research Centre, Ruakura Agricultural Centre, Hamilton, New Zealand

Introduction

The relatively recent expansion of deer farming around the world has necessitated the development of appropriate handling facilities and practices. Deer belonging to the genus *Cervus* are the most widely distributed and numerously farmed species and include red deer (*C. elaphus*) and wapiti (elk) (*C. elaphus nelsoni*). The next most common farmed species is the fallow (*Dama dama*). Smaller numbers of other species are farmed, including rusa (*C. timorensis*), sika (*C. nipon*), chital (*Axis axis*) and white-tail (*Odocoileus virginianus*).

The importance of accommodating the unique behavioural characteristics of deer into the design of safe and efficient handling facilities is well recognized (Kilgour and Dalton, 1984). In the absence of carefully controlled scientific studies, facility design has been based largely on general farmer knowledge of deer behaviour and trial and error evaluation of facilities. Nevertheless, this approach has yielded facilities that work well. This is a continually evolving process and some of the designs presented here have only just been developed.

The range of systems for handling are described in this chapter. The effectiveness of these systems and their use have been systematically assessed with reference to fundamental aspects of deer behaviour. Although there are variations in the behaviours of deer belonging to different species, the main one influencing handling relates to the degree of flightiness. The handling facilities and procedures for the less flighty species (red, wapiti and rusa) tend to be somewhat different from the more flighty ones (fallow, chital and white-tail) (English, 1992; Haigh, 1992, 1999; Woodford and Dunning, 1992). Any important distinctions are specifically mentioned.

Basic Behaviour Patterns of Deer

Taming, sensory capacities, physical agility, social organization and learning ability are among those aspects of the basic behavioural responses of animals that need to be considered in the design of handling facilities. Apart from social organization, there are relatively few scientific studies of these responses and their influence on animal handling. Available information is derived from observational studies of animals in the wild or during farming operations and published reports (where available).

Sensory and physical capacities

The position of the eyes on the head is similar to cattle and sheep and it can be assumed that the visual abilities of deer would be similar to these animals (Prince, 1977). Thus, deer are likely to have a wide visual field (about 300° with a blind spot to the rear and at ground level to the front of the animal when the head is raised (Hutson, 1985a) and good depth perception in a small area of binocular vision some short distance in front of the head). Inability of several deer species to detect bot flies hovering below the nose supports the notion that there is a blind spot in this area (Anderson, 1975).

Like other ruminants, deer detect moving objects readily but do not respond well to static objects (Cadman, 1966; McNally, 1977). Cattle and sheep have dichromatic colour vision but the extent to which deer can discriminate colours is not known. Deer are active at night and during the change of light, so it must be assumed that they have excellent vision under low light conditions. See Chapters 4 and 16 for more information on vision.

The avoidance by deer of humans approaching downwind and the use of gland secretions to mark trees and trails (McNally, 1977) indicate that deer have an acute sense of smell. Their sense of hearing is also well developed. Recorded calls of stags played to groups of hinds advances the breeding season by about 6 days (McComb, 1987). Observations by farmers that deer respond to the 'silent' dog whistle indicate that the effective hearing range extends to about 20 kHz or more.

Physical agilities

In contrast with the traditional farm animal species, deer are very agile. Most species have little difficulty in clearing barriers of 2–2.5 m in height from a standing position and can accelerate almost instantaneously from 0 to 50 km/h (Drew and Kelly, 1975). When presented with an obstacle, the natural tendency for deer is to lower the head and go under or push through it rather than go over the top (Clift et al., 1985). Within their natural home ranges, deer prefer to move along well-defined trails that are used by successive generations (Cadman, 1966).

Social behaviour

In the wild, deer species that are commonly farmed are group living, with highly organized social structures. Females associate with their dams and remain in matrilineal groupings throughout their lives (Clutton-Brock et al., 1982). Males disperse from the females and live in bachelor herds throughout the year except during the breeding season (rut), when individual stags move into the areas occupied by the females. In favourable habitats groups comprise up to 100 or more animals (Staines, 1974).

Dominance hierarchies are a feature of both male and female groupings (Clutton-Brock et al., 1982). A wide range of agonistic behaviours are used in the establishment and maintenance of these hierarchies. With males, antlers are the primary means of offence and defence. Threat gestures include lowering the head, directing the antlers towards the opponent and lateral body positioning. Animals with the largest body size and antlers tend to have the highest social ranking. During the annual velvet antler growing season and prior to antler hardening, stags use a range of other threat behaviours. These include head-high threats, which precede strikes by the forefeet or rearing on the hind feet and boxing with the forefeet, kicking with the rear feet, biting or biting threats where the head is tilted slightly, upper lips are raised and a hissing sound is made or the tongue may be protruding and accompanied by teeth-grinding (Lincoln et al., 1970; Bartos, 1985). Appeasement is indicated by an outstretched neck posture, turning the head away or movement away from the aggressor. Agonistic interactions between females are similar to those seen in males during the velvet-growing season (Haigh and Hudson, 1993).

The intensity of aggressive interactions between males is strongly influenced by hormonal status. During the late winter and spring, when the old antlers are cast and a new set is growing, testosterone levels are low. The stags are least aggressive during this period. The shedding of velvet and hardening of the antler corresponds to rising testosterone levels which remain elevated during the rut (Darling, 1937). Aggressiveness is highest during periods of high testosterone levels, especially where animals rejoin in groups after the breeding season.

Females move away from the matriarchal groups to give birth (Darling, 1937). The behaviour of the newborn in the first few weeks of life forms the basis of the group lifestyle and some of the responses to stresses in later life. After suckling, the newborn moves away from its mother and lies down in available shelter (Darling, 1937). This behaviour is characteristic of lying-out type species (Lent, 1974) and persists for the first 2–3 weeks of life. For the first few days of this period the calf freezes in response to disturbances (e.g. approaching human) (Kelly and Drew, 1976). Thereafter, a strong flight response to threatening stimuli is shown. At about 3 weeks of age the calf begins to follow its mother. It is at this time that the mother returns to the matriarchal group with the young and the foundations for leader–follower and herding behaviour typical of adult deer are established.

Flight, domestication, taming and learning

Wild deer are well known for their large flight distance and strong flight response. Wild-caught animals and deer bred in captivity habituate rapidly to human presence and the flight distance reduces to 30–50 m or less (Clutton-Brock and Guinness, 1975; Blaxter et al., 1988). Flight distances are less for animals confined in yards, when approached in familiar vehicles rather than on foot and where handlers are associated with the feeding of supplements. The flight distance reduces to zero for animals reared by hand from birth. Hand-reared male deer should not be kept as they become extremely aggressive toward humans during the breeding season (Gilbert, 1974). Hand-reared roe deer are especially aggressive (Hemmer, 1988). Animals with a zero flight distance cannot be moved readily other than by attracting with food.

The flight responses of a herd are strongly influenced by the behaviour of the lead animals. When disturbed from a distance, deer orient their heads toward the intruder and bunch together (Humphries et al., 1990; Bullock et al., 1993). Upon closer approach by the intruder, the lead animals (or most flighty ones) begin to move away from the source of the disturbance and the remainder of the group follow. If the intruder remains stationary, the leaders usually circle from the front of the group, around the outside facing back toward the intruder and then back up the centre of the mob to the front again (L. Matthews, unpublished observations). If the intruder approaches quickly the group will scatter in all directions. The flight distance for a group is reduced when extremely flighty individuals are removed.

There have been no formal studies of changes in the flightiness of deer over the generations as they have become increasingly domesticated (Kelly et al., 1984), although techniques for measuring temperament and stress are being developed (Matthews et al., 1994; Pollard et al., 1994b; Carragher et al., 1997). Nevertheless, in New Zealand it has become apparent that farmed deer are much less reactive to handling now than when deer farming began 40 years ago using wild-caught stock. Both genetic selection for less flighty animals and improved methods of training deer to handling have no doubt contributed to this effect. Regular gentle contact with the stock and hand-feeding of animals (particularly after weaning) seem to be particularly important factors in the taming of deer. Hand-reared animals have been trained to come to a call over distances of 1 km. Wild animals quickly learn to avoid areas which are frequently subject to disturbance by helicopters or humans. Farmed deer learned in one trial to avoid areas where they had been handled aversively (Pollard et al., 1994a). In addition, deer quickly learn to recognize a regular handler but react with flight to unfamiliar handlers (Bull, 1996).

Behaviour Relevant to Facility Design

Farm layout

The layout of the farm is centred around the need to confine the herd and control the feeding, breeding and handling of various subgroups of animals. A farm typically consists of a series of fields (paddocks) linked to a handling facility via a central laneway.

The agility of deer and the need to control their seasonal activities have had a major influence on the design of farm fences and laneways. Perimeter and race fences 2 m in height will discourage most animals from attempting to leap over, as deer seemingly are unable to see the top wire. Under pressure, animals have been known to jump over 2 m fences. Outside the breeding season and when food supply is not restricted, internal fences of between 1.2 and 1.5 m in height will normally contain animals.

Full-height fences are best constructed from 13-wire high-tensile netting with vertical stay wires at 150 mm or 300 mm spacings. Ideally, the fine wires should be about 100 mm apart near the ground, increasing to about 180 mm at the top. The closer spacing at the bottom prevents escapes and entanglement of smaller animals. An even finer mesh size is required to contain newborns.

Electric fencing is being used increasingly on deer farms. Internal fences comprising about four wires to a height of 1.5 m provide a convenient movable structure that will control most animals. Electrified outriggers about 600 mm above the ground will reduce fence-pacing and agonistic interactions between stags held in adjacent paddocks. Naïve animals should be introduced to electric fences in the presence of trained deer. Visual barriers (e.g. trees) along fence lines may also reduce fence-pacing or aggression between animals in neighbouring enclosures.

The ease of movement of deer into and out of paddocks is influenced by the position of the gateways. Deer tend to move more readily uphill than down, so gateways are best located at the tops of rises. On flat areas gates work well if sited near the corners. Short-wing fences leading to the gateways assist in funnelling deer from larger fields. Gate widths should be a minimum of about 3.5 m wide to avoid undue constriction of the flow of the herd.

Laneways linking the paddocks and the handling complex assist the movement of stock around the farm. The design of these raceways has been influenced by several behavioural characteristics of deer. Their natural tendency to bunch (Humphries et al., 1990) and move at speed is facilitated in raceways that are relatively wide, 5 m being a minimum. Wider races (up to 20 m) are more suitable for large herds and can also be used for drafting, since the handler can move past deer without encroaching on the flight zone. It has been suggested that deer move more readily along laneways incorporating curves (Haigh and Hudson, 1993). This has not been scientifically evaluated, and many farmers find that deer move just as readily along straight laneways.

It is important that straight sections do not lead directly into yards, as this creates an apparent dead-end, causing deer to baulk. Deer flow readily into yards if they are turned through a curve on the final approach. This effect is achieved by setting the yards off to one side of the race. Deer tend to 'cut the corners' when moving into and out of yards. Impacts with the walls at corners (which cause bruising) can be reduced by avoiding the use of sharp turns or smoothing the corners.

On occasions, deer approach yards on the run. This, coupled with their poor vision of stationary objects, increases the risk of collision with fences. The visibility of wire fences can be increased by affixing vertical wooden battens some 20–40 m out from the yards and solid boarding (or hessian or plastic mesh) over the final 20 m (Fig. 17.1). It is important to have a gradual increase in visibility of the fences at the yard entrance to avoid creating the impression of a dead-end.

Yard complex

Deer yard complexes serve four main functions – to hold, draft (sort), close handle and load out (or receive) deer. Close handling requires varying degrees of restraint depending on the procedure, the species (or breed) of deer and the individual animal. Low levels of restraint for practices such as oral administration (drenching) of anthelmintics or ear tagging can be applied manually in small working pens. Higher levels of restraint, which are required for artificial insemination or removal of antlers, can be applied in a cradle or purpose-built deer handler.

Layout and construction

There are innumerable variations in facility layout and construction that permit basic farming operations to be carried out easily. In the past, yard designs were based on a completely enclosed complex containing a central drafting area with several holding pens off it. Near the drafting pen were smaller working pens (in which animals can be manually restrained), a restrainer, a weighing platform and a race leading to a loading ramp (Yerex and Spiers, 1990). Nowadays, farmed deer are generally less flighty and facilities and handling procedures a little more typical of those used with cattle are being utilized. Thus, the indoor facilities for red deer are typically very

Fig. 17.1.   Laneway fencing on approach to yards.

well lit and well ventilated, and groups of animals are typically handled by one person working in the pens at ground level. Other changes in modern facilities include the greater provision of outdoor holding areas and the elimination of the central drafting pen. The layout of the indoor area is now more comparable to cattle facilities, with small groups of deer brought into larger holding areas (indoors or outdoors) and then drafted off into a number of smaller pens within the facility (Fig. 17.2). Drafting is undertaken in pens by cutting animals out by hand, as is common with cattle.

The design may also incorporate pens which contain one or two centrally mounted swinging gates (Fig. 17.2). In these rooms pen size can be adjusted readily to assist with restraint and drafting, and the gates are useful for pushing animals toward load-out ramps or other facilities such as restrainers or weighing scales.

The behaviour of deer is critical in relation to the design and use of these yarding facilities. In contrast with other livestock, the need to reduce flightiness has been at least as influential as the need for containment in the design and construction of deer yards. Walls are built to a

height of 2.25 m (2.6 m for fallow or 3 m for wapiti) to deter leaping, and are smooth-sided with no sharp corners, to reduce injury and bruising in these fast-reacting animals. Secure footing is provided by concrete or other compacted, free-draining flooring material (e.g. sand).

In the past, two common design features were used to assist in reducing flightiness. Yard walls were solidly or completely close-boarded on the assumption that deer unable to see into neighbouring pens would be less reactive to handling or other disturbances in those pens. However, recent informal observations suggest that deer remain more settled when there is a clear view of other nearby animals, approaching handlers or activity that produces unfamiliar noises (Pearse, 1992). Thus, wall designs are open-boarded (75 mm spaces for fallow and red deer, 150 mm for wapiti) or mesh (50 mm × 50 mm) above a height of about 1.2 m (for red and wapiti) or 1 m (for fallow) (Fletcher, 1991; Yerex, 1991). Walls can even be open-boarded right down to ground level (provided the gaps between the boards are no more than about 50 mm up to a height of about 1.2 m). Note, it is preferable to place the boards horizontally.

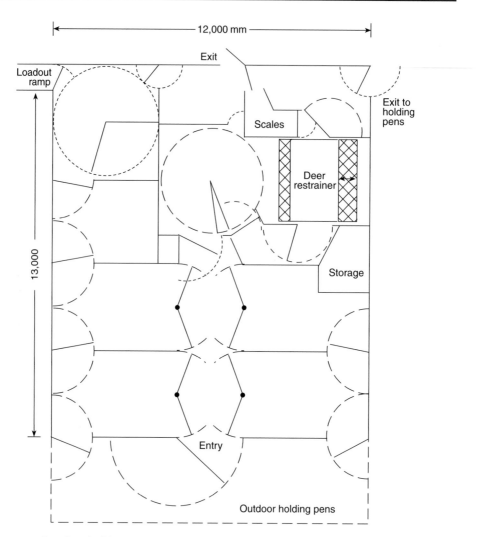

Fig. 17.2. Modern deer shed layout, incorporating large, small and adjustable-sized pens, working areas with one or two circular-swinging doors and a hydraulic deer restrainer with swinging push-up doors at its entrance.

This allows a handler, in an emergency, to climb the walls easily and prevents flickering of light between the boards as animals move (which can occur with vertically oriented boards) and disruption to the flow of deer. In close or high-pressure working areas, the walls should be solid-boarded from the floor to 2.5 m (or 3 m for wapiti).

Another common procedure that has been used in the past to reduce flightiness is to reduce the level of illumination in the pens (Kelly *et al.*, 1984). Control of light levels is usually achieved by constructing a roof over the yard complex and fitting artificial lights with dimmer controls. Although there have been no scientific studies of the effect of illumination on flightiness, practical experience indicates that fractious animals and flighty species such as fallow or white-tail are easier to handle under low light conditions, but more tractable deer can be handled without difficulty in well-lit facilities.

There is some scientific evidence to suggest that red deer may be less disturbed if handled or penned under low-light conditions (Pollard and

Littlejohn, 1994). High light levels have the advantage of allowing more easy inspection of animals for husbandry purposes. In addition, the level of arousal or aggressiveness of deer can be more readily detected should evasive actions be required. Flightiness is also influenced by the space available in a pen. Low ceilings (2 m for red deer and 2.5 m for wapiti) discourage leaping and boxing with the forelimbs and small group sizes reduce flightiness. Smaller space allowances may provide for improved welfare when animals are kept indoors on a long-term basis (Hanlon *et al.*, 1994), although on a short-term basis pen size is not an issue in terms of animal well-being (Pollard and Littlejohn, 1996).

Pens and races

The behaviour of deer in groups, especially when disturbed or handled, has a major influence on the design of pens and races. The ideal number and size of pens varies with the number of animals in the herd. Outdoor pens suitable for groups of up to 100 animals should provide 2–6 m^2 per animal depending on size and breed. Ideally, larger numbers should not be herded together at one time in a single pen. Large groups are usually broken down into smaller groups in outside holding pens soon after yarding. For ease of handling and to reduce stress, these groups should comprise no more than about 25 animals. Large stags and animals housed for longer periods or overnight require a minimum of about 1 m^2 per animal.

In the working area, practical experience has shown that deer are less likely to trample one another and are easiest to move amongst when group size is limited to five or six animals. To facilitate close handling and manual restraint, a space allowance of up to 0.5 m^2/100 kg animal is ideal. Thus, working pens typically measure 1.5 m wide × 1.5 m or 2.0 m long. The walls of the working pen should be solid-sided from floor to ceiling to reduce the possibility of injuries. A variation of this design that allows for rapid handling and the creation of variable-sized pens is based on a large pen which can be divided into segments by one or two swing gates mounted on a central pole (Fig. 17.2).

Traditionally, raceways have not been used as working areas for deer because not all animals remain settled long enough for the whole group to be processed. New systems that take advantage of several unique behavioural characteristics of deer have been developed and are eminently suitable for carrying out routine operations such as drenching, vaccinating and tagging in facilities that resemble those used for cattle.

One such system shown in Fig. 17.3 utilizes a U-shaped raceway (N. Cudby and L. Cooney, unpublished data). The wings of the U are 3–4 m long and the base is about 2 m long (Fig. 17.3a). Deer enter the 700 mm-wide lane through an offset race linked with the holding pens. The operator works the deer from the inside of the U. The inside race wall (Fig. 17.3b) is solid-sided to 1 m (or 1.2 m if the race is not on the same level as the operator's floor level) and above that it is curtained to 2 m. If the side wall is 1.2 m high, a catwalk needs to be positioned about 200 mm above floor level. The curtain is positioned to hang to the outside of the catwalk and is closed while animals are being loaded into the race; the operator accesses the deer through slits in the curtain and works on the inside of the curtain. A ceiling consisting of pipes, open boards or mesh is placed at 2 m height to discourage jumping. The outside wall is solid to 2 m or more. Weighing and drafting of animals in the U-race can be achieved from strategically placed weigh-scales and drafting gates, respectively. About two 100 kg animals/m of race can be handled in the U-shaped race at one time. Sliding doors located at intervals along the race can be used to facilitate animal control and drafting. Ideally, such doors should incorporate a see-through section about 1.2 m from the floor to allow animals to see each other (thereby reducing the likelihood that animals will turn around in the race). One way to position the U-race inside a deer facility is shown in Fig. 17.3c.

This type of race can also be used with wapiti, in which case the curtains can be replaced with pipe railings (running horizontally) and the ceiling may need to be raised a little.

A similar design functions well with fallow deer (Cash, 1987). In this case, the race is straight, with the walls and ceiling forming a tunnel, and light at one end is used to attract animals into the enclosure. The fallow tunnel-race is about 900 mm high × 310 mm wide.

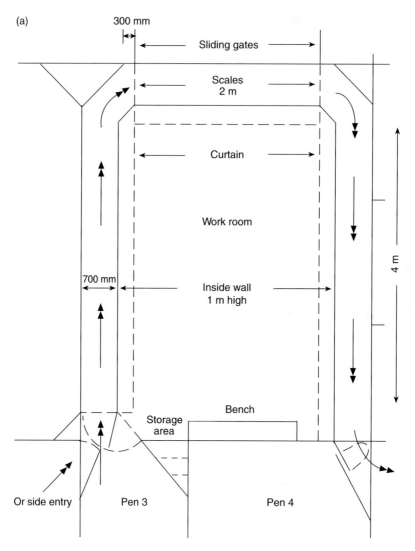

Fig. 17.3. (a) A modern handling race for use with groups of up to 25 red deer. Additional sliding gates can be positioned along the race to facilitate control of the animals. Drafting can be carried out from the race by placing additional gates in the outside wall leading to neighbouring pens. (b) The cross-section and dimensions of the race. Note that the curtain hangs 300 mm out from the low side wall, thereby allowing the operator easy movement and access when working the deer. If the raceway is not set below the shed floor (as shown) then the shorter side wall needs to be 1.2 m high. (c) The U-race is shown in relation to holding pens inside a deer handling facility. (*Continued.*)

Operator access to the animals is via a 160 mm gap at the top of the inside wall or in the ceiling. A facility for handling white-tail is described by Haigh (1995).

Races are used in several other parts of the yarding complex. Their design varies depending on whether they are used for moving groups of animals (e.g. from holding to drafting pens,

work pens or load-out ramps) or for moving single animals (e.g. on to weighing platforms or crushes). Deer move better as a group and this can be facilitated by constructing races at least 1.5 m wide so that animals can move two or three abreast (Grigor *et al.*, 1997b). Wider races should be used with larger mobs. Races for moving single animals should be 600–700 mm

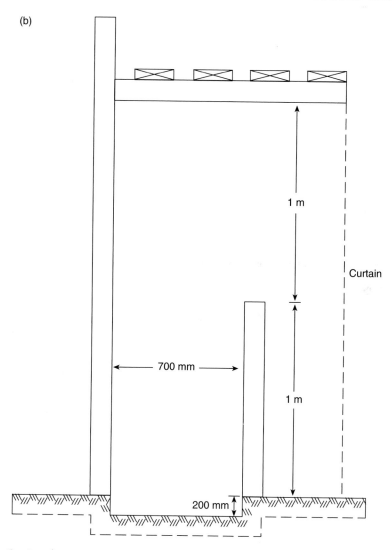

(b)

1 m

Curtain

700 mm

1 m

200 mm

Fig. 17.3. *Continued.*

wide to prevent animals turning around easily. Wider races may be required for stags with a large antler spread.

As has been shown with sheep (Hutson and Hitchcock, 1978), it seems that deer move better in races with solid sides at least part-way up the walls (Lee, 1992). These would limit distractions and also reduce the potential for trapping legs and causing injury. Entry to races can be encouraged by the use of forcing gates which swing around behind the animals. Whereas the need for such gates may indicate design faults in a red deer handling system, they are an essential component of safe systems for handling wapiti.

In most respects the general principles of facility design are similar for red deer and wapiti (apart from obvious differences in dimensions of enclosures and robustness of construction). Notwithstanding the greater size and strength of wapiti, the tendency for these animals to move less freely than red deer when in close proximity to handlers has led to the development of wapiti-specific pen and race features that facilitate animal flow and human safety. The most advanced designs (Fig. 17.4) are comprised of a

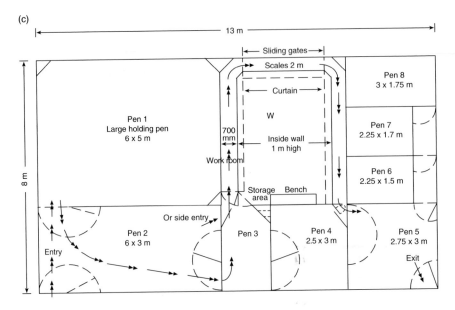

Fig. 17.3. *Continued.*

series of enclosures that function both as pens and raceways (Thorliefson *et al.*, 1997). These facilities are particularly suitable for less domesticated deer. As with red deer, the smallest enclosure units accommodate two to five animals (2.5–7 m²). These units are so designed that one of the side 'walls' is comprised of two swinging doors, and sliding doors are positioned at either end. The doors can be opened up to construct much larger pens. Reluctant animals can be moved with sliding or swing doors that can be pushed up behind animals. Within the yarding complex, gate widths of 1.2 m or more prevent animals from jamming during movement.

Walkways along the top of the pen walls provide quick access from one part of the facility to another. However, it is not desirable to handle animals from such heights as they often become too flighty and difficult to control. In raceways, animals are best worked from ground level.

Practical experience and scientific evidence (Grigor *et al.*, 1997b) shows that animals move more readily into races that incorporate curves, or where animals have a clear view of an exit. These features are particularly important in encouraging the movement of lead or single animals, especially where the race terminates in

a dead end. Grigor *et al.* (1997b) showed that deer, once in the raceway, move no more rapidly along curved races than straight laneways. However, practical experience suggests that movement through races or several pens adjacent to one another is achieved most readily if baffles, lanes or doors are arranged so as to create a zigzag pathway. Ideally, long, straight races should be avoided. Where this is impractical, straight sections should be kept as short as possible and incorporate a corner or curve as close as possible to areas that terminate in a dead end. There has been no systematic study of the effect of degree of curvature on ease of movement, but anecdotal evidence suggests that corners of up to 90° and higher are effective. One report maintains that moving deer around a 135° corner at the intersection of the race and restrainer assists animal flow (Goble, 1991). Thus, curves should immediately precede entry to enclosed yards and slaughter plants, restrainers, weighing platforms, load-out ramps and transport vehicles.

There have been no scientific studies of the effect of floor slope on deer movement. Practical experience during mustering and transport indicates that deer move uphill readily but are hesitant on downward slopes of 20° or more. Given appropriate facilities, deer unload at

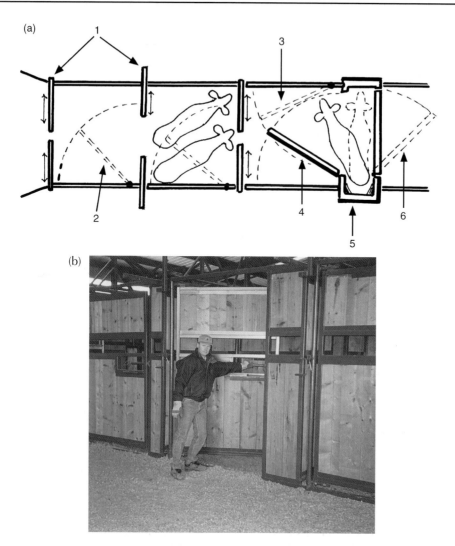

Fig. 17.4. (a) Elk (wapiti) handling facility. On both wapiti and red deer farms in Canada, producers have developed this design which eliminates the need for a single-file race. Wapiti stay calmer because two animals stand side by side as they move through the system. All fences, gates and walls in the facility are solid. 1. Biparting sliding gates separate the compartments; 2. A pusher gate is used to move the animals to the next compartment; 3. Pusher gate to urge the animal into the padded stall; 4. Side entrance gate on the padded stall; 5. Padded stall which holds the animal for palpation, injections and artificial insemination; 6. Exit gate from stall. The padded stall shown in this diagram fits snugly but does not squeeze the animal and has numerous small doors on both the side and the rear for access to the animal. Velveting is carried out in a conventional deer restrainer which is shown in Fig. 17.5. (b) Dan Sych in Alberta, Canada, demonstrates the operation of a pusher gate. All handling is carried out from ground level.

speed from transport vehicles. Ramps at least as wide as the exit door on transporters (typically 1.2 m) and preferably wider assist unloading. Swinging doors situated along the ramp can be used to push from behind animals reluctant to enter the transporter.

Restraining devices

Restraining devices are an important component in handling complexes where a high degree of animal control is required. As with most other aspects of the deer facilities, there are a great many different restrainer designs or procedures and they are being refined continually (Haigh, 1999). Restrainer designs vary according to degree of restraint required, size of animal and behavioural characteristics of individual animals, breeds or species.

Flighty species (e.g. fallow, chital) or fractious animals invariably require high levels of restraint for most procedures requiring close handling. For less flighty species (and smaller animals) many simple operations, such as drenching, can be carried out with low levels of restraint, e.g. manually or as provided between the animals themselves when small groups (four to six deer are contained within small working pens (2–3 m²)). A second handler located near one wall can be helpful in such situations to prevent deer circling. Deer which do not stand readily when immediately next to the handler will frequently allow human contact if a second animal is manoeuvred between the handler and target animal. Animals can be discouraged from backing up by placing a hand at the rear of the deer.

Other procedures, such as shaving and injecting for disease testing, can be performed most easily under moderate levels of restraint (e.g. in the U-shaped race as described earlier) or behind a swinging door offset from a solid wall. Reduced light levels, low ceilings and working from a slightly elevated position (e.g. from a 0.5–1 m-high catwalk) can reduce fractiousness during handling. Alternatively, larger animals or breeds (wapiti) can be restrained in narrow stalls. In some designs the stall is created by closing sliding doors fore and aft of the animal standing in a race and by narrowing a movable section of one of the race walls.

In other systems the stall is located at the end of, and perpendicular to, the raceway (Fig. 17.4). The nearside 'wall' of the stall doubles as a swinging door which is opened to allow the animal to enter. In some designs the walls are padded and, when closed, are only 45 cm apart. This provides slight but firm pressure on the animal and prevents turning. In other designs, the propensity for deer to turn when isolated in narrow compartments is accommodated by leaving a much wider gap between the walls. Openings in the walls, either between pipes or in the form of removable panels, provide access ways to the restrained animal.

A high degree of restraint is required for some procedures, e.g. removal of antlers, artificial insemination, and this can be provided by drop-floor cradles and hydraulic handlers which use both mechanical and psychological aids to inhibit or prevent flight or jumping responses. Prior hand-feeding of animals in the restrainer can reduce fractiousness during handling (Grandin et al., 1995). Secure mechanical restraint is provided by application of pressure alone or a combination of pressure together with removal of the animals' footing.

The most sophisticated, successful and versatile devices are called hydraulic workrooms or hydraulic deer handlers (Hutching, 1993), which can apply pressure, reduce footing and place a barrier over the animal's back. Typically, in its open state, the hydraulic handler consists of two well-padded walls about 1.4 m apart and 2–3 m long (see Fig. 17.5). One or more deer are walked into the pen-like space between the walls and then one of the walls is moved hydraulically toward the other to enclose the animal(s). The handler operates the device and works the deer from a raised platform. In earlier versions this platform was affixed to the moving wall, but the latest versions allow much higher acceleration of the moving wall and the operator works from the non-moving side. The walls can be adjusted up and down, which allows the opportunity to partially lift the animal off the floor and thereby reduce its flightiness.

In addition, the vertical angle of the walls can be adjusted to accommodate different body shapes and, on later models, the moving wall can be angled in the horizontal plane to form a wedge-shaped enclosure (which further restricts movement). On many models the walls incorporate removable panels which allow additional access to the animal. Several animals (with similar-sized bodies) can be accommodated one behind the other in the larger work room restrainers at one time, thereby increasing the efficiency of handling and reducing stress on the animals (since the animals are not isolated visually from one another). Further efficiencies are provided by the dual-roomed type of hydraulic restrainer, which comprises two outer fixed walls and a central moving wall common to both rooms. While deer are being restrained and worked on one side, the other room is opened and loaded (Hutching, 1993).

(a)

Fig. 17.5. (a) A hydraulically operated deer handler shown in its open position; this creates a small padded room about 1.4 × 2.6 m long which is suitable for holding one or two deer. The curtains are closed during loading but can be opened after the animal(s) has (have) been restrained. The operator works from the small platform and restrains an animal by operating the moving wall, and applying steady pressure to the animal aligned between the two walls. (*Continued overleaf.*)

The drop-floor cradle utilizes a combination of pressure and removal of footing to restrain deer (Haigh, 1999). The cradle has a removable floor and Y-shaped walls, one of which is (usually) movable. The deer enters the cradle when the floor is upraised. When the animal is in position, the movable wall is operated to narrow the walls and contain the animal, the floor is released and the animal's weight is taken by its thorax and abdomen. The sides of the cradle can be padded to reduce the potential for bruising.

With both cradles and workrooms, additional immobilization can be achieved by placing additional pressure at the withers (Haigh, 1999) or by using a head halter.

Transport

Practical experience gained in New Zealand and elsewhere in transporting hundreds of thousands of deer annually has led to the development of guidelines for the humane transportation of these animals (MAFF, 1989; AWAC, 1994). Recently published scientific studies, as reported below, support the view that deer can be transported safely and without undue distress provided the guidelines are followed. Transporters need to be specially constructed to take account of the physical abilities and behavioural requirements of deer.

Transport crate design

The pressure on animals is relatively high in transport crates, therefore as with close handling facilities, they need to be smooth-sided and any openings should be less than 50 mm wide if situated within 1.2 m of the floor.

The temperature inside transporters laden with deer is about 5° higher than the ambient temperature when the vehicle is moving (Harcourt, 1995). When stationary, the internal temperature rises quickly to much higher

Fig. 17.5. *Continued.* (b) A drop-floor cradle for restraining deer.

levels. Thus, ventilation slots about 100 mm wide are essential and should be located around the sides of the crate at a height of 1.4–1.6 m above the floor. Two such slots high on the crate walls are even better. There should be a 250 mm gap between the top of the internal walls of the pens and the roof to allow fresh air to circulate. Gaps for ventilation located closer to the floor should be covered with a 50 × 50 mm grid. Stoppages during transport need to be kept to a minimum to prevent excessive heat build-up in the pens and hyperthermia in the deer (Harcourt, 1995).

A ceiling is required to contain the animals during transit, and its height should be less than about 1.5 m to discourage aggressive behaviour and rearing. A grid (25 × 25 mm) supported 25–50 mm off the truck's deck is an appropriate flooring material for shorter journeys. On longer journeys a soft underfoot material will assist in preventing injuries to the feet.

Entrance ways to the transporter and pens should be at least 1.2 m wide to assist animal flow, and the doors need to be positioned so that they can be used to push up behind animals that baulk. It is useful to have the pens on the transporter arranged as modules with dimensions of about 1.2 m × 2.0 m (2.4 m^2) and connected to adjacent pens via swinging doors which can be opened to increase pen size. Each module is suitable for carrying six 100 kg liveweight animals. Larger animals can be accommodated in larger pens or by reducing the group size in the standard module.

Practical experience indicates that the ideal group size is six animals (100 kg weight), with a maximum of about eight. Grigor et al. (1998b) showed that smaller groups ($n = 5$) were less disturbed during travel than larger groups ($n = 10$). There have been several scientific studies that provide information on the effects of stocking density on the responses of deer to travel (Jago et al., 1993, 1997; Waas et al., 1997; Grigor et al., 1998b). The densities examined ranged from about 0.4 to 1.2 m^2/ 100 kg animal equivalent. Transport was found to be somewhat stressful in all studies, as measured by increases in plasma cortisol concentrations. Variations in space allowance had little additional effect on animal comfort or stress. At lower space allowances, animals were less able to orient in the preferred directions (parallel or perpendicular to the direction of travel), but this did not adversely affect their ability to maintain balance or avoid falling down.

A variety of other environmental and animal factors influence the ease of handling and animal well-being during transport. In a survey of industry statistics, Jago et al. (1996) reported that bruising was more likely for animals transported distances greater than 200 km, for smaller or less fat deer, and for males transported during the breeding season or held overnight in lairage in late winter. Subsequent scientific studies have confirmed that there are increases in bruising rates, muscle damage (Jago et al., 1997) and stress (Waas et al., 1999) with increasing distance travelled, but the magnitude of the effects is small, and there is little effect on levels of dehydration of journey times up to 6 h (Grigor et al., 1997a, 1998d; Jago et al., 1997). Biochemical signs of dehydration, as measured by increases in plasma sodium levels, become apparent 11–20 h following water withdrawal (Hargreaves and Matthews, 1995; Waas et al., 1999).

Journey parameters other than distance or time travelled have more important effects on animal well-being. Deer are more prone to losing their footing in the first few minutes of a journey (Jago et al., 1997) and on steep, winding sections of highway (Jago et al., 1997; Grigor et al., 1998b). Thus, careful driving is required at this time. In addition, there appears to be a greater physical challenge to animals at the rear of transporters as heart rates and

plasma lactate concentrations for animals in the middle and rear pens are higher than for those at the front (Waas et al., 1997). Agonistic interactions between deer on trucks are more common between recently mixed animals, and those differing greatly in body weight (Jago et al., 1997). Therefore, animals allocated to the same pen on transporters should be familiar with each other and of similar sizes. Practical experience indicates that animals of different ages or gender should not be mixed in pens. Further, stags during the rut or with velvet antlers longer than 60 mm should not be transported. Stags with hard antlers should be transported singly in pens, or preferably have the antlers removed before transport.

Behavioural measures of aversion and physiological measures of stress indicate that the process of transportation is a relatively mild stressor (in comparison with physical restraint) (Carragher et al., 1997; Waas et al., 1997; Grigor et al., 1998a). Thus, if best practices are followed, animal well-being should not be unduly compromised by transport and associated handling.

Lairage and Slaughter

In some countries deer are slaughtered in the field or in slaughter plants on farms, and this can be a satisfactory way to avoid undue stress (Smith and Dobson, 1990), particularly in animals not familiar with close handling or selected for tractability. With field slaughter, great care needs to be taken to ensure that appropriate standards of hygiene are met (Vaarala and Korkeala, 1999). In countries where large numbers of animals are farmed, most deer are transported and slaughtered in commercial abattoirs: this process is probably more suitable for the less flighty species and those animals that have become accustomed to farming routines.

In New Zealand the slaughterhouses are purpose-built for deer alone, but in other countries several different species may be processed through the same facility. Ultimate muscle pH after slaughter is often used as an indicator of meat quality, with stress contributing to the development of high pH (poor quality) meat (Lawrie, 1985). Typically, the pH values for deer slaughtered at commercial premises are low and within the range considered indicative of good-quality meat (Smith and Dobson, 1990; Jago et al., 1993, 1997; Grigor et al., 1997c; Pollard et al., 1999).

Practical experience indicates that the principles for good handling and facility design (as outlined above for farm facilities) apply equally well to slaughter plants. However, a number of features of lairage require special consideration, since the deer are often held for much longer periods indoors and there is a requirement to maintain a steady flow of animals to the stunning box.

Overnight lairage may increase the rate of bruising slightly at specific times of the year (e.g. during winter in males) (Jago et al., 1996), but appears to have little other adverse physiological effects on deer. Grigor et al. (1997c) showed that lairage for up to 18 h did not lead to dehydration, depletion of muscle glycogen or high pH. There is some evidence that lairage is associated with recovery from transport, as the activity of enzymes indicative of muscle damage declines with time in lairage (Jago et al., 1993; Grigor et al., 1997c) and activity levels return to pre-journey patterns (Grigor et al., 1997c). Nevertheless, measures of other behaviours indicate that longer periods in lairage may be less than ideal as the frequency of agonistic interactions increases with time in lairage (Pollard et al., 1999). Penning deer next to other species, particularly pigs, is undesirable as it leads to higher levels of aggression and heart rates in the deer (Abeyesinghe et al., 1997).

The design of the race leading to the stunning box is critical for achieving a steady flow and avoiding stress in the deer. Holding pens and raceways should be designed so that each animal can remain in direct or visual contact with at least one familiar companion right up until entry to the stunning box. The use of appropriately positioned swing doors and curves and illumination of the stunning box are the best ways to encourage movement of reluctant animals without causing undue distress. Electric goads should not be used on deer. A low ceiling constructed to permit access by the stunner will discourage rearing. In the interests of animal well-being, the shorter the time the animals are left in the stunning pen the better (Grigor et al., 1999).

Behaviour Relevant to Facility Use

The fundamental behaviour and stress reactions of deer have important implications not only for facility design but also for the efficiency and ease of handling the animals. These behaviours will be examined in the context of mustering, yarding and restraint. Although deer are flighty by nature, recent studies (Ingram et al., 1994, 1997; Carragher et al., 1997) have shown that common handling procedures (such as yarding and drafting) are not particularly stressful. Further, upon release back to pasture, the animals' maintenance activities and stress hormone concentrations return rapidly to normal levels (Ingram et al., 1994, 1997, 1999; Carragher et al., 1997).

Mustering and movement

Basic behaviours relevant to efficient mustering and movement include the degree of flightiness and familiarity with the environment, and the flocking and leader–follower tendencies of deer.

Extremely flighty animals or those unfamiliar with the farm environment are difficult to direct and are best moved by passive methods. This simply involves leaving open gates to laneways, fresh pasture or handling facilities for animals to move without encouragement from handlers. Most animals soon become familiar with the farm layout and handlers. For these tamer deer, either of two active methods of mustering are more appropriate and efficient. The simplest one utilizes the good learning abilities of deer and their tendency to follow a leader. Animals are trained to follow or approach a handler in response to visual or auditory cues. Untrained animals in the group will readily follow the leaders to new pastures or through handling facilities. Training can be carried out most readily during the hand-rearing of calves or the feeding of supplements to weaners.

The second and more common technique utilizes the natural antipredator responses of flocking and flight, together with the following tendency. As a handler approaches a herd, the animals bunch and direct their eyes and ears toward the intruder (Fig. 17.6). When the handler is near the boundary of the flight zone a proportion of the animals turn and face away

Fig. 17.6. Herding behaviour of red deer in response to a human intruder.

(Fig. 17.6). With further advances by the handler and gradual penetration of the flight zone the leaders will begin to move and the well-developed follower response stimulates the remainder of the group to follow.

For calm and coordinated movement it is important that the handler moves steadily but slowly and remains near the boundary of the flight zone. Occasionally, the lead animals will baulk during mustering, thereby causing the remainder of the herd to stop. To re-establish synchronized forward movement of the herd it is important for handlers to coordinate their movements with those of the lead animals. At the moment when the leader(s) has (have) returned to the front of the mob (often after having circled the group and moving back up through the centre) and is/are facing in the required direction, the handler should move forward into the flight zone. As the animals move into smaller enclosures (field to raceway to yards) their flight distance decreases and speed of travel changes, and frequent positional adjustments are required by the handler. Familiarity with the handlers and the use of one or two people help reduce flightiness. In addition,

machines (motorcycles, trucks and helicopters) and other animals (dogs and horses) that are familiar to the deer can be used successfully in the mustering process.

Sudden or deep penetration of the flight zone induces strong and less cohesive flight of the herd. In extreme cases, panic behaviour characterized by excessive and disoriented flight (Mills and Faure, 1990) may occur, thereby increasing the risk of animals running into fences and sustaining injuries. Panic responses are more likely in particularly young animals with little handling experience; flighty individuals especially if held singly, in small groups away from the normal social group, or when exposed to unfamiliar handlers, objects or noises (Bull, 1996). Unusual visual stimuli induce stronger and more sustained flight and fear responses than unfamiliar auditory stimuli (Herbold et al., 1992; Hodgetts et al., 1998). Through the process of social facilitation (Mills and Faure, 1990), other normally calm individuals may show similar panic behaviours. In cases where the majority of the herd is extremely disturbed, handling of the animals should be discontinued for several hours. Practical experience indicates that the flightiness of young animals during handling can be reduced by the use of older, quieter animals as leaders (Pollard *et al.*, 1992), by habituating them to yards prior to handling (particularly in conjunction with feeding in the handling areas) and by mustering and moving them as a group for several hours in a large field prior to any close handling (C. Brown and L. Matthews, unpublished observations). Herrmann (1991) has shown that the provision of artificial shelters in the paddocks assists in reducing the incidence of abnormal behaviours following disturbances, although Hodgetts *et al.*'s (1998) data did not support this finding.

Previous negative handling experiences can adversely influence the ease of handling of deer. Groups kept in fields near handling facilities are more flighty and pace up and down fence lines furthest from the yards (Diverio *et al.*, 1993, 1996), particularly in the first 3–4 h after handling (Matthews *et al.*, 1990). The avoidance of the handling area and repetitive pacing suggest that some aspects of the handling process are aversive. This is supported by anecdotal observations that some animals become increasingly difficult to collect from pasture and drive toward the handling facility after repeated handling. The aversive aspects most likely result from periods of visual isolation, physical restraint or particular handling procedures (Matthews and Cook, 1991; Pollard *et al.*, 1993; Price *et al.*, 1993; Grigor *et al.*, 1998a). Minimizing stress from these factors will improve animal flow. In addition, the use of food rewards after handling and familiarization with the handling process facilitates movement in sheep (Hutson, 1985a, b) and also seems to work with deer.

Yarding

An ideal group size for intensive handling of deer in yards is about six animals. Large mobs are usually broken down into smaller units in two steps. First, groups of about 25 animals are drafted off while the herd is contained in outdoor holding pens close to the yards. These groups are then run indoors and given access to as many holding pens as possible. The animals tend to settle out into small-sized groups in the various pens, which are then secured.

Drafting (sorting) techniques are similar to those used with cattle (Kilgour and Dalton, 1984). The handler enters the flight zone to induce movement and then moves forward or backward of the point of balance behind the shoulder to direct animals backward or forward, respectively. The use of short lengths of plastic pipe to extend the arms and solid wooden shields are useful in manipulating the point of balance. Apart from stags during the rut, deer should not be kept in isolation from their herdmates.

The procedures for moving animals between various sections of the yard complex are similar to those used in mustering. In well-designed facilities passive techniques work well on most occasions. Individuals or groups move readily if the doors ahead of the animals are left open – typically the handler does not need to apply pressure from behind the deer. Active techniques involving manipulation of the flight response are sometimes required, especially when moving lead animals into unfamiliar or dead-ended areas. As mentioned earlier, the use of curved entranceways assists movement. However, some animals show no flight response and may need to be pushed from behind with swing gates or shields. This situation occurs more frequently with tamer animals or wapiti breeds (Thorliefsen *et al.*, 1997). Shields are particularly effective in providing protection from strikes by the deer's front or hind feet.

Deer become reluctant to leave holding pens if they have had prior experience of aversive events in other parts of the handling facility (Grigor *et al.*, 1998a). Behavioural (Pollard *et al.*, 1994a; Grigor *et al.*, 1998a) and physiological measures (Pollard *et al.*, 1993; Carragher *et al.*, 1997; Waas *et al.*, 1997) indicate that the rank ordering of the relative aversiveness of various events is: drop-floor cradle; transport; social isolation; human proximity. J. Ingram and L. Matthews, unpublished data support other studies (Pollard *et al.*, 1993; Hanlon *et al.*, 1995) which show that mixing of unfamiliar deer is a highly aversive event.

Adverse experiences have less effect on the readiness with which deer move along raceways (Matthews *et al.*, 1990; Stafford and Mesken, 1992; Grigor *et al.*, 1997b). Similarly, speed of movement along a race is not influenced by the light levels in the area ahead (Grigor *et al.*, 1997b, 1998c), although practical experience suggests it is difficult to move animals toward pens or on to transporters if there are bright shafts of light directed toward the deer. Overall then, the speed of movement of deer in races is influenced more by the strength of the flight response to the handler than the attractiveness (or aversiveness) of areas ahead of the animal.

Practical experience indicates that familiarity with the handling facility may help to reduce the aversiveness of handling. Animals fed supplements indoors or simply run through the facility a number of times are less flighty. Price et al. (1993) measured the heart rates of deer in yards and found that stress levels declined with increasing familiarity with the handling areas. They also recorded heart rate changes when handlers entered the holding pens. Interestingly, there were no differences in heart rate responses between familiar and unfamiliar handlers. In contrast, Bull (1996) noted in red deer at pasture that the approach of a handler wearing unfamiliar clothes was associated with a higher frequency of behaviours indicative of disturbance (more movement and less grooming) than when the handler wore familiar clothes. Interestingly, the colour of clothing influences deer vigilance and the effect varies with time of day (Bull, 1996). Deer are most alert to white clothing at night and to red in the morning.

In yarding situations, the animals are frequently worked from well within their flight zones, thereby increasing the likelihood of exaggerated flight responses, stress, threats and aggression. Maintenance of dominant behaviour by the handler over mobs (but especially stags) eases handling and lessens the risk of injury to handlers. Holding a hand high above the head, working from slightly above the deer (e.g. on catwalks) or confronting an animal with a shield are useful techniques for increasing the apparent size and dominance of the handler. Because hand-reared deer have a decreased flight response to humans (Blaxter et al., 1988), it can be difficult to assert dominance over particularly aggressive deer (e.g. females with newborns and males during the rut). It is advisable to cull hand-reared males before sexual maturity. Working the animals from outside the pens using swing gates and sliding doors reduces flightiness and aggression, especially in fallow and wapiti.

Repeated handling results in weight loss or lack of growth in deer (J. Ingram and

L. Matthews, unpublished data). A number of techniques assist in reducing stress. Unfamiliar animals, or animals of different ages, sizes, sex and species should not be mixed (Hanlon et al., 1995; Jago et al., 1997). Individuals and small groups should remain in visual contact with other deer and be able to see and hear handlers (by talking quietly) approach. Unfamiliar and loud noises should be avoided (Hodgetts et al., 1998). Ideally, stags should not be handled in the rut. If rutting stags need to be handled there is less chance of injury to themselves and handlers if they have been de-antlered, are moved in small groups only and are penned individually. Immature deer are particularly flighty and are best handled as little as possible. Any handling that is necessary should take place in the holding and working pens rather than in restrainers and races which are designed for older animals. Animals appear to be easier to handle if permitted to settle for an hour or so after mustering and drafting. We have shown that, although handling is somewhat stressful, animals settle, both physiologically and behaviourally, soon after release to pasture (Carragher et al., 1997). Thus, animals not required indoors or any others showing signs of excessive stress should be released. With continued handling, stressed animals may lie down. Attempts to move them are usually unsuccessful. Electric prodders should not be used. Such animals should be left in the company of others to stand before handling again. Some animals remain consistently flighty or aggressive throughout the year. In the interests of animal and handler welfare these deer should be culled.

Restraint

With red deer, procedures requiring relatively little restraint (e.g. ear tagging, drenching) have traditionally been carried out in working pens under direct manual restraint, augmented by shields and doors. There has been a move away from these methods as farmers have developed more suitable handling races for use with red, fallow and wapiti deer which ensure greater handler safety and also seem to be less stressful for deer. Procedures requiring higher levels of restraint (e.g. antler removal) are usually performed in a deer handler or under chemical restraint (Haigh and Hudson, 1993).

The basic behaviour patterns of deer can be used to advantage in several ways in animal restraint. Red deer and wapiti can be encouraged to enter handling races and restrainers by the use of curved entrances and by giving the lead animal a clear view of an exit. Single animals move readily to restrainers if they are drafted off individually from the holding pen and run through curving laneways. The tendency of fallow, white tail and chital deer to move readily from dark to light is used to advantage by dimming the light in the holding pen and lighting the entrance to the handling race or crush (Langridge, 1992; Haigh, 1999). Other sources of light shining into darkened pens need to be covered, as fallow deer will persist in jumping toward these. In the handling races, animals remain settled as they are in visual and physical contact with other deer. Reducing visual stimulation either by darkening the environment or obscuring the eyes with a cloth tends to quieten animals (Jones and Price, 1992). Positioning a ceiling just above the heads of the animals prevents any jumping (Lee, 1992).

The concept of optimal pressure (Grandin, 1993) is relevant to the restraint of deer in crushes. Informal observations made while assessing the effects of analgesia (Matthews et al., 1992) indicate that deer remain settled in padded crushes for up to 60 min provided the shoulders of the animal are well restrained. Animals struggle in response to pressures that are too high or low, or when the shoulders are well forward in the crush. Experience has shown that wapiti remain settled standing in squeeze races where the width of the race is narrowed but little pressure is exerted on the animal (Lee, 1992).

Conclusion

Deer such as red and wapiti are rapidly becoming less flighty the longer they are farmed. This is most probably a result of three interacting factors: (i) improved genetics for tractability; (ii) better methods for training animals to handling; and (iii) the development of handling facilities more appropriate to the unique behavioural characteristics of each species.

As a result of these processes, lower levels of force are required in order to move or restrain the animals, which in turn improves deer well-being during handling. No doubt handling systems will continue to evolve, with benefits for both animal and operator safety. Areas of research that will hasten this process include a greater scientific understanding of genetic factors underlying good animal temperament and of developmental processes determining rapid habituation of deer to humans and handling.

References

Abeyesinghe, S.M., Goddard, P.J. and Cockram, M.S. (1997) The behavioural and physiological responses of farmed red deer (Cervus elaphus) penned adjacent to other species. Applied Animal Behaviour Science 55 (1–2), 163–175.

Anderson, J.R. (1975) The behaviour of nose bot flies (Cephenemyia apicata and C. jellisoni) when attacking black-tailed deer (Odocoileus hemionus columbianus) and the resulting reactions of the deer. Canadian Journal of Zoology 53, 977–992.

AWAC (1994) Code of recommendations and minimum standards for the welfare of animals transported within New Zealand. Code of Animal Welfare, No. 15, Ministry of Agriculture, Wellington, New Zealand.

Bartos, L. (1985) Social activity and the antler cycle in red deer stags. Royal Society of New Zealand Bulletin 22, 269–272.

Blaxter, K.L., Kay, R.N.B., Sharman, G.A.M., Cunningham, J.M.M., Eadie, J. and Hamilton W.J. (1988) Farming the Red Deer. Rowett Research Institute and Hill Farming Research Organisation, Department of Agriculture and Fisheries for Scotland, Edinburgh, UK.

Bull, R.L. (1996) The behavioural responses of farmed red deer (Cervus elaphus) to handling. Unpublished MSc thesis, University of Waikato, New Zealand, 81 pp.

Bullock, D.J., Kerridge, F.J., Hanlon, A. and Arnold, R.W. (1993) Short-term responses of deer to recreational disturbances in two deer parks. *Journal of Zoology* 230 (2), 327–332.

Cadman, W.A. (1966) The fallow deer. *Forestry Commission Leaflet* no. 52. HMSO, London.

Carragher, J.F., Ingram, J.R. and Matthews, L.R. (1997) Effect of yarding and handling procedures on stress responses of free-ranging red deer (*Cervus elaphus*). *Applied Animal Behaviour Science,* 51, 143–158.

Cash, R. (1987) Fallow handling systems: simplicity and reliability are vital. *The Deer Farmer* 39, 29–31.

Clift, T.R., Challacombe, J. and Dyce, P.E. (1985) Electric fencing for fallow deer. *Royal Society of New Zealand Bulletin* 22, 363–365.

Clutton-Brock, T.H. and Guinness, F.E. (1975) Behaviour of red deer (*Cervus elaphus* L.) at calving time. *Behaviour* 55, 287–300.

Clutton-Brock, T.H., Guinness, F.E. and Albon, S.D. (1982) *Red Deer: Behaviour and Ecology of Two Sexes.* Edinburgh University Press, Edinburgh, UK.

Darling, F.F. (1937) *A Herd of Red Deer.* Oxford University Press, London.

Diverio, S., Goddard, P.J., Gordon, I.J. and Elston, D.A. (1993) The effect of management practices on stress in farmed red deer (*Cervus elaphus*) and its modulation by long acting neuroleptics (LANS): behavioural responses. *Applied Animal Behaviour Science* 36, 363–376.

Diverio, S., Goddard, P.J. and Gordon, I.J. (1996) Physiological responses of farmed red deer to management practices and their modulation by long-acting neuroleptics. *Journal of Agricultural Science* 126 (2), 211–220.

Drew, K.R. and Kelly, R.W. (1975) Handling deer run in confined areas. *Proceedings of the New Zealand Society of Animal Production* 35, 213–218.

English, A.W. (1992) Management strategies for farmed Chital deer. In: Brown, R.D. (ed.) *The Biology of Deer.* Springer-Verlag, New York, pp. 189–196.

Fletcher, T.J. (1991) Deer. In: Anderson, R.S. and Edney, A.T.B. (eds) *Practical Animal Handling.* Pergamon Press, Oxford, UK, pp. 57–66.

Gilbert, B.K. (1974) The influence of foster rearing on adult social behaviour in fallow deer (*Dama dama*). In: Geist, V. and Walther, F. (eds) *The Behaviour of Ungulates and its Relation to Management.* IUCN Publication New Series No. 24, Morges, Switzerland, pp. 247–273.

Goble, K. (1991) Growing, yarding and harvesting of velvet. *Proceedings of the 43rd Ruakura Farmers' Conference.* Hamilton, New Zealand, pp. 146–148.

Grandin, T. (1993) Facility design in relation to behavior, stress and bruising. *Proceedings of the New Zealand Society of Animal Production* 53, 175–178.

Grandin, T., Rooney, M.B. and Phillips, M. (1995) Conditioning of Nyala (*Tragelaphus angassii*) to blood sampling in a crate with positive reinforcement. *Zoo Biology* 14, 261–273.

Grigor, P.N., Goddard, P.J., Cockram, M.S., Rennie, S.C. and MacDonald, A.J. (1997a) The effects of some factors associated with transportation on the behavioural and physiological reactions of farmed red deer. *Applied Animal Behaviour Science* 52 (1–2), 179–189.

Grigor, P.N., Goddard, P.J. and Littlewood, C.A. (1997b) The movement of farmed red deer through raceways. *Applied Animal Behaviour Science* 52 (1–2), 171–178.

Grigor, P.N., Goddard, P.J., Macdonald, A.J., Brown, S.N., Fawcett, A.R., Deakin, D.W. and Warriss, P.D. (1997c) Effects of the duration of lairage following transportation on the behaviour and physiology of farmed red deer. *Veterinary Record* 140 (1), 8–12.

Grigor, P.N., Goddard, P.J. and Littlewood, C.A. (1998a) The relative aversiveness to farmed red deer of transport, physical restraint, human proximity and social isolation. *Applied Animal Behaviour Science* 56 (2–4), 255–262.

Grigor, P.N., Goddard, P.J. and Littlewood, C.A. (1998b) The behavioural and physiological reactions of farmed red deer to transport: effects of sex, group size, space allowance and vehicular motion. *Applied Animal Behaviour Science* 56 (2–4), 281–295.

Grigor, P.N., Goddard, P.J., Littlewood, C.A. and Deakin, D.W. (1998c) Pre-transport loading of farmed red deer: effects of previous overnight housing environment, vehicle illumination and shape of loading race. *Veterinary Record* 142 (11), 265–268.

Grigor, P.N., Goddard, P.J., Littlewood, C.A. and MacDonald, A.J. (1998d) The behavioural and physiological reactions of farmed red deer to transport: effects of road type and journey time. *Applied Animal Behaviour Science* 56 (2–4), 263–279.

Grigor, P.N., Goddard, P.J., Littlewood, C.A., Warriss, P.D. and Brown, S.N. (1999) Effects of preslaughter handling on the behaviour, blood biochemistry and carcases of farmed red deer. *Veterinary Record* 144 (9), 223–227.

Haigh, J.C. (1992) Requirements for managing farmed deer. In: Brown, R.D. (ed.) *The Biology of Deer.* Springer-Verlag, New York, pp. 159–172.

Haigh, J.C. (1995) A handling system for white-tailed deer (*Odocoileus virginianus*). *Journal of Zoo Wildlife Medicine* 26 (2), 321–326.

Haigh, J.C. (1999) The use of chutes for ungulate restraint. In: Fowler, M.E. and Miller, R.E. (eds) *Zoo and Wildlife Medicine. Current Therapy 4.* W.B. Saunders, Philadelphia, pp. 657–662.

Haigh, J.C. and Hudson R.J. (1993) *Farming Wapiti and Red Deer.* Mosby, Toronto, pp. 67–98.

Hanlon, A.J., Rhind, S.M., Reid, H.W., Burrells, C., Lawrence, A.B., Milne, J.A. and McMillen, S.R. (1994) Relationship between immune response, live-weight gain, behaviour and adrenal function in red deer (*Cervus elaphus*) calves derived from wild and farmed stock, maintained at two housing densities. *Applied Animal Behaviour Science* 41 (3–4), 243–255.

Hanlon, A.J., Rhind, S.M., Reid, H.W., Burrells, C. and Lawrence, A.B. (1995) Effects of repeated changes in group composition on immune response, behaviour, adrenal activity and live-weight gain in farmed red deer yearlings. *Applied Animal Behaviour Science* 44 (1), 57–64.

Harcourt, R. (1995) Temperatures rising. *The Deer Farmer* 125, 41–42.

Hargreaves, A.L. and Matthews, L.R. (1995) The effect of water deprivation and subsequent access to water on plasma electrolytes, haematocrit and behaviour in red deer. *Livestock Production Science* 42, 73–79.

Hemmer, H. (1988) Ethological aspects of deer farming. In: Reid, H.W. (ed.) *The Management and Health of Farmed Deer.* Kluwer, Dordrecht, Netherlands, pp. 129–138.

Herbold, Von H., Sachentrunk, F., Wagner, S. and Willing, R. (1992) Einfluss anthropogener stoerreize auf die herzfrequenz von Rotwild (*Cervus elaphus*) und Rehwild (*Capreolus capreolus*). *Zeitschrift Jagdwissenschaft* 38, 145–159.

Herrmann, H.J. (1991) Aspects of the welfare of farmed red deer (*Cervus elaphus*) with results of a preliminary study of two types of environmental enrichment. Unpublished Masters thesis, University of Edinburgh, UK.

Hodgetts, B.V., Waas, J.R. and Matthews, L.R. (1998) The effects of visual and auditory disturbance on the behaviour of red deer (*Cervus elaphus*) at pasture with and without shelter. *Applied Animal Behaviour Science* 55 (3–4), 337–351.

Humphries, R.E., Smith, R.H. and Sibley, R.M. (1990) Effects of human disturbance on the welfare of park fallow deer. *Applied Animal Behaviour Science* 28, 302 (abstract).

Hutching, B. (1993) Deer handlers. *The Deer Farmer* 105, 39–47.

Hutson, G.D. (1985a) Sheep and cattle handling facilities. In: Moore, B.L. and Chenoweth, P.J. (eds) *Grazing Animal Welfare.* Australian Veterinary Association, Queensland, pp. 124–136.

Hutson, G.D. (1985b) The influence of barley food rewards on sheep movement through a handling system. *Applied Animal Behaviour Science* 14, 263–273.

Hutson, G.D. and Hitchcock, D.K. (1978) The movement of sheep around corners. *Applied Animal Ethology* 4, 349–355.

Ingram, J.R., Matthews, L.R. and McDonald, R.M. (1994) A stress free blood sampling technique for free-ranging animals. *Proceedings of the New Zealand Society of Animal Production* 54, 39–42.

Ingram, J.R., Matthews, L.R., Carragher, J.F. and Schaare, P. (1997) Plasma cortisol responses to remote adrenocorticotropic hormone (ACTH) infusion in free-ranging red deer (*Cervus elaphus*). *Domestic Animal Endocrinology* 14, 63–71.

Ingram, J.R., Crockford, J.N. and Matthews, L.R. (1999) Ultradian, circadian and seasonal rhythms in cortisol secretion and adrenal responsiveness to ACTH and yarding in unrestrained red deer (*Cervus elaphus*) stags. *Journal of Endocrinology* 162, 289–300.

Jago, J.G., Matthews, L.R., Hargreaves, A.L. and van Eeken, F. (1993) Preslaughter handling of red deer: implications for welfare and carcass quality. In: Wilson, P.R. (ed.) *Proceedings of a Deer Course of Veterinarians,* No. 10. Deer Branch of the New Zealand Veterinary Association, New Zealand, pp. 27–39.

Jago, J.G., Hargreaves, A.L., Harcourt, R.G. and Matthews, L.R. (1996) Risk factors associated with bruising in red deer at a commercial slaughter plant. *Meat Science* 44 (3), 181–191.

Jago, J.G., Harcourt, R.G. and Matthews, L.R. (1997) The effect of road-type and distance transported on behaviour, physiology and carcass quality of farmed red deer (*Cervus elaphus*). *Applied Animal Behaviour Science* 51, 129–141.

Jones, A.R. and Price, S.E. (1992) Measurement of heart rate as an additional means of ascertaining the effect of disturbance in park deer. In: Bullock, D.J. and Goldspink, C.R. (eds) *Management, Welfare and Conservation of Park Deer.* Universities Federation of Animal Welfare, Potters Bar, UK, pp. 95–107.

Kelly, R.W. and Drew, K.R. (1976) Shelter seeking and sucking behaviour of the red deer calf (*Cervus elaphus*) in a farmed situation. *Applied Animal Ethology* 2, 101–111.

Kelly, R.W., Fennessy, P.F., Moore, G.H., Drew, K.R. and Bray, A.R. (1984) Management, nutrition, and reproductive performance of farmed deer in New Zealand. In: Wemmer, C.M. (ed.) *Biology and Management of the Cervidae*. Smithsonian Institution Press, Washington, DC, pp. 450–460.

Kilgour, R. and Dalton, C. (1984) *Livestock Behaviour: a Practical Guide*. Methuen, Auckland, New Zealand.

Langridge, M. (1992) Establishing a fallow deer farm: basic principles. In: Asher, G.W. and Langridge, M. (eds) *Progressive Fallow Deer Farming: Farm Development and Management Guide*. Ruakura Agricultural Centre, New Zealand, pp. 17–28.

Lawrie, R.A. (1985) Chemical and biochemical constitution of muscle. In: Lawrie, R.A. (ed.) *Meat Science*. Pergamon Press, Oxford, UK, pp. 43–73.

Lee, A. (1992) Steiner's user-friendly yards make the difference. *The Deer Farmer* 95, 25–26.

Lent, P.C. (1974) Mother-infant relationships in ungulates. In: Geist, V. and Walther, F. (eds) *The Behaviour of Ungulates and its Relation to Management*. IUCN Publication New Series No. 24, Morges, Switzerland, pp. 14–55.

Lincoln, G.A., Youngson, R.W. and Short, R.V. (1970) The social and sexual behaviour of the Red Deer stag. *Journal of Reproduction and Fertility* (Suppl.) 11, 71–103.

MAFF (1989) *Guidelines for the Transport of Farmed Deer*. MAFF Publications, Lion House, Alnwick, UK.

Matthews, L.R. and Cook, C.J. (1991) Deer welfare research: Ruakura findings. *Proceedings of a Deer Course for Veterinarians*, No. 8. New Zealand Veterinary Association, pp.120–127.

Matthews, L.R., Cook, C.J. and Asher, G.W. (1990) Behavioural and physiological responses to management practices in red deer stags. *Proceedings of a Deer Course for Veterinarians*, No. 7. New Zealand Veterinary Association, pp. 74–85.

Matthews, L.R., Ingram, J.R., Cook, C.J., Bremner, K.J. and Kirton, P.G. (1992) Induction and assessment of velvet analgesia. *Proceedings of a Deer Course for Veterinarians*, No. 9. New Zealand Veterinary Association, pp. 69–76.

Matthews, L.R., Carragher, J.F. and Ingram, J.R. (1994) Post-velveting stress in free-ranging red deer. In: Wilson, P.R. (ed.) *Proceedings of a Deer Course for Veterinarians*, No 11. Deer Branch of the New Zealand Veterinary Association, New Zealand, pp. 138–146.

McComb, K. (1987) Roaring by red deer stags advances the date of oestrus in hinds. *Nature* 330, 648–649.

McNally, L. (1977) The senses of deer. *Deer* 4, 134–137.

Mills, A.D. and Faure, J.M. (1990) Panic and hysteria in domestic fowl: a review. In: Zyan, R. and Dantzer, R. (eds) *Social Stress in Domestic Animals*. Kluwer Academic Publishers, Dordrecht, Netherlands, pp. 248–272.

Pearse, A.J. (1992) Farming of wapiti and wapiti hybrids in New Zealand. In: Brown, R.D. (ed.) *The Biology of Deer*. Springer-Verlag, New York, pp. 173–179.

Pollard, J.C. and Littlejohn, R.P. (1994) Behavioural effects of light conditions on red deer in a holding pen. *Applied Animal Behaviour Science* 41 (1–2), 127–134.

Pollard, J.C. and Littlejohn, R.P. (1996) The effects of pen size on the behaviour of farmed red deer stags confined in yards. *Applied Animal Behaviour Science* 47 (3–4), 247–253.

Pollard, J.C., Littlejohn, R.P. and Suttie, J.M. (1992) Behaviour and weight change of red deer calves during different weaning procedures. *Applied Animal Behaviour Science* 35, 23–33.

Pollard, J.C., Littlejohn, R.P. and Suttie, J.M. (1993) Effects of isolation and mixing of social groups on heart rate and behaviour of red deer stags. *Applied Animal Behaviour Science* 38 (3–4), 311–322.

Pollard, J.C., Littlejohn, R.P. and Suttie, J.M. (1994a) Responses of red deer to restraint in a y-maze preference test. *Applied Animal Behaviour Science* 39 (1), 63–71.

Pollard, J.C., Littlejohn, R.P. and Webster, J.R. (1994b) Quantification of temperament in weaned deer calves of two genotypes (*Cervus elaphus* and *Cercus elaphus* × *Elaphurus davidianus* hybrids). *Applied Animal Behaviour Science* 41 (3–4), 229–241.

Pollard, J.C., Stevenson-Barry, J.M. and Littlejohn, R.P. (1999) Factors affecting behaviour, bruising and pHµ in deer slaughter premises. *Proceedings of the New Zealand Society of Animal Production* 59, 148–151.

Price, S., Sibly, R.M. and Davies, M.H. (1993) Effects of behaviour and handling on heart rate in farmed red deer. *Applied Animal Behaviour Science* 37, 111–123.

Prince, J.H. (1977) The eye and vision. In: Swenson, M.J. (ed.) *Dukes' Physiology of Domestic Animals*. Cornell University Press, UK, pp. 696–712.

Smith, R.F. and Dobson, H. (1990) Effect of pre-slaughter experience on behaviour, plasma cortisol and muscle pH in farmed red deer. *Veterinary Record* 126 (7), 155–158.

Stafford, K.J. and Mesken, A. (1992) Electro-immobilisation in red deer. *Proceedings of a Deer Course for Veterinarians,* No. 9. New Zealand Veterinary Association, pp. 56–68.

Staines, B.W. (1974) A review of factors affecting deer dispersion and their relevance to management. *Mammalian Review* 4, 79–91.

Thorliefson, I., Pearse, A. and Friedel, B. (1997) *Elk Farming Handbook.* North American Elk Breeders Association, Canada.

Vaarala, A.M. and Korkeala, H.J. (1999) Microbiological contamination of reindeer carcasses in different reindeer slaughterhouses. *Journal of Food Protection* 62 (2), 152–155.

Waas, J.R., Ingram, J.R. and Matthews, L.R. (1997) Physiological responses of red deer (*Cervus elaphus*) to conditions experienced during road transport. *Physiology and Behavior* 61(6), 931–938.

Waas, J.R., Ingram, J.R. and Matthews, L.R. (1999) Real-time physiological responses of red deer to translocations. *Journal of Wildlife Management* 63 (4), 1152–1162.

Woodford, K.B. and Dunning, A. (1992) Production cycles and characteristics of Rusa deer in Queensland, Australia. In: Brown, R.D. (ed.) *The Biology of Deer.* Springer-Verlag, New York, pp. 197–202.

Yerex, D. (1991) *Wapiti Behind the Wire: the Role of Wapiti in Deer Farming in New Zealand.* GP Books, Wellington, New Zealand.

Yerex, D. and Spiers, I. (1990) *Modern Deer Farm Management.* GP Books, Wellington, New Zealand.

Further Reading

Australian Code of Practice for Farming Deer in Western Australia http://www.dlgrd.wa.gov.au

England Deer Welfare Code http://www.defra.gov.uk

New South Wales Department of Primary Industry http://www.agric.nsw.gov.au

New Zealand Deer Welfare Codes http://www.biosecurity.govt (type 'deer welfare' into the search box).

North American Deer Farmers Association http://www.nafda.org

Ontario Elk Breeders Association http://www.oeba.ca

Pollard, J.C. and Wilson, P.R. (2002) Welfare of farmed deer in New Zealand. 1. Management practices. *New Zealand Veterinary Journal* 50, 214–220.

Saskatchewan Agriculture and Food http://www.agr.gov.sk.ca

The Deer Farmers Magazine http://www.deerfarmer.co.au

18 Poultry Handling and Transport

Claire A. Weeks

University of Bristol, Department of Clinical Veterinary Science, Langford, Bristol, BS40 5DU, UK; Claire.weeks@bristol.ac.uk

Introduction

More countries produce poultry intensively and on a larger scale than previously. More than 50 thousand million birds, mostly broilers, were slaughtered in 2004, and laying hen numbers are about 5500 million (Watt Poultry Statistical Yearbook, 2005). The capacity of poultry houses has increased, along with larger processing plants and faster line speeds. All birds are handled and transported at least twice. There is a need for this to be carried out both efficiently and with due regard for the welfare of the individual bird.

Farm animal welfare is currently a political priority in many countries (Blokhuis, 2004), with legislation covering many issues, including the welfare of animals in transit. Often, improving welfare benefits both meat quality and performance. Recent research has been aimed at quantifying the relative stressfulness and aversiveness to birds of multiple concurrent stressors during handling and transport. There have been developments in automated handling systems. The importance of training humans to handle birds more efficiently and humanely is also widely recognized.

Rearing Systems and On-farm Handling

Chicks

Careful handling and transport of both hatching eggs and newly hatched chicks are important for subsequent performance (Meijerhof, 1997).

Control of humidity as well as temperature is particularly important for eggs. Newly hatched chicks are usually tipped from incubator trays on to a series of conveyors, creating accelerations of up to 882 m/s, with several drops of 7–55 cm in height (Knowles *et al.*, 2004), before passing through automatic counters into lightweight, disposable containers perforated with ventilation holes for transport (see Figs 18.1 and 18.2). These authors also found that, in passing from one conveyor to another, if the belt speeds differed by more than 0.4 m/s, few chicks were able to remain standing.

Manual sexing and sorting of layer strains is routine and often used for meat birds. Automatic sexing before hatching is being developed with the potential to reduce handling stress (Phelps *et al.*, 2003). The yolk sac reserves enable chicks to be transported for 24 h or even longer with low mortality. Chick containers are transported by either truck or aircraft, generally in controlled environments to maintain uniformly warm yet well-ventilated conditions. Optimum temperature for chicks at normal stocking density in transport containers is 24–26°C (Meijerhof, 1997). On arrival, the chicks may be gently tipped out or removed manually.

Broilers, turkeys and ducks

Most poultry intensively reared for meat are placed in mixed or single-sex groups of up to 60,000 birds in environmentally controlled, dimly lit (3–20 lux)

Fig. 18.1. Automated chick-handling. Photograph courtesy of Dr Andrew Butterworth.

Fig. 18.2. Automated chick-counting into transport crate. Photograph courtesy of Dr Andrew Butterworth.

houses. Birds live at stocking densities of up to 45 kg/m^2 at slaughter age on a deep litter of wood shavings, straw or similar material, with automated provision of food and water. However, there are proposals within the European Union (EU) (2005) to limit stocking densities to a maximum of 38 kg/m^2, with lower rates (30 kg/m^2) to improve welfare of broilers housed in less well managed and equipped accommodation, as is already the case in Sweden (Berg and Algers, 2004). Although primarily using bird-based measures of welfare, the Directive also requires farmers to

keep records, to inspect birds twice daily and to meet minimal housing standards.

Several producers of turkeys and broilers now adopt lighting programmes (reviewed by Buyse *et al.*, 1996) and feeding programmes (Su *et al.*, 1999), which may be a requirement of assurance and marketing schemes. By reducing early growth rate, the lighting and feeding programmes may decrease the incidence and severity of leg problems, 'flip-overs', ascites and other consequences of genetic selection for excessively fast growth and food conversion efficiency.

In hot climates and systems such as organic, birds may be grown more slowly in naturally ventilated and lit pole barns or with access outside. Catching birds in such systems is invariably harder, because the birds are more active and have more space.

Most poultry meat is grown on contract or as part of an integrated system; so processors usually own and are responsible for the transport system. They provide specialist catching teams to depopulate the houses. Birds are calmer and less affected by the catching process if they are handled in darkness (Duncan, 1989), thus catching in the early hours in very dim light is common. Blue lights are useful with turkeys (Siegel, personal communication). Broilers are caught by one leg, inverted and carried in bunches of three or four per hand to the waiting crates or modules (Gerrits *et al.*, 1985).

Inversion may cause fear, as indicated by prolonged TI (tonic immobility) responses (Zulkifli *et al.*, 2000). To avoid dislocated hips and other injuries, handling by both legs is preferable, as is maintaining the bird upright (Gerrits *et al.*, 1985; Parry, 1989). Appropriate handling techniques for all species are given in Anderson and Edney (1991).

Methods of catching have been reviewed by Mitchell and Kettlewell (2004). Particularly with heavier birds, loose-crate systems are laborious and thus are decreasing in popularity. Changing to modular systems, comprising a unit of compartments or sliding drawers that can be moved by forklift on and off the transport vehicle and right into the poultry house or lairage, may reduce dead on arrivals to about one-third of previous levels (Aitken, 1985; Stuart, 1985). However, care needs to be exercised when loading the topmost drawers, particularly with heavy birds and turkeys. To prevent injuries to the head and wings when the drawers are closed, there should be a gap of 4.25–4.50 cm between the top of the drawer and the rack to prevent birds from being caught.

Manual catching is exhausting, repetitive and dirty for the humans employed (Bayliss and Hinton, 1990). Up to 1500 broilers may be caught per man-hour in shifts of 5 h (Metheringham, 1996). Significant welfare benefits such as reduced heart rates and catching damage were recorded for turkeys that could be driven into a modular system in comparison with three other systems in which they were manually caught and loaded (Prescott et al., 2000).

In many countries, dump module systems are used. To unload the broilers at the processing plant, the module is tipped and the chickens slide down the chute on to a wide conveyor belt. These systems work well for lightweight broilers, but they have increased levels of broken wings when used with heavy broilers. Unpublished industry data from Australia showed that on heavy birds, 1% of them had broken wings from the dumping process. This did not occur in the small roaster chickens. Another interesting finding was that 74% of the birds with broken wings were females. Birds with the longest slides from the top decks of the module have the most broken wings.

Further unpublished data from a US processing plant showed reduced broken wings to 1% in heavy chickens. The new systems have an additional conveyor so that the birds on the upper decks fall on an upper conveyor, while birds on the lower decks fall on to a lower conveyor. This greatly reduces the distance birds on the upper deck have to slide out to a conveyor.

Skeletal defects are a serious welfare issue and cause substantial economic losses in broilers, broiler breeders, ducks and turkeys (reviewed by Mench, 2004). The most common defect is lameness, which results in significant changes to behaviour and, in particular, the number of visits to the feeders is reduced in proportion to the degree of walking disability (Weeks and Kestin, 1997). This implies a cost to the bird of lameness, which may be attributed to pain (McGeown et al., 1999; Danbury et al., 2000).

Thus, most catching and handling procedures are likely to cause pain, especially inverting and carrying birds by the leg(s) and shackling. UK legislation based on an EC Directive would also regard the more severely affected birds as

being unfit to travel (MAFF, 1997). Severe clinical lameness following transport of male breeding turkeys with dyschondroplasia has also been reported (Wyers et al., 1991).

The husbandry and welfare of ducks in European systems has recently been considered (Rodenburg et al., 2005), but there appears to be no literature considering the catching and transportation of ducks or geese, despite over 2000 million ducks and 500 million geese being slaughtered annually worldwide.

Pullets

Laying hens may be reared in cages or on litter and will usually undergo transportation at about 18 weeks of age when they reach point of lay; they are then taken to the egg production farm. Perforated plastic crates – generally with solid floors – are widely used, particularly when the rearer is responsible for delivery and may be using general-purpose vehicles. Narrow, modular systems that can be loaded and unloaded directly into cages and wheeled on to the transporter are favoured by professionals using dedicated vehicles. These predominate in the USA and are increasing in popularity in Europe with decreasing labour availability.

The other main system is crates built as permanent fixtures on the bed of the lorry with a central ventilation channel. Hinged openings to the outside are used to load and unload the birds, which have therefore to be carried out of and into their housing. They may also be passed in handfuls from person to person.

Pullets are relatively valuable birds with good plumage (insulation) and are resilient to transport and handling stressors. Egg producers require them to arrive in good condition so they tend to be handled and loaded carefully. The vehicles are usually generously ventilated with air gaps above the floor and below the roof, air inlets in the headboard and either roof fans or a central ventilation channel the length of the trailer, including a slot in the roof.

Hens

The majority of laying hens are still kept in battery cages. Numbers in alternative systems such as percheries, aviaries or on free range have increased steadily over the past 10 years in northern Europe, where some countries have outlawed, and the EU is likely to ban in 2012, the conventional cage. After a productive year, hens are caught and transported – as 'spent' hens – to the slaughterhouse. Spent hens are generally purchased 'off farm'. Their low economic value reduces the care taken in handling and the investment in transport systems. Loose crates predominate, despite the labour costs of handling and cleaning them. Many types of vehicle may therefore be used, but dedicated trucks with central ventilation and side-curtains are common.

Hens are removed from cages either individually or in groups of two or three by pulling them out by one leg, despite recommendations to handle poultry by two legs (e.g. UK Codes of Recommendation). They may be struck against the cage entrance or food trough during removal, and in furnished cages the perches may form an obstruction. Hens may also hit cages or roof supports as they are carried down the narrow aisles of a battery house (Knowles, 1991, unpublished PhD thesis).

Depopulation is very labour-intensive, with bunches of inverted hens commonly being passed along a human chain. Kristensen et al. (2001) evaluated a modular system for depopulating battery cages and found a significant reduction in the time each bird was handled, from 64.5 to 4.5 s. Compared with manual handling, there was no difference in the proportion of damaged birds in the small trial, but the catchers preferred the modular system.

Hens in many alternative systems are difficult to catch, tending to crowd and pile up at the end of aisles and thus creating a potential for suffocation, or flap and fly, with the risk of injury to the catchers. In some systems the back of a tier is beyond arms' reach, resulting in birds having to be goaded or driven out, which is time-consuming and hazardous (L.J. Wilkins, 1992, personal communication). Ease of depopulation is therefore an important consideration to build into the design of any housing system.

A direct comparison of different catching and carrying methods for end-of-lay hens showed that plasma corticosterone concentrations were significantly higher when they were removed from their cages three at a time and carried in an inverted position from the house than when

they were removed singly and crated before removal from the house (Knowles and Broom, 1993). However, all hens in experimental handling treatments had high concentrations of corticosterone in comparison with the control birds, which were removed individually and gently from their cages in an upright position (Knowles and Broom, 1993).

Scott and Moran (1993) found that the fear levels of laying hens carried for 20 m on a flat-belt conveyor were lower than those of hens carried the same distance in an inverted position by hand or on a processing shackle. However, in the absence of a non-inverted control in this study, it is difficult to know whether the reduced fear was a consequence of upright conveyance or some other fear-reducing property of the flat-bed conveyor. As with broilers, well-designed automated handling devices would seem to have the potential to reduce trauma and fear.

Automated catching and handling systems

Automated catching and handling systems (reviewed by Scott, 1993) have great potential for reducing injury and distress for birds and humans alike (Lacy and Czarick, 1998; Knierim and Gocke, 2003). Several are now commercially available (Moran and Berry, 1992; Mitchell and Kettlewell, 2004; see Fig. 18.3). Uptake by the industry is increasing as the systems become more reliable and the difficulty of recruiting human catchers in some countries increases.

An early study revealed that the heart rate of broilers caught by an automatic combine returned to base levels more quickly than the heart rate of broilers caught manually (Duncan et al., 1986). Scott and Moran (1992) found significant increases in loss of balance, wing flapping and alarm calls by hens conveyed up or down slopes rather than horizontally. In particular, drops from one conveyor belt to another must be avoided as they tend to cause injuries and wing-flapping.

Although Ekstrand (1998) did not observe the birds during catching, this is the probable cause of differences found in carcasses examined after slaughter, where twice as many wing fractures and significantly more bruising – mainly of the wings – were seen in mechanically caught birds, but insignificant differences in birds dead

Fig. 18.3. Poultry-catching machine. The rotating heads with soft rubber fingers are mounted on the end of a long, pivoting boom. This provides the operator with more precise control because he/she can control both sideways and forward movement of the machine. To enable the operator to observe chickens more closely during pick-up, he/she stands next to the pick-up head and operates the machine with a controller that he/she carries.

on arrival (DOA) compared with manual catching. In the USA, improvements in catching machine technology during 2005 and 2006 have reduced the incidence of broken wings. Unpublished industry data indicate that machine-caught and hand-caught birds have similar incidences of broken wings.

Gracey (1986) noted DOAs averaged 0.2% for mechanically caught flocks compared with 0.3–0.6% for manually caught broilers. In the USA, where there are extremely hot temperatures of up to 38°C, mechanical catching had higher DOAs because it took longer than manual catching. One innovative company solved this problem by using two catching machines.

Other surveys have recorded similar levels of DOAs for mechanically harvested broilers compared with manually caught birds, or higher levels in spring (Knierim and Gocke, 2003; Nijdam et al., 2005a). It is likely that in both these studies the increased mortality was due to heat stress: Knierim and Gocke (2003) noted uneven stocking rates for machine-filled compartments and Nijdam et al. (2005a) suggested that particular climatic conditions were associated with the higher losses.

Compared with almost 88,000 manually caught broilers, Knierim and Gocke (2003) recorded significantly reduced injuries (3.1 versus 4.4% of birds) in over 108,000 mechanically caught broilers, particularly in respect of leg injuries, thus confirming the results of an earlier study by Lacy and Czarick (1998) and results of field trials with the Easyload system that found a halving of catching damage from 4–6% manually to 1–3% using the machine (*Poultry International*, 1998). However, in a 2001 field study at a commercial slaughterhouse, Nijdam et al. (2005a) found no differences in the percentages of bruising, meat quality or corticosterone levels between catching methods, indicating similar levels of stress.

It therefore appears that, with experience and careful management, mechanical and automated systems can perform at least as well and generally better than manual handling. Indeed, during the year of their survey, Knierim and Gocke (2003) noted a reduction in loading time by a third and significantly decreased rates of injury, which might in part be attributed to a reduction in speed of the conveyor from 1.4–1.6 to 0.8–1.0 m/s.

Effects of rearing experience on response to commercial catching and transportation

Birds exposed to outdoor environments and low stocking densities are generally less fearful than birds that experience a less stimulating environment (Grigor et al., 1995; Sanotra et al., 1998; Scott et al., 1998). The use of enrichment stimuli such as novel objects or even video stimuli during the rearing period can reduce the underlying fearfulness of domestic fowl, as revealed by their responses in a range of laboratory tests (Jones and Waddington, 1992; Jones, 1996). Nicol (1992) reported reduced levels of fear after transportation in broiler chickens that had been reared in an enriched environment. However, Scott et al. (1998) found no effect of enrichment when TI durations were measured after a 74-min journey, despite the fact that environmental enrichment reduced TI durations in birds caught and tested immediately.

Possibly, the combined effects of commercial-type handling and transportation are so great that minor reductions in underlying fearfulness or propensity to react to stressors are simply masked. Although Reed et al. (1993) found reduced levels of jumping and flapping behaviour (resulting in reduced trauma and injury at depopulation) in end-of-lay hens that had previously experienced environmental enrichment and regular handling, Kannan and Mench (1997) detected no beneficial effects of prior handling on the plasma corticosterone concentrations of broiler chickens subjected to commercial-type handling at 7 weeks of age. Nicol (1992) found that prior gentle handling by humans either had no effect or, in combination with physical enrichment, actually resulted in an increased fear response after transportation.

Although the potential for alterations to the rearing environment to modify the responses of birds to handling and transportation seems limited, there is an emerging consensus that fear or stress reactions can be modified by changes in commercial handling procedures per se. Jones (1992) found the TI response of both broilers and hens was reduced by gentle handling.

Kannan and Mench (1997) confirmed this result in broiler chickens that were subjected to a 2-min handling treatment and then returned to their home pens, where plasma corticosterone was sampled at hourly intervals for the following 4 h.

Birds that had received upright handling had lower plasma corticosterone concentrations than birds that had been inverted either individually or in groups of three. However, the effects of handling treatment were masked when the broilers were crated after the 2-min handling period.

This suggests that crating itself can be stressful, as found previously by Beuving (1980), but in apparent contrast to the results of a study with end-of-lay hens (Knowles and Broom, 1993). It should be noted, however, that in Knowles and Broom's study the alternative to crating was inverted conveyance down a narrow aisle rather than return to the home pen.

Consequences of Transportation

Transportation is an extremely stressful process for commercial poultry. Having lived in relatively uniform environments they are suddenly exposed to multiple changes that include being handled – as discussed above – and withdrawal of food and water. They experience stimuli that may be new, such as motion with vibration and impacts, or of greater intensity and more varied than previously – such as daylight, noise, overcrowding and temperature extremes.

There are economic and welfare benefits for minimizing these stressors. The potentially adverse consequences of transportation include physical, physiological and behavioural changes. The greater the duration of exposure to stressors, the greater the integrated stress for the bird. The resistance of birds to handling (Zulkifli et al., 2000) and to transportation stressors (Kolb and Seehawer, 2001) may be enhanced by the addition of ascorbic acid (vitamin C) to the drinking water.

Death

Death is the most obvious and easily recorded indicator of intolerable stress. Global figures for dead on arrivals (DOAs) are unknown. Published figures in Europe range up to 0.57% (Bingham, 1986; Bayliss and Hinton, 1990; Gregory and Austin 1992; Warriss et al., 1992a; Ekstrand, 1998), with a recent survey of 59 million broilers at one UK plant averaging 0.13% (Warriss et al., 2005). Using a figure of 0.3%

indicates that, worldwide, some 150 million birds die annually between farm and factory, representing a vast economic loss to the industry and a larger cost in terms of bird welfare.

If some birds are sufficiently stressed to die, many more will be stressed close to their capacity to survive. Surveys (Warriss *et al.*, 1992a) have shown that, the longer the journey time – especially > 4 h – and the longer the time between farm and slaughter, the greater the mortality rate and that time influenced mortality rates more than distance travelled (see Fig. 18.4).

Typical times in transit are unreported in most countries, but vary considerably. A survey of four UK broiler processing plants by Warriss *et al.* (1990) found average time from loading to unloading was 3.6 h, with a maximum of 12.8 h. Time in transit for 90% of turkeys was under 5 h, with a maximum of 10.2 h (Warriss and Brown, 1996). Other factors that increased broiler mortality were: (i) the length of waiting time in the holding area at the processing plant; (ii) older birds (during summer months); and (iii) arrival at the processing plant in the afternoon or evening rather than the morning (Bayliss and Hinton, 1990).

Several risk factors associated with mortality and bruising in transit have been statistically modelled, using data from 1907 Dutch and German broilers flocks slaughtered at a single Dutch plant during 2000 and 2001 (Nijdam *et al.*, 2004). The authors concluded that reducing stocking rate (OR = 1.09 for each additional bird in a compartment), transport time (OR = 1.06 per 15 min) and lairage time (OR = 1.03 per 15 min) would have a significant impact on broiler welfare

and reduce DOAs. Increased risk of mortality was also associated with: (i) high (> 15°C) or low (≤ 5°C) ambient temperature; (ii) daytime rather than night-time transport; (iii) larger flock size; (iv) heavier mean body weight; and (v) an interaction between ambient temperature and transport time. The catching company and breed also had a significant effect on risk of DOAs.

Thermal stress

Heat stress is thought to be the major contributor to both deaths (attributed to 40% of DOAs by Bayliss and Hinton, 1990) and overall transit stress. The recent survey in the UK by Warriss *et al.* (2005) indicated that mortality increased when ambient temperature rose above 17°C, with DOAs almost seven times greater when air temperatures were above 23°C. Cold-stress problems are seldom reported for broilers; however, cyanosis damage in turkeys in Canada was found to be associated with prolonged journey times (> 8 h) and sub-zero ambient temperatures (Mallia *et al.*, 2000). This study illustrates that the degree of thermal stress experienced by birds in transit depends on the duration and intensity of heat and cold stressors.

Side-curtains are used as protection against precipitation, wind, solar radiation and increasingly to hide the stock from public view. However, even in winter, these often restrict ventilation too much and excessive heat and moisture levels build up around the birds (Mitchell *et al.*, 1992; Webster *et al.*, 1992; Kettlewell *et al.*, 1993). The risk of

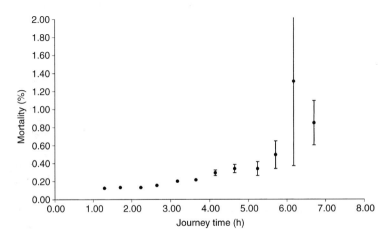

Fig. 18.4. Broiler death rates in relation to time in transit (from Warriss *et al.*, 1992a). Figure reproduced with permission of Carfax Publishing.

DOAs tends to be associated with high environmental temperatures and the stationary parts of the transportation process (i.e. loading, unloading and waiting at the factory (Ritz et al., 2005)).

Several studies in the temperate UK climate have measured the thermal environment in different parts of the loads of pullets, hens and broilers during winter and summer conditions and on stationary and moving vehicles of varying design (Webster et al., 1992; Kettlewell et al., 1993; Weeks et al., 1997). These have shown that most vehicles used for transporting poultry that are naturally ventilated do not provide a uniform thermal environment. Studies of the aerodynamics of full-size and scale models of one design of vehicle, including a trailer, have shown that, when moving, air predominantly enters at the lower rear end of the vehicle and moves forward to exit at the front (Baker et al., 1996; Hoxey et al., 1996).

Thus different parts of the load of poultry within the vehicle are over- or under-ventilated, and birds at the main inlet point are also susceptible to becoming wet unless protected. There are large differences between conditions on moving and on stationary vehicles, again primarily due to ventilation and to speed of air movement. For example, Weeks et al. (1997) calculated that average air speeds immediately surrounding the birds in moving vehicles varied between 0.9 and 2.4 m/s, with maxima of 6.0 m/s. In certain positions there was virtually no air movement to dissipate the body heat produced by the birds. Conditions in thermal 'hot' or 'cold' spots are frequently detrimental to welfare and may lead to deaths that can be correlated with climatic conditions and excessive or inadequate ventilation (Hunter et al., 1997).

This knowledge of the thermal conditions experienced by live birds in road transit and the physiological consequences of these (Mitchell and Kettlewell, 1994) leads to the inescapable conclusion that controlled and uniform ventilation is essential (Weeks et al., 1997; Mitchell and Kettlewell, 1998; Kettlewell et al., 2000). Heat losses from model chickens among live birds in vehicles fitted with both side-curtains and roof-mounted inlet fans were generally in the comfortable range, with little variation between areas of the load (Weeks et al., 1997). These authors suggested air speeds within bird crates or modules should be maintained between 0.3

and 1.0 m/s, except in extremely hot weather. Ventilation requirement is 100–600 m^3/h for typical commercial loads.

Kettlewell et al. (2000) proposed – for uniformly ventilated loads in temperate conditions (up to 20°C) – that 2.2 m^3/h/kg of chickens was sufficient. Birds can also become heat-stressed during loading of the trailer. In the USA, where the temperatures are often $\geq$ 30°C, a trailer with a bank of fans is parked alongside the truck while it is being loaded. Providing fan ventilation while loading is especially helpful to prevent heat stress in heavy birds.

It is strongly recommended that all vehicles be fitted with several temperature probes placed in close proximity to the birds. Temperatures should be both recorded and linked to an in-cab monitoring and alarm system. As a guide, that should be modified according to individual loads and vehicle designs; Weeks et al. (1997) indicated broilers and pullets transported at 10–15°C and poorly feathered end-of-lay birds at 22–28°C were likely to be thermally comfortable at the usual high stocking densities. In hot weather, the direct and indirect heating effects of solar radiation should be avoided by transportation at night or early in the day and parking in the shade. Research knowledge has now been transferred to commercially operated, fan-ventilated, temperature-controlled vehicles with sensors within the load and these are similar to the 'Concept 2000' vehicle (BBSRC, 2002).

Trauma

Osteoporosis in laying hens is almost ubiquitous, and a recent survey found 26–55% of laying hens had sustained fractures during production and 4–25% during depopulation, depending on housing system (Sandilands et al., 2005). Other data showed that up to 78% of birds in free-range systems had old breaks, principally to the keel (breastbone) (Wilkins et al., 2004). Previously, the problem of weak bones was thought to be primarily confined to caged birds (Gregory et al., 1990; Knowles and Broom, 1990; Keutgen et al., 1999). Two-leg handling substantially reduces bone breakage (Gregory and Wilkins, 1992). Thus hens in all systems need to be handled with great care to avoid creating new fractures. Unhealed old breaks are likely to be painful during handling.

Injuries to broilers from careless handling and catching include bruising to the breast, wings and legs, as well as fractures and dislocations, especially of the legs of heavier birds carried by a single leg (see reviews by Knowles and Broom, 1990; Nicol and Scott, 1990). It has been estimated that 25% of broilers processed in the USA may have such bruising damage (Farsaie et al., 1983).

Recently improved practices in the USA indicate that these injuries to the legs can be almost eliminated if a bird is carried less than 3 m by one leg. The catcher kneels down and loads his left hand with a bird between each finger. The birds remain on the ground until both hands are full. When the hands are full, the catcher rises and walks a short distance to the transport module. When this procedure is carried out by a skilled crew and only one leg is held in each space between the fingers, there are no breast bruises and very little leg damage (T. Grandin, personal communication, 2006).

Observations at the slaughter plant indicate that rough handling during shackling is a major cause of leg bruising (T. Grandin, personal communication, 2006). Post-slaughter rejection rates due to trauma damage range from 5 to 30% (Jepersen, 1982; Gerrits et al., 1985; Kettlewell and Turner, 1985). In a survey of downgrading at a turkey abattoir, McEwen and Barbut (1992) found substantial levels of bruised drums: fewer leg and breast scratches were seen where birds had clipped toenails, and spur-clipping reduced back scratches. They found no effect on injuries of truck design or stocking density, but increased half-wing trim and bruising of drums was associated with length of time spent on the truck. In their model, Nijdam et al. (2004) found low ambient temperatures (< 5°C) increased the risk of bruising, as did transport during the day and in summer.

Damage to muscle cells results in an increase in the concentration of creatine kinase (CK) in blood plasma, so it may be possible to use CK concentrations as a quantitative index of injury (Mitchell et al., 1992). Apart from muscle damage, broilers also sustain a worrying number of broken bones and dislocations during the catching and transportation process. Gregory and Wilkins (1990) found 3% of broilers had complete fractures before stunning at the processing plant, and 4.5% had dislocated femurs. The dislocated femurs were probably caused by swinging the birds by one leg before placing them in the crate,

as the chance of a broiler suffering a dislocation of this type increases with its bodyweight.

The incidence of dislocations recorded in live birds before stunning may not be a true reflection of the problem, since many dislocations result in fatal haemorrhaging. Indeed, when 1324 DOA broilers from six UK plants were examined, it was found that 27% had dislocated femurs (Gregory and Wilkins, 1992). These data indicate that the physical injury that occurs during manual catching and loading is a severe welfare problem for all poultry, and should accelerate the search for more humane alternative methods.

Fatigue

Many birds arrive at the slaughterhouse in an apparently exhausted state. It has been argued that the dehydration and depletion of body glycogen stores (Warriss et al., 1988), which occurs when broilers are subjected to food deprivation and simulated commercial transport, may produce a sensation of fatigue in birds (Warriss et al., 1993). However, progressive immobility is also a correlate of increasing fearfulness (Jones, 1996) and a response to painful (nociceptive) stimulation. This immobility may be related to learned helplessness, and could be an important 'cut-off' response in transported poultry.

Sherwin et al. (1993) investigated the effects of fasting and transportation on measures of fear and fatigue in broilers. Broilers subjected to either food deprivation for 10 h or a journey of 6 h, or both, were compared with control birds, which were neither fasted nor transported. When the behaviour of birds was monitored in their home pens after the treatments had ended it was found that both fasted and transported birds were more active than controls, showing less lying behaviour for at least 12 h, suggesting they were not particularly fatigued.

Hunger and thirst

Newly hatched chicks are not provided with food and water until they reach the rearing unit. During transportation, which may last up to 2 days on international journeys, chicks are thus completely reliant on yolk sac metabolism. Warriss et al. (1992b) found that chicks deprived in this way for 48 h weighed 16.5 g less than control chicks, which had access to food and water within 6 h of hatching, and also showed both physiological and behavioural signs of dehydration and thirst.

The fasting regime imposed on broilers before transportation may have important implications for welfare. At the very least, it affects the amount of weight lost, which occurs after 4–6 h of fasting at a rate of 0.2–0.5%/h as birds begin to metabolize body tissue (Veerkamp, 1986). Fasting is also aversive (Nicol and Scott, 1990) and may also increase stress (see below).

Broilers transported for 2, 4 or 6 h after feed withdrawal for 1–10 h had similar live and carcass weights to untransported controls (Warriss et al., 1993). However, transport significantly reduced liver weight and liver glycogen concentration. Glycogen depletion of the biceps muscle increased progressively with journey time, which could have reflected the muscular effort involved in maintaining balance in a moving vehicle. There was also evidence from this study that transported broilers were becoming dehydrated.

Withdrawal of food, or both food and water, for 24 h resulted in a 10% drop in live weight (0.43%/h), of which 41% was loss in carcass weight (Knowles et al., 1995). There was no evidence of significant dehydration in this study, nor in a survey of 800 broilers at two plants (Knowles et al., 1996).

A more recent study (Nijdam et al., 2005b) also noted significantly reduced bodyweight (0.42%/h) at slaughter in broilers that had had their feed withdrawn for 10 h before transport for 3 h plus 1 h lairage, compared with those that had had access to feed until transported, which had minimal weight loss. Changes in blood metabolites indicated stress and negative energy balance.

Physiological changes indicative of stress

Further physiological changes are associated with the crating and moving of poultry after they have been caught. Increased glucagon and plasma corticosterone (Freeman et al., 1984) and increased heterophil:lymphocyte ratios (Mitchell et al., 1992) have been found in broilers transported for 2–3 h. In a small study, Bedanova et al. (2006) reported significantly elevated haemoglobin and reduced erythrocyte counts in heavy

(3 kg) broilers crated at 105 cm^2/kg compared with those at 115 cm^2/kg and uncrated controls; they also found increased heterophil:lymphocyte ratios compared with the controls.

Thus, handling and crating per se appear to be stressful, possibly in part due to thermal stress, as the effects are more marked with further space restriction. The additional stress of transport under experimental conditions is variable. Broom et al., 1990 (cited in Knowles and Broom, 1990) found that the plasma corticosterone concentrations of hens rose significantly after crating, but found no subsequent difference between hens that had been left crated but stationary for 2 h and hens that had been transported in a van for 2 h. Duncan (1989) found that plasma corticosterone in broilers was significantly higher if they had been transported for 40 min after crating than if they had been left stationary in the crate in the same vehicle for 40 min.

Fear and aversion

Cashman et al. (1989) assessed the TI duration of nearly 700 broilers on arrival at four commercial processing plants. The overall mean TI duration was 12.6 min, a level comparable to that reported after exposure to high-intensity electric shock treatment (Gallup, 1973). Journey duration had the most significant effect on duration of TI. A strong positive linear relationship indicated that the birds' fear levels were primarily determined by transportation, and not just by the catching and loading procedures.

When a similar study was conducted with 300 laying hens, there was no evidence of a positive linear relationship between journey duration and the duration of TI (Mills and Nicol, 1990). However, the TI durations of hens undergoing short journeys were higher than those found for broilers. Differences in catching procedures for broilers and battery hens are a possible explanation, with handling effects probably contributing to the TI response of hens undergoing short journeys (Mills and Nicol, 1990).

Transportation involves simultaneous exposure to many factors, including noise, motion, heat and crowding. The birds' experience of some of these separate factors has been examined by tests of preference and aversion. Mean sound levels on animal transportation lorries typically fall within the range 95–103 dbA. The response of poultry to the specific noise encountered during transportation has not been assessed. It is generally thought that vibration is likely to be more aversive than noise.

Vibration

The fundamental frequency of most trucks used for poultry transport is 1–2 Hz, with a secondary peak at 10 Hz that coincides with the resonance frequency of poultry viscera (Scott, 1994). More detailed measurements by Randall et al. (1996) found resonant frequencies of broilers were around 15 Hz when sitting and 4 Hz when standing. Experimentally, passive avoidance procedures have been used to examine responses to vibration.

For example, Randall et al. (1997) used a wide range of horizontal and vertical vibrations, imposed for a 2-h period. They showed that both vertical and horizontal vibrations were aversive to broiler chickens, although a greater sensitivity to vertical vibration by a factor of between 1.3 and 2.5 was found at all acceleration levels. Aversion tended to increase with acceleration magnitude (0–5 m/s^2) and to decrease with increasing frequency (0–10 Hz) of the motion. As chickens find vibration < 5 Hz particularly aversive, Randall et al. (1997) concluded that the resonant frequencies of 1–5 Hz found on transporters are undesirable.

Although animals can reduce the effects of vibration by moving and by skeletal muscle tone, the scope for this in broilers with leg problems at high stocking density is very limited. Thus, evidence suggests that vibration does adversely affect the birds and should be reduced – for example, by using air suspension. Appropriate methodology to compare aversiveness of concurrent stressors during transport is being developed, initially using thermal and vibrational stressors (e.g. MacCalium et al., 2003).

Post-transport Handling and Environment

Thermal conditions at the end of the journey must be considered, as it can take 2–3 h to manually unload pullets. Broilers and spent hens may also have to wait at the processing plant (Warriss et al., 1990), either on the vehicle or unloaded in modules or stacks of crates. In both instances, a well-designed lairage is preferable to remaining outside exposed to the elements. It is important that the birds themselves receive adequate ventilation – measurements in two lairages found air movements around the stacks of modules > 1 m/s but < 0.1 m/s adjacent to the broilers (Quinn et al., 1998). Temperatures and humidities among the birds consequently rose rapidly to give rise to conditions of heat stress within 1 h in both winter and summer.

Average body temperatures rose by 0.3°C in the first hour of lairage and by 0.1°C thereafter for the 4 h of measurement (Warriss et al., 1999). The model birds used by Webster et al. (1992) and Weeks et al. (1997) indicated that conditions of substantial heat and cold stress were frequently experienced by hens and broilers in lairage during loading and unloading. Thus, the duration of such times needs to be kept to a minimum of preferably < 1 h.

A controlled environment providing adequate ventilation, while avoiding excessive wind and air movement on to the birds, is highly desirable. There should also be sufficient space around each module or stack for effective air exchange and flow. Monitoring of the condition of birds and their environment in lairage is as necessary then as during the journey. In practical terms, birds observed to be panting will become progressively dehydrated and increasingly heat-stressed. In some cases, the stress of transport may lead to PSE (pale, soft, exudative) meat in broilers similar to that in pigs. Showering heat-stressed broilers with water appears to be effective in reducing PSE and in improving breast meat quality (Guarnieri et al., 2004).

Following arrival at the processing plant, most broiler chickens and end-of-lay hens are manually removed from containers, although a mechanical unloading system has recently been developed and refined to improve broiler welfare (Tinker et al., 2005). Where electrical stunning is used, live birds are suspended by their legs from shackles for conveyance to the bath.

Many birds react to this potentially painful procedure by struggling, flapping their wings and attempting to right themselves. This can lead to injury and reduces the chance that the bird will be effectively stunned prior to slaughter. Jones and Satterlee (1997) reported that covering broilers' heads with a hood immediately before shackling reduced struggling in comparison with non-hooded controls. A smaller reduction in struggling was also obtained when the birds were fitted with transparent hoods, leading Jones et al. (1998a) to suggest that the effect was due both to the tactile properties of the hoods and to their ability to impair patterned vision.

Observations in US plants show that providing a breast rub made from strips of smooth conveyor belting will also reduce struggling and flapping. Reducing ambient light intensity by itself reduced struggling in broilers shackled in groups of three (Jones et al., 1998b), but not in broilers shackled individually (Jones et al., 1998a). The authors argue that the welfare benefits of reduced injury would outweigh any possible increases in fear suggested by the increased immobility observed when hoods are fitted. Fitting birds with hoods would not be practical on most commercial processing lines, but it may be possible to design a system for a slaughterhouse based on the principles of reduced light intensity, mild tactile contact and interference with clear vision.

Bird welfare would be greatly improved if the labour-intensive, stressful and often painful procedure of removing them from the containers and hanging them on shackles could be eliminated. Controlled atmosphere (gas) stunning of chickens and turkeys is now a commercial reality, with welfare and meat quality benefits such as reduced breast muscle haemorrhaging and bone breaks (Raj et al., 1997; Hoen and Lankhaar, 1999), and may also be used for ducks, which can be difficult to stun electrically (Raj et al., 1998). Automation of shackling is also being investigated (e.g. Lee, 2001), which will be easier with gas-stunned birds than conscious ones that may flap, struggle and experience pain when shackled (Sparrey and Kettlewell, 1994).

Training

The benefits for animal welfare of training stockpeople and handlers is increasingly recognized (Hester, 2005), with specific benefits from the alteration of attitudes (Hemsworth, 2003) and in handling and transport (Broom, 2005). Incentive programmes are also effective in reducing damage to birds. Providing incentive pay to employees will greatly reduce the incidence of broken wings during catching of broilers. In US plants, broken wings averaged 5–6%. The implementation of both incentive pay and auditing by restaurant company customers reduced broken wings to 1% or less in lightweight birds and < 3% in jumbo (3–3.5 kg) heavy birds. Canadian data from a large poultry company indicated that continuous auditing of catching has maintained the broken wing percentage of 1% for 2.1–2.6 kg chickens and < 1.5% for birds weighing 2.8–3.0 kg.

Conclusions

Systems of housing, catching, handling, transport and lairage that are more humane both for poultry and for human workers need further development. In many cases, mechanization is less stressful and more efficient. Improved techniques for depopulating houses need to be developed and considered as part of the system design. Machine harvesting of broilers is now more reliable and should be more widely adopted.

Scientific evidence demonstrates increasing stress and mortality in all classes of poultry as transportation time, holding time and feed and water deprivation time increase. Thermal stress is a major component of overall stress. Control of ventilation during transit should be provided to reduce this, as well as reducing the duration of all stages of transportation.

Hygiene regulations can conflict with animal welfare. EC hygiene regulations have forced the closure of several slaughterhouses in the UK and spent hens now have longer journeys.

Improved designs of vehicles are becoming commercially available. In the future, it may be possible to inspect birds in transit by accessing on-board information records of the conditions experienced on the journey thus far. The ability to access such information is a prerequisite for the sensible enforcement of welfare legislation or farm assurance standards.

References

Aitken, G. (1985) Poultry meat inspection as a commercial asset. *State Veterinary Journal* 39, 136–140.

Anderson, R.S. and Edney, A.T.B. (eds) (1991) *Practical Animal Handling*. Pergamon Press, Oxford, UK.

Baker, C.J., Dalley, S., Yang, X., Kettlewell, P. and Hoxey, R. (1996) An investigation of the aerodynamic and ventilation characteristics of poultry transport vehicles. 2. Wind tunnel experiments. *Journal of Agricultural Engineering Research* 65, 97–113.

Bayliss, P.A. and Hinton, M.H. (1990) Transportation of poultry with special reference to mortality rates. *Applied Animal Behaviour Science* 28, 93–118.

BBSRC (2002) *Science and Animal Welfare*. Biotechnology and Biological Sciences Research Council, Polaris House, North Star Avenue, Swindon, UK, p. 41 (also at http://www.bbsrc.ac.uk).

Bedanova, I., Voslarova, E., Vecerek, V., Pistekova, V. and Chloupek, P. (2006) Effects of reduction in floor space during crating on haematological indices in broilers. *Berliner un Munchener Tierartliche Wochenschrift* 119, 17–21.

Berg, C. and Algers, B. (2004) Using welfare outcomes to control intensification: the Swedish model. In: Weeks, C.A. and Butterworth, A. (eds) *Measuring and Auditing Broiler Welfare*. CAB International, Wallingford, UK, pp. 223–230.

Beuving, G. (1980) Corticosteroids in laying hens. In: Moss, R. (ed.) *The Laying Hen and its Environment*. Martinus Nijhoff, The Hague, Netherlands, pp. 65–82.

Bingham, A.N. (1986) Automation of broiler harvesting. *Poultry International* 25, 41–42.

Blokhuis, H.J. (2004) Recent developments in European and International welfare regulations. *World's Poultry Science Journal* 60 (4), 469–477.

Broom, D.M. (2005) The effects of land transport on animal welfare. *Revue Scientifique et Technique; Office International des Epizooties* 24 (2), 683–691.

Buyse, J., Simons, P.C.M., Boshouwers, F.M.G. and Decuypere, E. (1996) Effect of intermittent lighting, light intensity and source on the performance and welfare of broilers. *World's Poultry Science Journal* 52, 121–130.

Cashman, P.J., Nicol, C.J. and Jones, R.B. (1989) Effects of transportation on the tonic immobility fear reactions of broilers. *British Poultry Science* 30, 211–221.

Danbury, T.C., Weeks, C.A., Chambers, J.P., Waterman-Pearson, A.E. and Kestin, S.C. (2000) Preferential selection of the analgesic drug carprofen by lame broiler chickens. *Veterinary Record* 146, 307–311.

Duncan, I.J.H. (1989) The assessment of welfare during the handling and transport of broilers. In: Faure, J.M. and Mills, A.D. (eds) *The Proceedings of the Third European Symposium on Poultry Welfare*. French Branch of the World's Poultry Science Association, Tours, France, pp. 93–108.

Duncan, I.J.H., Slee, G., Kettlewell, P.J., Berry, P. and Carlisle, A.J. (1986) A comparison of the effects of harvesting broiler chickens by machine and by hand. *British Poultry Science* 27, 109–114.

Ekstrand, C. (1998) An observational cohort study of the effects of catching method on carcase rejection rates in broilers. *Animal Welfare* 7, 87–96.

EU (2005) *Proposal for a Council Directive Laying Down Minimum Rules for the Protection of Chickens Kept for Meat Production*. 2005 per 0099 (CNS) (also at http:www://europa.eu.int).

Farsaie, A., Carr, L.E. and Wabeck, C.J. (1983) Mechanical harvest of broilers. *American Society of Agricultural Engineers* 26, 1650–1653.

Freeman, B.M., Kettlewell, P.J., Manning, A.C.C. and Berry, P.S. (1984) The stress of transportation for broilers. *Veterinary Record* 114, 286–287.

Gallup, G.G. (1973) Tonic immobility in chickens: is a stimulus that signals shock more aversive than the receipt of shock? *Animal Learning and Behaviour* 2, 145–147.

Gerrits, A.R., de Kning, K. and Mighels, A. (1985) Catching Broilers. *Poultry* 1 (5) 20–23.

Gracey, J.F. (1986) *Meat Hygiene*, 8th edn. Baillière Tindall, London, pp. 455–458.

Gregory, N.G. and Austin, S.D (1992) Causes of trauma in broilers arriving dead at processing plants. *Veterinary Record* 131, 501–503.

Gregory, N.G. and Wilkins, L.J. (1990) Broken bones in chickens, 2. Effect of stunning and processing in broilers. *British Poultry Science* 31, 53–58.

Gregory, N.G. and Wilkins, L.J. (1992) Skeletal damage and bone defects during catching and processing. In: *Bone Biology and Skeletal Disorders in Poultry*, 23rd Poultry Science Symposium. World's Poultry Science Association, Edinburgh, UK.

Gregory, N.G., Wilkins, L.J., Eleperuha, S.D., Ballantyne, A.J. and Overfield, N.D. (1990) Broken bones in domestic fowls: effect of husbandry system and stunning method in end of lay hens. *British Poultry Science* 31, 59–70.

Grigor, P.N., Hughes, B.O. and Appleby, M.C. (1995) Effects of regular handling and exposure to an outside area on subsequent fearfulness and dispersal in domestic hens. *Applied Animal Behaviour Science* 44, 47–55.

Guarnieri, P.D., Soares, A.L., Olivio, R., Schneider, J.P., Macedo, R.M., Ida, E.I. and Shimokomaki, M. (2004) Preslaughter handling with water shower inhibits PSE (pale, soft, exudative) broiler breast meat in a commercial plant. Biochemical and ultrastructural observations. *Journal of Food Biochemistry* 28 (4), 269–277.

Hemsworth, P.H. (2003) Human–animal interactions in livestock production. *Applied Animal Behaviour Science* 81 (3), 185–198.

Hester, P.Y. (2005) Impact of science and management on the welfare of egg laying strains of hens. *Poultry Science* 84 (5), 687–696.

Hoen, T and Lankhaar, J. (1999) Controlled atmosphere stunning of poultry. *Poultry Science* 78 (2), 287–289.

Hoxey, R.P., Kettlewell, P.J., Meehan, A.M., Baker, C.J. and Yang, X. (1996) An investigation of the aerodynamic and ventilation characteristics of poultry transport vehicles. I. Full-scale measurements. *Journal of Agricultural Engineering Research* 65, 77–83.

Hunter, R.R., Mitchell, M.A. and Matheu, C. (1997) Distribution of 'dead on arrivals' within the bio-load on commercial broiler transporters: correlation with climatic conditions and ventilation regimen. *British Poultry Science* 38, S7–S9.

Jepersen, M. (1982) Injuries during catching and handling of broilers. In: Moss, R. (ed.) *Transport of Animals Intended for Breeding, Production and Slaughter.* Martinus Nijhoff, The Hague, Netherlands, pp. 39–44.

Jones, R.B. (1992) The nature of handling immediately prior to test affects tonic immobility fear reactions in laying hens and broilers. *Applied Animal Behaviour Science* 34, 247–254.

Jones, R.B. (1996) Fear and adaptability in poultry: insights, implications and imperatives. *World's Poultry Science Journal* 52,131–174.

Jones, R.B. and Satterlee, D.G. (1997) Restricted visual input reduces struggling in shackled broiler chickens. *Applied Animal Behaviour Science* 52 (1–2), 109–117.

Jones, R.B. and Waddington, D. (1992) Modification of fear in domestic chicks *Gallus gallus domesticus* via regular handling and early environmental enrichment. *Animal Behaviour* 43, 1021–1034.

Jones, R.B., Hagedorn, T.K. and Satterlee, D.G. (1998a) Adoption of immobility by shackled broiler chickens: effects of light intensity and diverse hooding devices. *Applied Animal Behaviour Science* 55, 327–335.

Jones, R.B., Satterlee, D.G. and Cadd, G.G. (1998b) Struggling responses of broiler chickens shackled in groups on a moving line: effects of light intensity, hoods and 'curtains'. *Applied Animal Behaviour Science* 58, 341–352.

Kannan, G. and Mench, J.A. (1997) Prior handling does not significantly reduce the stress response to pre-slaughter handling in broiler chickens. *Applied Animal Behaviour Science* 51, 87–99.

Kettlewell, P.J. and Turner, M.J.B. (1985) A review of broiler chicken catching and transport systems. *Journal of Agricultural Engineering Research* 31, 93–114.

Kettlewell, P., Mitchell, M. and Meehan, A. (1993) The distribution of thermal loads within poultry transport vehicles. *Agricultural Engineer* 48, 26–30.

Kettlewell, P.J., Hoxey, R.P. and Mitchell, M.A. (2000) Heat produced by broiler chickens in a commercial transport vehicle. *Journal of Agricultural Engineering Research* 75, 315–326.

Keutgen, H., Wurm, S. and Ueberschar, S. (1999) Pathologic changes in end-of-lay hens with regards to different housing systems. *Deutsche Tierarztliche Wochenschrift* 106, 127–133.

Knierim, U. and Gocke, A. (2003) Effect of catching broilers by hand or machine on rates of injuries and dead-on-arrivals. *Animal Welfare* 12 (1), 63–73.

Knowles, T.G. and Broom, D.M. (1990) The handling and transport of broilers and spent hens. *Applied Animal Behaviour Science* 28, 75–91.

Knowles, T.G. and Broom, B.M. (1993) Effect of catching method on the concentration of plasma corticosterone in end-of-lay battery hens. *Veterinary Record* 133, 527–528.

Knowles, T.G., Warriss, P.D., Brown, S.N., Edwards, J.E. and Mitchell, M.A. (1995) Response of broilers to deprivation of food and water for 24 hours. *British Veterinary Journal* 151, 197–202.

Knowles, T.G., Ball, R.C., Warriss, P.D. and Edwards, J.E. (1996) A survey to investigate potential dehydration in slaughtered broiler chickens. *British Veterinary Journal* 152, 307–314.

Knowles, T.G., Brown, S.N., Warriss, P.D., Butterworth, A. and Hewitt, L. (2004) Welfare aspects of chick handling in broiler and laying hen hatcheries. *Animal Welfare* 13 (4), 409–418.

Kolb, E. and Seehawer, J. (2001) Significance and application of ascorbic acid in poultry. *Archiv fur Gerflugelkunde* 65, 106–113.

Kristensen, H.H., Berry, P.S. and Tinker, D.B. (2001) Depopulation systems for spent hens – a preliminary evaluation in the United Kingdom. *Journal of Applied Poultry Research* 10 (2), 172–177.

Lacy, M.P. and Czarick, M. (1998) Mechanical harvesting of broilers. *Poultry Science* 77, 1794–1797.

Lee, K.M. (2001) Design criteria for developing an automated live-bird transfer system. *IEEE Transactions on Robotics and Automation* 17 (4), 483–490.

MacCalium, J.M., Abeyesinghe, S.M., White, R.P. and Wathes, C.M. (2003) A continuous-choice assessment of the domestic fowl's aversion to concurrent transport stressors. *Animal Welfare* 12, 95–107.

MAFF (1997) *The Welfare of Animals (Transport) Order, 1997.* The Stationery Office Ltd, London.

Mallia, J.G., Vaillancourt, J.P., Martin, S.W. and McEwen, S.A. (2000) Risk factors for abattoir condemnation of turkey carcasses due to cyanosis in southern Ontario. *Poultry Science* 79, 831–837.

McEwen, S.A. and Barbut, S. (1992) Survey of turkey downgrading at slaughter – carcass defects and associations with transport, toenail trimming, and type of bird. *Poultry Science* 71 (7), 1107–1115.

McGeown, D., Danbury, T.C., Waterman-Pearson, A.E. and Kestin, S.C. (1999) The effect of carprofen on lameness in broiler chickens. *Veterinary Record* 144, 668–671.

Meijerhof, R. (1997) The importance of egg and chick transportation. *World Poultry* 13 (11), 17–18.

Mench, J. (2004) Lameness. In: Weeks, C.A. and Butterworth, A. (eds) *Measuring and Auditing Broiler Welfare.* CAB International, Wallingford, UK, pp. 3–18.

Metheringham, J. (1996) Poultry in transit – a cause for concern? *British Veterinary Journal* 152, 247–250.

Mills, D.S. and Nicol, C.J. (1990) Tonic immobility in spent hens after catching and transport. *Veterinary Record* 126, 210–212.

Mitchell, M.A. and Kettlewell, P.J. (1994) Road transportation of broiler chickens – induction of physiological stress. *World's Poultry Science Journal* 50, 57–59.

Mitchell, M.A. and Kettlewell, P.J. (1998) Physiological stress and welfare of broiler chickens in transit: solutions, not problems! *Poultry Science* 77, 1803–1814.

Mitchell, M.A. and Kettlewell, P.J. (2004) Transport and handling. In: Weeks, C.A. and Butterworth, A. (eds) *Measuring and Auditing Broiler Welfare.* CAB International, Wallingford, UK pp. 145–160.

Mitchell, M.A., Kettlewell, P.J. and Maxwell, M.H. (1992) Indicators of physiological stress in broiler chickens during road transportation. *Animal Welfare* 1, 92–103.

Moran, P. and Berry, P.S. (1992) Mechanised broiler harvesting. *Farm Buildings and Engineering* 91, 24–27.

Nicol, C.J. (1992) Effects of environmental enrichment and gentle handling on behaviour and fear responses of transported broilers. *Applied Animal Behaviour Science* 33, 367–380.

Nicol, C.J. and Scott, G.B. (1990) Transport of broiler chickens. *Applied Animal Behaviour Science* 28, 57–73.

Nijdam, E., Arens, P., Lambooij, E., Decuypere, E. and Stegman, J.A. (2004) Factors influencing bruises and mortality of broilers during catching, transport and lairage. *Poultry Science* 83 (9), 1610–1615.

Nijdam, E., Delezie, E., Lambooij, E., Nabuurs, M.J.A., Decuypere, E. and Stegman, J.A. (2005a) Processing, products and food safety – comparison of bruises and mortality, stress parameters and meat quality in manually and mechanically caught broilers. *Poultry Science* 84 (3), 467–474.

Nijdam, E., Delezie, E., Lambooij, E., Nabuurs, M.J.A., Decuypere, E. and Stegman, J.A. (2005b) Feed withdrawal of broilers before transport changes plasma hormone and metabolite concentrations. *Poultry Science* 84 (7), 1146–1152.

Parry, R.T. (1989) Technological developments in pre-slaughter handling and processing. In: Mead, G.C. (ed.) *Processing of Poultry.* Elsevier, Amsterdam, pp. 65–101.

Phelps, P., Bhutada, A., Bryan, S., Chalker, A., Ferrell, B., Neuman, S., Ricks, S., Tran, H. and Butt, T. (2003) Automated identification of male layer chicks prior to hatch. *World's Poultry Science Journal* 59 (1), 33–38.

Poultry International (1998) At last – fully automated livebird harvesting. March, 44–48.

Prescott, N.B., Berry, P.S., Haslam, S. and Tinker, D.B. (2000) Catching and crating turkeys: effects on carcass damage, heart rate and other welfare parameters. *Journal of Applied Poultry Research* 9, 424–432.

Quinn, A.D., Kettlewell, P.J., Mitchell, M.A. and Knowles, T. (1998) Air movement and thermal microclimates observed in poultry lairages. *British Poultry Science* 39, 469–476.

Raj, A.B.M., Wilkins, L.J., Richardson, R.I., Johnson, S.P. and Wotton, S.B. (1997) Carcase and meat quality in broilers either killed with a gas mixture or stunned with an electric current under commercial processing conditions. *British Poultry Science* 38, 169–174.

Raj, A.B.M., Richardson, R.I., Wilkins, L.J. and Wotton, S.B. (1998) Carcase and meat quality in ducks killed with either gas mixtures or an electric current under commercial processing conditions. *British Poultry Science* 39, 404–407.

Randall, J.M., Cove, M.T. and White, R.P. (1996) Resonant frequencies of broiler chickens. *Animal Science* 62, 369–374.

Randall, J.M., Duggan, J.A., Alami, M.A. and White, R.P. (1997) Frequency weightings for the aversion of broiler chickens to horizontal and vertical vibration. *Journal of Agricultural Engineering Research* 68, 387–397.

Reed, H.J., Wilkins, L.J., Austin, S.D. and Gregory, N.G. (1993) The effects of environmental enrichment during rearing on fear reactions and depopulation trauma in adult caged hens. *Applied Animal Behaviour Science* 36, 39–46.

Ritz, C.W., Webster, A.B. and Czarick, M. (2005) Evaluation of hot weather and incidence of mortality associated with broiler live haul. *Journal of Applied Poultry Research* 14 (3), 594–602.

Rodenburg, T.B., Bracket, M.B.M., Berk, J., Cooper, J., Faure, J.M., Guémené, D., Guy, G., Harlander, A., Jones, T., Knierim, U., Kuhnt, K., Pingel, H., Reiter, K., Servière, J. and Ruis, M.A.W. (2005) Welfare of ducks in European duck husbandry systems. *World's Poultry Science Journal* 51, 633–646.

Sandilands, V., Sparks, N., Wilson, S. and Nevison, I. (2005) Laying hens at depopulation: the impact of the production system on bird welfare. *British Poultry Abstracts* 1, 23–24.

Sanotra, G.S., Vestergaard, K.S. and Thomsen, M.G. (1998) The effect of stocking density on walking ability, tonic immobility, and the development of tibial dyschondroplasia in broiler chicks. *Poultry Science* 77, 1844–1845.

Scott, G.B. (1993) Poultry handling: a review of mechanical devices and their effect on bird welfare. *World's Poultry Science Journal* 49, 44–57.

Scott, G.B. (1994) Effects of short-term whole-body vibration on animals with particular reference to poultry. *World's Poultry Science Journal* 50, 25–38.

Scott, G.B. and Moran, P. (1992) Behavioural responses of laying hens to carriage on horizontal and inclined conveyors. *Animal Welfare* 1, 269–277.

Scott, G.B. and Moran, P. (1993) Fear levels in laying hens carried by hand and by mechanical conveyors. *Applied Animal Behaviour Science* 36, 337–345.

Scott, G.B., Connell, B.J. and Lambe, N.R. (1998) The fear levels after transport of hens from cages and a free-range system. *Poultry Science* 77, 62–66.

Sherwin, C.M., Kestin, S.C., Nicol, C.J., Knowles, T.G., Brown, S.N., Reed, H.J. and Warriss, P.D. (1993) Variation in behavioural indices of fearfulness and fatigue in transported broilers. *British Veterinary Journal* 149, 571–578.

Sparrey, J.M. and Kettlewell, P.J. (1994) Shackling of poultry – is it a welfare problem? *World's Poultry Science Journal* 50, 167–176.

Stuart, C. (1985) Ways to reduce downgrading. *World Poultry Science* 41, 16–17.

Su, G., Sørensen, P. and Kestin, S.C. (1999) Meal feeding is more effective than early feed restriction at reducing the prevalence of leg weakness in broiler chicken. *Poultry Science* 78, 949–955.

Tinker, D., Berry, P., White, R., Prescott, N., Welch, S. and Lankhaar, J. (2005) Improvement in the welfare of broilers by changes to a mechanical unloading system. *Journal of Applied Poultry Research* 14 (2), 330–337.

Veerkamp, C.H. (1986) The influence of fasting and transport on yields of broilers. *Poultry Science* 57, 619–627.

Warriss, P.D. and Brown, S.N. (1996) Time spent by turkeys in transit to processing plants. *Veterinary Record* 139, 72–73.

Warriss, P.D., Kestin, S.C., Brown, S.N. and Bevis, E.A. (1988) Depletion of glycogen reserves in fasting broiler chickens. *British Poultry Science* 29, 149–154.

Warriss, P.D., Bevis, E.A. and Brown, S.N. (1990) Time spent by broiler chickens in transit to processing plants. *Veterinary Record* 127, 617–619.

Warriss, P.D., Bevis, E.A., Brown, S.N. and Edwards, J.E. (1992a) Longer journeys to processing plants are associated with higher mortality in broiler chickens. *British Poultry Science* 33, 201–206.

Warriss, P.D., Kestin, S.C. and Edwards, J.E. (1992b) Responses of newly hatched chicks to inanition. *Veterinary Record* 130, 49–53.

Warriss, P.D., Kestin, S.C., Brown, S.N., Knowles, T.G., Wilkins, L.J., Edwards, J.E., Austin, S.D. and Nicol, C.J. (1993) The depletion of glycogen stores and indices of dehydration in transported broilers. *British Veterinary Journal* 149, 391–398.

Warriss, P.D., Knowles, T.G., Brown, S.N., Edwards, J.E., Kettlewell, P.J., Mitchell, M.A. and Baxter, C.A. (1999) Effects of lairage time on body temperature and glycogen reserves of broiler chickens held in transport modules. *Veterinary Record* 145, 218–222.

Warriss, P.D., Pagazaurtundua, A. and Brown, S.N. (2005) Relationship between maximum daily temperature and mortality of broiler chickens during transport and lairage. *British Poultry Science* 46, 647–651.

Watt Poultry Statistical Yearbook (2005) Watt Publishing Co., Petersfield, UK.

Webster, A.J.F., Tuddenham, A., Saville, C.A. and Scott, G.A. (1992) Thermal stress on chickens in transit. *British Poultry Science* 34, 267–277.

Weeks, C.A. and Kestin, S.C. (1997) The effect of leg weakness on the behaviour of broilers. In: Koene, P. and Blokhuis, H.J. (eds) *Proceedings of the 5th European Symposium on Poultry Welfare*, Wageningen,

Netherlands, pp. 117–118 (published for Working Group IX of the European Federation of the World's Poultry Science Association, ISBN 90-6754-486-8).

Weeks, C.A., Webster, A.J.F. and Wyld, H.M. (1997) Vehicle design and thermal comfort of poultry in transit. *British Poultry Science* 38, 464–474.

Wilkins, L.J., Brown, S.N., Zimmerman, P.H., Leeb, C. and Nicol, C.J. (2004) Investigation of palpation as a method for determining the prevalence of keel and furculum damage in laying hens. *Veterinary Record* 155, 547–549.

Wyers, M., Cherel, Y. and Plassiart, G. (1991) Late clinical expression of lameness related to associated osteomyelitis and tibial dyschondroplasia in male breeding turkeys. *Avian Diseases* 35, 408–414.

Zulkifli, I., Norma, M.T.C., Chong, C.H and Loh, T.C. (2000) Heterophil to lymphocyte ratio and tonic immobility reactions to preslaughter handling in broiler chickens treated with ascorbic acid. *Poultry Science* 79, 402–406.

19 Stress Physiology of Animals During Transport

Toby Knowles* and Paul Warriss

*School of Veterinary Science, University of Bristol, Bristol;
Toby.Knowles@bristol.ac.uk

Introduction

There is an increasing public interest in, and concern for, the welfare of livestock during transport. The majority of people now live in towns and cities and are no longer in day-to-day contact with farm animals. They are relatively unfamiliar with the animals and the methods of husbandry under which they are kept, and to a large extent have an idealized picture of farming and animal production.

However, there is one point in most animal production systems that is commonly open to public view – when the animals are transported. However, although necessary, transport is generally an exceptionally stressful episode in the life of the animal and one that is sometimes far removed from an idealized picture of animal welfare.

Therefore, increasing public concern for animals during transport has spurred research into their welfare: research which has attempted to quantify the severity of the stress imposed by the various stages involved in transport and to identify acceptable conditions and methods to minimize the adverse effects of transport.

Most work has concentrated on quantifying and ameliorating the effects of road transport, as this is the major mode of animal transport, and it is road transport that is the main theme of this chapter. However, there have also been research programmes targeted at other forms of transport: the export of cull sheep from Australia

to the Middle East by ship can result in exceptionally high mortality rates during the sea journey. This has prompted the Australian government to fund research into the problem. An introduction to the literature covering this research, and that of shipping cattle by sea, can be found in Norris (2005).

A limited number of livestock, usually only those of high value, are transported by air. Recommendations for the transportation of live animals by air are detailed in the IATA (2005) Live Animal Regulations, which are updated annually to take account of the latest research findings. These regulations are enforced by the European Union (EU), the USA and many other countries for the air transport of all live animals.

The assessment of the welfare of animals during transport in any sort of objective and scientific way requires the measurement of something, in a quantifiable and repeatable manner. Broom (1986) defined an animal's welfare as 'the state of an individual with regard to its attempts to cope with its environment'. Within this definition, an animal attempts to maintain homeostasis through physiological and behavioural changes, and it follows that the greater the behavioural or physiological changes that are required, the more an animal is having to do to cope with the situation or environment, and the poorer its welfare is likely to be.

This approach provides a working basis by which welfare can be judged and is very much in line with the clinical biochemical approach to the diagnosis of disease in both human and veterinary medicine. In clinical biochemistry, one or a number of measured biochemical or haematological variables in an individual are compared to population norms in order to identify specific disorders (see, for example, Farver, 1997). This sounds fairly straightforward; however, welfare itself is not an objective, measurable thing but is an entirely human concept and as such cannot escape a high degree of subjective interpretation. Even when we can communicate with the animal that is being assessed – that is, within our own species – there arise differences in the assessment of the welfare of individuals owing to differences between 'assessees' and in the opinions and backgrounds of the assessors, and these are not the only source of variability.

As a whole, society's idea of what is acceptable human welfare has changed over time. How much more difficult it is then, to try to 'second guess' the welfare of an animal with which we cannot communicate and which is unlikely to view or interpret its situation in anything approaching our own, human terms and, furthermore, for a range of people then to come to an overall agreement on the level of its welfare?

Thus, the idea of measuring the magnitude of the behavioural and/or physiological adjustments that an animal has to make to cope with its environment provides a useful structure underpinning the assessment of an animal's welfare. However well scientifically founded the measurements, their interpretation cannot escape a high degree of subjective interpretation. We might ask 'what is an unacceptable level of mortality?', when, however well animals are transported, there will always be some deaths. We can measure increasing 'hunger' and dehydration in an animal by changes in blood biochemicals, but how hungry or thirsty can that animal be allowed to become before the situation is unacceptable, when the biochemical changes that are observed increase linearly over time? During mating, play or hunting, many of the biochemical variables that are commonly used as measures of animal welfare reach extreme values, but most people would not consider the welfare of an animal in these situations to be impaired.

The remainder of this chapter gives an introduction to the main physiological variables that have been used to assess the stress imposed on animals by transport. As far as possible these appear in functional groups. That is, they have been grouped as indicators of the various effects that are of interest: food deprivation, dehydration, muscular effort, etc. Following this we summarize the best practice relating to research to date. Some further details of species-specific research on transport can be found in other chapters, but additionally there are a number of reviews in the scientific literature listed in Box 19.1.

Physiological Variables

For a healthy, rested animal of a given species there is a range of values for each biochemical and haematological variable within which the level of each measure for any individual would normally be expected to fall. The distribution of values found in a healthy, rested population usually forms the familiar, bell-shaped, Gaussian distribution, except for the values of enzymes, for which the distribution is positively skewed, having a greater number of higher values. Published veterinary reference ranges for variables are quoted as the range of values within which 95% of the population would be expected to fall. These limits are the 2.5 and 97.5 percentiles of any distribution and are approximately equivalent to ± two standard deviations about the mean when the variable does have a Gaussian distribution.

Figure 19.1 shows the frequency distributions of plasma albumin levels and of the enzyme creatine kinase (CK) from control samples obtained from cattle in a study by Knowles et al. (1999a). The distribution of albumin values is very close to the Gaussian curve which is superimposed on the graph, whilst the distribution of CK values is far from Gaussian. The 2.5 and 97.5 centiles of the albumin values are 34.9 and 45.4 g/l, respectively, and for CK 58.6 and 302.4 U/l. The mean and standard deviation provide a useful summary of the albumin data, and the percentiles are close to the mean ± two standard deviations. This is not the case for the distribution of the CK values, which are strongly right-skewed.

Published reference ranges are useful in the diagnoses of a wide variety of diseases and can be useful for evaluating hypo- and hyperthermia, the degree of dehydration and, to a lesser extent, the degree of hunger arising during

Box 19.1. Published reviews from the scientific literature covering the road transport of livestock.

Adult cattle

Eicher, S.D. (2001) Transportation of cattle in the dairy industry: current research and future directions. *Journal of Dairy Science* 84 (E Suppl.), E19–E23.

Fike, K. and Spire, M.F. (2006) Transportation of cattle. *Veterinary Clinics of North America Food Animal Practice* 22, 305–320.

Knowles, T.G. (1999) A review of the road transport of cattle. *Veterinary Record* 144, 197–201.

Swanson, J.C. and Morrow-Tesh, J. (2001) Cattle transport historical, research and future perspectives. *Journal of Animal Science* 79 (E Suppl.), E102–E109.

Tarrant, P.V. (1990) Transportation of cattle by road. *Applied Animal Behaviour Science* 28, 153–170.

Warriss, P.D. (1990) The handling of cattle pre-slaughter and its effects on carcass meat quality. *Applied Animal Behaviour Science* 28, 171–186.

Calves

Knowles, T.G. (1995) A review of post-transport mortality among younger calves. *Veterinary Record* 137, 406–407.

Trunkfield, H.R. and Broom, D.M. (1990) The welfare of calves during handling and transport. *Applied Animal Behaviour Science* 28, 135–152.

Sheep

Knowles, T.G. (1998) A review of the road transport of sheep. *Veterinary Record* 143, 212–219.

Pigs

Tarrant, P.V. (1989) The effects of handling, transport, slaughter and chilling on meat quality and yield in pigs – a review. *Irish Journal of Food Science and Technology* 13, 79–107.

Warriss, P.D. (1987) The effect of time and conditions of transport and lairage on pig meat quality. In: Tarrant, P.V., Eikelenboom, G. and Monin, G. (eds) *Evaluation and Control of Meat Quality in Pigs*. Martinus Nijhoff Publishers, Dordrecht, Netherlands, pp. 245–264.

Warriss, P.D. (1998a) The welfare of slaughter pigs during transport. *Animal Welfare* 7, 365–381.

Warriss, P.D. (1998b) Choosing appropriate space allowances for slaughter pigs transported by road: a review. *Veterinary Record* 142, 494–454.

Cattle, sheep and goats

Wythes, J.R. and Morris, D.G. (1994) *Literature Review of Welfare Aspects and Carcass Quality Effects in the Transport of Cattle, Sheep and Goats (Parts A, B and C)*. Report prepared by Queensland Livestock and Meat Authority for Meat Research Corporation, Queensland Livestock and Meat Authority, Australia.

Sheep and pigs

Hall, S.J.G. and Bradshaw, R.H. (1998) Welfare aspects of the transport by road of sheep and pigs. *Journal of Applied Animal Welfare Science* 1, 235–254.

Deer

Weeks, C.A. (2000) Transport of deer: a review with particular relevance to red deer (*Cervus elaphus*). *Animal Welfare* 9, 63–74.

transport. However, it should be remembered that most of the physiological changes seen during transport are due to the action of normal homeostatic mechanisms taking place within a healthy population of animals in response to the variety of different stressors (see Table 19.1).

Thus, clinical reference ranges are of limited use in evaluating welfare during transport, as the animals being transported are generally all healthy. So, care should be taken in drawing any conclusions from comparisons with published normal ranges.

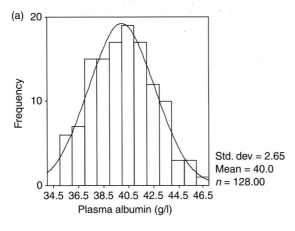

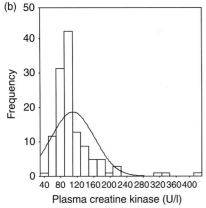

Fig. 19.1 The frequency histograms of (a) plasma albumin and (b) creatine kinase (CK) from rested cattle from a study by Knowles *et al.* (1999a). The Gaussian curves for distributions with the same means and standard deviations as the data are superimposed.

Table 19.1. Commonly used physiological indicators of stress during transport.

Stressor	Physiological variable
Measured in blood	
Food deprivation	↑ FFA, ↑ β-OHB, ↓ glucose, ↑ urea
Dehydration	↑ osmolality, ↑ total protein, ↑ albumin, ↑ PCV
Physical exertion	↑ CK, ↑ lactate
Fear/arousal	↑ cortisol, ↑ PCV
Motion sickness	↑ vasopressin
Other measures	
Fear/arousal and physical exertion	↑ heart rate, ↑ respiration rate
Hypothermia/ hyperthermia	body temperature, skin temperature

What is really of interest is the change of a variable over time within an individual animal, as this is an indication of the scale of the response that an animal is mounting in order to cope. And because of the inherent variability between individuals, measurements are best taken at the level of the individual over time. The following is an overview of some physiological variables that can give some insight into how an animal is coping with a given situation. However, instead of moving directly to descriptions of individual biochemical and haematological markers, we start with the ultimate indicator of the inability of an animal to cope.

Mortality

Mortality is a useful indicator of physiological stress. When an animal dies during transport it is because its physiological mechanisms have failed to maintain homeostasis. That transport is stressful is most readily quantified by the increased mortality that accompanies it. On average, in a given time period, a greater number of livestock will die if they are transported than if they are not. Of course, the animals that die initially are often those that are weaker, and the overall mortality rates during road transport for most livestock are usually only fractions of a percentage point. This means that the rate can only be accurately estimated when large enough numbers are surveyed. However, increased mortality is interpreted by most people as an indicator of poor welfare and of the stressful nature of transport, in a way that many other types of measures are not so easily agreed upon.

If an increase in mortality is seen, even if mortality is only occurring amongst weaker animals, it is an indication that conditions are harsher and that the animals that do survive are probably facing a greater challenge. If a large enough number of animals is being considered, mortality rate is also a useful measure of the relative 'stressfulness' of methods of handling and transport. For instance, Warriss et al. (1992) surveyed journey times and mortality rate amongst broilers transported to slaughter and found a strong relationship between the two variables. The results showed a marked non-linear increase in bird mortality for journeys greater than 4 h, strongly suggesting that transport for longer than 4 h was undesirable.

However, results like these are more easily obtained from broilers, as they have the highest mortality rates amongst the commonly transported types of livestock (the high mortality rates in broilers are due in part to the heavy commercial selection for improved growth rate and feed conversion to slaughter at 42 days, which has rather disregarded bird viability much beyond this age. The same trend can be seen occurring in pigs, which are under similar selection pressures but have a longer reproductive cycle and are not yet at the same level of selection).

Liveweight, β-hydroxybutyrate (β-OHB), free fatty acids and liver glycogen as indicators of fasting

Transport can involve extended periods without food or water and, as a consequence, there is an initial loss of live weight which is predominantly due to loss of gut fill; approximately 7% of body weight in ruminants and 4% in pigs is lost during the first 18–24 h. Generally, weight loss during transport is accelerated compared with when an animal is simply deprived of food and water and not transported. In ruminants, the main loss of gut fill takes place during the first 18–20 h of transport. Loss of gut fill in itself is unlikely to be directly deleterious to the animal. There is, however, an approximately linear loss of body weight, measured as a decrease in carcass weight, which is due to dehydration and to the use of body reserves. This can be measured as the rate of loss of carcass weight and, within species, has been found to show quite large variation across different studies, these differences being due to the condition of the animals, the environment and the conditions of transport.

Once an animal is deprived of food and water, it has to rely on its body reserves to buffer it, until it can feed and drink again. The main energy store in the body is in the form of lipids, and by far the most important of these are triacylglycerols (or triglycerides), which can also provide thermal insulation. Triacylglycerols are mobilized by breaking them down into the constituent glycerol and fatty acids. These non-esterified fatty acids (NEFA) or free fatty acids (FFA) are transported in the blood bound to proteins. Triacylglycerols can be synthesized by many types of cell, but most synthesis takes place in the liver, adipose tissue and the small intestine.

Lipolysis, the mobilization of body fat, is under hormonal control. As an animal fasts, much less glucose is available from the gut or glycogen reserves, and this results in decreased levels of glucose in the blood plasma. This, in turn, leads to hormone changes, increased glucagon levels and decreased insulin levels, which trigger hormone-sensitive lipase to break down adipose triacylglycerols, which are hydrolysed to FFA and glycerol. FFA can be utilized directly by most tissues and, as an insoluble lipid, is bound to albumin in the plasma for transport around the body, whilst glycerol is transported dissolved in plasma water. Thus, during starvation, levels of FFA rise in the plasma, whilst actions that promote FFA synthesis suppress lipolysis and plasma FFA levels are not elevated.

The liver holds a reserve of glycogen and, during the first day of fasting, this reserve diminishes rapidly. Levels of liver glycogen can be measured by biopsy or at slaughter. Changes in liver weight can also be used as a measure of the use of these reserves. There are also reserves of glycogen within the skeletal muscle that tend to be conserved, even after several days of fasting.

During fasting, the usual metabolic pathways are modified and greater amounts of ketones are produced from FFA in the liver. Very high levels of FFA are damaging to tissues. The liver converts these to ketones. One of the main ketones is β-hydroxybutyrate (β-OHB) (or 3-hydroxybutyrate (3-OHB)). Because FFA can be utilized by most tissues, it was not clear until recently why ketones were produced. It now appears, however, that many tissues more easily utilize β-OHB than FFA. In fact, in some species such as man, ketones form the main energy source for the brain during fasting.

This is not the case with the sheep or pig, where the brain still relies on glucose as the main energy source. Ketones are the main fuel of resting skeletal muscle during short-term fasting but, during long-term starvation or exercise, FFA become the main energy source. There is a biological limit to the amount of FFA that can be present in the plasma, as all FFA have to be bound to albumin for transport. Levels of ketones in the plasma are not restricted in this way, which is important, as levels of plasma albumin decrease during fasting, thus reducing the amount of FFA that can be transported.

During exercise, glucose, ketones and FFA are all used as fuel. After strenuous exercise, ketone oxidation by muscles is reduced and this leads to an increase in plasma levels of FFA and β-OHB that may be several times higher than pre-exercise levels. However, for several minutes immediately after exercise, levels may fall momentarily below pre-exercise levels as the metabolism adjusts.

Plasma osmolality, total protein, albumin and packed cell volume as indicators of dehydration

Water is essential to all of the processes that take place within the body, accounting for 60% of the total body weight of most domestic animals. However, adipose tissue contains little water, and 'fat' animals such as fattened lambs and pigs will contain a lower percentage of water. Total body water is considered, physiologically, to be made up of the extracellular fluid (ECF) volume and intracellular fluid (ICF) volume, where the ECF is all fluid outside the cells. Fluid present in the gut is sometimes considered as part of the ECF.

In ruminants, the forestomach may contain a substantial amount of fluid: up to 30–60 l in adult cattle, which during periods of water deprivation can act as a buffer in maintaining effective circulating volume. During periods of inadequate water uptake, the water losses are balanced proportionately between the ICF and non-gut ECF, thus electrolyte balance between the two is maintained. Packed cell volume (PCV), total plasma protein and plasma albumin are convenient and simple measures of dehydration.

PCV is the percentage of the blood volume occupied by cells (predominantly the red blood cells), the remainder of the volume being fluid. Thus, as long as there is no loss or gain of cells, PCV is a measure of the plasma volume. However, many species have a reserve of red blood cells in the spleen that are readily released in response to excitement and stressors, so it is useful to use total plasma protein and albumin levels in conjunction with PCV when assessing levels of hydration.

The assumption is made that the total amount of protein present in the plasma remains the same. Both total plasma protein and plasma albumin should show the same type of change if the change is due to dehydration and not to a dietary effect. It should be noted that the percentage changes in protein and PCV will not be the same for a given loss of plasma volume, e.g. a 50% plasma volume deficit would result in a 100% increase in protein but only, perhaps, a 40% increase in PCV.

Osmolality can be used as a further, simple measure of plasma water content as it is a colligative property and therefore includes all solute species. As a rough guide to the extent of dehydration, clinical signs are usually apparent when 4–6% of total body weight of 'effective' (not including fluid in the gut) total body water has been lost; moderate dehydration is when

8–10% has been lost and severe dehydration is said to occur when losses are greater than 12% (Carlson, 1997).

Heart rate, respiration rate, plasma cortisol and glucose as indicators of a general reaction to stress

An initial response to stress is the release of the hormones adrenaline and noradrenaline into the bloodstream from the adrenal glands. Noradrenaline is also released from sympathetic nerve endings, where it can act directly. The release of these hormones causes an acute increase in heart rate and blood pressure and stimulates hepatic glycogenolysis. This leads to an increased availability of glucose and a rise in plasma glucose levels within minutes. The effects of these hormones provide a useful measure of stress, but the hormones themselves have a rather short half-life in the blood stream and direct measurement of them is problematic.

In the slightly longer term, an animal's response to stress is mediated mainly through the hypothalamus–pituitary system, a system in which neural and endocrine control systems are integrated in such a highly complex and inter-dependent manner that it is only described superficially here. Glucocorticoid hormones, pro-duced in and released from the cortex of the adrenal glands in response to an extremely wide range of stimuli/stressors, play a major role in mediating the physiological response. Cortisol is the central glucocorticoid in mammalian farm species and corticosterone in avian species. The pathway leading to the release and control of cortisol acts through the hypothalamus, pitu-itary and the adrenal cortex and is summarized in Fig. 19.2.

The glucocorticoids play a major role in glu-cose metabolism, inhibition of protein synthesis, initiation of proteolysis and the modulation of immunological mediators such as lymphokines and as mediators of inflammatory reactions, caus-ing anti-inflammatory effects. Because of the role of the brain in the release of glucocorticoids, they are widely interpreted as a measure of an animal's psychological perception of a situation, in addition to the extent of its physiological reaction.

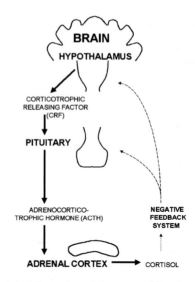

Fig.19.2. The main pathways controlling the release of cortisol.

Creatine kinase, muscle glycogen and lactate as indicators of physical activity

The enzyme creatine kinase (CK, also referred to as creatine phosphokinase (CPK)) is present in muscle, where it makes adenosine triphosphate (ATP) available for use in muscle contraction by the phosphorylation of adenosine diphosphate (ADP) from creatine phosphate. It appears in the circulating plasma as a result of tissue dam-age and is relatively organ-specific, occurring as three isoenzymic forms with an additional fourth variant that derives from mitochondria. Identification of the relative levels of isoenzymes present in the blood allows determination of the tissue which is the source – and the relative extent to which damage has occurred. During exercise there is increasing CK3 (the main isoenzyme present in muscle and also known as CK-MM) activity present in the blood as it leaks from the cells of skeletal muscle. Lactate dehydrogenase (LDH) has also been used as a measure of mus-cle damage; however, LDH activity is high in various tissues throughout the body and mea-surements cannot be so organ-specific.

During exercise, the main fuels for muscular contraction are glucose and fatty acids from the blood. There is also an intramuscular carbohydrate reserve in the form of glycogen, and it is when muscle glycogen stores are depleted that exhaustion has been shown to set in. The main extramuscular carbohydrate source is glycogen in the liver. Reserves of muscle and liver glycogen may be measured by biopsy but, as most transport of animals is to slaughter, it is usually assayed in muscle and liver sampled immediately after slaughter. Metabolism of glucose can take place aerobically or anaerobically, in which latter case there may be a gradual build-up of lactate.

Lipids can be metabolized only aerobically. At the start of fairly intense exercise, metabolism is mainly anaerobic. If the exercise is not too strenuous, aerobic metabolism of glucose and lipids takes over and lactate production decreases. The harder the exercise is, the higher is the percentage use of glucose over lipid, thus lactate production is closely correlated with the intensity of exercise and may be seen as increased levels of lactate in muscle and in plasma.

The degree to which the reserves of muscle glycogen are depleted at the time of slaughter has an effect on the post-mortem changes that take place in the muscle. If glycogen reserves have been depleted to any great extent, the muscle produces meat of an inferior quality that looks dark, tends to have a less acceptable eating quality and is more prone to microbial spoilage, partly because it has a higher pH (is less acidic). Meat with this quality problem is commonly referred to as 'dark, firm and dry', or DFD, and is most prevalent amongst cattle and pigs, being less common in sheep.

Urea

Any process which increases protein catabolism will tend to result in increased levels of plasma urea. Thus, levels of urea increase in response to stress, when levels of cortisol increase, and they will also rise as a result of food deprivation.

Vasopressin as an indicator of travel sickness

The major role of the hormone vasopressin is to regulate body water homeostasis, in regard to the relative osmolality of the ECF and ICF, by affecting reabsorption. Its release is mainly triggered by increased plasma osmolality and it acts by causing water retention. However, its release

is relatively insensitive to changes in plasma volume, thus it normally plays little role in the maintenance of overall water balance (i.e. how much water is in the whole animal). Increased levels of vasopressin have been shown to be of use as an indicator of nausea and vomiting. Two types of vasopressin occur: most mammals produce arginine vasopressin, while the pig produces only lysine vasopressin.

We have been able to discuss only briefly the main biochemical and haematological indicators that have been used to evaluate animal welfare during transport. For more in-depth information the reader should refer to one of the many textbooks dealing with clinical veterinary biochemistry. Kaneko *et al.* (1997) provide a comprehensive reference.

Immune response

One of the well-known effects of transport is that it can modify the susceptibility of animals to infection. This is through the effects of stress on the animal's immune system, which is modified primarily by the increased levels of circulating corticosteroid hormones, but also by other hormones such as vasopressin and oxytocin. At a superficial level, changes may be seen in counts of circulating lymphocytes, neutrophils, leucocytes and eosinophils. Direct counts of these cells or ratios of counts have been commonly used to assess welfare during, and after, transportation. However, the changes that can be seen in immune status and their relationship with transport are too complex to describe adequately here. For a more comprehensive recent overview see Broom (2006).

The Physiological Responses of Cattle, Sheep and Pigs to Transport

Cattle

Mortality rates amongst cattle transported by road are generally much lower than those of other forms of livestock. To a large extent this is because the care with which animals are transported and the attention paid to their welfare is in proportion to the value of the individual animal (Hails, 1978). Over the first 18–24 h of transport loss of body weight can range from 3–11%. This is mostly due to loss of gut fill. Loss of carcass weight increases approximately linearly with transport time and has variously been reported to range from < 1% to 8% over 48 h.

Access to water can reduce both loss of body weight and loss of carcass weight (Warriss, 1990). After 24 h of transport there is an increase in plasma levels of β-OHB, FFA and osmolality, total protein and albumin indicative of mobilization of food reserves and increasing dehydration (Tarrant et al., 1992; Warriss et al., 1995; Knowles et al., 1999a).

After 24–31 h of transport, levels of plasma cortisol, glucose and CK are elevated. In sheep these variables generally return to pre-transport levels after approximately 9 h of transport, but in cattle they tend to remain elevated or to steadily increase. Additionally, in cattle there is a gradual depletion of muscle glycogen (Knowles et al., 1999a) and an associated increase in the pH of the meat (Tarrant et al., 1992). These changes arise because cattle prefer to stand during transport as they are relatively heavy animals, and lying can produce considerable pressure on the parts of the body in contact with the floor of the vehicle, especially during a rough journey. The act of lying down and rising is difficult on a moving lorry at the stocking densities used for transport and there is a risk of being trampled or fallen upon.

The changes seen in blood variables indicate that there is some physical effort involved in remaining standing and having to maintain balance against the motion of the vehicle. Despite the dangers and discomfort involved with lying, towards the end of the first 24 h of transport some cattle do lie down (Tarrant et al., 1992; Knowles et al., 1999a). This could be because of the physical effort involved in standing, although the physiological changes seen do not indicate excessive physical demand. Knowles et al. (1999a) hypothesized that the animals could possibly be in need of sleep, as those animals that did lie down displayed higher levels of plasma cortisol. Raised levels of plasma cortisol are associated with sleep deprivation in humans (Leproult et al., 1997).

Knowles et al. (1999a) offered water to cattle onboard lorries for 1 h following 14 h of transport within the UK. They found that < 60% of animals drank, few drank fully and that activity levels rose whilst the vehicle was motionless, leading them to conclude that the stop merely prolonged transport and further exhausted the animals rather than providing any recovery. Warriss et al. (1995) found that it took cattle 5 days to recover the live weight lost during 15 h of transport. Knowles et al. (1999a) found little difference in the pattern of recovery following either 14, 21, 26 or 31 h of transport. Levels of plasma β-OHB, FFA, urea and glucose had recovered to pre-transport levels after 24 h in lairage with food and water freely available, as had levels of plasma cortisol. Levels of indicators of hydration took up to 72 h to return to pre-transport levels, whilst full pre-transport live weight had not been recovered even after 72 h of lairage.

Based on both the physiological indicators of fatigue and dehydration and the behaviour of the animals, both Tarrant et al. (1992) and Knowles et al. (1999a) suggest a maximum continuous transport time of no longer than 24 h for cattle. Knowles et al. (1999a) recommend a mid-transport lairage period of, ideally, 24 h with food and water available, to allow recovery from the physical demands of transport. They considered that short, mid-transport stops were unlikely to provide reasonable opportunity for rest or recovery.

However, including a lairage stop of any length provides an opportunity for cattle from different sources to exchange pathogens. Experience in the USA has shown that 200–300 kg cattle will suffer fewer post-transport health problems if they are transported for a complete 32-h journey without any lairage stops. Whether the increased health problems are due to exposure to novel pathogens or to the inadequacy of the lairage conditions, essentially extending the stress of transport, is not known (Grandin, 1997).

Recently, workers have used measurements of the down-regulation of glucocorticoid and β-adrenergic receptors on the surface of lymphocytes to monitor the levels of stress suffered by young cattle (300 kg) during transport (Odore et al., 2004). Using indices related to immune function in this way could also be useful in other species. Another measure of stress in cattle is the antioxidant capacity of the serum (Chirase et al., 2004; Pregel et al., 2005).

In 207-kg beef steers transported for 3105 km, there was significantly lower serum antioxidant capacity compared to that at the start of the trip (Chirase et al., 2004).

Another possible measure of stress is explained by Dixit et al. (2001). They found that transport stress affected adrenocorticotropin secretion from lymphocytes after a 14-h trip. Dixit et al. (2001) conclude that ACTH secretion from lymphocytes could be used to assess stress, because a long trip increased ACTH secretion from lymphocytes, while a short, 30-min trip had little effect.

Tarrant et al. (1992) studied the effects of three stocking densities on 600 kg cattle that were transported for 24 h. Following transport they found that levels of plasma CK, cortisol and glucose had increased with increasing stocking density, as had the amount of bruising on the carcasses, indicative of increased physical and psychological stress and poorer welfare. They concluded that stocking densities above 550 kg/m were unacceptable for this size of animal on long journeys.

These results run counter to the popularly held belief within the industry that packing animals in tightly helps support them and prevents them from being jolted and bruised. Too high a stocking density was found to prevent the animals from holding a proper footing, by overly restricting their movement. The highest stocking density, and the one that was found to be unacceptable, was that which would normally be considered to represent a full load – the maximum number of animals that could be held in a pen and the gate still easily closed (see also Chapter 9, this volume).

Young calves (cattle less than 1 month of age)

Neonatal animals are generally less well adapted to cope with transport and are more vulnerable than adult animals. The long-distance transport of very young cattle is common and usually takes place within days or weeks of birth, whilst the animal is still unweaned and is fully dependent on milk. Calf mortality *during* transport tends to be low; however, mortality rates *following* transport can be high, usually as a result of disease (Knowles, 1995).

In a large-scale survey of calf mortality and husbandry within the UK, Leech et al. (1968) estimated the mortality of transported calves to be 160% that of calves that had remained on their farm of birth. Mortality of calves transported below 1 month of age remained markedly above that of home-bred calves until 2 months after purchase. In calves < 1 month old, various authors have reported a strong negative correlation between mortality/morbidity and age when first transported (Knowles, 1995). In addition to the age at which calves are transported, the length of time that marketing takes is also important. Mormede et al. (1982) found less post-transport disease amongst calves whose marketing took only 13 rather than 37 h.

The reactivity of the adrenal glands to ACTH increases with age and is not fully developed in the calf (Hartmann et al., 1973). Several authors report that the increase in plasma cortisol usually seen in response to transport is not present in young calves (Knowles, 1995). Neither do calves show the usual increase in heart rate and plasma glucose levels (Knowles et al., 1997). These authors concluded that calves were unable to respond to the stress of transport because of their immaturity, and that the lack of a cortisol response was not because they were relatively 'unstressed' by the process of transportation. Using measurements of rectal temperature, Knowles et al. (1997, 1999b) found that, when transported during cold weather, calves found difficulty in maintaining body temperature during transport and regulating it afterwards. Loss of live weight was greater in the cold.

During and immediately after long-distance transport, calf hauliers within Europe prefer to feed a glucose and electrolyte solution rather than milk replacer, as they report that this reduces the incidence of diarrhoea. Knowles et al. (1997, 1999b) found that feeding electrolyte during transport of 19–24 h provided only little benefit in terms of rehydration and improvements in levels of plasma metabolites, and so recommended that it was best to complete the journey without the disruption and stress of feeding.

Liquid feeding of unweaned calves requires the observation, and often the handling, of each individual animal. It also requires attention to hygienic presentation of the feed, which has to be made up to the correct temperature and solution strength in order to avoid digestive problems.

There was some evidence from the study of Knowles *et al.* (1999b) that feeding just cold water during transport was detrimental to the calves.

If there is sufficient room for them to do so, calves spend much of the time lying down during road transport. Knowles *et al.* (1997) found that calves spent approximately 50% of the time lying during 24 h of transport. During cold weather, the proportion of the journey spent lying increased to 80–90% (Knowles *et al.*, 1999b). Following transport for 24 h, Knowles *et al.* (1999b) reported that most of the commonly measured physiological variables had returned to pre-transport values after 24 h of lairage and feeding – except for live weight and levels of plasma CK, which took up to 7 days to recover.

Overall, present evidence indicates that young calves should not be transported until they are at least over the age of 1 month, but further work is required to confirm that this age limit should not be further extended. The only exception to this recommendation would be transport for a short distance to a calf-rearing facility. If they are to be transported, then it is best to keep the marketing time to a minimum, to avoid feed/rest stops if transport is for no longer than 24 h, to avoid exposure of the calves to cold and to avoid cross-contamination of animals from different sources. The animals should be well bedded – especially in cold weather – and transported at a stocking density that allows enough space for them all to lie down.

Sheep

The mortality rate amongst slaughter lambs transported by road within the UK has been estimated at 0.018% (Knowles *et al.*, 1994a), at 0.10% within South Africa (Henning 1993) and at between 0.74 to 1.63% within Queensland, Australia (Shorthose and Wythes, 1988). In the UK, those lambs going direct from farm to slaughterhouse have an estimated mortality rate of 0.007% compared with 0.031% for those passing through a live auction market (Knowles *et al.*, 1994a). Occasionally, mass deaths within single loads of sheep are reported. These are most often associated with a combination of high ambient temperatures and reduced ventilation on a stationary lorry. In many countries

there is a trade in cull sheep. There is anecdotal evidence that the mortality rates amongst these relatively infirm, low-value animals can be high during transport.

The loss of live weight during transport has been well documented in lambs. Wythes and Morris (1994) averaged the results from eight pieces of work and found live-weight losses of 3, 5, 7.5, 11, 12 and 14% over 6, 12, 24, 48, 72 and 96 h, respectively, with food withdrawal alone; however, losses as high as 20% after just 72 h have been reported by Horton *et al.* (1996), when also deprived of water. They also found that following transport, food intake was depressed. Combining data from various sources, Wythes and Morris (1994) found the average rate of loss of carcass weight to be 1.7% per day over 4 days when lambs were deprived of only feed and not transported, with a range of 1.3–2.3%/day.

During periods of fasting and transport of up to 72 h, plasma levels of β-OHB have been found to increase linearly at a rate of approximately 0.006 mmol/l/h (Warriss *et al.*, 1989a; Knowles *et al.*, 1995). Levels of plasma FFA tend to rise linearly with periods of fasting and transport at a rate of approximately 20 μmol/l/h, but peak and flatten out, with no further increase, between 18–24 h, whilst levels of plasma urea increase approximately linearly by 30–50% during 24 h of transport (Knowles *et al.*, 1995, 1998).

When sheep were held without food or water for 48 h at temperatures up to 35°C, Parrott *et al.* (1996) found little evidence of dehydration from measurements of plasma osmolality, but they did find evidence that the sheep were unable to maintain water balance if they consumed feed. Sheep transported for up to 24 h in the summer in the UK showed no signs of dehydration as measured by plasma total protein, albumin and osmolality. However, sheep transported across France for 24 h, during which daytime temperatures rose above 20°C, showed signs of dehydration, with increases in plasma total protein, albumin and osmolality of approximately 10, 12 and 5%, respectively (Knowles *et al.*, 1996).

In accord with Parrott *et al.* (1996), Knowles *et al.* (1996) noted that feeding during and after transport tended to disrupt water balance. This has important implications for the length of mid-transport lairage stops, as after short periods of food and water deprivation sheep are primarily interested in eating and do not drink readily or immediately (Knowles *et al.*, 1994b). A lairage stop of just 1 h, as is presently required for transport of > 14 h within Europe, is sufficient for the animals to eat but not to drink, so animals may be reloaded after having consumed a high dry-matter feed, but no water. A minimum mid-transport lairage time of 8 h has been recommended (Knowles, 1998).

Measurements of heart rate, plasma cortisol, glucose and CK have shown that it is the initial stages of transport that are most stressful to sheep (Knowles *et al.*, 1995). Heart rate peaks at loading and there is a rise in cortisol, glucose and CK levels at loading, but after 9 h of transport these variables have generally returned to approximate basal levels, and the only measurable changes seen are then due to the effects of feed and water deprivation, which can be exaggerated by the conditions of transport. However, the conditions of transport are important. Sheep that are loaded at too high a stocking density to be able to lie down readily show elevated levels of plasma CK, indicative of physical fatigue caused by having to remain standing (Knowles *et al.*, 1998).

As long as they are fit, loaded at an appropriate stocking density, the ambient temperature is not extreme and the load is properly ventilated, sheep appear to cope reasonably well during transport. However, Horton *et al.* (1996) reported that, after passing through a live auction market, lambs transported for 72 h without food or water – whilst not differing in terms of performance or blood metabolites from animals simply deprived of food and water for 72 h – suffered in terms of compromised general health. This was probably a result of confinement on the lorry and exposure to unfamiliar animals and pathogens, combined with the effects of deprivation and transport per se.

After transport, the recovery of physiological variables to pre-transport levels appears to take place in three stages (Knowles *et al.*, 1993). After 24 h of lairage, with food and water, variables usually associated with short-term stress and the variables associated with dehydration had returned to normal levels. After 96 h there had been a well-defined recovery in live weight, and levels of most of the metabolites measured

had returned to normal levels. At 144 h of lairage a fuller recovery had taken place, levels of creatine kinase had fallen and all variables had stabilized.

Where appropriate it is always preferable, and generally makes better economic sense, to transport carcasses rather than live animals. Transport is stressful and transport times should be kept to a minimum. After 9 h of transport the changes in physiological variables with time tend to be linear and are of little help in determining a maximum acceptable transport time. Behavioural studies of motivation to feed have shown that sheep will begin to work for food after 10–12 h of deprivation.

At present, all the evidence taken together points to an acceptable maximum journey time in the region of 24 h when transport is continuous and when food and water are not available. If a lairage stop is included in a journey then it should be for a minimum of 8 h, with both food and water continuously available. However, a lairage stop does increase the chance of cross-infection between animals from different sources, and animals stressed by the process of transport will already tend to be immunologically compromised and vulnerable.

Pigs

The mortality rate amongst pigs transported to slaughter within the UK has changed little over 20 years, and is estimated to be 0.061%, with a further 0.01% of pigs dying in the lairage pens before slaughter (Warriss and Brown, 1994). However, there are marked differences in mortality rates between different countries. The rates are particularly higher in countries where the slaughter pig population contains a large proportion of genes from stress-susceptible breeds such as the Pietrain and Belgian Landrace. Estimates range from 0.3–0.5% in Belgium and Germany (see Warriss, 1998a). Murray (2000) gives a mortality figure for pigs transported in Alberta, Canada, in 1996 of 0.14%, of which half was attributed to the presence of the Halothane gene. In the USA pigs grown to 130 kg also have higher death losses (Chapter 1).

The other major factor influencing the mortality of pigs during transit is ambient temperature. Pigs are sensitive to high temperatures because they are poorly adapted to lose heat unless allowed to wallow, a behaviour not possible during transport. The relationship between mortality and ambient temperature is curvilinear. This is illustrated in Fig. 19.3 with data for the UK from Warriss and Brown (1994), which show that there was a marked increase in mortality when average monthly temperatures rose above 15°C.

Because pigs find it hard to thermoregulate when confined in vehicles during transport, they are susceptible to heat stress. Measuring animals' temperature is therefore a potential way of monitoring stress suffered during transport. Warriss et al. (2006) used a thermal imaging camera to record the temperature of the ears of transported pigs at slaughter. Ear temperature was significantly correlated with the pigs' core temperature, measured as the temperature of the blood lost at exsanguination. Moreover, ear temperature was positively correlated to serum CK activity, and blood temperature was positively correlated to serum cortisol concentration.

Thermal images could therefore be useful in assessing the physiological state of pigs non-invasively and so monitoring their welfare. Other factors of importance in determining mortality are the time of last feed before loading, vehicle deck, stocking density and possibly journey time. Pigs fed too soon (< 4 h) before transport are more likely to die, as are those carried on the bottom deck, at higher densities and for longer. However, the evidence for the latter is contradictory.

Pigs find simulated transport aversive (Ingram et al., 1983), particularly the vibration associated with it (Stephens et al., 1985) and if they have recently eaten a large meal. Because pigs may vomit during transport (Bradshaw and Hall, 1996; Riches et al., 1996) and show increased circulating levels of vasopressin, a hormone associated with feelings of motion sickness in humans, part of their aversion may be attributable to similar feelings of sickness.

That pigs find at least some aspects of transport psychologically stressful is evidenced by increases in plasma adrenaline (Dalin et al., 1993), indicating stimulation of the sympatho–adrenal system, and of cortisol (see, e.g. Dantzer, 1982), with corresponding depletion of adrenal ascorbic acid (Warriss et al., 1983), indicating stimulation of the hypophyseal–adrenal axis. They may also find it physically stressful, based on elevations of circulating activities of the enzyme CK (Honkavaara, 1989).

The physical stress they experience will be determined by the comfort and length of the journey. It is likely to be greater if vibration levels are higher. Modern vehicles with air suspension and driven on smooth roads will provide more comfort than older vehicles with traditional spring suspension systems driven on more poorly surfaced roads. Physical stress and the associated fatigue are likely to be higher if pigs stand, rather than lie down, during the journey. There is some debate about whether pigs prefer to stand or lie down. The available evidence has been reviewed by Warriss (1998b), who suggested that it pointed to the view that pigs preferred to stand on short journeys in which the conditions made it uncomfortable to lie down. These conditions could be excessive vibration or uncomfortable flooring, perhaps because of inadequate bedding. But, under comfortable conditions, many – if not all – pigs would lie down if given sufficient space, especially on longer journeys.

What constitutes sufficient space was also discussed by Warriss (1998b). It is equivalent to a stocking density of not higher than about 235–250 kg/m² for normal slaughter pigs weighing between 90 and 100 kg. For smaller pigs, the space requirement would be expected to be slightly greater and for larger pigs slightly less. EC Directive 95/29/EC requires that the

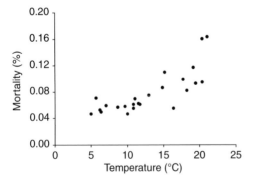

Fig. 19.3. The relationship between average monthly ambient temperature and the mortality of pigs transported to slaughter.

loading density for pigs of around 100 kg should not exceed 235 kg/m^2. Also, the Directive recognizes that this density may be too high under certain conditions.

Breed of pig, size and physical condition of the animals – or weather and journey time – may mean that the space allowed has to be increased by up to 20%. Pigs carried at very high stocking densities show increased circulating levels of CK (Warriss et al., 1998). The provision of appropriate amounts of space is especially important with longer journeys.

The physiological state of an animal at slaughter often affects subsequent lean meat quality. Thus, muscle glycogen depletion can lead to elevated ultimate muscle pH in the meat. Longer transport sometimes results in muscle glycogen depletion and more meat with high ultimate pH (see Malmfors, 1982; Warriss, 1987). This is seen as a higher prevalence of dark, firm, dry (DFD) pork. Rigor mortis occurs when the levels of adenosine triphosphate (ATP) in the muscles after death are exhausted because of failure of regeneration from creatine phosphate, adenosine diphosphate and glycogen. Factors that affect the concentrations of glycogen and creatine phosphate in the muscles at death can therefore influence the rate of development of rigor in carcasses. These factors are often associated with pre-slaughter stress. Warriss et al. (2003) showed that earlier rigor development in commercially slaughtered pigs was associated with higher concentrations of cortisol, lactate and CK in the blood lost at exsanguination. This suggested that the average rates of rigor development, or the proportion of carcasses in full rigor, could be used to monitor pre-slaughter stress, including that caused by transport in groups of slaughter pigs.

During transport, pigs are deprived of food. They are often also deprived of water, although current EU legislation (EC Directive 95/29/EC) prescribes that pigs transported for more than 8 h must have continuous access to drinking water. There is, however, evidence that pigs drink only very small amounts of water (Lambooy, 1983; Lambooy et al., 1985). Some studies (Warriss et al., 1983) have indicated that pigs can become dehydrated after only short

journeys, but data from other work (Becker et al., 1989) have not supported this. Food deprivation leads to losses in live and carcass weights (Warriss, 1985), liver weight and liver glycogen (Warriss and Bevis, 1987) and muscle glycogen (Warriss et al., 1989b). These are undesirable. However, some period of food withdrawal before transport is desirable to minimize mortality and, in the case of slaughter pigs, to facilitate hygienic carcass dressing. Four hours has been recommended (Warriss, 1994), but this may be too short a time, based on observations of vomiting during transport, although the ideal period of withdrawal is not clear (Warriss, 1998a).

Many slaughter pigs are mixed with unfamiliar animals when they are assembled for sending to slaughter. This usually leads to fighting, particularly between dominant individuals. The consequences are elevations in circulating cortisol, CK and lactate and depletion of muscle glycogen (Warriss, 1996). Mixing pigs is undesirable from the points of view of both welfare and meat quality. Nevertheless, in the UK about 40% of pigs show some evidence of fighting, and 5–10% show evidence of severe fighting.

A complicating factor in the interpretation of changes in the levels of stress indicators in the blood of pigs may be the social status of different individuals. Hicks et al. (1998) measured the effects of a 4-h transport, heat stress and cold stress on weanling pigs (4 weeks old). There were interactions between the effects of treatment and social status on albumin/globulin ratio, natural killer cell cytotoxicity, lymphocyte proliferation and cortisol concentrations. After transport, subordinate piglets tended to have higher serum cortisol and lower albumin/globulin ratios, compared with piglets of higher social status.

Acknowledgements

Permission to reproduce copyrighted material for Fig. 19.3 was kindly granted by *The Veterinary Record*, 7 Mansfield Street, London. The authors thank Julie Edwards for help with the preparation of the figures.

References

Becker, B.A., Mayes, H.F., Hahn, G.L., Nienaber, J.A., Jesse, G.W., Anderson, M.E., Heymann, H. and Hedrick, H.B. (1989) Effect of fasting and transportation on various physiological parameters and meat quality of slaughter hogs. *Journal of Animal Science* 67, 334–341.

Bradshaw, R.H. and Hall, S.J.G. (1996) Incidence of travel sickness in pigs. *Veterinary Record* 139, 503 (letter).

Broom, D.M. (1986) Indicators of poor welfare. *British Veterinary Journal* 142, 524–526.

Broom, D.M. (2006) Behaviour and welfare in relation to pathology. *Applied Animal Behaviour Science* 97, 73–83.

Carlson, G.P. (1997) Fluid, electrolyte, and acid–base balance. In: Kaneko, J.J., Harvey, J.W. and Bruss, M.L. (eds) *Clinical Biochemistry of Domestic Animals*, 5th edn. Academic Press, San Diego, California, pp. 485–516.

Chirase, N.K., Greene, L.W., Purdy, C.W., Loan, R.V., Aovemann, B.W., Parker, D.E., Walborg, E.F., Stevens, D.E., Xu, Y. and Klaunig, J.E. (2004) Effect of transport stress on respiratory disease, serum antioxidant status and serum concentrations of lipid peroxidation biomarkers in beef cattle. *American Journal of Veterinary Research* 65, 860–864.

Dalin, A.M., Magnusson, U., Häggendal, A. and Nyberg, L. (1993) The effects of transport stress on plasma levels of catecholamines, cortisol, corticosteroid-binding globulin, blood cell count and lymphocyte proliferation in pigs. *Acta Veterinaria Scandinavica* 34, 59–68.

Dantzer, R. (1982) Research on farm animal transport in France: a survey. In: Moss, R. (ed.) *Transport of Animals Intended for Breeding, Production and Slaughter*. Martinus Nijhoff, The Hague, Netherlands, pp. 218–230.

Dixit, V.D., Morohrens, M. and Parvizi, N. (2001) Transport stress modulates adrenocorticotropin secretion from peripheral bovine lymphocytes. *Journal of Animal Science* 79, 729–734.

Eicher, S.D. (2001) Transportation of cattle in the dairy industry: current research and future directions. *Journal of Dairy Science* 84 (E Suppl.), E19–E23.

Farver, T.B. (1997) Concepts of normality in clinical biochemistry. In: Kaneko, J.J., Harvey, J.W. and Bruss, M.L. (eds) *Clinical Biochemistry of Domestic Animals*, 5th edn. Academic Press, San Diego, California, pp. 1–19.

Fike, K. and Spire, M.F. (2006) Transportation of cattle. *Veterinary Clinics of North America Food Animal Practice* 22, 305–320.

Grandin, T. (1997) Assessment of stress during handling and transport. *Journal of Animal Science* 75, 249–257.

Hails, M.R. (1978) Transport stress in animals: a review. *Animal Regulation Studies* 1, 289–343.

Hall, S.J.G. and Bradshaw, R.H. (1998) Welfare aspects of the transport by road of sheep and pigs. *Journal of Applied Animal Welfare Science* 1, 235–254.

Hartmann, H., Meyer, H., Steinbach, G., Deschner, F. and Kreutzer, B. (1973) Allgemeines adaptations-syndrom (seleye) beim kalb. 1. Normalverhalten der blutbildwerte sowie des glukose- und 11 -OHKS-blutspiegels. *Archiv fur Experimentelle Veterinarmedizin* 27, 811–823.

Henning, P.A. (1993) Transportation of animals by road for slaughter in South Africa. In: *Proceedings of the 4th International Symposium on Livestock Environment, American Society of Agricultural Engineers, 6–9 July 1993*, pp. 536–541.

Hicks, T.A., McGlone, J.J., Whisnant, C.S., Kattesh, H.G. and Norman, R.L. (1998) Behavioural, endocrine, immune, and performance measures for pigs exposed to acute stress. *Journal of Animal Science* 76, 474–483.

Honkavaara, M. (1989) Influence of selection phase, fasting and transport on porcine stress and on the development of PSE pork. *Journal of Agricultural Science in Finland* 61, 415–423.

Horton, G.M.J., Baldwin, J.A., Emanuele, S.M., Wohlt, J.E. and McDowell, L.R. (1996) Performance and blood chemistry in lambs following fasting and transport. *Animal Science* 62, 49–56.

IATA (2005) *Live Animal Regulations* (32nd edn). International Air Transport Association, Montreal, Canada (http://www.iata.org/).

Ingram, D.L., Sharman, D.F. and Stephens, D.B. (1983) Apparatus for investigating the effects of vibration and noise on pigs using operant-conditioning. *Journal of Physiology* 343, 2–3.

Kaneko, J.J., Harvey, J.W. and Bruss, M.L. (1997) (eds) *Clinical Biochemistry of Domestic Animals*, 5th edn. Academic Press, San Diego, California.

Knowles, T.G. (1995) A review of post-transport mortality among younger calves. *Veterinary Record* 137, 406–407.

Knowles, T.G. (1998) A review of the road transport of slaughter sheep. *Veterinary Record* 143, 212–219.

Knowles, T.G., Warriss, P.D., Brown, S.N., Kestin, S.C., Rhind, S.M., Edwards, J.E., Anil, M.H. and Dolan, S.K. (1993) Long-distance transport of lambs and the time needed for subsequent recovery. *Veterinary Record* 133, 287–293.

Knowles, T.G., Maunder, D.H.L. and Warriss, P.D. (1994a) Factors affecting the mortality of lambs in transit to or in lairage at a slaughterhouse, and reasons for carcass condemnation. *Veterinary Record* 135, 109–111.

Knowles, T.G., Warriss, P.D., Brown, S.N. and Kestin, S.C. (1994b) Long-distance transport of export lambs. *Veterinary Record* 134, 107–110.

Knowles, T.G., Brown, S.N., Warriss, P.D., Phillips, A.J., Dolan, S.K., Hunt, P., Ford, J.E., Edwards, J.E. and Watkins, P.E. (1995) The effects on sheep of transport by road for up to twenty-four hours. *Veterinary Record* 136, 431–438.

Knowles, T.G., Warriss, P.D., Brown, S.N., Kestin, S.C., Edwards, J.E., Perry, A.M., Watkins, P.E. and Phillips, A.J. (1996) The effects of feeding, watering and resting intervals on lambs exported by road and ferry to France. *Veterinary Record* 139, 335–339.

Knowles, T.G., Warriss, P.D., Brown, S.N., Edwards, J.E., Watkins, P.E. and Phillips, A.J. (1997) Effects on calves less than one month old of feeding or not feeding them during road transport of up to 24 hours. *Veterinary Record* 140, 116–124.

Knowles, T.G., Warriss, P.D., Brown, S.N. and Edwards, J.E. (1998) The effects of stocking density during the road transport of lambs. *Veterinary Record* 142, 503–509.

Knowles, T.G., Warriss, P.D., Brown, S.N. and Edwards, J.E. (1999a) The effect of transporting cattle by road for up to 31 hours. *Veterinary Record* 145, 575–582.

Knowles, T.G., Brown, S.N., Edwards, J.E., Phillips, A.J. and Warriss, P.D. (1999b) The effect on young calves of a one hour feeding stop during a 19-hour road journey. *Veterinary Record* 144, 687–692.

Lambooy, E. (1983) Watering pigs during road transport through Europe. *Fleischwirtshaft* 63, 1456–1458.

Lambooy, E., Garssen, G.J., Walston, P., Mateman, G. and Merkus, G.S.M. (1985) Transport of pigs by car for 2 days: some aspects of watering and loading density. *Livestock Production Science* 13, 289–299.

Leech, F.B., Macrae, W.D. and Menzies, D.W. (1968) *Calf Wastage and Husbandry in Britain*. 1962–63. HMSO, London.

Leproult, R., Copinschi,G., Buxton, O. and VanCauter, E. (1997) Sleep loss results in an elevation of cortisol levels the next evening. *Sleep* 20, 865–870.

Malmfors, G. (1982) Studies on some factors affecting pig meat quality. In: *Proceedings of the 28th European Meeting of Meat Research Workers*, Indtituto del Frio, Madrid, September 1982, pp. 21–23.

Mormede, P., Soissons, J., Bluthe, R-M., Raoult, J., Legraff, G., Levieux, D. and Dantzer, R. (1982) Effect of transportation on blood serum composition, disease incidence and production traits in young calves. Influence of journey duration. *Annales de Recherches Vétérinaires* 13, 369–384.

Murray, A.C. (2000) Reducing losses from farm gate to packer. *Advances in Pork Production* 11, 175–180.

Norris, R.T. (2005) Transport of animals by sea. *Revue Scientifique et Technique – Office International des Epizooties* 24, 673–681.

Odore, R., D 'Angelo, A., Badino, P., Bellino, C., Pagliasso, S. and Re, G. (2004) Road transportation affects blood hormone levels and lymphatic glucocorticoid and p-adrenergic receptor concentrations in calves. *The Veterinary Journal* 168, 213–214.

Parrott, R.F., Lloyd, D.M. and Goode, J.A. (1996) Stress hormone responses of sheep to food and water deprivation at high and low ambient temperatures. *Animal Welfare* 5, 45–56.

Pregel, P., Bollo, E. and Cannizo, F.T. (2005) Antioxidant capacity as a reliable marker of stress in dairy calves transported by road. *Veterinary Record* 156, 53–54.

Riches, H.L., Guise, H.J. and Penny, R.H.C. (1996) A national survey of transport conditions for pigs. *Pig Journal* 38, 8–18.

Shorthose, W.R. and Wythes, J.R. (1988) Transport of sheep and cattle. *Proceedings of the 34th International Congress of Meat Science and Technology*, Brisbane, Australia, pp. 122–129.

Stephens, D.B., Bailey, K.J., Sharman, D.F. and Ingram, D.L. (1985) An analysis of some behavioural effects of the vibration and noise components of transport in pigs. *Quarterly Journal of Experimental Physiology* 70, 211–217.

Swanson, J.C. and Morrow-Tesh, J. (2001) Cattle transport historical, research and future perspectives. *Journal of Animal Science* 79 (E Suppl.), E102–E109.

Tarrant, P.V., Kenny, F.J., Harrington, D. and Murphy, M. (1992) Long-distance transportation of steers to slaughter, effect of stocking density on physiology, behavior and carcass quality. *Livestock Production Science* 30, 223–238.

Warriss, P.D. (1985) Marketing losses causes by fasting and transport during the preslaughter handling of pigs. *Pig News and Information* 6, 155–157.

Warriss, P.D. (1987) The effect of time and conditions of transport and lairage on pig meat quality. In: Tarrant, P.V., Eikelenboom, G. and Monin, G. (eds) *Evaluation and Control of Meat Quality in Pigs.* Martinus Nijhoff Publishers, Dordrecht, Netherlands, pp. 245–264.

Warriss, P.D. (1990) The handling of cattle pre-slaughter and its effects on carcass meat quality. *Applied Animal Behaviour Science* 28, 171–186.

Warriss, P.D. (1994) Ante-mortem handling of pigs. In: Cole, D.J.A., Wiseman, J. and Varley, M.A. (eds) *Principles of Pig Science.* Nottingham University Press, Nottingham, UK, pp. 425–432.

Warriss, P.D. (1996) The consequences of fighting between mixed groups of unfamiliar pigs before slaughter. *Meat Focus International* 5, 89–92.

Warriss, P.D. (1998a) The welfare of slaughter pigs during transport. *Animal Welfare* 7, 365–381.

Warriss, P.D. (1998b) Choosing appropriate space allowances for slaughter pigs transported by road: a review. *Veterinary Record* 142, 449–454.

Warriss, P.D. and Bevis, E.A. (1987) Liver glycogen in slaughtered pigs and estimated time of fasting before slaughter. *British Veterinary Journal* 143, 354–360.

Warriss, P.D. and Brown, S.N. (1994) A survey of mortality in slaughter pigs during transport and lairage. *Veterinary Record* 134, 513–515.

Warriss, P.D., Dudley, C.P. and Brown, S.N. (1983) Reduction in carcass yield in transported pigs. *Journal of the Science of Food and Agriculture* 34, 351–356.

Warriss, P.D., Bevis, E.A., Brown, S.N. and Ashby, J.G. (1989a) An examination of potential indices of fasting time in commercially slaughtered sheep. *British Veterinary Journal* 145, 242–248.

Warriss, P.D., Bevis, E.A. and Ekins, P.J. (1989b) The relationships between glycogen stores and muscle ultimate pH in commercially slaughtered pigs. *British Veterinary Journal* 145, 378–383.

Warriss, P.D., Bevis, E.A., Brown, S.N. and Edwards, J.E. (1992) Longer journeys to processing plants are associated with higher mortality in broiler chickens. *British Poultry Science* 33, 201–206.

Warriss, P.D., Brown, S.N., Knowles, T.G., Kestin, S.C., Edwards, J.E., Dolan, S.K. and Phillips, A.J. (1995) The effects on cattle of transport by road for up to fifteen hours. *Veterinary Record* 136, 319–323.

Warriss, P.D., Brown, S.N., Knowles, T.G., Edwards, J.E., Kettlewell, P.J. and Guise, H.J. (1998) The effect of stocking density in transit on the carcass quality and welfare of slaughter pigs: 2. Results from the analysis of blood and meat samples. *Meat Science* 50, 447–456.

Warriss, P.D., Brown, S.N. and Knowles, T.G. (2003) Measurements of the degree of development of rigor mortis as an indicator of stress in slaughtered pigs. *Veterinary Record* 153, 739–742.

Warriss, P.D., Pope, S.J., Brown, S.N., Wilkins, L.J. and Knowles, T.G. (2006) Estimating the body temperature of groups of pigs by thermal imaging. *Veterinary Record* 158, 331–334.

Wythes, J.R. and Morris, D.G. (1994) *Literature Review of Welfare Aspects and Carcass Quality Effects in the Transport of Cattle, Sheep and Goats (Parts A, B and C).* Report prepared by Queensland Livestock and Meat Authority for Meat Research Corporation. Queensland Livestock and Meat Authority, Australia.

20   Handling and Welfare of Livestock in Slaughter Plants

Temple Grandin

Department of Animal Sciences, Colorado State University, Fort Collins, Colorado, USA

Introduction

Animal welfare is a global issue. The OIE (World Organization for Animal Health) now has guidelines for animal welfare during slaughter. These guidelines cover handling, pre-slaughter and stunning methods, and are minimum standards. Welfare standards mandated by large meat buyers are usually more strict.

Gentle handling in well-designed facilities will minimize stress levels, improve efficiency and maintain good meat quality. Cattle that become excited during restraint may have tougher meat and more borderline dark-cutters (Voisinet *et al.*, 1997; King *et al.*, 2006). Rough handling or poorly designed equipment is detrimental to both animal welfare and meat quality. Quiet handling and a great reduction in electric prod use in the stunning race resulted in a 10% reduction of pale, soft, exudative (PSE) pork. Careful handling prior to slaughter helps to maintain meat quality in pigs (Barton-Gade, 1989; Barton-Gade *et al.*, 1993; van der Wal, 1997; Faucitano, 1998; Meisinger, 1999; Hambrecht *et al.*, 2004, 2005; Kuchenmeister *et al.*, 2005).

Progressive slaughter-plant managers recognize the importance of good handling practices. Constant monitoring of handler performance is required to maintain high standards of animal welfare. Safety for plant employees is another benefit of keeping animals calm during handling. This is especially important when large animals such as cattle, bison, elk, camels or water buffalo are handled.

How Stressful is Slaughter?

Numerous studies (Grandin, 1997b) have been conducted to determine the relative stressfulness of different husbandry and slaughter procedures. Measurements of cortisol (stress hormone) is the most common method of evaluating handling stresses. One must remember that cortisol is a time-dependent measure. It takes approximately 15–20 min for it to reach its peak value after an animal is stressed (Lay *et al.*, 1992, 1998). Evaluations of handling and slaughter stress will be more accurate if behavioural reactions, heart rate and other blood chemistries are also measured. Adrenaline and noradrenaline have limited value in measuring slaughtering stress, because both captive-bolt and electrical stunning trigger massive releases (Warrington, 1974; Pearson *et al.*, 1977; van der Wal, 1978). If the stunning method is applied properly, the animal will be unconscious when the hormone release occurs and there will be no discomfort.

Absolute comparisons of cortisol levels between studies must be done with great caution. Cortisol levels can vary greatly between individual animals (Ray *et al.*, 1972). Cattle that show signs of behavioural excitement and have a faster exit speed when released from a race usually have higher levels than calm animals (King *et al.*, 2006). A review of many studies done with extensively raised cattle unaccustomed to close human contact indicates that cortisol levels in cattle fall into three basic categories: (i) resting baseline levels; (ii) levels provoked by being held in a race or restraint in a headgate (bail) for blood testing; and (iii) excessive levels that are double or triple the farm restraint levels (Grandin, 1997a).

Baseline levels vary from a low of 2 ng/ml (Alan and Dobson, 1986) to 9 ng/ml (Mitchell *et al.*, 1988). Beef cattle on a research station that had become accustomed to being handled for different experiments had cortisol levels that ranged from 10 ng/ml in calm animals to 15 ng/ml in the more excitable individuals (King *et al.*, 2006). Restraining extensively raised semiwild cattle for blood testing under farm conditions elicits cortisol readings of 25–33 ng/ml in steers (Zavy *et al.*, 1992), 63 ng/ml in steers and cows (Mitchell *et al.*, 1988), 27 ng/ml in steers (Ray *et al.*, 1972) and 24–46 ng/ml in weaner calves (Crookshank *et al.*, 1979). In Brahman and Brahman-cross cattle, cortisol values ranged from 30 to 35 ng/ml after 20 min of restraint in a squeeze chute (Lay *et al.*, 1992, 1998). The levels were 12 ng/ml after 5 min of restraint and rose to 23 ng/ml after 10.5 min of restraint (Lay *et al.*, 1998).

In some studies, cortisol levels were expressed in nmol/ml by multiplication by 0.36. Tame animals that are trained to lead often have baseline cortisol levels when they are handled for veterinary procedures. When brought to a slaughter plant, tame draught animals often show little or no behavioural signs of agitation.

Slaughter stress similar to on-farm handling

When slaughtering is carried out carefully, cortisol levels in cattle can be substantially lower compared with on-farm handling of extensively raised cattle. Tume and Shaw (1992) reported that steers and heifers slaughtered in a small research abattoir had average cortisol levels of only 15 ng/ml, and cattle slaughtered in a commercial slaughter plant had levels similar to those of on-farm handling of extensively raised cattle. β-Endorphin levels, which are another indicator of stress, were not significantly different between the two

groups. Research by Gerry Means at the Lethbridge Research Centre in Alberta, Canada, indicates that β-endorphin rises in response to pain, whereas cortisol is affected by psychological stress.

For commercial slaughter of extensively raised cattle with captive-bolt stunning, the following average cortisol values have been recorded: 45 ng/ml (Dunn, 1990), 25–42 ng/ml (Mitchell *et al.*, 1988), 44.28 ng/ml (Tume and Shaw, 1992) and 24 ng/ml (Ewbank *et al.*, 1992). A French study with intensively raised young fed bulls had cortisol levels of 21 mg/ml at slaughter (Mounier *et al.*, 2006). When things go wrong, the stress levels increase greatly. Cockram and Corley (1991) reported a median value of 63 ng/ml. One animal had a high of 162 ng/ml.

The slaughter plant observed by Cockram and Corley (1991) had a poorly designed forcing pen and slippery floors. About 38% of the cattle slipped after exiting the holding pens and 28% slipped just before entering the race. Cortisol levels also increased when delays increased waiting time in the single-file race. This was the only study where vocalizations shortly before stunning were not correlated with cortisol levels. This can probably be partly explained by earlier stress caused by the slippery floors. Ewbank *et al.* (1992) reported the lowest average value. This may be explained by excellent handling before entering the stunning box. Cortisol levels are lower and stress is reduced when cattle were transported and slaughtered with their penmates (Mounier *et al.*, 2006).

Detrimental effects of poor handling of cattle

Ewbank *et al.* (1992) found a high correlation between cortisol levels and handling problems in the stunning box. Use of a poorly designed head-restraint device, which greatly increased behavioural agitation and the time required to restrain the animal, resulted in cortisol levels jumping from 24 to 51 ng/ml. In the worst case, the level increased to 96 ng/ml. Cattle slaughtered in a badly designed restraining pen that turned them upside down had average values of 93 ng/ml (Dunn, 1990). Very few sexually mature bulls have been studied, though Cockram and Corley (1991) had a few in their study. Sexually mature bulls have much lower cortisol levels than steers, cows or heifers (Tennessen *et al.*, 1984).

Stress in sheep and pigs

Less clear-cut results have been obtained in sheep. Slaughter in a quiet research abattoir resulted in much lower average levels (40 ng/ml) compared with a large noisy commercial plant that had dogs (61 ng/ml) (Pearson *et al.*, 1977). Shearing and other on-farm handling procedures provoked similar or slightly greater stress levels of 73 ng/ml (Hargreaves and Hutson, 1990), 72 ng/ml (Kilgour and de Langen, 1970) and 60 ng/ml (Fulkerson and Jamieson, 1982). Ninety to 120 min of restraint and isolation stress will increase cortisol levels to 80 to 100 ng/ml in sheep (Apple *et al.*, 1993; Rivalland *et al.*, 2007). Baseline levels were 22 ng/ml.

Electric goads are very stressful for pigs. Numerous shocks from an electric goad significantly raised heart rate compared with animals that had not been shocked (Brundige *et al.*, 1998; Kuchenmeister *et al.*, 2005). Multiple shocks from an electric prod while moving pigs up a ramp and through alleys significantly increased the percentage of non-ambulatory pigs (Benjamin *et al.*, 2001). Blood lactate levels were also significantly higher: high blood lactate is strongly correlated with poor-quality pork (Hambrecht *et al.*, 2004).

Methods for stress reduction

These studies indicate that careful slaughter can be less stressful than on-farm restraint and handling. There is a need to improve practices and equipment. Pre-slaughter stress can greatly exceed farm stress levels when poorly designed equipment is used. When good facilities are combined with well-trained personnel, cattle and sheep can be induced to move through the entire system with no signs of behavioural agitation. The author has observed cattle entering a stunning restrainer like cows in a milking centre. Moving pigs through a high-speed, 1000 animals/h slaughter plant in a calm manner is very difficult. Some large, 1000/h plants have installed two stunning systems to reduce stress. Pigs will be easier to handle and move to the stunning area if they have been rested 1–3 h in the lairage. Resting pigs also

reduces the incidence of PSE and helps to preserve pork quality (Milligan *et al.*, 1998; Perez *et al.*, 2002; Warriss, 2003; Kim *et al.*, 2004).

Every extra handling procedure causes increased stress and bruises. Elimination of unnecessary procedures at the slaughter plant will reduce stress. Bray *et al.* (1989) found that multiple stresses, such as washing, fasting and shearing, have a cumulative, deleterious effect on meat quality. Pigs should be sorted and tattooed prior to leaving the farm. Weighing of live animals at the slaughter plant can be eliminated if the marketing system is on a carcass weight basis.

Washing of cattle and sheep in Australia and New Zealand caused additional stress (Walker *et al.*, 1999). Kilgour (1988) found that washing sheep was a very stressful procedure. Washing had little effect on carcass cleanliness unless the sheep were very dirty (Glover and Davidson, 1977). Washing also increased bruising and DFD meat, with a high ultimate pH (Petersen, 1977; Geesink *et al.*, 2001).

Electric prods should be replaced with other driving aids, such as a flag made from plasticized cloth (see Fig. 20.1). This works especially well for moving pigs out of pens and down alleys.

Farm, environmental and genetic effects on animal handling

Producers must avoid breeding excitable, nervous animals that are difficult to handle. There are problems with very excitable pigs that are almost impossible to handle quietly in a high-speed slaughter plant (Grandin, 1991a). These pigs constantly back up in the race and have excessive flocking behaviour. Shea-Moore (1998) has measured a highly significant difference in the behaviour of high-lean and fat-type pigs. High-lean pigs were more fearful and explored an open field less than did the fat type. The lean pigs also baulked more and took three times longer to walk down an alley with contrasting shadows. Further research showed that high-lean-gain pigs got into significantly more fights than low-lean-gain pigs (Busse and Shea-Moore, 1999).

Breeders need to select lean pigs that have a calmer temperament. Other researchers have observed that pigs from certain farms are more difficult to drive (Hunter *et al.*, 1994). Providing extra environmental stimulation in swine confinement

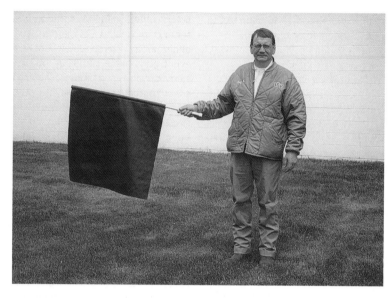

Fig. 20.1. Flag made from plasticized cloth for moving pigs out of pens and down alleys. The animals move quickly away from the rustling sound.

buildings – such as rubber hoses to chew on and people walking in the pens – will produce calmer pigs that drive more easily (Grandin, 1989; Pedersen *et al.*, 1993). Playing a radio in the fattening pens at a reasonable volume level will prevent excessive startle reactions to noises such as a door-slamming.

It is especially important to provide environmental stimulation for pigs that will be transported a short distance to the slaughter plant. Short-transit-time animals are often more difficult to handle because they have not had sufficient time to calm down during the trip. The author has also observed that heavy 125 kg pigs that had been fed excessive amounts of the beta agonist ractopamine were more likely to become non-ambulatory during handling. Producers who have high percentages of non-ambulatory pigs should receive financial penalties.

Accustoming animals to handling

The author has observed that pigs from excitable genetic lines will be easier to handle at the slaughter plant if they are trained to be accustomed to a person walking through their pens: 10–15 s per pen every day for the entire finishing period works well. In facilities where large numbers of pigs are housed in each pen, the time should be 10–15 s per 50 pigs. The person should quietly walk through each pen in a different random direction each day to teach the pigs to rise quietly and flow around them – the person should not just stand in the pen. This trains the animals to approach and chew boots instead of driving.

Geverink *et al.* (1998) also reported that pigs that had been walked in the aisles during finishing on the farm were easier to handle. In another experiment, moving pigs out of their pens 1 month prior to slaughter improved their willingness to move (Abbott *et al.*, 1997). Walking the pens of fattening pigs as little as once or twice a week to get pigs accustomed to people moving through them made them easier to drive at the slaughter plant (Brown *et al.*, 2006).

Further observations in cattle slaughter plants indicated that some extensively raised cattle had never seen a person on foot. When they arrived at the slaughter plant they were difficult and dangerous to handle. This problem is most likely to occur in cattle with European Continental genetics (Grandin and Deesing, 1998). Stress would be reduced if ranchers worked to accustom their cattle to both people on foot and people on horses. At one feedlot, excessively wild cattle had 30% dark-cutters. For more information on the benefits of getting cattle accustomed to handling, see Chapter 5, this volume.

In developing countries where the animals are accustomed to close contact with people, the animals will usually remain calm at the slaughter plant. The author has observed that most animals do not react when they see other animals stunned or bled: they do not appear to understand what is happening. However, dismemberment or seeing the head separated from the body may cause an animal to become agitated.

Objective Evaluation of Animal Welfare

A study by von Wenzlawowlez et al. (1999) showed that many plants did not have procedures that fulfilled European animal welfare requirements. As discussed in Chapter 1, this volume, the use of an objective scoring system to continuously measure the quality of animal handling and stunning is recommended (Grandin, 1998a, b, 2001a, 2005a, b). The five major critical control points for monitoring welfare at slaughter plants are: (i) percentage of animals stunned correctly on the first attempt; (ii) percentage of animals that remain insensible throughout the slaughter process; (iii) percentage of animals that vocalize during handling or stunning; (iv) percentage that slip or fall during handling or stunning; and (v) percentage of animals prodded with an electric prod.

Vocalization scoring

Vocalizations in cattle and pigs are highly correlated with physiological stress measurements (Dunn, 1990; Warriss et al., 1994; White et al., 1995). A higher percentage of cattle vocalized when they were inverted on to their backs for ritual slaughter compared with upright restraint (Dunn, 1990). When freeze-branding was compared with hot-iron branding, 23% of the hot-iron cattle vocalized and only 3% of the freeze-branded vocalized (Watts and Stookey, 1998).

Isolation also increased cattle vocalization (Watts and Stookey, 1998). Grandin (1998a) observed 1125 cattle in six commercial slaughter plants during handling. The cattle were scored as

being either a vocalizer or a non-vocalizer. Nine per cent vocalized (112 cattle), and all except two vocalized in direct response to an aversive event, such as electric prodding, excessive pressure from a restraint device, slipping or falling or missed stuns. Further observations have shown that leaving a bovine alone in the stunning box for too long may cause it to vocalize. In well-run slaughter plants with calm handling, 3% or less of the cattle will vocalize during handling and stunning (Grandin, 1998b). Watts and Stookey (1998) suggested that vocalization in cattle may be especially useful for measuring severe stress.

The level of squealing in pigs is correlated with meat quality problems (Warriss et al., 1994). Pig squeals can also be measured with a sound meter or by determining the percentage of time pigs are quiet in the stunning pen, restrainer, race and crowd pen. As each pig is stunned, score the entire stunning area as either quiet or heard a squeal. For each stunning interval, score yes or no on squealing. Weary et al. (1998) found that, as pain increased, the frequency of high-pitched pig squeals also increased.

Vocalization scoring will work well on cattle and pigs, but it should not be used with sheep. Sheep moving quietly through a race will vocalize. Vocalization scoring should be done during actual handling and stunning, because animals standing in the yards sometimes vocalize to each other. Vocalization scoring should be interpreted as a statistical property of a group of animals, instead of using it to evaluate the condition of individual animals (Watts and Stookey, 1998).

Vocalization scoring will not work if an electric current is used to immobilize an animal, because it prevents the animal from vocalizing. Electro-immobilization should not be used on conscious animals, because it is highly aversive (Grandin et al., 1986; Pascoe, 1986; Rushen, 1986). Electro-immobilization must not be confused with electrical stunning, which induces instantaneous unconsciousness.

Best Practice at the Slaughterhouse

Use of scientifically researched stunning methods

Slaughter plants should use stunning methods that have been verified by scientific testing.

There are a number of reviews on humane stunning methods (Grandin, 1980d, 1985/86, 1994; Eikelenboom, 1983; UFAW, 1987; Gregory, 1988, 1992, 2004, 2007; Devine et al., 1993; Gregory, 2004; OIE, 2005a, b). Many people that first view slaughter are concerned that stunned animals are conscious when reflexes cause the legs to move. The legs may move vigorously on a properly stunned, insensible animal. Eye reflexes, vocalizations and rhythmic breathing must be absent. When the animal is suspended upside down on the shackle, the head should hang straight down and the neck should be straight and limp; there should be no signs of an arched back. Sensible animals will attempt to right themselves by arching their backs and raising their heads. Information on troubleshooting problems with electrical or captive bolt stunning can be found in Grandin (2001b, 2002).

Reduction of noise

Observations in many slaughter plants indicate that noisy equipment and people yelling increase levels of excitement and stress. The author has observed improvements in handling and calmer cattle and pigs after a noise problem was corrected. Clanging and banging noises should be eliminated by rubber pads on gates, and the shackle return should be designed to prevent sudden impact noise. Spensley et al. (1995) found that novel noises ranging from 80 to 89 dB increased heart rate in pigs. Intermittent noises were more disturbing to pigs than continuous noise (Talling et al., 1998). Air exhausts on pneumatically powered gates should be piped outside or muffled, and hydraulic systems should be engineered to minimize noise. High-pitched sounds from hydraulic systems are very disturbing to cattle.

Cattle held overnight in a noisy yard close to the unloading ramp were more active and had more bruising compared with cattle in quieter pens (Eldridge, 1988). Grandin (1980a) discussed the use of music to help mask disturbing noises. The cattle became accustomed to the music in the holding pens and it provided a familiar sound when the animals approach noisy equipment. The type of building construction will greatly influence noise levels.

Plants with high concrete ceilings and precast concrete walls have more echoes and noise than plants built from foam-core insulation board. In one plant with precast concrete walls, the sound level was 93 dB at the restrainer when the equipment was running, and the pigs were quiet. Pigs tended to become agitated because the sound increased from 88 dB in the lairage to 93 dB as the pigs approached. In a survey of 24 plants, the pork plant with the quietest handling and low pig vocalization rates had been engineered to reduce noise (Grandin, 1998b).

Lighting and distractions

Animals have a tendency to move from a darker place to a brighter one (van Putten and Elshof, 1978; Grandin, 1982, 1996; Tanida et al., 1996). Races, stunning boxes and restrainers must be illuminated with indirect, shadow-free light. Lamps can be used to attract animals into a dark race or stunbox. Shadows, sparkling reflections, seeing people up ahead or a small chain hanging in a race will cause animals to baulk. Moving existing lamps will often eliminate reflections on wet floors on shiny metal.

Grandin (1996, 1998c, 2005a; Grandin and Johnson, 2005) reviewed all the little distractions that cause baulking. Quiet handling and minimal usage of electric prods is impossible if animals baulk at distractions. To find the distractions that cause baulking, a person needs to get down and look up a race from the animal's eye level. When animals are calm, they will look right at distractions that cause baulking (see Fig. 20.2).

One of the worst distractions is air blowing down a race into the faces of approaching animals. Distractions can also have time-of-day effects. Shadows may cause a problem when the sun is out and animals may move more easily at night. Small distractions that are easy to fix can ruin the efficiency of the best system. When troubleshooting handling problems, one must be very observant to determine whether the problem is caused by a basic design mistake in the system or by a little distraction or lighting problem that is easy to fix. Sometimes there can be more than one distraction in a system that must be removed.

Elimination of distractions is essential to make it possible to reduce electric prod use. Many plants that were audited for animal welfare by a

Fig. 20.2. A calm animal will look right at a sparkling reflection or other distraction that makes it baulk.

customer had to find and eliminate distractions so they could pass an audit in electric prod use (Grandin, 2005b).

Does blood upset livestock?

Many people interested in the welfare of livestock are concerned about animals seeing or smelling blood. Cattle will baulk and sniff spots of blood on the floor (Grandin, 1980a, d); washing blood off facilitates movement. Baulking may be a reaction to novelty. A piece of paper thrown in the race or stunning box elicits a similar response. Cattle will baulk and sometimes refuse to enter a stunning box or restrainer if the ventilation system blows blood smells into their faces at the stunning-box entrance. They will enter more easily if an exhaust fan is used to create a localized zone of negative air pressure. This will suck smells away from cattle as they approach the stunning-box entrance.

Observations in Jewish ritual (kosher) slaughter plants indicate that cattle will readily walk into a restraining box that is covered with blood. In kosher slaughter, the throat of a fully conscious animal is cut with a razor-sharp knife. The cattle will calmly place their heads into the head restraint device and some animals will lick blood or drink it. Kosher slaughter can proceed very calmly with few signs of behavioural agitation if the restraining box is operated gently (Grandin, 1992, 1994).

However, if an animal becomes very agitated and frenzied during restraint, subsequent animals often become agitated. An entire slaughter day can turn into a continuous chain reaction of excited animals. The next day, after the equipment has been washed, the animals will be calm. The excited animals may be smelling an alarm pheromone from the blood of severely stressed cattle. Blood from relatively low-stressed cattle may have little effect. However, blood from severely stressed animals that have shown signs of behavioural agitation for several minutes may elicit a fear response. Eible-Eibesfeldt (1970) observed that, if a rat is killed instantly in a trap, the trap can be used again. The trap will be ineffective if it injures and fails to kill instantly.

Research with pigs and cattle indicates that there are stress pheromones in saliva and urine. Vieville-Thomas and Signoret (1992) and Boissy *et al.* (1998) reported that pigs and cattle tend to avoid places or objects stained with urine from a stressed animal. The stressor must be applied for 15–20 min to induce the effect. In the cattle experiment, the animals were given repeated shocks for 15 min. Rats showed a fear response to the blood of rats and mice that had been killed with carbon dioxide (CO_2) (Stevens and Gerzog-Thomas, 1977). Carbon dioxide causes secretion of adrenocortical steroids (Woodbury *et al.*, 1958).

Observations by the author of CO_2 euthanasia of mice indicated that they gasped and frantically attempted to escape; both guinea pig and human blood had little effect on rats (Hornbuckle and Beall, 1974; Stevens and Gerzog-Thomas, 1977). Possibly, this was due to less stress in these blood donors. Carbon dioxide at high concentrations does not produce excitement in guinea pigs (Hyde, 1962). Stevens and Gerzog-Thomas (1977) also found the alarm substance in the blood and muscle, whereas rat and mouse brain tissue did not provoke a fear response (Stevens and Gerzog-Thomas, 1977). In a choice test, rats avoided muscle and blood,

but there was no difference between brain tissue and water (Stevens and Saplikoski, 1973).

Design and Management of Lairages and Stockyards

Different countries have specific requirements; for example, truck size will determine the size of each holding pen. Space and facilities must also be designed for specialized functions, such as weighing, sorting and animal identification. Long, narrow pens are recommended in stockyards and lairages in slaughter plants (Kilgour, 1971; Grandin, 1979, 1980e, 1991b). One advantage of long, narrow pens is more efficient animal movement: animals enter through one end and leave through the other. To eliminate 90° corners, the pens can be constructed on a 60–80° angle (Fig. 20.3). Long, narrow pens maximize lineal fence length in relation to floor area, which may help to reduce stress (Kilgour, 1978; Grandin, 1980c, e). Cattle and pigs prefer to lie along the fence line (Stricklin *et al.*, 1979; Grandin, 1980c). Observations indicate that long, narrow pens may also help to reduce fighting (Kilgour, 1976).

Minimum space requirements for holding fattened, feedlot steers for less than 24 h are 1.6 m^2 for hornless cattle and 1.85 m^2 for horned cattle (Grandin, 1979; Midwest Plan Service,

Fig. 20.3.   Long, narrow pens on a 60° angle.

1987), and for slaughter-weight pigs and lambs 0.5 m^2. During warm weather, pigs require more space. Wild, extensively raised cattle may require additional space. However, providing too much space may increase stress, because wild cattle tend to pace in a large pen. Enough space must be provided to allow all animals to lie down at the same time. To avoid bunching and trampling, 25 m is the maximum recommended length for each holding pen, unless block gates are installed to keep groups separated. Shorter pens are usually recommended. Pen and alley width recommendations can be found in Grandin (1991b).

Restraint by tying the legs or confinement in a container that does not allow the animal to turn around must not be used as a means of holding animals in lairage before slaughter. Sheep tied by the legs for 2 h and isolated from other sheep had very high cortisol levels – 100 ng/ml (Apple et al., 1993). This is 30–40% higher than the stress of shearing or slaughter (Pearson et al., 1977; Hargreaves and Hutson, 1990).

Avoidance of mixing strange animals

To reduce stress, prevent fighting and preserve meat quality, strange animals should not be mixed shortly before slaughter (Grandin, 1983a; Tennessen et al., 1984; Barton-Gade, 1985). Solid pen walls between holding pens prevent fighting through the fences. Solid fences in lairage and stockyard pens are especially important if wildlife, such as deer, elk or buffalo, are handled.

Pigs present some practical problems in keeping them separated. In the USA, pigs are transported in trucks with a capacity of over 200 animals. However, they are fattened in much smaller groups on many farms, and some large farms fatten pigs in groups of 200 or more. Observations at US slaughter plants indicate that mixing 200 pigs from three or four farms resulted in less fighting than mixing six to 40 pigs. One advantage of the larger group is that an attacked pig has an opportunity to escape. To determine whether mixing strange pigs is causing a welfare problem, the amount of scratches and wounds can be quantified. Price and Tennessen (1981) found a tendency towards more dark, firm, dry (DFD) carcasses – and hence more stress – when small groups of seven bulls were mixed, compared with larger groups of 21 bulls.

Fighting between slaughter-weight pigs can be reduced by having mature boars present in the holding pens (Grandin and Bruning, 1992). Another method for reducing fighting is to feed pigs excess dietary tryptophan for 5 days before slaughter (Warner et al., 1992). Further information on agonistic behaviour in pigs can be found in a review by Petherick and Blackshaw (1987).

When strange bulls are mixed, physical activity during fighting increases DFD meat. The installation of either steel bars or an electric grid over the holding pens at the abattoir prevented dark cutting in bulls (Kenny and Tarrant, 1987). These devices prevent mounting. The electric grid should be used only with animals that have been fattened in pens equipped with an electric grid. In Sweden and other countries where small numbers of bulls are fattened, individual pens are recommended at the abattoir (Puolanne and Aalto, 1981).

In small European slaughter plants, the holding area consists of a series of single-file races which lead to the stunner. Bulls are unloaded directly into the races and each bull is kept separated by guillotine gates. This system is recommended in situations where a farmer markets a few intensively raised bulls at a time from each fattening pen. Using a single-file race as lairage should not be considered for wild, extensively raised cattle. If the entire pen is marketed at one time, the animals can be penned together in a group at the plant. Another common European design is tie-up stalls with halters. Tie-up stalls should be used only for animals that are trained to lead.

Flooring and fencing design

Floors must have a non-slip surface (Stevens and Lyons, 1977; Grandin, 1983a; OIE, 2005b). In three out of 11 (17%) beef plants surveyed, slippery floors caused cattle to become agitated (Grandin, 1998a). Smooth floors and slipping increase levels of stress (Cockram and Corley, 1991). For cattle, concrete floors should have deep, 2.5 cm V grooves in a 20 cm square or diamond pattern. For pig and sheep abattoirs, the wet concrete should be imprinted with a stamp constructed from expanded steel mesh. The expanded steel should have a 3.8 cm-long opening (Grandin, 1982). A broom finish prevents slipping when new (Applegate *et al.*, 1988), but practical experience has shown that it quickly wears out and the animals will fall down.

Some abattoirs in developing countries have smooth floors because it is intended that the animals fall down. This is a cruel practice that should be eliminated. These plants may need to build a single-file race if they are handling extensively raised wild cattle or sheep. Concrete slats may be used in holding pens, but the drive alleys should have a solid concrete floor. Precast slats for cattle or swine confinement buildings will work. The slats should have a grooved surface. Slats or gratings used in pig and sheep facilities should face in the proper direction to prevent the animals from seeing light coming up through the floor.

Animals will baulk at sudden changes in floor texture or colour. Flooring surfaces should be uniform in appearance and free from puddles. In facilities that are washed, install concrete kerbs between the pens to prevent water in one pen flowing into another. Drains should be located outside the areas where animals walk. Livestock will baulk at drains or metal plates across an alley (Grandin, 1987). Flooring should not move or jiggle when animals walk on it: flooring that moves causes pigs to baulk (Kilgour, 1988). Lighting should be even and diffuse, to reduce shadows. Lamps can be used to encourage animals to enter races (Grandin, 1982). Additional information on the effects of vision and lighting on livestock handling is given in Chapters 5, 10, 14 and 15, this volume.

Bruising

Cattle and sheep can have bruised meat even though the hide is undamaged. Bruises can occur up until the moment of bleeding. Meischke and Horder (1976) determined that stunned cattle could be bruised when they were ejected from the stunning box. Pigs are slightly less susceptible to bruising, but meat quality deteriorates when they become excited or hot. Lean pigs bruise more easily than fatter pigs. Edges with a small diameter will cause severe bruises. Steel angles or I-beams should not be used in the construction of pens or races, as animals bumping against the edges will bruise.

Round-pipe posts and fence rails are recommended. Surfaces coming into contact with animals should be smooth and rounded (Stevens and Lyons, 1977; Grandin, 1980b); exposed sharp ends of pipes should be bevelled to prevent gouging.

Areas that have completely solid fences should have all posts and structural parts on the outside away from the animals. An animal rubbing against a smooth, flat metal surface will not bruise. All gates should be equipped with tiebacks to prevent them from swinging into the alley. Guillotine gates should be counterweighted and padded on the bottom with conveyor belting or large-diameter hose (Grandin, 1983a). Bruising on pigs can be prevented by cutting 45 cm off the bottom of guillotine gates and replacing the bottom portion of the gate with a curtain of flexible rubber belting. The pigs think the curtain is solid and few attempt to go through it.

Races for cattle and sheep

Single-file races work well for cattle and sheep, because they are animals that will naturally walk in single file. Cattle will walk in single file when moving from pasture to pasture. Races are required for low-stress handling of extensively raised cattle and sheep that have had little contact with people. They are not required if animals are trained to lead with a halter or if they are sufficiently tame to allow people to touch them without moving away.

In small abattoirs in developing countries, a tame animal can be led to the slaughter area. Stunning can be performed while the animal is standing. All races should have solid outer fences to prevent the animals from seeing people, vehicles, moving equipment or other distractions outside the fence. Animal entry into the race can sometimes be facilitated if solid shields are installed to prevent approaching animals from seeing people standing by the race. A curved, single-file race is especially recommended for the moving of cattle (Rider *et al.*, 1974; Grandin, 1980e; Figs 20.4 and 20.5).

An inside radius of 5 m is ideal for cattle (additional recommendations are given in Chapter 7, this volume). Walkways for the handler should run alongside the race, and the use

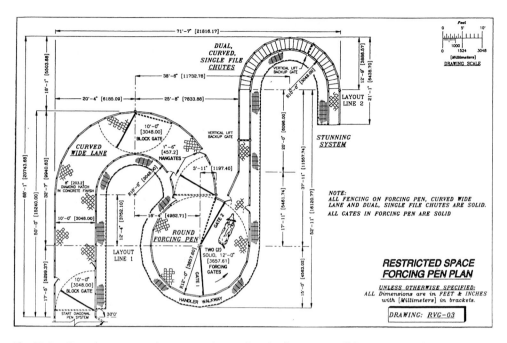

Fig. 20.4. Curved race system for extensively raised cattle; this system will fit in a restricted space.

Fig. 20.5. Curved cattle race with solid sides at a slaughter plant; this design improves cattle movement.

of overhead walkways should be avoided. In slaughter plants with restricted space, a serpentine race system can be used (Grandin, 1984). A race system at an abattoir must be long enough to ensure a continuous flow of animals to the stunner, but not be so long that animals become stressed waiting in line. The hold-down rack over the top of a single-file race must not touch the backs of the animals. Animals tend to panic if the hold-down bars press on their backs.

Handling systems for pigs

Hartung *et al.* (1997) found that pigs were less stressed in a very short 3.5-m race compared with in an 11-m race. German plants run at slower speeds than US plants, and a 3.5 m-long race may cause more stress in a plant running 800 pigs per hour because the short race makes it more difficult to keep up with the line. In plants slaughtering 240 or fewer pigs per hour, stunning them with electric tongs in groups on the floor was less stressful than a double, single-file race (Warriss *et al.*, 1994). In larger plants, floor stunning with tongs tends to get rough and careless. In a small plant, pigs stunned in small groups on the floor had better meat quality (Stuier and Olsen, 1999). Two races are sometimes built side by side, because the pigs will enter more easily (Grandin, 1982). The outer walls of the race are solid, but the inner fence between the two races is constructed from bars. This enables the pigs to see each other and promotes following behaviour.

Researchers in Denmark have developed an excellent handling system for quietly moving groups of four or five pigs into a CO_2 chamber (Barton-Gade and Christenson, 1999; Fig. 20.6). A major advantage of this handling system is that use of the electric goad can be completely eliminated. The group gas-stunning system has been installed in many large pork slaughter plants in North America and Europe, and handling has been improved because the single-file race has been eliminated.

From a handling standpoint, the biggest potential problem with gas-stunning of pigs is use of an undersized system where the gondolas are overloaded. Pigs in the gondolas must have enough room so they can comfortably stand and not be on top of each other. There are concerns about the humaneness of CO_2 in certain breeds and genetic lines of pigs. Forslid (1987) and Ring (1988) reported that CO_2 was humane for Yorkshire and Landrace pigs. Grandin (1988a) reported that certain breeds, such as Hampshire, had a very bad reaction and became highly agitated prior to the onset of unconsciousness. Dodman (1977) also reported variations in a pig's reaction to CO_2. Genetic selection is one solution to this problem.

The work by Forslid (1987) measured electrical responses from the brain of purebred Yorkshires. Forslid's experiments need to be conducted in other genetic types of pigs. Even if CO_2 is mildly aversive to certain pig genotypes, their overall welfare may be improved if the single-file race is eliminated. Raj *et al.* (1997) found that mixing 30% CO_2 with 60% argon was more humane than 90% CO_2. If the pigs struggle violently or attempt to climb the walls of the gondola before they collapse and lose sensibility, the gas mixture is not acceptable.

Fig. 20.6. Danish group handling system for moving groups of five pigs into a CO_2 chamber.

Crowd pens for all species

Round crowd pens leading up to a single-file race are very efficient for all species because, as the animals go around the circle, they think they are going back to where they came from (Grandin, 1998b). The recommended radius for round crowd pens is 3.5 m for cattle, 1.83–2.5 m for pigs and 2.4 m for sheep. For all species, solid sides are recommended on both the race and the crowd pen leading to the race (Rider *et al.*, 1974; Brockway, 1975; Grandin, 1980c, 1982). For operator safety, gates for personnel must be constructed so that they can escape charging cattle. The crowd gate should also be solid to prevent animals from turning back. Wild animals tend to be calmer in facilities with solid sides. The crowd pen must be constructed on a level floor. Animals will pile up if the crowd pen is on a ramp (see Chapter 7, this volume, for further information).

Crowd pens with a funnel design work well for cattle and sheep, but very poorly for pigs. A crowd pen for pigs must be designed with an abrupt entrance to the race to prevent jamming. Hoenderken (1976) devised a crowd pen constructed like a series of stair steps, which varied in width from one pig wide to two pigs wide and three pigs wide. This design works fairly well when pigs are handled rapidly in batches, but it works poorly in continuous-flow systems. A round crowd pen in which two crowd gates continually revolve and with an abrupt entrance to the single-file race is being successfully used in several US pig slaughter plants. Another design is a single, offset step equal to the width of one pig. It prevents jamming at the entrance to a single-file race (Grandin, 1982, 1987).

Jamming can be further prevented by installing an entrance restricter at single-file race entrances. The entrance of the single-file race should provide only 5 mm on each side of each pig. A double race should also have a single offset step to prevent jamming.

The paramount handling problem is overloading of the crowd pen and the alley leading up to it. Cattle and pigs will move more easily if the crowd pen is filled half full and they are moved in small bunches. Tina Widowski and her colleagues at the University of Guelph, Ontario, Canada, conducted a research study that quantified the advantage of filling the crowd pen half full (T. Widowski, personal communication, 2006) (see Fig. 20.7). Another mistake is

Fig. 20.7. Round crowd pen for cattle. A round crowd pen is efficient for all species because animals think they are going back to where they came from. Cattle and pigs move easily when the pen is half full, as shown in this photograph. Photograph by T. Grandin from Grandin (1998b).

in attempting to forcibly push animals with powered crowd gates. The author has observed many plants where powered crowd gates were abused. Sheep can be moved in large groups due to their intense following behaviour.

Unloading

More than one truck unloading ramp is usually required to facilitate prompt unloading (Weyman, 1987). During warm weather, prompt unloading is essential, because heat rapidly builds up in stationary vehicles. In some facilities, unloading pens will be required. These pens enable animals to be unloaded promptly prior to sorting, weighing or identification checking. After one or more procedures have been performed, the animals move to a holding pen. Facilities used for unloading only should be 2.5–3.0 m wide to provide the animals with a clear exit into the alley (Grandin, 1980c, 1991b).

One big problem in some developing countries is that they may have no unloading ramps. Animals are jumped out of vehicles and may break legs. Ramps can be easily built from either steel or concrete by local welders or masons. People skilled in welding and masonry are readily available in all developing countries. In most countries, two ramps will be needed, one for large trucks and one for smaller pickups (ute) trucks. Stationary ramps are easier for local people to build than complicated devices such as hydraulic tailgate lifts on trucks, or ramps that move up and down.

Ramps and slopes

Ideally, an abattoir stockyard should be built at truck-deck level to eliminate ramps for both unloading and movement to the stunner. This is especially important for pigs. The maximum angle for non-adjustable livestock unloading ramps is 20–25°. If possible, the ramp to the stunner should not exceed 10° for pigs, 15° for cattle and 20° for sheep. Ramp angles to the stunner should be more gradual than the maximum angles that will work for loading trucks. To reduce the possibility of falls, unloading ramps should have a flat deck at the top. This provides a level surface for animals to walk on when they first step off the truck (Stevens and Lyons, 1977; Grandin, 1979; Agriculture Canada, 1984). This same principle also applies to ramps to the stunner. A level portion facilitates animal entry into the restrainer or stunning box.

Grooved stair steps are recommended on concrete ramps (United States Department of Agriculture, 1967; Grandin, 1980c, 1991b). They are easier to walk on after the ramp becomes worn or dirty. The dimensions for cattle and other large animals are a 10-cm rise and a 30-cm or longer tread width. Further recommendations are given in Chapter 7, this volume and Grandin (1991b). However, in new, clean facilities, small pigs showed no preference between stair steps and closely spaced cleats (Phillips et al., 1987, 1989). For slaughter-weight pigs, cleats should be spaced 15 cm apart (Warriss et al., 1991). For cattle, the spacing should be 20 cm of space in between the cleats. There were very little differences in pig stress levels when they were loaded and unloaded with either a ramp or a hydraulic tailgate lift (Brown et al., 2005).

Design and Maintenance of Restraint, Stunning and Slaughtering Areas

Restraint methods that cause suffering

The OIE (2005a) guidelines for animal welfare state that methods of restraint that cause avoidable suffering should not be used. Some of the worst methods that should never be used are:

- Suspending or hoisting animals (other than poultry) by the feet or legs.
- Mechanical clamping of the legs of mammals.
- Breaking legs.
- Cutting leg tendons.
- Blinding by poking out eyes.
- Severing the spinal cord with a puntilla (dagger) that paralyses the animal.
- Electrical immobilization with currents that are not sufficient to cause loss of sensibility.

Conveyor restrainer systems

One of the first modern systems was the 'V' conveyor restrainer for pigs (Regensburger, 1940).

This consists of two obliquely angled conveyors that form a V. Pigs ride with their legs protruding through the space at the bottom of the V. The V restrainer is a comfortable system for pigs with round, plump bodies and for sheep (Grandin, 1980c). Pressure against the sides of the pig will cause it to relax (Grandin et al., 1989). However, the V restrainer is not suitable for restraining calves or extremely heavily muscled pigs with large, overdeveloped hams (Lambooy, 1986). The V pinches the large hams and the slender forequarters are not supported. Some of the very lean, long pigs are also not supported properly.

Researchers at the University of Connecticut, USA, developed a laboratory prototype for a double-rail restrainer system to replace V conveyor restrainers (Westervelt et al., 1976; Giger et al., 1977). Calves and sheep are supported under the belly and the brisket by two moving rails. This research demonstrated that animals restrained in this manner were under minimal stress. Sheep and calves rode quietly on the restrainer and seldom struggled. The space between the rails provides a space for the animal's brisket and prevents uncomfortable pressure on the sternum. The prototype was a major step forward in humane restrainer design, but many components still had to be developed to create a system that would operate under commercial conditions.

In 1986, the first double-rail restrainer was designed and installed in a large commercial calf and sheep slaughter plant by Grandin Livestock Handling Systems and Clayton H. Landis in Souderton, Pennsylvania, USA (Grandin, 1988b, 1991c; Figs 20.8 and 20.9). In the early 1990s, the Stork Company in the Netherlands developed a restrainer where pigs ride on a moving-centre conveyor.

In 1989, the first double-rail restrainer was installed in a large cattle slaughter plant by Grandin Livestock Handling Systems and Swilley Equipment, Logan, Iowa, USA (Grandin, 1991c, 1995). Half the cattle in North America are now handled in this system, and the system has been installed in 25 plants. The double-rail restrainer has many advantages compared with the V restrainer (Grandin, 1983b). Stunning is easier and more accurate, because the operator can stand 28 cm closer to the animal. Cattle enter more easily because they can walk in with their legs in a natural position.

Proper design is essential for smooth, humane operation. Incoming cattle must not be able to see light coming up from under the restrainer. It must have a false floor below the restrained animal's feet, to provide incoming cattle with the appearance of a solid floor to walk on. To keep cattle calm they must be fully restrained and settled down on the conveyor before they emerge from under the hold-down rack. Their back feet must be off the entrance ramp before they see out from beneath the hold-down rack. If the hold-down is too short, the cattle are more likely to become agitated. The principle is to block the animal's vision until it is fully restrained.

Moving animals easily into restrainers

Animals should enter a restrainer easily. If they baulk, figure out why they are baulking instead

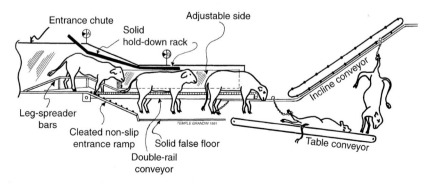

Fig. 20.8. Diagram of double-rail, centre-track restrainer.

Fig. 20.9. Large steer in the double-rail, centre-track restrainer. Note that the conveyor is shaped to fit the animal's brisket.

of resorting to increased electric prodding. Lighting problems and the distractions discussed in Grandin (1996, 1998b; Chapter 7, this volume) can cause animals to refuse to enter. In one plant, adding a light at the restrainer entrance reduced baulking and use of the electric prod. The percentage of cattle vocalizing due to electric prodding was reduced from 8 to 0%. Designers and animal handlers should learn to use behaviour principles in controlling an animal. Blocking an animal's view of people with a piece of rubber belting can keep even wild cattle quiet.

Details of design are very important in animal handling systems. A 0.5 m difference in the length of a shield to block an animal's vision can be the difference between calm animals and agitated, frightened animals. Cattle stay calm in the restrainer because the next animal in line can see an animal directly in front of it. Watts and Stookey (1998) found that cattle vocalized less in a race when they could see another animal in front of them less than 1 m away.

Conventional stunning boxes

A common mistake is to build stunning boxes too wide. A 76 cm-wide stunning box will hold all cattle with the exception of some of the largest bulls. Stunning boxes must have non-slip floors to allow the animal to stand without slipping.

In a conventional cattle stunning box, stunning accuracy can be greatly improved by the use of a yoke to hold the head. Yokes and automatic head restraints for cattle have been developed in Australia (CSIRO, 1989; Buhot et al., 1992), New Zealand and England. Ewbank et al. (1992) found that cattle had higher stress levels when their heads were restrained. The system they observed was poorly designed and lacked a rear pusher gate. Forcing the animal's head into the restraint was difficult and took an average of 32 s.

To minimize stress, the yoke must be designed so that the animal will enter it willingly, and it must be stunned immediately after the head is caught. Australian head-restraint equipment with a rear pusher gate works well (CSIRO, 1989). The rear pusher gate eliminates prodding. Lamps can also be used to encourage cattle to hold their heads up for stunning. New Zealand researchers have devised a humane system for electrical stunning of cattle while they are held in a head restraint (Gregory, 1993). Specifications for proper electrode placement are given in Cook et al. (1991). In countries where the animals are tame, a stunning box may not be needed. An animal that is trained to lead can be stunned while it is standing, having no restraint except a halter and a lead rope.

Restraint for ritual slaughter

Some Muslim religious authorities will allow head-only electrical stunning for halal slaughter, but Jewish (kosher) slaughter is always conducted on conscious animals. In some countries, such as the USA, it is legal to suspend live animals by one back leg for ritual slaughter. This cruel practice is very dangerous. Replacement of shackling and hoisting by a humane restraint device will greatly reduce accidents (Grandin, 1990). In Europe, Canada and Australia, the use of humane restraining devices is required. When ritual slaughter is being evaluated from a welfare standpoint, the animal's reactions to restraint must be separated from its reactions to throat-cutting without stunning (Grandin, 1994; Grandin and Regenstein, 1994).

In small local abattoirs in developing countries no restraint devices are available, so the OIE (2005a) does allow rope-casting and the tying of legs for slaughter without stunning. These methods are most stressful for extensively raised animals unaccustomed to close contact with people. The time that the animal is restrained must be as brief as possible.

Slaughter without stunning can be carried out easily in sheep and goats when they are standing with a person straddling their back. The author strongly recommends that larger plants install more modern equipment. In Middle Eastern countries, sheep are moved through a V conveyer restrainer and the throat cut is made after the sheep is ejected from the restrainer into a level, moving-table conveyor. This system is superior to systems where sheep are tied up, but it would have poorer welfare than a system where the throat cut is performed while the animal is held upright.

The first restraining device for ritual slaughter was developed in Europe: the Weinberg casting pen consists of a narrow stall that slowly inverts the animal until it is lying on its back. It is less stressful than shackling and hoisting, but it is much more stressful than more modern, upright, restraint devices (Dunn, 1990). Animals restrained in the Weinberg casting pen had much higher levels of vocalizations and cortisol

(stress hormone) compared with cattle restrained in the upright restraining pen.

The Facomia pen is an improved rotating pen. It holds the animal's head and body more securely than the old-fashioned Weinberg. An adjustable side presses against the body to support its body during rotation. However, it is probably more stressful than the best upright restraint. Cattle resist inversion: inverted cattle twist their necks in an attempt to right their heads, and they may aspirate greater amounts of blood.

A major innovation in ritual restraint equipment was the ASPCA (American Society for the Prevention of Cruelty to Animals) pen (Marshall et al., 1963). It consists of a narrow stall with solid walls with an opening in the front for the animal's head. A lift under the belly prevents the animal from collapsing after the throat-cut (see Fig. 20.10).

Proper design and operation are essential (Grandin, 1992, 1994; Grandin and Regenstein, 1994). The belly lift should not lift the animal off the floor. Some older models of this pen cause excessive stress because the animal is lifted off the floor by the belly lift. A stop should be installed to restrict belly lift travel to 28 in (71 cm). The air or hydraulic pressure that operates the rear pusher gate should be reduced to prevent excessive pressure on the animal's rear. The head-holder must have a stop or a bracket to prevent excessive bending of the neck and a pressure-limiting device to prevent the application of excessive pressure. A 25 cm (10 in)-wide forehead fold-down bracket covered with rubber belting will help prevent discomfort to the animal.

Flow controls or speed reducers should be installed in the hydraulic or pneumatic system to prevent sudden jerky movement. All pneumatically or hydraulically powered devices must have pressure regulators that will limit the maximum pressure applied to an animal. Devices for holding the head usually require less pressure than devices for holding the body. Head-holding devices should have their own separate pressure control set at a lower pressure. If the animal vocalizes or strains when a restraint device is applied, it is too tight.

Further developments in ritual slaughter equipment are the use of a mechanical head holder on a V restrainer and ritual slaughter of

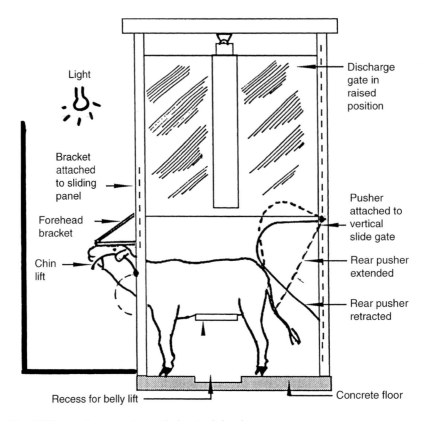

Fig. 20.10. ASPCA pen for restraining cattle for ritual slaughter.

calves on the double rail (Grandin, 1988b). The author has developed a head-holding device for the large-cattle double-rail restrainer (see Fig. 20.11). It is being used in a large commercial slaughter plant for both ritual slaughter and captive-bolt stunning. A slot in the forehead bracket is provided for captive-bolt stunning and it could be modified for electric stunning. The research team at the University of Connecticut, USA, has also developed a small, inexpensive restrainer to hold calves and sheep during ritual slaughter (Giger *et al.*, 1977). For larger calves, a miniature ASPCA pen has been installed in many plants.

Behavioural principles of restraint

During work on restraint systems at four different kosher slaughter plants, the author developed four behavioural principles of restraint.

- Block vision. The animal must see a lighted place to move into, but solid panels or curtains should be used to prevent it from seeing people.

- Slow steady movement. Parts of an apparatus that press against an animal must move with slow, steady movement. Sudden jerky motion scares. Hydraulic and pneumatic controls should be designed to work like a car accelerator to enable the operator to speed up or slow down movement of the device.

- Optimum pressure. A device must hold an animal tightly enough for it to feel held but not so tightly that it causes discomfort. Pinch points and sharp edges must be eliminated.

- Does not trigger righting reflex (fear of falling). The device should hold the animal in a comfortable upright position. If the animal slips or feels unbalanced, it may struggle. Box-type restrainers must have non-slip flooring.

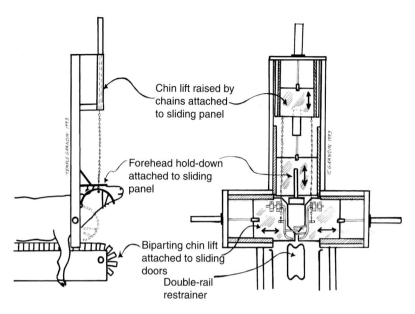

Fig. 20.11. Head-holding device for ritual slaughter or captive-bolt stunning.

Conclusions

Carefully conducted slaughter is less stressful than on-farm handling and restraint. When different systems are being evaluated, the variables of basic equipment design must be separated from rough handling and distractions that make animals baulk. Correcting these problems will usually improve animal movement in all types of equipment. Engineers who design equipment must pay close attention to design details. Layout mistakes, such as bending a race too sharply at the junction between the race and the forcing pen, will cause baulking. Pigs handled in a race system with layout mistakes had increased stress (Weeding et al., 1993).

Management must become increasingly sensitive to animal welfare. The single most important factor that determines how animals are treated is the attitude of management. New developments in equipment design will make quiet humane handling easier, but all systems must have good management to go with them. Sloppy practices that can be easily corrected by good management are a major cause of poor stunning.

A lack of maintenance was a primary cause of captive-bolt stunning failure (Grandin, 2002). In pigs, placing the electrodes in the wrong position and poor bleeding are the main causes of return to sensibility (Grandin, 2001b). Management must closely supervise both employee behaviour and equipment maintenance. The worldwide OIE standards for animal welfare are a minimum standard. Large meat-buying customers will often have stricter standards for their suppliers.

References

Abbott, T.A., Hunter, E.J., Guise, J.H. and Penny, R.H.C. (1997) The effect of experience of handling on pigs willingness to move. *Applied Animal Behaviour Science* 54, 371–375.

Agriculture Canada (1984) *Recommended Code of Practice for Care and Handling of Pigs*. Publication 1771/E, Agriculture Canada, Ottawa, Canada.

Alan, M.G.S. and Dobson, H. (1986) Effect of various veterinary procedures on plasma concentrations of cortisol, luteinizing hormone and prostaglandin E$_2$ metabolite in the cow. *Veterinary Record* 118, 7–10.

Apple, J.K., Minton, J.E., Parsons, K.M. and Unruh, J.A. (1993) Influence of repeated restraint and isolation stress and electrolyte administration on pituitary–adrenal secretions, electrolytes and other blood constituents of sheep. *Journal of Animal Science* 71, 71–77.

Applegate, A.L., Curtis, S.E., Groppel, J.L., McFarlane, J.M. and Widowski, T.M. (1988) Footing and gait of pigs on different concrete surfaces. *Journal of Animal Science* 66, 334–341.

Barton-Gade, P. (1985) Developments in the pre-slaughter handling of slaughter animals. In: *Proceedings, European Meeting of Meat Research Workers*, Albena, Bulgaria, pp. 1–6.

Barton-Gade, P. (1989) Pre-slaughter treatment and transportation in Denmark. In: *Proceedings, International Congress of Meat Science and Technology*, Copenhagen.

Barton-Gade, P. and Christenson, L. (1999) *Transportation and Pre-stun Handling CO$_2$ Systems*. Report No. 53.310, Danish Meat Institute, Roskilde, Denmark.

Barton-Gade, P., Blaabjerg, L. and Christensen, L. (1993) A new lairage system for slaughter pigs. *Meat Focus International* 2, 115–118.

Benjamin, M.E., Gonyou, H.W., Ivers, D.L., Richard, L.F., Jones, D.J., Wagner, J.R., Seneriz, R. and Anderson, D.B. (2001) Effect of handling method on the incidence of stress response in market swine in a model system. *Journal of Animal Sciences* 79 (Suppl. 1), 279 (abstract).

Boissy, A., Terlow, C. and LeNeindre, P. (1998) Presence of pheromones from stressed conspecifics increases reactivity to aversive events in cattle, evidence for the existence of alarm substances in urine. *Physiology and Behavior* 4, 489–495.

Bray, A.R., Graafhuis, A.E. and Chrystall, B.B. (1989) The cumulative effect of nutritional shearing and preslaughter washing stresses on the quality of lamb meat. *Meat Science* 25, 59–67.

Brockway, B. (1975) *Planning a Sheep Handling Unit*. Farm Buildings Centre, Kenilworth, UK.

Brown, J.A., Toth, E.L., Stanton, A.L., Lawlis, P. and Widowski, T. (2006) The effects of different frequencies of weekly human interaction on handling responses of market hogs. *Journal of Animal Science* 84 (Suppl. 1), 301 (abstract).

Brown, N.S., Knowles, T.G., Wilkins, L.J., Chadd, S.A. and Warriss, P.D. (2005) The response of pigs to being loaded or unloaded onto commercial animal transporters using three systems. *Veterinary Journal* 70, 91–100.

Brundige, L., Okeas, T., Doumit, M. and Zanella, A.J. (1998) Loading techniques and their effect on behavior and physiological responses of market pigs. *Journal of Animal Science* 76 (Suppl. 1), 99 (abstract).

Buhot, J.W., Egan, A.F., Wharton, R.A. and Bowtell, K.D. (1992) An automated system for dressing bovines. In: *38th Congress of Meat Science and Technology*. INRA, Thiex, St Gene Champanelle, France.

Busse, C.S. and Shea-Moore, M.M. (1999) Behavioral and physiological responses to transportation stress. *Journal of Animal Science* 77 (Suppl. 1), 147 (abstract).

Cockram, M.S. and Corley, K.T.T. (1991) Effect of pre-slaughter handling on the behaviour and blood composition of beef cattle. *British Veterinary Journal* 147, 444–454.

Cook, C.J., Devine, C.E., Gilbert, K.V., Tavener, A. and Day, A.M. (1991) Electroencephalograms in young bulls following upper cervical vertebrae-to-brisket stunning. *New Zealand Veterinary Journal* 39, 121–125.

Crookshank, H.R., Elissalde, M.H., White, R.G., Clanton, D.C. and Smalley, H.E. (1979) Effect of transportation and handling of calves on blood serum composition. *Journal of Animal Science* 48, 430–435.

CSIRO (1989) *Cattle Head Capture Unit, Meat Research Newsletter*. CSIRO Meat Research Laboratory, Cannon Hill, Brisbane, Queensland, Australia.

Devine, C.E., Cook, C.J., Maasland, S.A. and Gilbert, K.V. (1993) The humane slaughter of animals. In: *39th International Congress of Meat Science and Technology*, Agriculture-Canada, Lacombe Research Station, Alberta, Canada, pp. 223–228.

Dodman, N.H. (1977) Observations on the use of the Wernberg dip-lift carbon dioxide apparatus for pre-slaughter anesthesia of pigs. *British Veterinary Journal* 133, 71–80.

Dunn, C.S. (1990) Stress reactions of cattle undergoing ritual slaughter using two methods of restraint. *Veterinary Record* 126, 522–525.

Eible-Eibesfeldt, I. (1970) *Ethology: the Biology of Behavior*. Holt Rhinehart and Winston, New York, p. 236.

Eikelenboom, G. (ed.) (1983) *Stunning of Animals for Slaughter*. Martinus Nijhoff, Boston, Massachusetts.

Eldridge, G.A. (1988) The influence of abattoir lairage conditions on the behavior and bruising of cattle. In: *Proceedings, 34th International Congress of Meat Science and Technology*. Brisbane, Australia.

Ewbank, R., Parker, M.J. and Mason, C.W. (1992) Reactions of cattle to head restraint at stunning: a practical dilemma. *Animal Welfare* 1, 55–63.

Faucitano, L. (1998) Preslaughter stressors effect on pork: a review. *Muscle Foods* 9, 293–303.

Forslid, A. (1987) Transient neocortical, hippocampal and amygdaloid EEG silence induced by one minute infiltration of high concentration CO_2 in swine. *Acta Physiologica Scandinavica* 130, 1–10.

Fulkerson, W.J. and Jamieson, P.A. (1982) Patterns of cortisol release in sheep following administration of synthetic ACTH and imposition of various stressor agents. *Australian Journal of Biological Science* 35, 215–222.

Geesink, G.H., Mareko, M.H.D., Motto, J.D. and Bickerstaff, R. (2001) Effects of stress and high voltage electrical stimulation on tenderness in lamb *M. longissimus*. *Meat Science* 57, 265–271.

Geverink, N.A., Kappers, A., Van de Burgwal, E., Lambooij, E., Blokhuis, J.H. and Wiegant, V.M. (1998) Effects of regular moving and handling on the behavioral and physiological responses of pigs to pre-slaughter treatment and consequences for meat quality. *Journal of Animal Science* 76, 2080–2085.

Giger, W., Prince, R.P., Westervelt, R.G. and Kinsman, D.M. (1977) Equipment for low stress animal slaughter. *Transactions of the American Society of Agricultural Engineers* 20, 571–578.

Glover, A.F. and Davidson, R.M. (1977) *Report on Cutter Lamb Survey*. New Zealand Meat Producers Board, Wellington, New Zealand.

Grandin, T. (1979) Designing meat packing plant handling facilities for cattle and hogs. *Transactions of the American Society of Agricultural Engineers* 22, 912–917.

Grandin, T. (1980a) Livestock behavior as related to handling facility design. *International Journal of the Study of Animal Problems* 1, 33–52.

Grandin, T. (1980b) Bruises and carcass damage. *International Journal of the Study of Animal Problems* 1, 121–137.

Grandin, T. (1980c) Designs and specifications for livestock handling equipment in slaughter plants. *International Journal of the Study of Animal Problems* 1, 178–299.

Grandin, T. (1980d) Mechanical, electrical and anaesthetic stunning methods for livestock. *International Journal for the Study of Animal Problems* 1, 242–263.

Grandin, T. (1980e) Observations of cattle behavior applied to the design of cattle handling facilities. *Applied Animal Ethology* 6, 10–31.

Grandin, T. (1982) Pig behavior studies applied to slaughter plant design. *Applied Animal Ethology* 9, 141–151.

Grandin, T. (1983a) Welfare requirements of handling facilities. In: Baxter, S.H., Baxter, M.R. and MacCormack, J.A.D. (eds) *Farm Animal Housing and Welfare*. Martinus Nijhoff, Boston, Massachusetts, pp. 137–149.

Grandin, T. (1983b) *System for Handling Cattle in Large Slaughter Plants*. Paper No. 83–4506, American Society of Agricultural Engineers, St Joseph, Michigan.

Grandin, T. (1984) Race system for slaughter plants with 1.5 m radius curves. *Applied Animal Behaviour Science* 13, 295–299.

Grandin, T. (1985/1986) Cardiac arrest stunning of livestock and poultry. In: Fox, M.W. and Mickley, L.D. (eds) *Advances in Animal Welfare Science*. Martinus Nijhoff, Boston, Massachusetts, pp. 1–30.

Grandin, T. (1987) Animal handling. *Veterinary Clinics of North America Food Animal Practice* 3, 323–338.

Grandin, T. (1988a) Possible genetic effect on pig's reaction to CO_2 stunning. In: *Proceedings, 34th International Congress of Meat Science and Technology*. CSIRO Meat Research Laboratory, Cannon Hill, Queensland, Australia.

Grandin, T. (1988b) Double rail restrainer conveyor for livestock handling. *Journal of Agricultural Engineering Research* 41, 327–338.

Grandin, T. (1989) Effect of rearing environment and environmental enrichment on behavior and neural development of young pigs. Doctoral dissertation, University of Illinois, Urbana-Champaign, Illinois.

Grandin, T. (1990) Humanitarian aspects of shehitah in the United States. *Judaism* 39, 436–446.

Grandin, T. (1991a) Handling problems caused by excitable pigs. In: *Proceedings, 37th International Congress of Meat Science and Technology*. Federal Centre for Meat Research, Kulmbach, Germany.

Grandin, T. (1991b) Principles of abattoir design to improve animal welfare. In: Matthew, J. (ed.) *Progress in Agricultural Physics and Engineering*. CAB International, Wallingford, UK.

Grandin, T. (1991c) *Double Rail Restrainer for Handling Beef Cattle*. Paper No. 91–5004, American Society of Agricultural Engineers, St Joseph, Michigan.

Grandin, T. (1992) Observations of cattle restraint devices for stunning and slaughtering. *Animal Welfare* 1, 85–91.

Grandin, T. (1994) Euthanasia and slaughter of livestock. *Journal of the American Veterinary Medical Association* 204, 1354–1360.

Grandin, T. (1995) Restraint of livestock. In: *Proceedings of the Animal Behavior and the Design of Livestock and Poultry Systems International Conference*. Northeast Regional Agricultural Engineering Service, Cooperative Extension, Cornell University, Ithaca, New York, pp. 208–223.

Grandin, T. (1996) Factors that impede animal movement at slaughter plants. *Journal of the American Veterinary Medical Association* 209, 757–759.

Grandin, T. (1997a) Assessment of stress during handling and transport. *Journal of Animal Science* 75, 249–257.

Grandin, T. (1997b) *Survey of Handling and Stunning in Federally Inspected Beef, Pork, Veal and Sheep Slaughter Plants*. ARS Research Project No. 3602–32000–002–08G, United States Department of Agriculture, Washington, DC.

Grandin, T. (1998a) Objective scoring of animal handling and stunning practices in slaughter plants. *Journal of the American Veterinary Medical Association* 212, 36–93.

Grandin, T. (1998b) Solving livestock handling problems in slaughter plants. In: Gregory, N.G. (ed.) *Animal Welfare and Meat Science*. CAB International, Wallingford, UK.

Grandin, T. (1998c) The feasibility of using vocalization scoring as an indicator of poor welfare during slaughter. *Applied Animal Behaviour Science* 56, 121–128.

Grandin, T. (2001a) Welfare of cattle during slaughter and the prevention of nonambulatory (downer) cattle. *Journal of the American Veterinary Medical Association* 219, 1377–1382.

Grandin, T. (2001b) Solving return to sensibility after electrical stunning in commercial pork slaughter plants. *Journal of the American Veterinary Medical Association* 219, 608–611.

Grandin, T. (2002) Return to sensibility problems after penetrating captive bolt stunning of cattle in commercial beef slaughter plants. *Journal of the American Veterinary Medical Association* 221, 1258–1261.

Grandin, T. (2005a) Maintenance of good animal welfare standards in beef slaughter plants by use of auditory programs. *Journal of the American Veterinary Medical Association* 226, 370–373.

Grandin, T. (2005b) *Recommended Animal Handling Guidelines of Audit Guide*, 2005 edn. American Meat Institute Foundation, Washington, DC (http://www.animalhandling.org; http://www.grandin.com).

Grandin, T. and Bruning, J. (1992) Boar presence reduces fighting in slaughter weight pigs. *Applied Animal Behaviour Science* 33, 273–276.

Grandin, T. and Deesing, M. (1998) Genetics and behavior during restraint and herding. In: Grandin T. (ed.) *Genetics and the Behavior of Domestic Animals*. Academic Press, San Diego, California, pp. 113–144.

Grandin, T. and Johnson, C. (2005) *Animals in Translation*. Scribner, New York (now published by Harcourt).

Grandin, T. and Regenstein, J.M. (1994) Religious slaughter and animal welfare: a discussion for meat scientists. In: *Meat Focus International*. CAB International, Wallingford, UK, pp. 115–123.

Grandin, T., Curtis, S.E. and Widowski, T.M. (1986) Electro-immobilization *versus* mechanical restraint in an avoid–avoid choice test. *Journal of Animal Science* 62, 1469–1480.

Grandin, T., Dodman, N. and Shuster, L. (1989) Effect of naltrexone on relaxation induced by flank pressure in pigs. *Pharmacology, Biochemistry and Behavior* 33, 839–842.

Gregory, N.G. (1988) Humane slaughter. In: *34th International Congress of Meat Science and Technology*. CSIRO, Meat Research Laboratory, Brisbane, Australia.

Gregory, N.G. (1992) Pre-slaughter handling, stunning and slaughter. In: *38th International Congress of Meat Science and Technology*. INRA, Thiex, Genes Champanelle, France.

Gregory, N.G. (1993) Slaughter technology. electrical stunning in large cattle. *Meat Focus International* 2, 32–36.

Gregory, N.G. (2004) Recent concerns about stunning and slaughter. *Meat Science* 70, 481–491.

Gregory, N.G. (2007) *Animal Welfare and Meat Production*, 2nd edn. CAB International, Wallingford, UK.

Hambrecht, E., Eissen, J.J., Nooijent, R.I., Ducro, B.J., Smits, C.H., den Hartog, L.A. and Verstegen, M.W. (2004) Preslaughter stress and muscle energy determine pork quality at two commercial slaughter plants. *Journal of Animal Science* 82, 1401–1409.

Hambrecht, E., Eissen, J.J., Newman, D.J., Smiths, C.H., Verstegen, M.W. and den Hartog, L.A. (2005) Preslaughter handling effects pork quality and glycolytic potential in two muscles differing in fiber type composition. *Journal of Animal Sciences* 83, 900–907.

Hargreaves, A.L. and Hutson, G.D. (1990) The stress response in sheep during routine handling procedures. *Applid Animal Behaviour Science* 26, 83–90.

Hartung, V.J., Floss, M., Marahrens, M., Nowak, B. and Fedlhusen, F. (1997) Stress response of slaughter pigs in two different access systems to electrical stunning. *DTW Disch Tierarztl Wochenschr* 104 (2), 66–68.

Hornbuckle, P.A. and Beall, T. (1974) Escape reactions to the blood of selected mammals by rats. *Behavioral Biology* 12, 573–576.

Hoenderken, R. (1976) Improved system for guiding pigs for slaughter to the restrainer. *Die Fleischwirtschaft* 56 (6), 838–839.

Hornbuckle, P.A. and Beall, T. (1974) Escape reactions to the blood of selected mammals by rats. *Behavioral Biology* 12, 573–576.

Hunter, E.J., Weeding, C.M., Guise, H.J., Abbott, T.A. and Penny, R.H. (1994) Pig welfare and carcass quality: a comparison of the influence of slaughter handling systems in two abattoirs. *Veterinary Record* 135, 423–425.

Hyde, J.L. (1962) The use of solid carbon dioxide for producing short periods of anesthesia in guinea pigs. *American Journal of Veterinary Research* 23, 684–685.

Kenny, F.J. and Tarrant, P.V. (1987) The behavior of young Friesian bulls during social regrouping at an abattoir: influence of an overhead electrified wire grid. *Applied Animal Behaviour Science* 18, 233–246.

Kilgour, R. (1971) Animal handling in works: pertinent behavior studies. In: *Proceedings of the 13th Meat Industry Research Conference*, Hamilton, New Zealand, pp. 9–12.

Kilgour, R. (1976) The behaviour of farmed beef bulls. *New Zealand Journal of Agriculture* 132 (6), 31–33.

Kilgour, R. (1978) The application of animal behavior and the humane care of farm animals. *Journal of Animal Science* 46, 1479–1486.

Kilgour, R. (1988) Behavior in the pre-slaughter and slaughter environments. In: *Proceedings, International Congress of Meat Science and Technology, Part A*, Brisbane, Australia, pp. 130–138.

Kilgour, R. and de Langen, H. (1970) Stress in sheep from management practices. *Proceedings, New Zealand Society of Animal Production* 30, 65–76.

Kim, D.H., Seo, J.T., Ha, D.M. and Lee, C.Y. (2004) Effects of fasting and handling stress of market pigs on plasma concentrations of stress associated enzymes and carcass quality. *Journal of Animal Sciences* (Suppl. 1), 60 (abstract).

King, D.A., Schuchle-Pfeiffer, C.E., Randel, R.D., Welsh, T.H., Oliphant, R.A., Baird, B.E. *et al.* (2006) Influence of animal temperament and stress responsiveness on carcass quality and beef tenderness of feedlot cattle. *Meat Science* 74, 546–556.

Kuchenmeister, V., Kuhn, G. and Ender, K. (2005) Preslaughter handling of pigs and the effect on heart rate, meat quality, including tenderness and sarcoplasmic reticulum Ca^{2+} transport. *Meat Science* 71, 690–695.

Lambooy, E. (1986) Automatic electrical stunning of veal calves in a V-type restrainer. In: *Proceedings, 32nd European Meeting of Meat Research Workers*, Ghent, Belgium, Paper 2:2, pp. 77–80.

Lay, D.C., Friend, T.H., Randel, R.D., Bowers, C.L., Grissom, K.K. and Jenkins, O.C. (1992) Behavioral and physiological effects of freeze and hot iron branding on crossbred cattle. *Journal of Animal Science* 70, 330–336.

Lay, D.C., Friend, T.H., Randel, R.D., Bowers, C.L., Grissom, K.K., Neuendorff, D.A. and Jenkins, O.C. (1998) Effects of restricted nursing on physiological and behavioral reactions of Brahman calves to subsequent restraint and nursing. *Applied Animal Behaviour Science* 56, 109–119.

Marshall, M., Milburg, E.E. and Shultz, E.W. (1963) *Apparatus for Holding Cattle in Position for Humane Slaughtering*. US Patent 3,092,871, US Patent and Trademark Office, Washington, DC.

Meischke, H.R.C. and Horder, J.C. (1976) A knocking box effect on bruising in cattle. *Food Technology in Australia* 28, 369–371.

Meisinger, D. (1999) *A System for Assuring Pork Quality*. National Pork Producers Council, Des Moines, Iowa.

Midwest Plan Service (1987) *Structures and Environment Handbook*, 10th edn. Iowa State University, Ames, Iowa, p. 319.

Milligan, S.D., Ramsey, C.B., Miller, M.F., Kaster, C.S. and Thompson, L.D. (1998) Resting of pigs and hot-fat trimming and accelerated chilling of carcasses to improve meat quality. *Journal of Animal Science* 76, 74–86.

Mitchell, G., Hattingh, J. and Ganhao, M. (1988) Stress in cattle assessed after handling, transport and slaughter. *Veterinary Record* 123, 201–205.

Mounier, L., Dubroeveof, H., Andanson, S. and Veissier, L. (2006) Variations in meat pH of beef bulls in relation to conditions of transfer to slaughter and previous history of the animals. *Journal of Animal Science* 84, 1567–1576.

OIE (2005a) *Terrestrial Animal Health Code: Guidelines for the Slaughter of Animals for Human Consumption*. Organization mundial de la Sante Animals (World Organization and Animal Health, Paris (http://www.oie.int/eng/en_index.htm).

OIE (2005b) *Terrestrial Animal Health Code: Guidelines for the Land Transport of Animals*. Organization Mundial de la Santa Animals (World Organization for Animal Health, Paris).

Pascoe, P.J. (1986) Humaneness of electro-immobilization unit for cattle. *American Journal of Veterinary Research* 10, 2252–2256.

Pearson, A.J., Kilgour, R., de Langen, H. and Payne, E. (1977) Hormonal responses of lambs to trucking, handling and electric stunning. *New Zealand Society of Animal Production* 37, 243–249.

Pedersen, B.K., Curtis, S.E., Kelley, K.W. and Gonyou, H.W. (1993) Well being of growing finishing pigs: environmental enrichment and pen space allowance. In: Collins, E. and Boon, C. (eds) *Livestock Environment IV*. American Society of Agricultural Engineers, St Joseph, Michigan, pp. 43–150.

Perez, M.P., Palachio, J., Santolaria, M.P., del Acena, M.C., Chacon, G., Verde, M.T., Calvo, J.H., Zaragoza, M.P., Garcon, M. and Garcia Behrenguer, S. (2002) Influence of lairage time on some welfare and meat quality parameters in pigs. *Veterinary Research* 33, 239–250.

Petersen, G.V. (1977) Factors associated with wounds and bruises in lambs. *New Zealand Veterinary Journal* 26, 6–9.

Petherick, J.C. and Blackshaw, J.K. (1987) A review of the factors influencing aggressive and agonistic behavior in the domestic pig. *Australian Journal of Experimental Agriculture* 27, 605–611.

Phillips, P.A., Thompson, B.K. and Fraser, D. (1987) *Ramp Designs for Young Pigs*. Technical Paper No. 87–4511, American Society of Agricultural Engineers, St Joseph, Michigan.

Phillips, P.A., Thompson, B.K. and Fraser, D. (1989) The importance of cleat spacing in ramp design for young pigs. *Canadian Journal of Animal Science* 69, 483–486.

Price, M.A. and Tennessen, T. (1981) Preslaughter management and dark cutting in carcasses of young bulls. *Canadian Journal of Animal Science* 61, 205–208.

Puolanne, E. and Aalto, H. (1981) The incidence of dark cutting beef in young bulls in Finland. In: Hood, D.E. and Tarrant, P.V. (eds) *The Problem of Dark Cutting Beef*. Martinus Nijhoff, The Hague, Netherlands, pp. 462–475.

Raj, A.B., Johnson, S.P., Wotton, S.B. and McInstry, J.L. (1997) Welfare implications of gas stunning of pigs: the time to loss of somatosensory evolved potentials and spontaneous electrocorticogram of pigs during exposure to gasses. *Veterinary Journal* 153, 329–339.

Ray, D.E., Hansen, W.J., Theures, B. and Stott, G.H. (1972) Physical stress and corticoid levels in steers. *Proceedings, Western Section American Society of Animal Science* 23, 255–259.

Regensburger, R.W. (1940) *Hog Stunning Pen*. US Patent 2,185,949, US Patent and Trademark Office, Washington, DC.

Rider, A., Butchbaker, A.F. and Harp, S. (1974) *Beef Working, Sorting and Loading Facilities*. Technical Paper No. 74–4523, American Society of Agricultural Engineers, St Joseph, Michigan.

Ring, C. (1988) Two aspects of CO_2 stunning method for pigs: animal protection and meat quality. In: *34th International Congress of Meat Science and Technology*. CSIRO Meat Research Laboratory, Cannon Hill, Queensland, Australia.

Rivalland, E.T., Clarke, I.J., Turner, A.I., Pompolo, S. and Tilbrook, A.J. (2007) Isolation and restraint stress results in differential activation of corticotropin releasing hormone and argonine vasopressen neurons in sheep. *Neuroscience* (in press).

Rushen, J. (1986) Aversion of sheep to electro-immobilization and physical restraint. *Applied Animal Behaviour Science* 15, 315.

Shea-Moore, M. (1998) The effect of genotype on behavior in segregated early weaned pigs tested in an open field. *Journal of Animal Science* 76 (Suppl. 1), 100 (abstract).

Spensley, J.C., Wathes, C.M., Waron, N.K. and Lines, J.A. (1995) Behavioral and physiological responses of piglets to naturally occurring sounds. *Applied Animal Behaviour Science* 44, 277.

Stevens, D.A. and Gerzog-Thomas, D.A. (1977) Fright reactions in rats to conspecific tissue. *Physiology and Behavior* 18, 47–51.

Stevens, D.A. and Saplikoski, N.J. (1973) Rats reactions to conspecific muscle and blood evidence for alarm substances. *Behavioral Biology* 8, 75–82.

Stevens, R.A. and Lyons, D.J. (1977) *Livestock Bruising Project: Stockyard and Crate Design*. National Materials Handling Bureau, Department of Productivity, Canberra, Australia.

Stricklin, W.R., Graves, H.B. and Wilson, L.L. (1979) Some theoretical and observed relationships of fixed and portable spacing behavior in animals. *Applied Animal Ethology* 5, 201–214.

Stuier, S. and Olsen, E.V. (1999) Drip loss dependent on stress during lairage and stunning. In: *International Congress of Meat Science and Technology, Yokohama, Japan*, Paper No. 4-P29, pp. 302–303.

Talling, J.C., Waran, N.K., Wathes, C.M. and Lines, J.A. (1998) Sound avoidance by domestic pigs depends on characteristics of the signal. *Applied Animal Behaviour Science* 58, 255–266.

Tanida, H., Miura, A., Tanaka, T. and Yoshimoto, T. (1996) Behavioral responses of piglets to darkness and shadows. *Applied Animal Behaviour Science* 49, 173–183.

Tennessen, T., Price, M.A. and Berg, R.T. (1984) Comparative responses of bulls and steers to transportation. *Canadian Journal of Animal Science* 64, 333–338.

Tume, R.K. and Shaw, F.D. (1992) Beta-endorphin and cortisol concentrations in plasma of blood samples collected during exsanguination of cattle. *Meat Science* 31, 211–217.

UFAW (1987) *Proceedings, Symposium on Humane Slaughter of Animals.* Universities Federation Animal Welfare, Potters Bar, UK.

USDA (United States Department of Agriculture) (1967) *Improving Services and Facilities at Public Stockyards.* Agriculture Handbook 337, Packers and Stockyards Administration, United States Department of Agriculture, Washington, DC.

van der Wal, P.G. (1978) Chemical and physiological aspects of pig stunning in relation to meat quality: a review. *Meat Science* 2, 19–30.

van der Wal, P.G. (1997) Causes of variation in pork quality. *Meat Science* 46, 319–327.

van Putten, G. and Elshof, W.J. (1978) Observations of the effects of transport on the well being and lean quality of slaughter pigs. *Animal Regulation Studies* 1, 247–271.

Vieville-Thomas, R.K. and Signoret, J.P. (1992) Pheromonal transmission of aversive experiences in domestic pigs. *Journal of Chemical Endocrinology* 18, 1551.

Voisinet, B.D., Grandin, T., O'Connor, S.F., Tatum, J.D. and Deesing, M.J. (1997) *Bos indicus* cross feedlot cattle with excitable temperaments have tougher meat and a higher incidence of borderline dark cutters. *Meat Science* 46, 367–377.

von Wenzlawowlez, M., von Hollenben, K., Kofer, J. and Bostelmann, N. (1999) Welfare monitoring at slaughter plants in Styria (Austria). In: *International Congress of Meat Science and Technology, Yokomaha, Japan.* Paper No. 2-P3, pp. 64–65.

Walker, P., Warner, R., Weston, P. and Kerr, M. (1999) Effect of washing cattle pre-slaughter on glycogen levels of *m. semitendinosus* and *m. semimembranosus*. In: *International Congress of Meat Science and Technology*, Yokohama, Japan. Paper No. 4-P27, pp. 298–299.

Warner, R.D., Eldridge, G.A., Barnett, J.L., King, R.H. and Winfield, C.G. (1992) Preslaughter behavior and meat quality of pigs fed excess dietary tryptophan. In: *38th International Congress of Meat Science and Technology*. Department of Technology de la Viande, INRA, Thiex 63122, St Genes Chapanelle, France.

Warrington, P.D. (1974) Electrical stunning: a review of the literature. *Veterinary Bulletin* 44, 617–633.

Warriss, P.D. (2003) Optimal lairage times and conditions for slaughter pigs: a review. *Veterinary Record* 153, 170–176.

Warriss, P.D., Bevis, E.A., Edwards, J.E., Brown, S.N. and Knowles, T.G. (1991) Effects of angle of slope on the care with which pigs negotiate loading ramps. *Veterinary Record* 128, 419–421.

Warriss, P.D., Brown, S.N. and Adams, S.J.M. (1994) Relationship between subjective and objective assessment of stress at slaughter and meat quality in pigs. *Meat Science* 38, 329–340.

Watts, J.M. and Stookey, J.M. (1998) Effects of restraint and branding on rates and acoustic parameters of vocalization in beef cattle. *Applied Animal Behaviour Science* 62, 125–135.

Weary, D.M., Braithwaite, L.A. and Fraser, D. (1998) Vocal response to pain in piglets. *Applied Animal Behaviour Science* 56, 161–172.

Weeding, C.M., Hunter, D.J., Guise, H.J. and Penny, H.C. (1993) Effects of abattoir and slaughter handling systems on stress indicators in pig blood. *Veterinary Record* 133, 10–13.

Westervelt, R.G., Kinsman, D., Prince, R.P. and Giger, W. (1976) Physiological stress measurement during slaughter of calves and lambs. *Journal of Animal Science* 42, 831–834.

Weyman, G. (1987) *Unloading and Loading Facilities at Livestock Markets.* Council of National and Academic Awards, Environmental Studies, Hatfield Polytechnic, Canberra, Australia.

White, R.G., DeShazer, J.A., Tressler, C.J., Borcher, G.M., Davey, S., Waninge, A., Parkhurst, A.M., Milanuk, M.J. and Clems, E.T. (1995) Vocalizations and physiological response of pigs during castration with and without anesthetic. *Journal of Animal Science* 73, 381–386.

Woodbury, D.M., Rollins, L.T., Gardner, M.D., Hirschi, W.L., Hogan, J.R., Rallison, M.L., Tanner, G.S. and Brodie, D.A. (1958) Effects of carbon dioxide on brain excitability and electrolytes. *American Journal of Physiology* 190, 79–94.

Zavy, M.T., Juniewicz, P.E., Phillips, W.A. and Von Tungeln, D.L. (1992) Effects of initial restraint, weaning and transport stress on baseline and ACTH stimulated cortisol responses in beef calves of different genotypes. *American Journal of Veterinary Research* 53, 551–557.

21 Biosecurity for Animal Health and Food Safety

K.E. Belk, J.A. Scanga and T. Grandin

Department of Animal Sciences, Center for Red Meat Safety, Colorado State University, Fort Collins, Colorado, USA

Introduction

In the modern era of global trade and travel, multinational corporations, intensive livestock raising practices and terrorism, persistent threats to animal and public health from foreign animal diseases and emerging foodborne pathogens have elevated the importance of implementing sound biosecurity measures during livestock production. Alone, the economic consequences to the livestock and meat industry of an animal health crisis (either zoonotic or other animal diseases) merit serious effort towards their prevention; but the ramifications to the industry are vastly escalated when the biosecurity threat affects food safety and public health.

People who handle and transport animals must understand the importance of preventing the spread of animal disease. They are in a unique position to help control spread of diseases and they should understand biosecurity. More training is needed for transporters and truck drivers. In a Kansas study, only 38% of the trucking companies trained their employees to recognize foreign animal diseases (Fike and Spire, 2006).

In addition to the financial burden of morbidity and mortality, biosecurity crises particularly influence global trade in agricultural products. Because every $1 billion in US agricultural exports creates 15,000 US jobs (currently 885,000 jobs in the USA), and exports fully support one-third of all jobs in rural communities (USDA–FAS, 2005), destabilization of agricultural trade due to a biosecurity crisis has serious financial consequences. For example, due to a classical swine fever outbreak during 1997 and 1998, the Netherlands – a perennial pork-exporting power – was forced to reduce hog production by 20%, greatly reducing their exports as well and opening doors for other countries to capture a greater global market share.

Bovine spongiform encephalopathy (BSE) and the avian influenza A virus (H5N1, 'bird flu') recently have dramatically improved demand for pork, particularly in Asian countries (USDA–FAS, 2005), at the expense of the beef and poultry trades, respectively. Further outbreaks of foot-and-mouth disease in some countries (e.g. Brazil and Uruguay) periodically continue to impede market access for meat products exported from those countries, thereby favorably affecting the competitiveness of other exporting nations during those periods of time when product from the affected country is banned in export destinations.

Although BSE did not reduce domestic US demand for beef as it did in European countries following the epidemic in the UK, significant losses ensued in the US beef industry due to closure of export markets. Although some markets reopened to US beef quickly following detection of BSE (e.g. Mexico), the most critical markets – such as

Japan and Korea – have been severely affected. Randy Blach of Cattle-Fax estimated in 2004 that loss of US beef export markets had resulted in approximately US$13–15/cwt (live basis) price reductions for fed cattle, essentially costing the US beef industry about US$165–190 per head marketed (personal communication). The USDA Foreign Agricultural Service reported in 2004 that, while the US exported US$3.1 billion in beef during 2003, only 15% of those levels were realized in 2004 following detection of BSE.

Today, and aside from natural routes of transmission for animal or human pathogens, many policy-makers and scientists are concerned about acts of bioterrorism against the livestock and meat industry; a natural phenomenon given the events of 11 September 2001. The US General Accounting Office issued reports (USGAO, 2003) citing gaps in federal protection of agriculture and the food supply from bioterrorism but, to their credit, USDA appears to have made progress towards addressing many of the criticisms. None the less, it was noteworthy that Secretary Tommy G. Thompson of the US Department of Health and Human Services said, before leaving office, that 'one of our greatest potential vulnerabilities to terror attack is something we depend on every day – our food supply'.

This chapter explores the evolving and escalating needs, risks and methods for addressing issues associated with implementation of on-farm biosecurity, biosafety and biocontainment practices in the livestock industry.

Terminology

Before a coherent discussion of biosecurity can ensue, it is important to define the terminology to be used. A quick review of published information regarding biosecurity reveals that the term biosecurity itself, along with a number of related terms (i.e. agrosecurity, biosafety, biocontainment, biodefence), have differing meanings to differing individuals or groups (e.g. see Sandia National Laboratories, 2004), while few definitions for use of the terms are available to review. In fact, few well-known dictionaries even provide definitions for the word biosecurity and associated terms. Lack of standardized terminology

is particularly problematic when, for example, one considers the quantity of regulatory policies developed that use such terms to govern biosecurity, biosafety and biocontainment for a wide array of entities.

Kirkpatrick and Selk (2006) said that 'biosecurity is used to describe programs for preventing the introduction of pathogens considered potentially harmful to the health and well-being of the [beef cattle] herd'. Clearly, using their definition, biosecurity is a veterinary term addressing the prevention of disease introduction for purposes of the protection of animal health. Conversely, the National Research Council (2006) defined the term biosecurity to mean 'the policies and measures taken to minimize the risk of introducing an infectious pathogen into the human, agricultural animal and research animal populations'. Use of the latter definition clearly is broader, including protections for both animal and public health.

The Sunshine Project (2003) stated that the term biosafety 'shifts with diplomatic and scientific context, its two major usages relating to laboratory containment and to biotechnology hazards', and that 'biosecurity is a closely related term increasingly heard in arms control and in health and agriculture, but which also lacks a consistent usage'. This latter reference also stated that, 'on a very practical level, there may be differences between means to prevent an unintended release into the environment (sometimes referred to as "biosafety") and means to prevent abuse or theft (sometimes referred to as "biosecurity")'. Use of the term biosecurity in this context again has a meaning differing from those proposed by other authors.

Although not standardized, for purposes of discussion in this chapter, the term biosecurity is defined to mean: a series of management practices designed to minimize or prevent the exposure to, and introduction of, subpopulations of livestock to human and animal disease infectious agents. Such management practices may include, but not necessarily be limited to, testing and screening protocols for surveillance and process verification, isolation and quarantine, immunization, waste management, selective purchasing and monitoring.

The term biosecurity conveys the same meaning as the terms agrosecurity, agrodefence and biodefence. However, in this context, the term biosecurity does not convey that intentional diversion of human or animal infectious agents is prevented – as it may be applied by research laboratories; the terms bioterrorism or agroterrorism are used to reflect intentional diversion of human or animal infectious agents.

The term biosafety is defined to reflect a differing risk: the prevention of accidental release (loss of control) of human and animal disease infectious agents into the environment, or prevention of exposure of another subpopulation of livestock to an infectious agent. In this context, the term biosafety implies that there should be accountability for accidental dissemination of infectious agents. The term biocontainment then conveys a meaning similar to that of biosafety; a series of management strategies to prevent the spread of human or animal infectious agents within or among subpopulations of livestock (i.e. control of a detected infectious outbreak).

Therefore, in livestock production, a biosecurity plan may incorporate biosecurity (as defined), biosafety and biocontainment components. Clearly, the most thorough of plans would address all three facets to manage transmission of infectious agents.

Biosecurity and Government

Most countries have government agencies whose responsibility is to implement programmes that will prevent the spread of animal disease and protect the food supply from dangerous pathogens. Most programmes require both agricultural government agencies and human health agencies to work together.

In the USA, the Secretary of Agriculture (US Department of Agriculture, USDA) is delegated authority by Congress (by 45 acts) to ensure animal and plant health; these responsibilities are regulated by the Animal and Plant Health Inspection Service (APHIS). Furthermore, the Secretary of Agriculture also is authorized by the Federal Meat Inspection Act – via the Food Safety Inspection Service (FSIS) – to ensure protection of 'the health and welfare of consumers . . . by assuring that meat and meat food products distributed to them are wholesome, not adulterated, and [are] properly marked, labeled, and packaged'.

Thus, APHIS is responsible for the protection and promotion of US animal health, while FSIS is responsible for the protection and promotion of public health as it relates to the consumption of meat and poultry products. The Center for Veterinary Medicine (CVM), Food and Drug Administration (FDA), Department of Health and Human Services (USDHHS) is delegated authority to regulate the manufacture and distribution of food additives and drugs that will be given to animals.

Therefore, in practice, all three agencies – along with the US Customs Service of the Department of Homeland Security (USDHS) and state and local governments – would be expected to play roles with respect to biosecurity, biosafety and biocontainment in the US livestock industries. Additionally, in conjunction with the USDHHS and its associated agencies, all have been assigned accountability under Homeland Security Presidential Directive 9 (HSPD-9, 30 January 2004, http://www.whitehouse.gov/news/releases/2004/02/20040203-2.html) for establishing surveillance, monitoring, traceability and diagnostic capability for biocontainment during acts of bioterrorism affecting livestock, plants and foods.

As a function of the regulatory and service responsibilities delegated to the aforementioned US government agencies, several systems were implemented – and are constantly being improved – to address any biosecurity, biosafety, biocontainment or bioterrorism crisis. Three of these systems merit consideration in this chapter.

First, the National Animal Health Laboratory Network (NAHLN) was developed, which has established a series of federal and state resources to enhance detection of, and to enable a rapid and sufficient response to, animal health emergencies. The NAHLN has garnered the power of the US land-grant university system by enlisting state veterinary diagnostic laboratories into the network, substantially improving capacity and communication in the event of an animal health crisis.

Secondly, FDA and FSIS developed the Food Emergency Response Network (FERN): a laboratory network similar to that implemented to address animal health crises. The FERN system includes laboratories of the Centers for Disease Control and Prevention (CDC) and

FDA of USDHHS, FSIS of USDA, the US Environmental Protection Agency (a stand-alone federal agency), the US Department of Energy (USDOE) and the 50 US states. The system allows a tiered screening and confirmation-testing laboratory network that is responsible for: (i) analysing food samples implicated in threats, terrorist events or contamination; (ii) responding to large-scale emergencies involving food; and (iii) conducting ongoing programmes to monitor foods for presence of threat agents. The system provides for prevention of crises, preparedness and response in the event of a public health event, and recovery following an emergency.

Lastly, the USDA Farm Service Agency (previously unmentioned) developed a Systematic Tracking for Optimal Risk Management (STORM) homeland security database. The STORM system is a web-based application used to easily record, maintain and track incidents relating to emergency preparedness and homeland security; it provides the capability to report and document suspicious, unusual or threatening activities, which can then be immediately accessible to state and national offices.

The US federal government, along with many state and local governments, have routinely conducted simulations of emergency events to provoke debate, develop new ideas and generally improve US capabilities in the event of an animal health or food safety crisis. Among these were USDA's Crimson Sky, Crimson Guard and Crimson Winter.

In addition to government programmes, private industry has programmes for verification of the origin of meat and poultry. Some large meat-buying customers are demanding source verification, and innovative niche marketers of meat products are forming alliances with producers where source verification is one of their requirements.

Likewise, although perhaps in a fashion that may seem too passive, implementation of a National Animal Identification System (NAIS) in the USA now seems imminent. Secretary of Agriculture Mike Johanns announced in April of 2006 that USDA would incrementally implement the remaining goals of NAIS so that it would be operational by 2007 (USDA, 2006), and that the system would achieve full producer participation by 2009. The NAIS will provide capability to identify all premises that have had direct contact with an animal health threat within 48 h of discovery of the threat.

As such, the NAIS plan calls for every farm or ranch in the USA with livestock to be identified with a premise number (PID) and, eventually, every individual beef and equine animal (IAID) or swine, poultry and fish animal lot (ALID) will be given its own unique identification number using a national numbering system. Animal movement data will be captured and stored in database systems, allowing for national searching capabilities; by early 2007, USDA expects to have an Animal Trace Processing System (the 'Metadata System') in place for use in animal disease investigations.

Traceability of Livestock

An important component of biosecurity is the ability to track the movement of animals. In many countries, trace-back and identification of animals has been in place for many years. Countries such as Denmark and Australia, where meat export has been vital to their economy, were some of the first countries to mandate traceability. After the BSE outbreaks in Europe, government authorities put in place better programmes for tracking the movements of livestock. These programmes are continually evolving and use a range of identification methods, from simple tags to other more sophisticated computerized systems.

Animal Health

The World Organization for Animal Health (OIE), the governments of individual countries and states within a country maintain lists of animal diseases whose occurrence must be reported to veterinary and human health agencies. The diseases that are listed have great economic impact on livestock and poultry producers. Some of them are zoonotic and can be transmitted from animals to humans. An alphabetical list of Select Agents and Toxins that are regulated by USDA, USDHHS or a combination of the two is provided by Princeton University Environmental Health and Safety (Princeton

University, 2006), as per Final Select Agent Rules published on 18 March 2005. The list, except for exclusions, includes viruses, bacteria, fungi, toxins, genetic elements, recombinant nucleic acids and recombinant organisms (Princeton University, 2006) and is provided in Box 21.1.

Over the years, efforts to control animal health threats have prevented the introduction, or have resulted in eradication, of many animal diseases in the USA and other countries. For example, APHIS (9 CFR 71.3) lists foot-and-mouth disease, rinderpest, African swine fever, hog cholera, Teschen disease, contagious bovine pleuropneumonia, European fowl pest, dourine, contagious equine metritis, vesicular exanthema, screw worms, glanders and scabies in sheep as diseases that 'are not known to exist in the United States'.

Conversely, APHIS lists (9 CFR 71.3) equine piroplasmosis, bovine piroplasmosis or splenetic fever, scabies in cattle, pseudorabies, acute swine erysipelas, tuberculosis, Johne's disease, brucellosis, scrapie, bluetongue, anthrax, chlamydiosis, poultry disease caused by *Salmonella enteritidis* serotype Enteritidis and Newcastle disease as 'communicable' animal diseases 'endemic in the United States'.

Box 21.2 provides a list of animal diseases that, in an effort to standardize global animal health biosecurity, biosafety and biocontainment measures, are reportable when detected by member nations to the World Organization for Animal Health (OIE, 2006). Many of the diseases listed – that would be of concern to livestock producers – also are zoonotic.

Recent animal health outbreaks of concern in the USA have included highly pathogenic avian influenza (Texas), BSE (Washington, Texas, Alabama) and exotic Newcastle disease. Currently, due to the potential for mutation and transmission among humans – possibly leading to pandemic – officials globally are monitoring prevalence and movement of the H5N1 avian influenza A virus.

Foreign diseases that may present risk to US livestock operations, should measures to prevent their introduction fail, include those listed by APHIS as 'not known to exist in the United States'. However, in total, over 17 animal reproductive diseases and 12 diseases known to be transmitted faecally/orally (including foodborne pathogens) or through feed are generally identified by veterinarians as cause for concern in the USA. Respiratory disease in the US cattle herd accounts for over one-quarter of health-related deaths (Drovers Alert, 2006).

Biosecurity for beef and dairies

Biosecurity, biosafety and biocontainment on the farm and during livestock transportation are thus crucial, as opportunities for disease transmission continue to escalate (i.e. via global travel, industry concentration and/or bioterrorism). Several templates exist to assist livestock producers as they implement protocols for managing biosecurity, biosafety and biocontainment; most are available via internet web sites.

For example, scientists at Oklahoma State University (Kirkpatrick and Selk, 2006) published via the web a very good outline of requirements for achieving biosecurity in beef cattle operations. The authors stipulated that 'producers must make a decision on the level of risk they are willing to accept', and indicated that the level of biosecurity measures implemented on an individual operation may also depend on the scope, production target and potential 'economic consequences' of animal disease. Kirkpatrick and Selk (2006) recommended that each producer classify their individual operation's needs based on a six-point scale, where 1 reflected a closed, specific-pathogen-free herd while 6 reflected a farm/ranch that allowed entry of new animals with no medical records and no isolation. They listed 'components of a biosecurity program' and provided examples for such programmes.

To achieve true biosecurity on the farm, one of the apparent best recommended systems was published by the Bovine Alliance on Management and Nutrition (BAMN, 2001) for use in dairies; in fact, the programme also is provided electronically on the APHIS web site (http://www.aphis.usda.gov). The programme outlines four procedural steps that producers should take to develop an appropriate on-farm biosecurity programme, including: (i) identify products and their value; (ii) identify and prioritize risk; (iii) evaluate how diseases are introduced to the dairy; and (iv) implement a plan for disease control. Likewise, BAMN also published guidelines

Box 21.1. Selected agents/toxins regulated by USDA, USDHHS, or a combination of the two (overlap), via Final Rules published on 18 March 2005 (Princeton University, 2006).

Abrin (USDHHS)
African horse sickness virus (USDA, animal)
African swine fever virus (USDA, animal)
Akabane virus (USDA, animal)
Avian influenza virus (highly pathogenic)
     (USDA, animal)
Bacillus anthracis (overlap)
Bluetongue virus (exotic) (USDA, animal)
Botulinum neurotoxins (overlap)
Botulinum neurotoxin-producing strains of
     Clostridium (overlap)
Bovine spongiform encephalopathy
     agent (USDA, animal)
Brucella abortus (overlap)
Brucella melitensis (overlap)
Brucella suis (overlap)
Burkholderia mallei (formerly
     Pseudomonas mallei) (overlap)
Burkholderia pseudomallei (overlap)
Camel pox virus (USDA, animal)
Central European Tick-borne encephalitis
     virus (USDHHS)
Cercopithecine herpesvirus 1
     (Herpes B virus) (USDHHS)
Classical swine fever virus (USDA, animal)
Clostridium perfringens epsilon toxin (overlap)
Coccidioides immitis (overlap)
Coccidioides posadasii (USDHHS)
Conotoxins (USDHHS)
Cowdria ruminantium (Heartwater)
     (USDA, animal)
Coxiella burnetii (overlap)
Crimean-Congo haemorrhagic fever
     virus (USDHHS)
Diacetoxyscirpenol (USDHHS)
Eastern equine encephalitis virus (overlap)
Ebola viruses (USDHHS)
Far Eastern tick-borne encephalitis (USDHHS)
Flexal virus (USDHHS)
Foot-and-mouse disease virus (USDA, animal)
Francisella tularensis (overlap)
Goat pox virus (USDA, animal)
Guanarito virus (USDHHS)
Hendra virus (overlap)
Japanese encephalitis virus
     (USDA, animal)
Junin virus (USDHHS)
Kyasanur Forest disease (USDHHS)
Lassa fever virus (USDHHS)
Liberobacter africanus (USDA, plant)
Liberobacter asiaticus (USDA, plant)
Lumpy skin disease virus (USDA, animal)
Machupo virus (USDHHS)

Malignant catarrhal fever virus (Alcelaphine
     herpes virus type 1) (USDA, animal)
Marburg virus (USDHHS)
Menangle virus (USDA, animal)
Monkeypox virus (USDHHS)
Mycoplasma capricolum/M. F38/M. *mycoides
     capri* (contagious caprine pleuropneumonia)
     (USDA, animal)
Mycoplasma mycoides mycoides (contagious
     bovine pleuropneumonia) (USDA, animal)
Newcastle disease virus (velogenic)
     (USDA, animal)
Nipah virus (overlap)
Omsk Haemorrhagic Fever (USDHHS)
Peronosclerospora philippinensis (USDA, plant)
Peste des petits ruminants virus (USDA, animal)
Ralstonia solanacearum, race 3, biovar 2
     (USDA, plant)
Ricin (USDHHS)
Rickettsia prowazekii (USDHHS)
Rickettsia rickettsii (USDHHS)
Rift Valley fever virus (overlap)
Rinderpest virus (USDA, animal)
Russian spring and summer encephalitis
     (USDHHS)
Sabia virus (USDHHS)
Saxitoxin (USDHHS)
Sclerophthora rayssiae var. *zeae* (USDA, plant)
Sheep pox virus (USDA, animal)
Shiga-like ribosome inactivating proteins
     (USDHHS)
Shigatoxin (overlap)
Staphylococcal enterotoxins (overlap)
Swine vesicular disease virus (USDA, animal)
Synchytrium endobioticum (USDA, plant)
T-2 toxin (overlap)
Tetrodotoxin (USDHHS)
Variola major virus (Smallpox virus) (USDHHS)
Variola minor virus (Alastrim) (USDHHS)
Venezuelan equine encephalitis virus (overlap)
Vesicular stomatitis virus (exotic) (USDA, animal)
Xanthomonas oryzae pv. *oryzicola* (USDA, plant)
Xylella fastidiosa (citrus variegated chlorosis strain)
     (USDA, plant)
Yersinia pestis (USDHHS)
Nucleic acids that can produce infectious forms of
any of the select agent viruses, i.e. recombinant
nucleic acids that encode for the functional form(s)
of any of the select agent toxins if the nucleic acids
can be expressed *in vivo* or *in vitro*, or are in a
vector or recombinant host genome and can be
expressed *in vivo* or *in vitro*, select agents that have
been genetically modified.

Box 21.2. Animal diseases for which, when detected, the World Organization for Animal Health must be notified (OIE, 2006).

Multiple species diseases
Anthrax
Aujeszky's disease
Bluetongue
Brucellosis (*Brucella abortus*)
Brucellosis (*Brucella melitensis*)
Brucellosis (*Brucella suis*)
Crimean Congo haemorrhagic fever
Echinococcosis/hydatidosis
Foot-and-mouth disease
Heartwater
Japanese encephalitis
Leptospirosis
New World screw worm (*Cochliomyia hominivorax*)
Old World screw worm (*Chrysomya bezziana*)
Paratuberculosis
Q fever
Rabies
Rift Valley fever
Rinderpest
Trichinellosis
Tularaemia
Vesicular stomatitis
West Nile fever

Cattle diseases
Bovine anaplasmosis
Bovine babesiosis
Bovine genital campylobacteriosis
Bovine spongiform encephalopathy
Bovine tuberculosis
Bovine viral diarrhoea
Contagious bovine pleuropneumonia
Enzootic bovine leukosis
Haemorrhagic septicaemia
Infectious bovine rhinotracheitis/pustular vulvovaginitis
Lumpky skin disease
Malignant catarrhal fever
Theileriosis
Trichomonosis
Trypanosomosis (tsetse-transmitted)

Sheep and goat diseases
Caprine arthritis/encephalitis
Contagious agalactia
Contagious caprine pleuropneumonia
Enzootic abortion of ewes (ovine chlamydiosis)

Maedi-visna
Nairobi sheep disease
Ovine epididymitis (*Brucella ovis*)
Peste des petits ruminants
Salmonellosis (*S. abortusovis*)
Scrapie
Sheep pox and goat pox

Equine diseases
African horse sickness
Contagious equine metritis
Dourine
Equine encephalomyelitis (Eastern)
Equine encephalomyelitis (Western)
Equine infectious anaemia
Equine influenza
Equine piroplasmosis
Equine rhinopneumonitis
Equine viral arteritis
Glanders
Surra (*Trypanosoma evansi*)
Venezuelan equine encephalomyelitis
Swine diseases
African swine fever
Classical swine fever
Nipah virus encephalitis
Porcine cysticercosis
Porcine reproductive and respiratory syndrome
Swine vesicular disease
Transmissible gastroenteritis

Other diseases
Camelpox
Leishmaniasis

Avian diseases
Avian chlamydiosis
Avian infectious bronchitis
Avian infectious laryngotracheitis
Avian mycoplasmosis (*M. gallisepticum*)
Avian mycoplasmosis (*M. synoviae*)
Duck virus hepatitis
Fowl cholera
Fowl typhoid
Highly pathogenic avian influenza
Infectious bursal disease (Gumboro disease)
Marek's disease
Newcastle disease
Pullorum disease
Turkey rhinotracheitis

for producers to follow when considering bio-security related to feedstuffs.

In contrast to readily available biosecurity measures that may be taken on the farm to pre-vent introduction of infectious agents into livestock populations, very little information exists to assist producers with biosafety and biocontain-ment. In truth, management of the production of feed, along with controls for water and the envi-ronment, may be considered to be biosafety measures (prevention of accidental dissemina-tion of infectious agents), but such logic is less straightforward.

In fact, true biosafety and biocontainment measures on the farm would minimize the likelihood that serious infectious agents could be disseminated to another subpopulation, and pro-ducers are less likely to implement programmes addressing such objectives if legal liability or economic disincentives exist to do so. Clearly, if sound biosafety measures had existed in the UK during the foot-and-mouth disease outbreak in 2001 – in conjunction with sound biosecurity procedures on farms – the outbreak would not have escalated and affected the numbers of ani-mals that it ultimately did, because disease trans-mission would have been controlled.

Likewise, in the case of major animal disease outbreaks, it is likely that government agencies would impose biocontainment (e.g. quarantining or sacrificing, in certain cases, large numbers of livestock) measures on individual producers if those producers had not detected the outbreak or had not taken internal measures to prevent disease dissemination. It is clear that, if bioterrorism were to be detected, multiple agencies in the USA would be involved with containment of a resulting outbreak once detected.

Interestingly, particularly in light of the eco-nomic consequences of an animal disease out-break, available data suggest that biosecurity, biosafety and biocontainment procedures are not well implemented by US livestock produc-ers. For example, APHIS administers the National Animal Health Monitoring System (NAHMS); dur-ing completion of Dairy NAHMS 2002, APHIS col-lected information concerning biosecurity on US dairies (APHIS, 2004). Although a high percent-age of producers (> 92%) had implemented insect and rodent control procedures on farms, only slightly more than one-half of the produc-ers had maintained closed herds or controlled

access to cattle feed by other livestock and wild game, and less than 15% of all producers had mandated visitor sanitation requirements or did not allow visitors on the farm.

Interestingly, in light of current concerns regarding avian influenza, only 20% of produc-ers had implemented procedures to control birds. In fact, only 5.1% of dairies had developed 'written procedures specifically related to pre-venting the introduction and spread of disease', and over 75% of operations had performed no testing whatsoever to monitor animal disease prevalence or to verify adequate protection.

Biosecurity for pigs and poultry at large integrated companies

In contrast to the beef and dairy industry, fully integrated pork and poultry companies have detailed programmes for prevention of the introduction of disease into their herds or flocks (Seaman and Fangman, 2001; Nelson, 2004; British Columbia Poultry, 2005; Pescatore, 2006). Rigorous testing or isolation of incoming breeding animals is mandatory.

Rodents are controlled, and netting is used to keep wild birds out of the pig and poultry houses. Vehicles and people moving between different farms follow procedures to prevent disease transmission. Vehicles are washed and tyres are sprayed with disinfectant. Employees must wear freshly laundered clothes and visitors wear company-provided boots and coveralls. Truck drivers loading out market pigs are not allowed to enter the building where the pigs are raised: the pigs are brought out of the building by farm personnel.

Some companies require all employees and visitors to shower at the facility and wear com-pany clothes that are laundered at the site. Tools and other supplies coming into the farm have to be sprayed with disinfectant. The large integrated companies often have more strict biosecurity than some independent producers. A survey in Den-mark indicated that 10% of producers allowed truck drivers to enter the production area (Boklund et al., 2003–2004). This is a poor practice.

Animal health has always impacted the financial viability of livestock production opera-tions. But, today, efforts to expand export oppor-tunities and public safety concerns are vastly different than at any time in history; animal health issues – particularly zoonotic diseases – now have correspondingly greater impact on the economics of production. For these reasons, livestock producers must 'bite the bullet' and take seriously the need for implementation of biosecurity, biosafety and biocontainment principles on the farm/ranch.

Public Health

Human infection resulting from a foodborne pathogen is a very public, high-impact event that elicits outrage in modern society, and hence results in immediate and widespread effects on product demand and trade. Livestock are considered to be a major reservoir for pathogens. To address modern food safety concerns, research funding has historically targeted development of packing plant decontamination technologies (Hardin *et al.*, 1995; Reagan *et al.*, 1996; Dorsa, 1997). However, although efforts to develop preharvest biosafety controls for foodborne pathogens have received less funding emphasis, scientists are slowly developing such controls.

Proceedings of a National Cattlemen's Beef Association-sponsored *E. coli* Summit suggested that producers should strive to 'evaluate adoption of interventions or Good Management Practices (GMPs) that have been scientifically validated [and] that use a combination of interventions' (NCBA, 2003).

Efforts to control dissemination of *Escherichia coli* O157:H7 in the beef supply chain also reflect implementation of biosecurity, biosafety and biocontainment programmes targeting that organism, and provide a useful case study for assessing the effectiveness of preharvest efforts to control foodborne pathogens. The research literature suggests that initial infection or exposure of cattle to *E. coli* O157:H7 occurs early in life and from a varied array of sources, is abundant in the bovine production environment from that point forward and may be widespread on cow–calf ranches as well as in feedlots and dairies. Thus, most management practices that would theoretically be implemented on a livestock production operation to control foodborne pathogens would be considered to be biosafety and biocontainment procedures.

Recently, Childs *et al.* (2006) conducted a molecular study to characterize how *E. coli* O157:H7 is transmitted from livestock production channels through the system to contaminate meat products. Pulsed-field gel electrophoresis and multiplex-DNA marker gels yielded fingerprint patterns for bovine hide *E. coli* O157:H7 isolates obtained at packing plants; this allowed determination of genetic homology among samples collected from hides/colons and environmental samples. Isolates recovered from the hides of cattle at harvest were genetically related to isolates recovered from feedlot pen floors, cattle shipping trailers, packing plant pen water troughs and from the hides/colons of other animals.

Although much of the pathogenic contamination on cattle at packing plants was not sourced to preceding production facilities, the researchers found that feedlot, transportation trailer and packing plant holding pen environment – as well as cross-contamination events – contributed to contamination on harvest-ready cattle hides. In the livestock production environment, pathogens deposited at various production sites from previously housed cattle were allowed to persist, and a constant potential for contamination with foodborne human disease pathogens was sustained (Childs *et al.*, 2006).

Pathogen-shedding in Beef

Transportation and consolidation of livestock as they progress through production and marketing channels may further contribute to dissemination of foodborne pathogens. For example, beef calves that had been vaccinated and pre-weaned 13–29 days before transport shed fewer salmonellae compared with calves that had had no vaccinations and had been weaned 1 day before transport (Bach *et al.*, 2004).

In another study of 200 fed cattle in a commercial feedlot, Barham *et al.* (2002) provided data showing that prevalence of *E. coli* O157:H7 had declined between the feedlot and the packing plant, while the reverse was true for *Salmonella* spp.; transport trailer prevalence of *E. coli* O157:H7 and *Salmonella* spp. was 5.4 and 59%, respectively. The authors concluded that transportation may result in cattle stress that influences shedding of some foodborne human pathogens (e.g. *Salmonella* spp.); however, based on their results, prevalence of other pathogens may not be influenced by transportation stress.

Similarly, Beach *et al.* (2002) showed an increase in hide prevalence of *Salmonella* spp. following transport to a packing plant, while prevalence of *Campylobacter* on fed cattle hides decreased following transport. Those authors concluded that, for feedlot cattle, transportation did not affect shedding of *Salmonella* or *Campylobacter* – but did affect contamination on cattle hides at the time of harvest (Beach *et al.*, 2002). Other factors that may affect shedding of *E. coli* O157:H7 are: (i) the condition of feedlot pens. Cattle from muddy pens shed more *E. coli* than cattle from dry pens (Smith *et al.*, 2001a, b); and (ii) vehicles may also be a source of contamination. Wray *et al.* (1991) found that washing and cleaning of vehicles reduced *Salmonella* contamination from 22.6% of the vehicles to 6.5%. Reicks *et al.* (2007) found that clean trucks had a lower percentage of *Salmonella* positive samples compared with dirty trucks. Feedlot cattle transported in clean and dirty trucks had a similar prevalence of *Salmonella* on their hides after arrival at a slaughter plant.

Washing live feedlot cattle in an attempt to remove pathogens is highly stressful during cold weather and may increase the percentage of dark-cutters. The mud is stuck on like cement, and washing will not remove it. Some packing plants are effectively removing a portion of the pathogens by washing stunned and bled feedlot cattle. They have developed high-pressure wash cabinets and other equipment for the removal of mud. The best approach to the mud problem is better management of feedlot pens to reduce mud.

In new feedlot construction, the pens should be sloped 2–4%. However, washing of clean grass-fed cattle is effective for reducing contamination. To reduce stress and improve animal welfare, the handlers must avoid spraying high-pressure water in the animal's face. During freezing weather, washing of live cattle should be confined to their legs, and irritating disinfectants should not be used.

Based on these and other studies, biosafety and biocontainment methods during livestock production should most likely consider, and be extended through, transport of animals to their next destination. Contamination at the lairage is also a concern. In the UK, 50% of swabs from lairage pens tested positive for either *E. coli*, *Salmonella* or *Camphylobacter* (Small *et al.*, 2002, 2003).

Although epidemiological studies have not shown that management of environmental foodborne pathogen contamination will result in reduced prevalence on the exterior of animals and carcasses at the time of harvest (Rasmussen and Casey, 2001), such prerequisite management is necessary before the use of microbiological intervention systems during livestock production can be completely effective. Therefore, effectiveness of management and intervention practices in controlling environmental human pathogens found in livestock production environments must be evaluated as a system. Hussein *et al.* (2001) stated that, 'because of the complexity of the problem, the on-farm and off-farm factors affecting beef safety should be elucidated, their roles clearly defined and their additive impacts determined'.

Ransom *et al.* (2003) provided data indicating that 7.1% of pre-evisceration carcasses tested positive for *E. coli* O157 if feedlot pen-floor faecal prevalence was less than 20%, while pre-evisceration prevalence increased to 12.5% if feedlot pen-floor faecal prevalence was greater than 20%. Research results illustrate that, if biosafety and biocontainment procedures are to be implemented by livestock producers to control human foodborne pathogens, such measures must be specifically designed to reduce pathogen loads on exterior surfaces of animals before they enter packing plants for harvest.

Other interventions in reduction of pathogens in beef

A novel approach for reducing beef pathogen prevalence was proposed by Brett Finlay and Andrew Potter at the University of British Columbia and the University of Saskatchewan, respectively (Finlay, 2003), and should soon be available commercially. These researchers developed a vaccine that may prevent attachment of *E. coli* O157:H7 in the large intestine of cattle. Potter *et al.* (2004) showed that, when cattle were vaccinated with proteins (Tir and EspA) secreted by *E. coli* O157:H7, pathogen numbers and the number of cattle shedding *E. coli* O157 were both reduced.

To date, the most effective technology for reducing the prevalence of *E. coli* O157:H7 in fed cattle has been that which utilizes feed and/or water as the vehicle of decontamination. Anderson *et al.* (2000) and Callaway *et al.* (2002) proposed that sodium chlorate be used to reduce pathogen loads in livestock; sodium chlorate eliminated organisms reliant on nitrate reductase metabolism from the gastrointestinal tract of animals during research trials. However, at this point, federal US approval status for use of sodium chlorate as a feed or water ingredient for foodborne pathogen control is unclear. Notwithstanding, sodium chlorate may offer potential in future human foodborne pathogen biosafety and biocontainment systems.

Direct-Fed Microbial (DFM) products are approved for use in reducing pathogen prevalence in livestock. Brashears *et al.* (2003a) demonstrated a methodology for isolating competitive exclusion products (*Lactobacillus acidophilus* NPC-747) to be fed to cattle to inhibit growth and proliferation of *E. coli* O157:H7. Brashears *et al.* (2003b) also published findings of studies reporting on the effectiveness of such DFM feed additives. Younts-Dahl *et al.* (2004) reported that providing a diet containing *Lactobacillus acidophilus* NP-51 plus *Propionibacterium freudenreichii* to feedlot cattle was effective in reducing prevalence of *E. coli* O157; prevalence on hides of harvested cattle was reduced from 50.0% in control cattle to 25.0% in treated cattle in that study (Younts-Dahl *et al.*, 2004).

Since 1971, neomycin sulphate has been approved in the USA for use in treatment of cattle, horses, sheep, swine, goats, cats, turkeys, chickens, ducks and mink for bacterial enteritis. However, research results now exist suggesting that this chemical may also be effective in reducing prevalence of *E. coli* O157:H7 in the faeces of treated calves.

Woerner *et al.* (2006) recently reported results of a study that included feedlot cattle treatments of: (i) feeding neomycin sulfate; (ii) feeding *Lactobacillus acidophilus* NPC-747; or (iii) vaccination with an *E. coli* O157:H7 bacterin vaccine (produced by Fort Dodge Animal Health, Kansas City, USA), as well as all possible combinations of the treatments. Pathogen prevalence was highest on hides and in faeces of cattle receiving no treatment (control), whereas prevalence of *E. coli* O157 was lower on and in treated cattle.

Neomycin appeared to be the most effective single intervention, reducing prevalence levels in faeces and on hides by 45.8 and 31.8%, respectively, compared to controls. When used singularly, *Lactobacillus acidophilus* NPC-747 and vaccination were similar to each other in effectiveness, reducing pathogen prevalence (from control levels) by 17.6 and 20.3% on hide samples, respectively, and by 32.5 and 31.1%, respectively, in faeces (Woerner *et al.*, 2006).

The US National Cattlemen's Beef Association has published a series of 'Production Best Practices' for differing sectors of the beef industry as a primary result of the E. coli Summit (NCBA, 2003). One such set of Production Best Practices was developed for preharvest control of *E. coli*. The concept of implementing Production Best Practices to achieve biosafety and biocontainment for *E. coli* O157:H7 contamination of live cattle would essentially incorporate a combination of intervention strategies in conjunction with controlled management practices (Gansheroff and O'Brien, 2000).

A production facility that follows Production Best Practices should use microbiological testing to establish an *E. coli* O157:H7 prevalence history in animals and in the environment, and then use resulting data comparatively to verify effectiveness of biosafety and biocontainment management programmes.

Shedding of Pathogens in Pigs

Research has clearly shown that pigs resting in the lairage at the meat plant become internally contaminated with *Salmonella*. Necropsy data showed that on the farm, only 5.3% of the pigs were contaminated, but *Salmonella* contamination was found in 39.9% of the pigs in the lairage (Hurd *et al.*, 2001). Swab tests of the pork plant lairage indicated that it was heavily contaminated with *Salmonella*. At one plant there were 36 different *Salmonella* serovars and the ileocaecal lymph nodes became contaminated after a 2.5-h rest period (Rustagho *et al.*, 2003).

Better cleaning of the lairage pens will reduce *Salmonella* contamination but not eliminate it. In one study, cleaning reduced *Salmonella* contamination in the pens from 25% of the samples to 10% (Swanenburg *et al.*, 2001). In an attempt to reduce *Salmonella* contamination of pigs, Hurd *et al.* (2005) reduced the time resting in the lairage from 4 h to 15–45 min. Holding the pigs for a shorter period significantly reduced shedding of *Salmonella* but the short resting period had a very detrimental effect on meat quality (Hurd *et al.*, 2005). Resting pigs from 1–3 h in the lairage will reduce PSE (pale, soft, exudative) pork and improve meat quality (Milligan *et al.*, 1998; Perez *et al.*, 2002; Warriss, 2003).

To resolve this conflict between pork quality and contamination, a multifaceted approach will be required. Below are some suggestions.

1. Better cleaning of lairage pens. A small study by Schmidt *et al.* (2004) found that cleaning the lairage pens had mixed results in reducing contamination of the pigs, but was very effective in cleaning the pen floors. A further study (O'Connor *et al.*, 2006) showed that contamination in different parts of the lairage was highly variable. This may be a factor in explaining the mixed results from lairage cleaning. However, cattle placed in lairage pens that were positive for *E.coli* were eight times more likely to test positive on their hides (Grant Dewell, 2007, personal communication).

2. Shedding of *Salmonella* by pigs can be reduced by eliminating overnight holding of pigs in the lairage and maintaining 3-h maximum rest time.

3. Pigs from *Salmonella*-free and *Salmonella*-positive farms should be kept in separate parts of the lairage. The clean pigs should be processed first. Separation of free and positive pigs will help reduce contamination (Swanenburg *et al.*, 2001; Nowak *et al.*, 2007).

4. Washing and sanitizing of trucks will greatly reduce contamination. Rajkowski *et al.* (1998) found that cleaning and sanitizing trucks reduced contamination from 41.5% of the samples to 2.77%. To prevent pathogens from the meat plant from causing disease back on the farm, many companies have a rigorous programme of trailer sanitation. After washing and disinfection, the truck and trailer must be completely dry before the next load. Complete dryness is essential to eliminate the porcine reproductive and respiratory syndrome virus (PRRSV)(Dee *et al.*, 2004a, b). Jeff Hill (2006, personal communication) states that his company, Premium Standard Farms, dries trailers in a heated room with fans to ensure

complete dryness. Heating the interior of the trailer to 71°C for 30 min is effective against PRRSV (Dee *et al.*, 2005).

Basic Disease Control Principles on the Farm for All Species of Animals

1. Isolate all new breeding stock arriving at a farm in a separate area away from other animals. Consult with local veterinary authorities for the recommended isolation time.

2. When purchasing breeding stock, have the animals tested for diseases that are a problem in your geographical area. Consult with local veterinary authorities.

3. Rendering and dead-stock trucks should not be allowed to enter the farm. Bring dead stock to the outer perimeter of the farm for pick-up.

4. Livestock trucks should be kept clean and all visitors and drivers should have clean footwear. This is especially important if a visitor or vehicle is arriving from an area where many different animals have co-mingled, such as auction markets and slaughter plants.

5. Biosecurity sequence – for all species, most veterinarians recommend that sanitation and biosecurity measures need to be more strict for newborn animals and valuable breeding stock compared with those for animals being fed for meat. Below is a list of five phases, moving from the least contaminated to the most contaminated. Some veterinarians call this the biosecurity pyramid. Seaman and Fangman (2001) have a good description. Phase 1 is at the top, where measures to prevent the introduction of disease should be the most strict.

- Foundation herds and flocks:
  - Bull, boar or stallion studs that distribute large amounts of semen;
  - Specific pathogen-free herds or flocks;
  - Foundation genetic breeders, all species.
- Newborns and commercial multiplier herds and flocks:
  - Chick hatcheries;
  - Dairy calf-rearing facilities;
  - Sows with piglets;
  - Breeder flocks that produce chicks for commercial egg-laying or meat;
  - Multiplier herds that produce replacement commercial females for many farms.

- Commercial livestock used for breeding and production:
  o Beef cow herds on pasture;
  o Dry sows;
  o Lactating and dry dairy cows;
  o Sheep, goats or horses on pasture.
- Feedlots and production operations – with no breeding:
  o Beef feedlots;
  o Swine finishing (fattening) barns;
  o Lamb feedlots;
  o Poultry grow-out barns;
  o Laying hens.
- Areas of animal co-mingling – all species:
  o Auction markets;
  o Buying stations;
  o Slaughter plants;
  o Rendering plants;
  o Traditional 'wet' meat markets with different species co-mingling.

Flow of people and traffic should move from the least contaminated Phase 1 and 2 areas to the more contaminated Phase 3, 4 and 5 areas. Boots and clothes should be washed before a person can return to animals that are higher up on the pyramid. Providing boots or shoe covers for visitors is strongly recommended. Dee *et al.* (2004a, b) reported that providing disposable plastic boots and bleach footbaths were effective for preventing the mechanical transmission of PRRSV to pigs. On pig and poultry operations the rules are often very strict, and all visitors to the farms wear plastic boots and disposable coveralls. Some poultry and pork companies require 24–72 h of down time after visiting Phases 3, 4 or 6 before a person can return to Phases 1 and 2. The author has visited over ten

different integrated poultry and pork companies in the USA, Canada, Europe and other countries. The biosecurity pyramids in all of these companies were similar for the different phases of production.

6. After visiting areas where foot-and-mouth disease, African swine fever or other diseases that require notification are prevalent, follow all OIE, CDC (Center for Disease Control) and local government guidelines on down time, cleaning and disinfection before contacting animals. Foot-and-mouth disease transmission can be prevented by having contaminated people shower and change into clean, laundered clothes (Amass *et al.*, 2003).

7. Use clean technique for injections, ear tagging and surgical procedures. Milking equipment and pregnancy testing can also spread disease. Clean the milking machine after milking a cow with mastitis. Sleeves should be changed after palpating a cow that has a disease transmissible to other animals. Work with your local veterinary specialists and meat quality assurance teams to improve your practices.

8. Keep a log book for all visitors, truck drivers, employees and other people who come to your farm.

9. Consult with your local veterinarian and other skilled professional for specific biosecurity recommendations for your farm or ranch.

10. Producer groups, government officials and meat-buying customers must all work together to make sure that regulations and procedures that help prevent the introduction of diseases are obeyed. Diseases that may disrupt the economy must be prevented from entering areas that are free of these diseases.

References

Amass, S.F., Pacheco, J.M., Mason, P.W., Schneider, J.L., Alvarez, R.M., Clark, L.K. and Ragland, D. (2003) Procedures for preventing the transmission of foot-and-mouth disease virus to pigs and sheep by personnel in contact with infected animals. *Veterinary Record* 153, 137–140.
Anderson, R.C., Buckley, S.A., Kubena, L.F., Stanker, L.H., Harvey, R.B. and Nisbet, D.J. (2000) Bacterial effect of sodium chlorate on *Escherichia coli* O157:H7 and *Salmonella typhimurium* DT 104 in rumen contents *in vitro*. *Journal of Food Protection* 63, 1038–1042.
APHIS (2004) *Animal Disease Exclusion Practices on US Dairy Operations, 2002*. National Animal Health Monitoring System, Fort Collins, Colorado (http://www.aphis.usda.gov/vs/ceah/ncahs/nahms/dairy/dairy02/Dairy02_biosecurity.pdf).

Bach, S.J., McAllister, T.A., Mears, G.J. and Schwartzkopf-Genswein, K.S. (2004) Long-haul transport and lack of preconditioning increases fecal shedding of *Escherichia coli* and *Escherichia coli* O157:H7 by calves. *Journal of Food Protection* 67, 672–678.

BAMN (2001) *Biosecurity on Dairies*. Bovine Alliance on Management and Nutrition (http://www.aphis. usda.gov/vs/ceah/ncahs/nahms/dairy/bamn/BAMNBiosDairy01.pdf).

Barham, A.R., Barham, B.L., Johnson, A.K., Allen, D.M., Blanton, Jr, J.M. and Miller, M.F. (2002) Effects of the transportation of beef cattle from the feedyard to the packing plant on prevalence levels of *Escherichia coli* O157 and *Salmonella* spp. *Journal of Food Protection* 65, 280–283.

BC Poultry Industry (2005) *Biosecurity Manual*. British Columbia Chicken Marketing Board (http://www. bcchicken.ca).

Beach, J.C., Murano, E.A. and Acuff, G.R. (2002) Prevalence of *Salmonella* and *Campylobacter* in beef cattle from transport to slaughter. *Journal of Food Protection* 65, 1687–1693.

Boklund, A., Mortensen, S. and Houe, H. (2003–2004) Biosecurity in 121 Danish sow herds. *Acta Veterinaria Scandinavica* (Suppl.) 100, 5–14.

Brashears, M.M., Jaroni, D. and Trimble, J. (2003a) Isolation, selection and characterization of lactic acid bacteria for a competitive exclusion product to reduce shedding of *Escherichia coli* O157:H7 in cattle. *Journal of Food Protection* 66, 355–363.

Brashears, M.M., Galyean, M.L., Loneragan, G.H., Mann, J.E. and Killinger-Mann, K. (2003b) Prevalence of *Escherichia coli* O157:H7 and performance by beef feedlot cattle given *Lactobacillus* direct-fed microbials. *Journal of Food Protection* 66, 748–754.

Callaway, T.R., Anderson, R.C., Genovese, K.J., Poole, T.L., Anderson, T.J., Byrd, J.A., Kubena, L.F. and Nisbet, D.J. (2002) Sodium chlorate supplementation reduces *E. coli* O157:H7 populations in cattle. *Journal of Animal Science* 80, 1683–1689.

Childs, K.D., Simpson, C.A., Warren-Serna, W., Bellenger, G., Centrella, B., Bowling, R.A., Ruby, J., Stefanek, J., Vote, D.J., Choat, T., Belk, K.E., Scanga, J.A., Sofos, J.H. and Smith, G.C. (2006) Molecular characterization of *Escherichia coli* O157:H7 of hide contamination routes – feedlot to harvest. *Journal of Food Protection* 69 (9), 1240–1247.

Dee, S.A., Deen, J., Otake, S. and Pitjoan, C. (2004a) An experimental model to evaluate the role of transport vehicles as a source of transmission of porcine reproduction and respiratory syndrome virus for susceptible pigs. *Canadian Journal of Veterinary Research* 68, 19–26.

Dee, S., Deen, J., Burns, D., Douthit, G. and Pijoan, C. (2004b) An assessment of sanitation protocols for commercial transport vehicles contaminated with porcine reproductive and respiratory syndrome virus. *Canadian Journal of Veterinary Resarch* 68, 208–214.

Dee, S., Torremorell, M., Thompson, B., Deen, J. and Pijoan, C. (2005) An evaluation of thermo-assisted drying and decontamination for the elimination of porcine reproductive and respiratory syndrome virus from contaminated livestock transport vehicles. *Canadian Journal of Veterinary Records* 69, 58–63.

Dorsa, W.J. (1997) New and established carcass decontamination procedures commonly used in the beef processing industry. *Journal of Food Protection* 60, 1146–1151.

Drovers Alert (2006) Respiratory disease causes most death loss. *Drovers Magazine* (Lenexa, Kansas) 7, 19.

Fike, K. and Spire, M.F. (2006) Transportation of Cattle. *Veterinary Clinics Food Animal Practices* 22, 305–320.

Finlay, B.B. (2003) Pathogenic *E. coli*: from molecules to vaccine. Presented at the *2nd Governors Conference on Ensuring Meat Safety, E. coli O157:H7 Progress and Challenges*, University of Nebraska-Lincoln, Nebraska, 7 April 2003.

Gansheroff, L.J. and O'Brien, A.D. (2000) *Escherichia coli* O157:H7 in beef cattle presented for slaughter in the US: higher prevalence rates than previously estimated. *Proceedings of the Natural Academy of Sciences* 97, 2959–2961.

Hardin, M.D., Acuff, G.R., Lucia, L.M., Oman, J.S. and Savell, J.W. (1995) Comparison of methods for decontamination from beef carcass surfaces. *Journal of Food Protection* 58, 368–374.

Hurd, H.S., McKean, J.D., Wesley, J.V. and Karriker, L.A. (2001) The effect of lairage on *Salmonella* isolation from market swine. *Journal of Food Protection* 64, 939–944.

Hurd, H.S., Gailey, J.K., McKean, J.D. and Griffith, R.W. (2005) Variable abattoir conditions affect *Salmonella enterica* prevalence and meat quality in swine and pork. *Foodborne Pathogens and Disease* 2, 77–81.

Hussein, H.S., Lake, S.L. and Ringkob, T.P. (2001) Cattle as a reservoir of Shiga-like toxin-producing *Escherichia coli* including O157:H7 – pre-and post-harvest control measures to assure beef safety. *Professional Animal Science* 17, 1–16.

Kirkpatrick, J.G. and Selk, G. (2006) *Biosecurity in the Beef Cattle Operation.* Bulletin F-3281, Division of Agricultural Science and Natural Resources, Oklahoma State University, Oklahoma City, Oklahoma, pp. 1–4 (http://osuextra.okstate.edu/pdfs/F-3281web.pdf).

Milligan, S.D., Ramsey, C.B., Miller, M.F., Kaster, C.S. and Thompson, L.D. (1998) Resting of pigs and hot-fat trimming and accelerated chilling of carcasses to improve pork quality. *Journal of Animal Sciences* 76, 74–86.

National Research Council (2006) *Guidelines for the Humane Transportation of Research Animals.* Committee on Guidelines for the Humane Transportation of Laboratory Animals, The National Academies Press, Washington, DC, pp. 59–70.

NCBA (2003) *E. coli* O157:H7 solutions: the farm to table continuum. *Executive Summary of the Beef Industry E. coli Summit Meeting,* 7–8 January 2003.

Nelson, T.M. (2004) *Biosecurity for Poultry.* University of Maryland, Baltimore, Maryland (http://www.agric. umd.edu).

Nowak, B., Von Muffling, T., Chaunchom, S. and Hartung, J. (2007) *Salmonella* contamination in pigs at slaughter and on the farm: a field study using antibody ELISA test and a PCR technique. *International Journal of Food Microbiology* (in press).

O'Connor, A.M., Gailey, J., McKean, J.D. and Hurd, H.S. (2006) Quantity and distribution of *Salmonella* recovered from three swine lairage pens. *Journal of Food Protection* 69, 1717–1719.

OIE (2006) *Diseases Notifiable to the OIE.* World Organization for Animal Health (http://www.oie.int/eng/ maladies/en_classification.htm).

Perez, M.P., Palachio, J., Santolaria, M.P., del Acena, M.C., Chacon, G., Verde, M.T., Calvo, J.H., Zangoza, M.P., Gascon, M. and Garcia Bethenguer, S. (2002) Influence of lairage time on some welfare and meat quality parameters in pigs. *Veterinary Research* 33, 239–250.

Pescatore, A. (2006) *Biosecurity: What can a Grower Do?* University of Kentucky, Lexington, Kentucky (http://www.ca.uky.edu).

Potter, A.A., Klashinksy, S., Li, Y., Frey, E., Townsend, H., Rogan, D., Erickson, G., Hinkley, S., Klopfenstein, T., Moxley, R.A., Smith, D.R. and Finlay, B.B. (2004) Decreased shedding of *Escherichia coli* O157:H7 by cattle following vaccination with type III secreted proteins. *Vaccine* 22, 362–369.

Princeton University (2006) *II. Biohazardous Research Project Registration and Approval.* (http://web.princeton.edu/sites/ehs/biosafety/biosafetypage/section2.htm).

Rajkowski, K.T., Eblem, S. and Laubauch, C. (1998) Efficacy of washing and sanitizing trailers used for swine transport in reduction of *Salmonella* and *Escherichia* coli. *Journal of Food Protection* 61, 31–35.

Ransom, J.R., Sofos, J.N., Belk, K.E., Dewell, G.A., McCurdy, K.S., Smith, G.C. and Salman, M.D. (2003) Determining the prevalence of *Escherichia coli* O157 in cattle and beef from the feedlot to the cooler. In: *90th Annual Meeting of the International Association for Food Protection,* New Orleans, Louisiana, 175 pp.

Rasmussen, M.A. and Casey, T.A. (2001) Environmental and food safety aspects of *Escherichia coli* O157:H7 infections in cattle. *Critical Review of Microbiology* 27, 57–73.

Reagan, J.O., Acuff, G.R., Buege, D.R., Buyck, M.J., Dickson, J.S., Kastner, C.L., Marsden, J.L., Morgan, J.B., Ranzell Nickelson, I., Smith, G.C. and Sofos, J.N. (1996) Trimming and washing of beef carcasses as a method of improving the microbiological quality of meat. *Journal of Food Protection* 59, 751–756.

Reicks, A.L., Brashears, M.M., Adams, K.D., Brooks, J.C., Blanton, J.R. and Miller, M.F. (2007) Impact of transportation of feedlot cattle to the harvest facility on the prevalence of *Escherichia* D157, *Salmonella* and total aerobic microorganisms on hides. *Journal of Food Protection* 70, 17–21.

Rustagho, M.H., Hard, H.S., McKean, J.D., Ziemer, C.J., Gailey, J.K. and Leite, R.C. (2003) Preslaughter holding environment in pork plants is highly contaminated with *Salmonella enterica. Applied Environmental Microbiology* 69, 4489–4494.

Sandia National Laboratories (2004) *International Biosecurity Symposium: Securing High Consequence Pathogens and Toxins, Symposium Summary.* Sandia Report SAND2004-2109, pp. 1–40. Sandia National Laboratories, Albuquerque, New Mexico (http://www.biosecurity.sandia.gov/documents/Int-Symposium-Summary-Final.pdf).

Schmidt, P.L., O'Connor, A.M., McKean, J.D. and Hurd, H.S. (2004) The association between cleaning and disinfection of lairage pens and the prevalence of *Salmonella enterica* at swine harvest. *Journal of Food Protection* 67, 1384–1388.

Seaman, J.S. and Fangman, T.J. (2001) *Biosecurity for Today: Swine Operation.* University of Missouri, Missouri (http://www.muextension.edu).

Small, A., Reid, C.A., Avery, S.M., Karabasil, N., Crowley, C. and Bunci, S. (2002) Potential for spread of *Escherichia coli* O157, *Salmonella* and *campylobacter* in the lairage environment of abattoirs. *Journal of Food Protection* 65, 931–936.

Small, A., Reid, C.A. and Buncie, S. (2003) Conditions in lairages at abattoirs for ruminants in southwest England and *in vitro* survival of *Escherichia coli* O157, *Salmonella Kedovgou* and *Campylobacter jujuni* on lairage-related substrates. *Journal of Food Protection* 66, 570–575.

Smith, D., Blackford, M., Younts, S., Moxley, R., Gray, J., Hungerford, L., Milton, T. and Klopfenstein, T. (2001a) The relationship of the characteristics of feedlot pens to the percentage of cattle shedding *Escherichia coli* O157:H7 within the pen. *Nebraska Beef Report*, University of Nebraska, Omaha, Nebraska, pp. 81–83.

Smith, D., Blackford, M., Younts, S., Moxley, R., Gray, I., Hungerford, L., Milton, T. and Klopfenstein, T. (2001b) Ecological relationship between the prevalence of cattle shedding *Escherichia* coli O157:H7 and characteristics of the cattle or conditions in the feedlot pen. *Journal of Food Protection* 64, 1899–1903.

Sunshine Project (2003) *Biosafety, Biosecurity and Biological Weapons*. Backgrounder #11 (http://www.sunshine-project.org/).

Swanenburg, M., Urings, H.A., Keuzenkamp, D.A. and Snijders, J.M. (2001) *Salmonella* in the lairage of pig slaughterhouses. *Journal of Food Protection* 64, 12–16.

US General Accounting Office (2003) *Livestock and Poultry: World Markets and Trade*. Report GAO-03-06, Performance and Accountability series, Department of Agriculture, USDA Foreign Agricultural Service, November 2005 (http://www.fas.usda.gov/currwmt.asp).

USDA–FAS (2005) Livestock and poultry: world markets and trade. http://www.fas.usda.gov/currumt.asp

USDA (2006) *Johanns Releases National Animal Identification System Implementation Plan*. News Release No. 0120.06 (http://www.usda.gov/wps/portal/usdahome).

Warriss, P.D. (2003) Optimal lairage times and conditions for slaughter pigs: a review. *Veterinary Record* 153, 170–176.

Woerner, D.R., Ransom, J.R., Sofos, J.N., Scanga, J.A., Smith, G.C. and Belk, K.E. (2006) Cattle feedlot intervention treatments to reduce *Escherichia coli* O157 contamination. *Food Protection Trends* 26(6), 393–400.

Wray, C., Todd, N., McLaren, I.M. and Beedel, Y.F. (1991) The epidemiology of *Salmonella* in calves: the role of markets and vehicles. *Epidemiology and Infection*, 107, 521–525.

Younts-Dahl, S.M., Galyean, M.L., Loneragan, G.H., Elam, N.A. and Brashears, M.M. (2004) Dietary supplementation with *Lactobacillus*- and *Propionibacterium*-based direct-fed microbials and prevalence of *Escherichia coli* O157 in beef feedlot cattle and on hides at harvest. *Journal of Food Protection* 67, 889–893.

Useful Websites[1]

Animal Behaviour, Professional Societies and Journals

Animal Behavior Society
Excellent links and educational materials. Publish the *Journal of Animal Behavior*
www.animalbehavior.org

Association for the Study of Animal Behaviour
Links to many behaviour societies
http://asab.nottingham.ac.uk

Applied Animal Behavior Science
Access behaviour abstracts in this journal
www.scirus.com

International Society of Applied Ethology
Many researchers in farm animal behaviour belong to this society
www.appliedethology.org

Universities Federation Animal Welfare
Publishes the *Journal of Animal Welfare*
www.ufaw.org.uk

Journal of Applied Animal Welfare Sciences (JAAWS)
Archive of articles
www.psyeta.org/jaaws

Industry and Veterinary Organizations

American Meat Institute
Contains guidelines for humane slaughter and welfare audit forms
www.animalhandling.org

Animal Transportation Association
Contains regulations and transport information
www.aata.animaltransport.org

[1] If a web address does not work, type the name of the organization into a search engine.

IATA – International Air Transport Association
Contains international regulations on transport
http://www.iata.org

British Veterinary Medical Association
Publisher of *Veterinary Record*
www.bva.co.uk
www.bva.publications.com

American Veterinary Medical Association
Publisher of *Journal of the American Veterinary Medical Association*
www.avma.org
Index to veterinary associations all over the world
www.vetmedicine.about.com

Australian Meat Industry Council
Contains welfare guidelines
www.amic.org.au

Government websites

Agriculture, Fisheries and Forestry of Australia
Contains welfare and transport codes
www.affa.gov.au

European Union
Contains welfare and transport codes
www.europa.ev.int

OIE Organisation Mondiale de la Santé Animale (World Organisation for Animal Health)
Contains transport and humane slaughter codes
www.oie.int

Dept of Environment, Food and Rural Affairs in the UK
Contains welfare and transport codes
www.defra.gov.uk

Animal Plant Health Inspection Service of the US Department of Agriculture
www.aphis.usda.gov

Food Safety Inspection Service, US Department of Agriculture
Contains humane slaughter and food safety regulations
www.fsis.gov

National Agriculture Library, Animal Welfare Information Center, AWIC
Database of animal welfare information
www.nal.usda.gov/awic

Canadian Food Inspection Agency
Contains humane slaughter and food safety regulations
www.inspection.gc.ca

Instituto Nacional de Carnes
Montevideo, Uruguay
Animal welfare guidelines
www.inac.gub.uy

CSIRO – Searchable database of Australian research on agriculture and livestock
www.CSIRO.au

Information Sites on Behaviour, Transport and Welfare

Temple Grandin's website
Information on handling, humane slaughter transport and facility design
www.grandin.com

Netvet – Mosby's veterinary guide to the internet
Lots of links to livestock web pages
Type Netvet into Google or Yahoo
www.netvet.wusti.edu

Prairie Swine Centre in Canada
Research studies on pig behaviour
www.prairieswine.com

Pubmed National Library of Medicine
Search the scientific literature
Type Pubmed into Google

Elsevier Journals Search Engine
Search the scientific literature
www.scirus.com

Purdue Center for Food Animal Well-Being
Links to journals and animal behaviour research
Type the centre's name into Google

Western College of Veterinary Medicine Saskatchewan
Lots of links to behaviour research and information
Joe Stookey's site
www.usask.ca/weum/herdmed/appliedethology

Virtual Livestock Library
Lots of good links, run by the Animal Science Department, Oklahoma State University
Type Virtual Livestock Library into Google
www.ansci.okstate.edu/library

Non-governmental animal welfare organizations

Compassion in World Farming
www.ciwf.org.uk

Farm Animal Welfare Council
Publishes animal welfare report
www.fawc.org.uk

Humane Slaughter Association
Publishes training materials
www.hsa.org

Humane Society of the USA
www.hsus.org

People for the Ethical Treatment of Animals
www.peta.org

World Society for the Protection of Animals
www.wspa-international.org

Index